Edited by
Karen K. Gleason

CVD Polymers

Related Titles

Picart, C., Caruso, F., Voegel, J.-C. (eds.)

Layer-by-Layer Films for Biomedical Applications

2014

Print ISBN: 978-3-527-33589-3; also available in electronic formats

Narayan, J.

Nanostructured Semiconductor Thin Films

2014

Print ISBN: 978-3-527-33473-5; also available in electronic formats

Ariga, K. (ed.)

Organized Organic Ultrathin Films

Fundamentals and Applications

2013

Print ISBN: 978-3-527-32733-1 also available in electronic formats

Theato, P., Klok, H. (eds.)

Functional Polymers by Post-Polymerization Modification

Concepts, Guidelines, and Applications

2013

Print ISBN: 978-3-527-33115-4 also available in electronic formats

Decher, G., Schlenoff, J.B. (eds.)

Multilayer Thin Films

Sequential Assembly of Nanocomposite Materials

Second Edition

2012

Print ISBN: 978-3-527-31648-9 also available in electronic formats

Schlüter, D.A., Hawker, C., Sakamoto, J. (eds.)

Synthesis of Polymers

New Structures and Methods

2012

Print ISBN: 978-3-527-32757-7 also available in electronic formats

Pacchioni, G., Valeri, S. (eds.)

Oxide Ultrathin Films

Science and Technology

2012

Print ISBN: 978-3-527-33016-4 also available in electronic formats

Friedbacher, G., Bubert, H. (eds.)

Surface and Thin Film Analysis

A Compendium of Principles, Instrumentation, and Applications

Second Edition

2011

Print ISBN: 978-3-527-32047-9 also available in electronic formats

Knoll, W., Advincula, R.C. (eds.)

Functional Polymer Films

2 Volume Set

2011

Print ISBN: 978-3-527-32190-2; also available in electronic formats

Edited by Karen K. Gleason

CVD Polymers

Fabrication of Organic Surfaces and Devices

WILEY-VCH
Verlag GmbH & Co. KGaA

The Editor

Prof. Dr. Karen K. Gleason
Massachusetts Institute of Technology
Department of Chemical Engineering
77 Massachusetts Avenue
Cambridge, MA 02139
United States

Cover
We would like to thank Kenneth Lau from Drexel University, Philadelphia, USA for providing us with the images we used in the cover illustration.

Library of Congress Card No.: applied for

British Library Cataloguing-in-Publication Data
A catalogue record for this book is available from the British Library.

Bibliographic information published by the Deutsche Nationalbibliothek
The Deutsche Nationalbibliothek lists this publication in the Deutsche Nationalbibliografie; detailed bibliographic data are available on the Internet at <http://dnb.d-nb.de>.

Print ISBN: 978-3-527-33799-6
ePDF ISBN: 978-3-527-69026-8
ePub ISBN: 978-3-527-69028-2
Mobi ISBN: 978-3-527-69029-9
oBook ISBN: 978-3-527-69027-5

Cover Design Adam Design
Typesetting Laserwords Private Limited, Chennai, India
Printing and Binding Markono Print Media Pte Ltd, Singapore

Printed on acid-free paper

Contents

List of Contributors

Peter Baumann
Aixtron SE
Dornkaulstraße 2
Kaiserstraße 98
52134 Herzogenrath
Germany

Salmaan Baxamusa
Physical and Life Sciences
Directorate
Lawrence Livermore National
Laboratory
7000 East Avenue
Livermore, CA
USA

Stacey F. Bent
Stanford University
Department of Chemical
Engineering
Shriram Center
Stanford, CA 94305
USA

Magnus Bergkvist
SUNY College of Nanoscale
Science and Engineering
255 Fuller Road
Albany, NY 12203
USA

Vijay Jain Bharamaiah Jeevendra Kumar
SUNY Polytechnic Institute
Colleges of Nanoscale Science &
Engineering
255 Fuller Road
Albany, NY 12203
USA

Daniel D. Burkey
University of Connecticut
Chemical and Biomolecular
Engineering
191 Auditorium Road
Unit 3187 Storrs
CT 06269-3237
USA

Kenneth C. K. Cheng
University of Michigan
Chemical Engineering
Materials Science and
Engineering
Biointerfaces Institute
NCRC, B26, Rm. 133S
2800 Plymouth Road
Ann Arbor, MI 48109
USA

Anna Maria Coclite
Graz University of Technology
Institute of Solid State Physics
Petersgasse 16
8010 Graz
Austria

Mariadriana Creatore
Eindhoven University of Technology
Department of Applied Physics
P.O. Box 513
Groene Loper 19
5600 MB Eindhoven
The Netherlands

Melik C. Demirel
Pennsylvania State University
Engineering Science and Mechanics
Materials Research Institute
212 Earth and Engineering Science Building
University Park
PA 16802
USA

Xiaopei Deng
University of Michigan
Chemical Engineering
Materials Science and Engineering
Biointerfaces Institute
NCRC, B26, Rm. 133S
2800 Plymouth Road
Ann Arbor, MI 48109
USA

Markus Gersdorff
Aixtron SE
Dornkaulstraße 2
Kaiserstraße 98
52134 Herzogenrath
Germany

Karen K. Gleason
Massachusetts Institute of Technology
Department of Chemical Engineering
77 Massachusetts Avenue
Cambridge, MA 02139
USA

Malancha Gupta
University of Southern California
Chemical Engineering and Materials Science
925 Bloom Walk
HED 216
Los Angeles, CA 90089-1211
USA

Matthew J. Hancock
Veryst Engineering, LLC
47A Kearney Road
Needham, MA 02494
USA

Rachel M. Howden
Massachusetts Institute of Technology (MIT)
Department of Chemical Engineering
77 Massachusetts Avenue
Cambridge, MA 02139
USA

Sung Gap Im
Korea Advanced Institute of Science and Technology (KAIST)
Department of Chemical and Biomolecular Engineering and KI for NanoCentury
291 Daehak-ro (373-1 Guseong-dong)
Yuseong-gu
Daejeon 305-701
Republic of Korea

Bong Jun Kim
Korea Advanced Institute of Science and Technology (KAIST)
Department of Chemical and Biomolecular Engineering and KI for NanoCentury
291 Daehak-ro (373-1 Guseong-dong)
Yuseong-gu
Daejeon 305-701
Republic of Korea

Martin Kunat
Aixtron SE
Dornkaulstraße 2
Kaiserstraße 98
52134 Herzogenrath
Germany

Juergen Kreis
Aixtron SE
Dornkaulstraße 2
Kaiserstraße 98
52134 Herzogenrath
Germany

Joerg Lahann
University of Michigan
Chemical Engineering
Materials Science and Engineering
Biointerfaces Institute
NCRC, B26, Rm. 133S
2800 Plymouth Road
Ann Arbor, MI 48109
USA

Kenneth K. S. Lau
Drexel University
Department of Chemical and Biological Engineering
3141 Chestnut Street
Philadelphia, PA 19104
USA

Yu Mao
Oklahoma State University
Biosystems and Agricultural Engineering
111 Agricultural Hall
Stillwater, OK 74078-6016
USA

W. Shannan O'Shaughnessy
GVD Corporation
45 Spinelli Place
Cambridge, MA 02138
USA

Gozde Ozaydin Ince
Sabancı University
Faculty of Engineering and Natural Sciences
Materials Science and Nanoengineering
Orta Mahalle, Universite Caddesi No: 27
34956 Orhanlı Istanbul
Turkey

Alberto Perrotta
Eindhoven University of Technology
Department of Applied Physics
P.O. Box 513
Groene Loper 19
5600 MB Eindhoven
The Netherlands

and

Dutch Polymer Institute (DPI)
P.O. Box 902
J.F. Kennedylaan, 2
5600 AX Eindhoven
The Netherlands

Markus Schwambera
Aixtron SE
Dornkaulstraße 2
Kaiserstraße 98
52134 Herzogenrath
Germany

Hyejeong Seong
Korea Advanced Institute of
Science and Technology (KAIST)
Department of Chemical and
Biomolecular Engineering and KI
for NanoCentury
291 Daehak-ro (373-1
Guseong-dong)
Yuseong-gu
Daejeon 305-701
Republic of Korea

Wyatt E. Tenhaeff
University of Rochester
Department of Chemical
Engineering
249A Gavett Hall
500 Joseph C. Wilson Blvd.
Rochester, NY 14627
USA

Jose L. Yagüe
Massachusetts Institute of
Technology
Department of Chemical
Engineering
77 Massachusetts Avenue
Cambridge, MA 02139
USA

Rong Yang
Harvard Medical School
Children's Hospital Boston
Department of Anesthesiology
Division of Critical Care
Medicine
Laboratory for Biomaterials and
Drug Delivery
300 Longwood Avenue
Boston, MA 02115
USA

and

Massachusetts Institute of
Technology
Department of Chemical
Engineering
77 Massachusetts Avenue
Cambridge, MA 02139
USA

Youngmin Yoo
Korea Advanced Institute of
Science and Technology (KAIST)
Department of Chemical and
Biomolecular Engineering and KI
for NanoCentury
291 Daehak-ro (373-1
Guseong-dong)
Yuseong-gu
Daejeon 305-701
Republic of Korea

Jae Bem You
Korea Advanced Institute of Science and Technology (KAIST)
Department of Chemical and Biomolecular Engineering and KI for NanoCentury
291 Daehak-ro (373-1 Guseong-dong)
Yuseong-gu
Daejeon 305-701
Republic of Korea

Han Zhou
Stanford University
Department of Chemical Engineering
Shriram Center
Stanford, CA 94305
USA

1
Overview of Chemically Vapor Deposited (CVD) Polymers

Karen K. Gleason

1.1
Motivation and Characteristics

Chemical vapor deposition (CVD) is a powerful technology for surface engineering. When combined with the richness of organic chemistry, CVD enables polymeric coatings to be deposited without solvents [1–3]. The implantation of biomedical devices into humans, the stable functioning of printed circuit boards in harsh environments, and long-lasting, highly lubricious surfaces on industrial parts are just a few examples of the applications which employ CVD polymers. Research for CVD polymers has been undertaken in a diverse array of fields that include biotechnology, nanotechnology, optoelectronics, photonics, microfluidics, sensing, composites, and separations.

In CVD polymerization, gas phase monomers are converted directly to thin solid macromolecular films. By eliminating the need to dissolve macromolecules, CVD allows the synthesis of insoluble polymers and highly crosslinked organic networks. CVD also enables the polymerization of monomer units that undergo unwanted side reactions in solution and copolymerization of pairs of monomers that lack a common solvent.

CVD polymer films can be applied to nearly any substrate. Actually, for certain polymers and certain substrates, CVD polymerization can be the sole fabrication option. Low surface temperatures allow CVD polymers to be grown directly on fragile objects such as tissue paper and porous polymeric membranes. CVD is ideal for substrates that swell, dissolve, or otherwise degrade in solvents or for substrates that cannot withstand the high temperatures of "spray and bake" melt processing. This versatility is in contrast to, for example, self-assembled monolayers (SAMs) that are compatible only with specific surfaces, such as gold.

Monitoring during CVD polymerization allows for the *in situ* determination of growth rate and for film growth to be terminated when a precise thickness is reached. High deposition rates ($>1\ \mu m\ min^{-1}$) and film thickness ($>200\ \mu m$) have been reported, as have ultrathin pinhole-free CVD polymer layers (<10 nm). Such

CVD Polymers: Fabrication of Organic Surfaces and Devices, First Edition. Edited by Karen K. Gleason.

ultrathin pinhole-free films are difficult to achieve by solution processing because of de-wetting, liquid thinning, and surface tension effects. Ultrathin layers are also problematic/challenging in melt processing because of the large sizes of the polymer particles which are initially sprayed on.

1.1.1 Quality

For the highest quality polymer films, CVD is the method of choice, even for layers that can be formed by other means. Low impurity levels are possible for vapor deposited films since the low molecular weight reactants utilized can be purified to a high degree. In contrast, when a thin film is cast from a liquid phase solution of polymer dissolved in a solvent, a variety of unintended impurities and intentional additives are typically present, as sodium, excipients, and glidents. When dried, these species remain behind, reducing the purity of the polymer film. Additionally, the removal of the solvent is generally not fully complete and this residual solvent can have significant effects on the film properties, such as lowering the glass transition temperature due to plasticization effects.

1.1.2 Conformality

A uniform thickness of film over every contour of a geometric feature is termed conformal coverage. Conformal coverage is essential for substrates of practical interest, which have micro- and/or nano-scale topography. Conformality results from vapor phase reactants (i) that undergo isotropic diffusion and display a low probability of "sticking" during a single collision with the surface, or (ii) that undergo self-limiting surface reactions. In contrast, achieving conformality is challenging for solution processing. Surface tension can drive the formation of liquid bridges that prevent penetration of the coating into the geometric feature, leaving behind voids. Alternatively, liquid solutions can well up in the bottom of the features, display meniscus formation, and show thinning behavior on the side walls of the features.

Conformal organic functionalization finds utility for micro- and nano-structured surfaces for microelectronics, organic electronics, microelectrical and mechanical systems (MEMS), microfluidic devices, and energy storage and harvesting applications; the encapsulation of micro- and nano-particles; and the chemical modification of the internal surfaces of porous materials, including foams and membranes. The conformal coating of rough fibrous substrates, such as paper and textiles, results in a surface modification completely around each fiber, yet leaves the regions between fibers open, thus maintaining both high surface area and breathability. This capability opens the possibility for device formation on inexpensive and highly utilizable everyday substrates.

1.1.3 Durability

Durability is paramount in the majority of applications for polymeric coatings. The formation of insoluble crosslinked organic networks by CVD is one synthesis strategy for achieving durability. A second strategy is covalent bond formation at the interface, which grafts the CVD film to its underlying substrate. These two strategies have been employed both independently as well as in combination.

Using monomers with multiple propagation sites allows for the direct deposition of crosslinked organic thin film by CVD. For example, CVD biopassive films resulting from the use of a trivinyl monomer have displayed excellent dielectric performance under continuous electrical bias and soaking in saline solution for at over 9 years. Crosslinking can also improve mechanical properties. Moreover, if the level of crosslinking is below the percolation of rigidity limit, the films remain flexible.

To achieve grafted interfaces that can resist delamination, surface sites must be activated for reaction with the CVD monomers. Desirable active sites on the surface include unpaired electrons, also known as free radicals or dangling bonds, as well as specific chemical functional groups, particularly vinyl groups. The most desirable strategies for forming active surface sites are carried out in the CVD chamber immediately prior to film growth, allowing the overall process to remain as a single step and also allowing a high-quality interface formation in the controlled environment of the reactor.

1.1.4 Composition

The organic functional groups of CVD polymers provide systematic control over surface properties including, but not limited to, wettability, lubricity, adhesion, and biocompatibility. Through selection of the appropriate monomer(s), CVD films can be optimized for a variety of surface and bulk properties including chemical inertness, corrosion resistance, thermal and UV stability, mechanical stability, coefficient of friction, hydrophobicity, permeability to gases, low dielectric constant, electrical conductivity, and also for biocompatibility, biopassivation, biofouling resistance, and biostability.

The chemical specificity of the functional groups of CVD polymers is valuable for applications in sensing and separations. Additionally, the biological specificity that can be achieved with CVD polymers has broad potential in medicine and biotechnology, and for applications including biosensors, controlled drug release, implantable medical devices, tissue engineering, biometric surfaces, and medical diagnostics. Surfaces with more than one functional group, the so-called multifunctional surfaces, have been fabricated both by CVD copolymerization and via patterning strategies on multilayer CVD polymer stacks.

Functional organic moieties at the surface of CVD polymers can also be utilized for the specific binding of molecules and particles. Fluorescent dyes, biological growth factors, and antibodies have all been successfully attached to the surfaces of CVD polymers, as have inorganic nanoparticles including metals and quantum dots.

While, inorganic layers typically have fixed thickness, some organic materials display dynamic thickness variations in response to external changes in electrical fields, magnetic fields, temperature, pH, or during exposure to specific chemical or biological molecules. Thus, devices incorporating organic materials can be designed to change dimension for transducing chemical events into electrical or optical responses. To obtain this responsive behavior requires a high concentration of organic functional groups achievable by polymer CVD.

By simply changing the ratio of gases introduced into a CVD reactor over the course of the process, film composition can be graded or layered during growth upward from the substrate. Hence, the concentration of specific organic functional group displayed at the surface of the film can differ from the concentration in the underlying layer. This feature has made it possible to deposit antibiofouling layers where the key zwitterionic moiety is concentrated at the top surface of the film. The ability to independently optimize the properties of the underlying film from the density of the surface functionalization moieties is an important design feature of the CVD polymerization method.

1.2 Fundamentals and Mechanisms

CVD processes are controlled by the introduction rates and the types of the feed gases as well as the conditions inside the reaction chamber, including partial pressures, temperatures, electric fields, and photon fluxes [4–6]. Homopolymer CVD films result from feeding only one monomer, while CVD copolymers result from introducing a mixture of monomers. Additional species may be introduced along with the monomer(s) in order to (i) promote the desired chemical reactions, (ii) inhibit unwanted reactions, or (iii) simply to function as an inert diluent. For reactants, which are not gases, vapors are produced from volatile liquid or solid sources. In the CVD chamber, one or more chemical reactions yield a thin solid polymeric layer as the desired reaction product. Any by-products of a volatile reaction are pumped out of the CVD chamber. For the majority of CVD polymer process, the reactor chamber is operated at modest vacuum levels, but some operate at atmospheric pressure.

1.2.1 Gas Phase and Surface Reactions

Chemical reactions distinguish CVD from physical vapor deposition processes such as evaporation and sputtering. Physical vapor deposition is compatible with small organic molecules but is challenging for macromolecules, as the decomposition of a polymer often precedes its vaporization. Chemical reactions in CVD processes can occur both in the gas phase and at the surface. For the homogeneous gas phase reactions, the reaction kinetics and equilibria depend on the partial pressures of the reactants and the gas phase temperature. Similarly, the rates and equilibrium concentrations associated with the heterogeneous reactions are determined by surface concentrations and the surface temperature. The surface concentration of a reactant can approach, or even exceed, a monolayer. At these high surface concentrations, the film deposition rates can become quite fast, even though the partial pressure of the reactant in the gas phase may be quite low. In this situation, the surface can be viewed as a means to concentrate the reactant to liquid-like concentrations, albeit only in a very thin surface layer.

CVD processes comprise a series of chemical and physical steps. For a surface reaction, the reactants must diffuse through the gas phase to arrive at the surface, adsorb on the surface, and then undergo chemical reaction. Any volatile products must desorb from the surface and diffuse away. The overall rate of the CVD process can be controlled by any one of these individual steps. Understanding which step is rate limiting is essential for optimizing the CVD process. For example, if the surface reaction is rate limiting, raising the surface temperature while holding all the other conditions constant will increase the film deposition rate. However, CVD polymerization rates are often observed to decrease as the surface temperature rises. This behavior is consistent with the rate limiting step being the adsorption of the reactant to the surface.

Solid films formed by surface reactions are typically desired over films formed by the agglomeration of particles formed by gas phase reactions. The latter are powdery deposits that are often quite rough and lack cohesion. Solid particles impinging on the surface are also a potential source of pinhole defects.

1.2.2 The Monomer Saturation Ratio

Two important parameters for CVD polymerization are the partial pressure of the monomer, P_m, and the saturation pressure of the monomer at the temperature of interest, P_{sat}. The saturation ratio, S, is equal to P_m/P_{sat}, and determines (i) the concentration of monomer absorbed on the surface, (ii) the film growth rate, (iii) the number-average molecular weight of the polymer film formed, and (iv) the degree of conformality achieved. For example, monolayer coverage for an adsorbed vinyl monomer has been observed at $S = 0.35$.

For $S = 1$, a liquid will be in equilibrium with the monomer vapor. Formation of a liquid in the CVD system is typically undesirable as it may necessitate physical cleaning of the reactor or pumping the reactor out for a long time so that the residual condensate does not introduce subsequent run to run process variability. Condensation occurs at cold spots, including regions cooled by the expansion of a gas. Condensation can also result from temporary increases in P_m, which can occur when system valves open and close. To avoid condensation, CVD polymerization often operates with S significantly below unity. However, with care,

CVD polymer synthesis can be carried out at high values of S, resulting in both extremely high deposition rates and high molecular weight films.

1.2.3
Process Simplification and Substrate Independence

For polymers applied to a surface by any number of well-known solution-based methods including spin-coating, dip-coating, and ink-jet printing, a post-deposition curing step is often used for removing residual solvent and/or crosslinking the films. In this case, the overall process can be considered to have three steps: (i) synthesis of the bulk polymer from its monomer(s), (ii) thin film formation, and (iii) curing. In contrast, the synthesis of polymer chains from their monomers and the thin film formation occur in a single step in the CVD reactor. Since there is no residual solvent, no curing step is required after the deposition. Also, CVD can allow for *in situ* grafting of the film to its substrate and the formation of patterns in the same single step.

Solution-based techniques require that a substrate is wetted by the solution. Hence, different surface preparation protocols typically need to be developed and optimized for each type of substrate. This need for customization is absent in CVD because de-wetting effects are eliminated. Combining the dry processing conditions with low surface temperatures make CVD a substrate independent method. For instance, antibiofouling polymeric layers grow equivalently on silicon wafers and polymeric membranes that are placed side by side in a CVD chamber. Also, directly patterned vapor printed electrodes are formed using the identical process on a wide variety of textiles, plastics, and papers.

Another consideration is avoiding the dissolution of previously applied layers during the fabrication of multilayer stacks by solution processing. This requires identification of mutual exclusive solvents, a requirement which is obviated by the use of CVD.

1.3
Scale-Up and Commercialization

Much of the knowledge for scaling up the reactors for organic CVD processes is built upon the much larger and more mature field of CVD processes for inorganic materials [2, 4]. For the inorganic materials, CVD is well known for the repeatable synthesis of high-purity, defect-free films; for the capability to systematically tune film properties through variations in process parameters; and for the ability to be scaled-up and automated for the manufacture of uniform films on silicon wafers of ever larger diameter in the semiconductor industry. Understanding the chemical kinetics and fluid dynamics of CVD polymer reactors has allowed successful scale-up to batch reactors >1 m in width. Larger reactors offer the process throughput required for commercial applications.

The flexibility of iCVD (initiated chemical vapor deposition) polymers makes them compatible with substrates like plastics, papers, textiles, and membranes. Such flexible substrates are required for use with economical roll-to-roll processing. For commercialization, the all-dry nature of CVD has the potential to reduce environmental, health, and safety impacts associated with solvents as well as the economic costs associated with their use, recycling, or disposal.

In semiconductor manufacturing, multiple vacuum processing tools are integrated into clusters. These cluster tools are essential for producing high-quality interfaces because they allow wafers to be transferred between steps without(?) exposure to ambient contaminants. Similarly, the reactor designs for organic CVD are also amenable with integration with cluster tools, and hence offer the promise of hybrid organic–inorganic optoelectronic devices having high-quality interfaces.

At the scale of the laboratory bench, less investment in equipment is required to apply and cure polymer solutions to form thin films. However, at commercial scales, vacuum deposition is a proven and cost-effective technology. For example, large-scale roll-to-roll vacuum deposition processing is commonly utilized for applying barrier materials in the food packaging industry. In the semiconductor industry, spin track systems with automated wafer handling systems and a controlled humidity environment are of comparable cost to a CVD tool. However, for spin-on materials, additional capital investment is required for the highly controlled furnaces for curing the spin-on materials. Such curing equipment is not required for a single-step CVD process. While for many years, spin-on low dielectric constant materials were expected to be adopted by the semiconductor industry, current low dielectric constant materials are predominately manufactured by CVD. This choice was driven both by the superior performance of the CVD films and the higher than anticipated costs for spin-tracks, furnaces, and the high-purity specialty spin-on resins.

1.4 Process and Materials Chemistry

In many cases, CVD polymerization methods represent the direct translation of solution phase organic synthesis methods into heterogeneous processes for surface modification [4, 7]. The vast knowledge of traditional chain growth and step growth polymer synthesis provides insight into the chemical mechanisms responsible for the assembly of monomers into macromolecular units for achieving the properties desired in the resulting thin solid films. The use of each type of CVD process used, as discussed in following text, is determined by the polymerization mechanisms for the monomer(s) of interest and the resultant film properties desired.

1.4.1 Initiated CVD (iCVD) and Its Variants

The iCVD is a versatile method for synthesizing chain growth polymers. While iCVD has been predominately used for free radical polymerization, ionic polymerization has also been demonstrated [1, 8]. For the iCVD process, a thermal initiator is simultaneously introduced into the growth chamber along with the monomer(s). Heated filament wires suspended a few centimeters above the growth surface result in the selective thermal decomposition of the initiator. The resulting fragments react with the monomers adsorbed on the cooled substrate surface and polymerization ensues. Without the use of an initiator, film growth rates are quite low. Two variations of iCVD are *photo initiated chemical vapor deposition* (piCVD) and *initiated plasma enhanced chemical vapor deposition (iPECVD)*. For piCVD, a photosensitive initiator is decomposed to free radicals using a UV light. For iPECVD, low-power plasma excitation induces the decomposition of the initiator.

The iCVD method is compatible with vinyl monomers, including, but not limited to, acrylates, methacrylates, and styrenes. The highly desirable pendant functional groups of the monomers, such as epoxy, amine, or carboxylic acid moieties, can be fully retained in the iCVD films, even at high deposition rates. Full retention of the functional groups is essential for fabricating films that are responsive to changes in humidity, temperature, or pH. Most iCVD vinyl polymers are electrical insulators but using monomers with acetylenic bonds extends iCVD to conjugated semiconducting polymers. Monomers already available in commercial quantities at low cost are highly favored for commercial applications.

1.4.2 Plasma Enhanced CVD (PECVD)

The longstanding method of PECVD (plasma enhanced chemical vapor deposition) makes use of the partially ionized gas, known as plasma [4, 9, 10]. Electric fields necessary for ionization can be excited by a direct current, radio frequency, or a microwave frequency generator. The plasma is capable of monomer fragmentation that can self-initiate deposition. Thus, the introduction of an intentional initiator species is not required. Monomers for PECVD include those employed for iCVD as well as those that extend to other gases that react upon fragmentation. However, monomer fragmentation reduces the incorporation of desired functional groups into the PECVD films, and can also result in branching, undesirably high degrees of crosslinking, and "dangling bond" defects (e.g., an unsatisfied bond with an unpaired electron). For this reason, PECVD films often lack the regular repeat units found in conventional polymers. Lowering the plasma power and/or using process variations such as pulsed, atmospheric, or downstream PECVD, can reduce monomer fragmentation but often reduces the film growth rates as well. Dangling bond defects inside the film can lead to reactions with oxygen and water from the air to produce undesirable aging effects. The use of plasma excitation at the start of a deposition can be highly advantageous because subsequent reactions of the free radical sites with the film forming species can greatly improve adhesion.

1.4.3
Poly(p-xylylene) (PPX) and Its Derivatives ("Parylenes")

Interestingly, some CVD polymers, such as poly(*p*-xylylene) (PPX) and its various functional variants, commonly termed "parylenes" have no direct analog in solution chemistry [11–14]. Volatizing and pyrolyzing the solid [2,2]paracyclophane dimer produces a monomer that absorbs and polymerizes on a cooled surface. The dimer and several of its substituted forms are commercially available. The corresponding dielectric films, particularly parylene C, which contains one chlorine per monomer unit, have a long history of commercial use for the encapsulation of biomedical devices and for the protection of electronics. Several transition metals, metal salts, and organometallic complexes inhibit PPX deposition, and this inhibition effect has been exploited to achieve additive patterning. Deposition has also been demonstrated using custom synthesized dimers possessing a variety of functional groups including, amino, hydroxyl, anhydride, triflate, and trifluoroacetyl moieties. Using the same type of reactor as for PPX, the electroluminescent material, poly(phenyl vinylene), can be synthesized for integration into light emitted diodes.

1.4.4
Oxidative CVD (oCVD)

The oCVD (oxidative chemical vapor deposition) method enables the step growth of conjugated polymers that are typically either semiconductors or electrical conductors [15]. For oCVD, both the oxidant and monomer impinge on the substrate through the vapor phase. Adsorption and spontaneous surface reaction proceeds directly on the substrate without the need for heated filaments, plasma excitation, or UV exposure. oCVD poly(ethylenedioxythiophene) (PEDOT) displays both high conductivity and transparency as desired for the integration of an electrode material into flexible electronic devices. The conductivity, transparency, and work function of oCVD PEDOT can be systematically tuned. Additionally, oCVD PEDOT can be grafted to its substrate for producing novel organic/inorganic hybrid interfaces as well as for ensuring resistance to delamination. Sufficient adhesion is essential for high-resolution lithographic patterning.

1.4.5
Vapor Deposition Polymerization (VDP) and Molecular Layer Deposition (MLD)

Both VDP (vapor deposition polymerization) and MLD (molecular layer deposition) are useful for step growth polymers synthesized through condensation reactions [13, 16, 17]. For VDP, reactive bifunctional monomers are simultaneously coevaporated and then spontaneously polymerize on the substrate, releasing a condensation by-product. For example, the VDP of diamines with dianhydrides yields poly(amic acid) films by the elimination of water. Also, the condensation reaction of carbonyl chloride and amine groups, which releases HCl, has also proved useful for forming nylons and for Kevlar® by VDP.

Alternating the introduction of the difunctional precursors under conditions that lead to self-limited surface reactions, differentiates MLD from VDP. In analogy to the related method of atomic layer deposition (ALD) for inorganic materials, MLD excels at producing uniform, defect-free, conformal coatings with thicknesses as low as 5 nm. MLD is also useful for hybrid organic–inorganic alucone and zincone polymers. The growth rates of MLD films are limited by the number of reactive surface sites and the experimental time required to switch between the different precursors.

1.4.6 Additional Methods

Other CVD polymerization techniques which have been less commonly used to date include ring-open polymerization by free radical, anionic, and cationic mechanisms; ring-opening metathesis polymerization (ROMP); and polyaddition to from polypeptides [4, 15, 18].

Additional methods have been developed, which are closely allied to CVD. Here, not all of the reactants are introduced through the vapor phase. Examples include the pre-application of low volatility initiators or oxidants to a surface, typically from a solution, before exposure to volatile monomer(s). A broad range of synthesis strategies are opened up by the pre-application of less volatile reactants. For chain growth polymers, the pre-application of initiators is used for *vapor-phase assisted surface polymerization (VASP)* for living free-radical polymerization, *surface-initiated vapor deposition polymerization (SI-VDP)* for nitroxide-mediated synthesis, and *gas-phase assisted surface polymerization (GASP)* for atom transfer radical polymerization (ATRP). *Vapor phase polymerization (VPP)* uses the pre-application of oxidant for step growth polymerization of conjugated polymers. In each of these methods, the available supply of the pre-applied reactant is fixed. Additionally, the pre-application is a second processing step, which introduces additional cost and integration challenges.

1.5 Summary

CVD polymerization represents the intersection of all-dry and scalable microfabrication technology with the chemistry of functional and responsive organic materials. Unique applications enabled by CVD polymers derive from the ability to avoid high substrate temperatures and solvents. In a single step, vapor-phase monomers undergo selective reactions to produce high-purity, conformal, and durable polymeric layers. CVD polymer processes for functionalizing solid surfaces have only begun to be exploited for novel surface modification and device applications.

Acknowledgments

This work was supported by the MIT Institute for Soldier Nanotechnologies (ISN) under Contract DAAD-19-02D-0002 with the U.S. Army Research Office, the Masdar Institute, the Office of Naval Research. The National Science Foundation Scalable Nanomanufacturing Program, Shell, and Chervon, are also kindly acknowledged.

References[1)]

1. Coclite, A.M., Howden, R.M., Borrelli, D.C., Petruczok, C.D., Yang, R., Yagüe, J.L., Ugur, A., Chen, N., Lee, S., Jo, W.J., Liu, A., Wang, X., and Gleason, K.K. (2013) *Adv. Mater.*, **25**, 5392.
2. Reeja-Jayan, B. Kovacik, P., Yang R., Hossein S., Ugar A., Kim D.H., Petruczok C.D., Wang X., Liu A., and Gleason K.K. (2014) A route towards sustainability through engineered polymeric interfaces. *Adv. Mater. Interfaces*, **1**, 1400117, doi: 10.1002/admi.201400117.
3. Ozaydin-Ince, G., Coclite, A.M., and Gleason, K.K. (2012) *Rep. Prog. Phys.*, **75**, 016501.
4. Alf, M.E., Asatekin, A., Barr, M.C., Baxamusa, S.H., Chelawat, H., Ozaydin-Ince, G., Petruczok, C.D., Sreenivasan, R., Tenhaeff, W.E., Trujillo, N.J., Vaddiraju, S., Xu, J., and Gleason, K.K. (2010) *Adv. Mater.*, **22**, 1993.
5. Baxamusa, S.H., Im, S.G., and Gleason, K.K. (2009) *Phys. Chem. Chem. Phys.*, **11**, 5227.
6. Vedula, R., Kaza, S., and Desu, S.B. (2001). *Chemical Vapor Deposition of Polymers: Principles, Materials, and Applications*, (eds J.-H. Park and T.S. Sudarshan), Copyright © 2001 ASM International®, pp. 243–285 doi: 10.1361/chvd2001p243
7. Im, S.G. and Gleason, K.K. (2011) *AIChE J.*, **57**, 276.
8. Yagüe, J.L., Coclite, A.M., Petruczok, C., and Gleason, K.K. (2013) *Macromol. Chem. Phys.*, **214**, 302.
9. Yasuda, H. (1985) *Plasma Polymerization*, Academic Press.
10. D'Agostino, R. (1990) *Plasma Deposition, Treatment, and Etching of Polymers*, Academic Press.
11. Fortin, J.B. and Lu, T. (2004) *Chemical Vapor Deposition Polymerization: The Growth and Properties of Parylene Thin Films*, Kluwer Academic Publishers.
12. Chen, H. and Lahann, J. (2011) *Langmuir*, **27**, 34.
13. Sreenivasan, R. and Gleason, K.K. (2009) *Chem. Vap. Deposition*, **15**, 77.
14. Vaeth, K.M. and Jensen, K.F. (1997) *Appl. Phys. Lett.*, **71**, 2091.
15. Bhattacharyya, D., Howden, R.M., Borrelli, D.C., and Gleason, K.K. (2012) *J. Polym. Sci., Part B: Polym. Phys.*, **50**, 1329.
16. Zhou, H. and Bent, S.F. (2013) *J. Vac. Sci. Technol.*, 040801. doi: 10.1116/1.4804609
17. George, S.M., Yoon, B., and Dameron, A.A. (2009) *Acc. Chem. Res.*, **42**, 498.
18. Lahann, J. (2006) *Polymer Int.*, **1370** (55), 1361.

1) Only review articles and books have been cited. The reader is advised to consult these or the subsequent chapters in this volume for more detail and for citations to original research papers.

Part I
Fundamentals

CVD Polymers: Fabrication of Organic Surfaces and Devices, First Edition. Edited by Karen K. Gleason.

2
Growth Mechanism, Kinetics, and Molecular Weight

Kenneth K. S. Lau

2.1
Introduction

In the chemical vapor deposition (CVD) of polymers, as with any chemical synthesis or process, a detailed understanding of the fundamental phenomena underlying the overall procedure is highly beneficial toward enabling effective integration of the deposited materials in desired applications. The ability to control the factors influencing the polymer growth process is important in deriving the requisite polymer chemical makeup and physical morphology. This operating knowledge is also critical in achieving uniform thin films that conform to the substrate surface topology, which in many practical cases can be quite complex, for example, high aspect ratio features, narrow tortuous channels, or interconnected porous networks.

In this chapter, we will review the body of literature studying the growth behavior and kinetics of polymer CVD, specifically of initiated chemical vapor deposition (iCVD). The iCVD process as will be described in this chapter and in a large portion of this book is unique in enabling the synthesis of polymers by borrowing valuable processing knowledge from conventional liquid phase polymerization approaches. As a result, the iCVD technique has been highly successful in making a wide array of polymers and for a wide range of applications, and a deeper understanding of the iCVD process will naturally help facilitate rapid development of future applications. The growth behavior of other polymer CVD methods, including photo-initiated chemical vapor deposition (piCVD), oxidative chemical vapor deposition (oCVD), initiated plasma-enhanced chemical vapor deposition (iPECVD), parylene CVD, and molecular layer deposition (MLD), will be discussed in subsequent chapters.

Here, the general processing scheme of iCVD will first be described to highlight the features of this polymer CVD approach. Then, the kinetic and transport phenomena and growth mechanism related to the iCVD process will be discussed, particularly by detailing the growth behavior of several major classes of polymer chemistries, including fluorocarbon, organosilicon, acrylate and methacrylate, vinyl, and ring-opened polymers. The chapter will conclude by emphasizing

CVD Polymers: Fabrication of Organic Surfaces and Devices, First Edition. Edited by Karen K. Gleason.

the value of processing knowledge in effecting the structure and properties of resulting materials and devices.

2.2 iCVD Process

The iCVD process in general involves many components expected of any CVD process: (i) precursor selection, (ii) vapor delivery, (iii) reaction chamber, (iv) temperature control, (v) pressure control, (vi) exhaust management, (vii) vacuum pump system, and (viii) process monitoring and other peripherals. However, there are several important features that make iCVD unique. As illustrated by the process scheme shown in Figure 2.1, iCVD involves the following main steps:

- *Steps 1 and 2*: vapor delivery of two main precursors, a monomer (or two or more co-monomers) and a corresponding initiator, into the iCVD reaction chamber.
- *Step 3*: adsorption of monomer vapor onto the cooled substrate surface.
- *Step 4*: activation of initiator vapor by the array of resistively heated filament wires suspended above the substrate to form an activated initiator.
- *Step 5*: polymerization at the substrate surface to form a growing film, first by initiation through the reaction of the activated initiator with an adsorbed monomer molecule, subsequent propagation by the addition of more adsorbed monomer units, and finally termination by capping the active polymer chain ends.

a) *Precursor selection.* One of the unique features of iCVD polymerization is the use of precursors that are commonly used in liquid phase polymerization. For example, vinyl, acrylate, and methacrylate monomers can be polymerized using free radical initiators like organic peroxides, or ethylene oxide monomer can be polymerized using ionic initiators like boron trifluoride. Aside from these main precursors, other additives can be included, such as crosslinking agents.

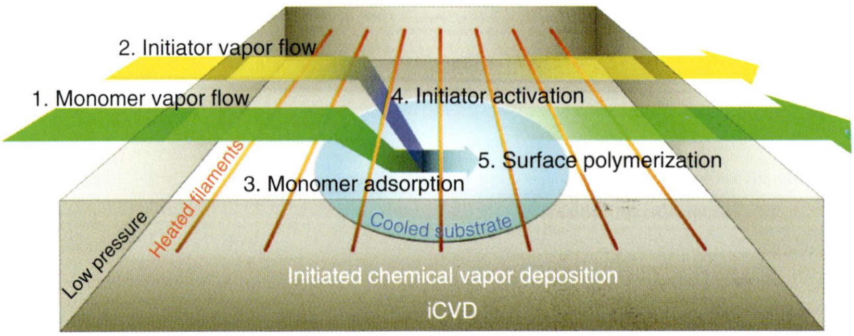

Figure 2.1 General process scheme of initiated chemical vapor deposition (iCVD).

b) *Vapor delivery.* Since most of the monomers and initiators are liquids, they generally need to be vaporized for delivery into the reaction chamber. This can be achieved through direct heating of source vessels containing neat liquid precursors to increase their vapor pressure. Heated mass flow controllers or needle valves can be used to accurately meter in the flow at set flow rates. As an alternative, a carrier gas can be used to help deliver the precursor flow.
c) *Reaction chamber.* Several different reactor configurations have been used for iCVD, including flat cylindrical, rectangular, tall barrel, rotating bed, and roll-to-roll systems. The main common feature in these reactors is the use of an array of filament wires on an electrically insulating support that is positioned a set distance from and parallel to the substrate surface to be coated. The vapor flow direction can either be parallel or perpendicular to the substrate and filament array plane.
d) *Temperature control.* There are two separate temperature regions in iCVD. First is the temperature of the heated filament wires (typically 200–400 °C), which can be controlled through the use of a power supply that maintains a constant power loading. Second is the temperature of the substrate (typically 0–30 °C), which can be controlled through backside contact with a stage maintained at a set temperature using a thermal fluid from a recirculating chiller/heater.
e) *Pressure control.* The iCVD polymerization process is carried out at low yet mild pressures (typically 0.1–10 Torr). Pressure in the reactor is generally controlled by a downstream throttle valve in a feedback loop with a pressure gauge and controller. As a practical consideration, the pressure is usually maintained such that the partial pressures of the monomer and initiator inside the reactor are below their saturation pressures to prevent liquid condensation on the substrate.
f) *Exhaust management.* Since iCVD is operated as a continuous, steady flow system, there will be gaseous vapors that leave the reactor. These include unreacted precursors given that precursor conversion is far from complete (typically <10%) and there may be specific chemical reactions that lead to gaseous by-products. These reactive vapors need to be scavenged from the exhaust stream prior to entering the vacuum pump system and this can be done through traps and scrubbers.
g) *Vacuum pump system.* To achieve the mildly low pressures, a roughing pump is typically sufficient, although depending on the size and intended use of the iCVD system, an additional roots blower can be added. For oil-based pump systems, fluorocarbon oils are preferred due to their inertness, but ideally a dry vacuum system is more desirable to eliminate oil backstreaming, reactor contamination, and potential oil degradation from reaction with exhaust gases.
h) *Process monitoring/other peripherals.* To monitor the growing polymer film thickness, laser interferometry or ellipsometry can be used for the case where the substrate has a reflective surface. In other cases, a quartz crystal microbalance (QCM) can be used to obtain real-time film growth rate during deposition. Other diagnostic tools, like a downstream gas phase Fourier

transform infrared (FTIR) spectroscopy or mass spectrometer system, can be added to determine the gas phase species and concentration in the iCVD reactor.

As is apparent, iCVD is similar to hot wire chemical vapor deposition (HWCVD) or hot filament chemical vapor deposition (HFCVD) in that heated filaments are used to enable reaction pathways for deposition. However, the key difference with iCVD is the deliberate use of a separate initiator that allows much lower filament temperatures for activation; iCVD is typically in the range of 200–400 °C while HWCVD is generally in excess of 800 °C. As a result of the much lower activation temperature, the monomer precursor structure is protected from thermal degradation and only the initiator is thermally activated selectively. This highly defined reaction pathway leads ultimately to the formation of correspondingly well-defined polymers, and it is therefore not too surprising that iCVD has been extremely successful in making a wide variety of polymers. Table 2.1 gives a comprehensive list of iCVD polymers derived to date from different sets of monomers and initiators that include fluorocarbons [1–20], organosilicons [14, 20–40], acrylates [37, 38, 40–89], methacrylates [15, 38, 44, 45, 49–51, 53, 54, 59, 66, 68, 69, 73, 80–86, 89–155], styrenes [42, 43, 114, 156–162], vinyls [46, 47, 54, 55, 76, 157, 160–174], crosslinked polymers [41, 46, 47, 49, 50, 55–58, 61, 62, 67, 72, 75–79, 87–90, 95–98, 115, 121, 129, 134–140, 142, 143, 147, 148, 153, 167, 168, 170, 174–179], and ring-opened polymers [37, 180–182].

2.3 Kinetics and Growth Mechanism

The ability of iCVD to synthesize such a large number of polymers stems in large from the processing knowledge gained over the course of its development as a polymer CVD technique. With the iCVD process, there are several key process parameters that can be externally controlled – precursor selection, precursor flow rates, reactor pressure, filament temperature, and substrate temperature. Here, we will discuss relevant kinetic and parametric studies for different classes of polymers to understand the underlying factors and mechanisms dictating polymer growth behavior in iCVD. The discussions of these studies are arranged in more or less a chronological order to further understand the rationale of iCVD's process development.

2.3.1 Fluorocarbon Polymers

One of the very first polymers synthesized by iCVD is polytetrafluoroethylene (PTFE), more commonly known by its DuPont trademark as Teflon. The general iCVD reaction pathway in making PTFE uses hexafluoropropylene oxide (HFPO;

Table 2.1 iCVD polymer chemistries.[a)]

Polymer class	Monomer/precursor[b)]	Initiator[c)]	References[d), e), f)]
Fluorocarbon	**HFPO**	—	**[1]**, [2], [3], [4], [5], [6], [7], [8], [9], [10], [11]
	—	**PFBSF**	[12], [13], [14], [15], [16], [17], [18]
	—	**PFOSF**	**[19]**
	HFPO + D_3	—	[20]
Organosilicon	$\mathbf{D_3}$	—	**[21]**
	$\mathbf{D_3}$ + HFPO	—	[20]
	$\mathbf{D_4}$	—	**[21]**, **[22]**
	HVDS	**TBPO**	**[23]**, [24]
	HVDS + V_3D_3	—	**[25]**, [26]
	$\mathbf{V_3D_3}$	**PFOSF**	[14], **[27]**, [28], [29]
	—	**TBPO**	[26], **[30]**, [31], [32], [33], [34], [35], [36], **[37]**
	$\mathbf{V_3D_3}$ + HVDS	**TBPO**	**[25]**, [26]
	$\mathbf{V_4D_4}$	**TBPO**	[38], [39], [40]
Acrylate	**AA** + NIPAAm + EGDA	**TBPO**	[41]
	DEAAm + EGDA	**TBPO**	[41]
	DEAAm + DMAAm +	**TBPO**	[41]
	DEAEA	**TAPO**	**[42]**, **[43]**
	DFHA + GMA	**TBPO**	[44], [45]
	DMAAm + MAH + DEGDVE	**TBPO**	[46], [47]
	DMAAm + DEAAm +	**TBPO**	[41]
	DMAAm + NIPAAm +	**TBPO**	[41]
	EA	**TAPO**	**[48]**
	EA + MAA	**TBPO**	[49], [50]
	MCA	**TBPO**	[51]
	MCA + MAA	**TBPO**	[51]
	nBA	**TAPO**	**[48]**, **[52]**
	—	**TBPO**	[38, 53, 54]
	nBA + DEGDVE	**TBPO**	[55]
	nHA	**TAPO**	**[48]**
	nPA	**TAPO**	**[48]**
	nPrA	**TAPO**	**[48]**
	NIPAAm	**TBPO**	**[37]**, [56], [57], [58], [59], [60]
	NIPAAm + DEGDVE	**TBPO**	[56], [57], [58]
	NIPAAm + EGDA	**TBPO**	[41], [61]
	NIPAAm + EGDMA	**TBPO**	[62]
	NIPAAm + AA + EGDA	**TBPO**	[41]
	NIPAAm + DMAAm + EGDA	**TBPO**	[41]
	PFDA	**TBPO**	[37], [40], [54], [59], **[63]**, [64], **[65]**, [66], [67], [68], [69], [70], [71], [72], [73], [74], [75], [76]
	PFDA + DVB	**TBPO**	[72], [75], [77]
	PFDA + EGDA	**TBPO**	[78], [79]
	PFDA + HEMA	**TBPO**	[80], [81], [82], [83], [84]

(continued overleaf)

Table 2.1 (*Continued*)

Polymer class	Monomer/precursor[b]	Initiator[c]	References[d), e), f)]
	PFDA + MAA	**TBPO**	[85], [86]
	PrgA	**TBPO**	[87]
	PrgA + EGDA	**TBPO**	[87]
	tBA	**TBPO**	[88]
	tBA + DEGDVE	**TBPO**	[55], [88], [89]
Methacrylate	**CHMA**	**TAPO**	[90]
	—	**TBPO**	**[91]**, **[92]**, [93], [94]
	—	**TBPOB**	[94]
	CHMA + EGDMA	**TAPO**	[90]
	—	**TBPO**	[95]
	DMAEMA + EGDA	**TBPO**	[96]
	DMAEMA + EGDMA	**TBPO**	[97], [98]
	DPAEMA	**TBPO**	[99]
	FMA	**TBPO**	**[100]**
	GMA	**TAPO**	[15], [90], [101], [102]
	—	**TBPO**	[73], **[91]**, **[103]**, [104], **[105]**, [106], [107], [108], [109], [110], [111], [112], **[113]**, [114], **[115]**, [116], [117], [118], [119], [120]
	GMA + DFHA	**TBPO**	[44], [45]
	GMA + EGDA	**TBPO**	[121]
	GMA + PFEMA	**TBPO**	[45]
	HEMA	**TAPO**	[122], [123], [124], [125], [126], [127], [128]
	—	**TBPO**	[38], [53], [54], [59], [66], [68], [129], [130], [131], [132], [133]
	HEMA + EGDA	**TBPO**	[89], **[129]**, [134], [135], [136], [137]
	HEMA + EGDMA	**TBPO**	[138], [139]
	HEMA + PFDA	**TBPO**	[80], [81], [82], [83], [84]
	HEMA + PFM	**TBPO**	[66]
	HEMA + PFM + EGDA	**TBPO**	[140]
	HFDMA	**TBPO**	[141]
	MAA	**TBPO**	[51], [131]
	MAA + EA	**TBPO**	[49], [50]
	MAA + EGDA	**TBPO**	[142]
	MAA + EGDMA	**TBPO**	[49], [50], [143]
	MAA + MCA	**TBPO**	[51]
	MAA + PFDA	**TBPO**	[85], [86]
	MAA + PFEMA	**TBPO**	[44], [45]
	MMA	**TBPO**	[144], **[145]**, **[146]**
	nBMA	**TBPO**	[54]
	neoPMA	**TBPO**	[147], **[148]**, **[149]**
	neoPMA + EGDA	**TBPO**	[147]
	oNBMA	**TBPO**	[69], [131], [150]
	PFEMA	**TBPO**	[151], [152]
	PFEMA + GMA	**TBPO**	[45]

(continued overleaf)

Table 2.1 (*Continued*)

Polymer class	Monomer/precursor[b)]	Initiator[c)]	References[d), e), f)]
	PFEMA + MAA	**TBPO**	[45], [44]
	PFM	**TBPO**	[38], [54], [69], [153]
	PFM + EGDA	**TBPO**	[153]
	PFM + HEMA	**TBPO**	[66]
	PFM + HEMA + EGDA	**TBPO**	[140]
	PrgMA	**TBPO**	[154], [155]
Styrene	**4AS**	**TBPO**	[114], [156]
	4AS + MAH	**TBPO**	[157]
	DMAMS	**TAPO**	**[42]**, **[43]**, [158]
	—	**TBPO**	[159]
	S + MAH	**TBPO**	**[160]**, **[161]**, [162]
Other vinyls	**4VP**	**TBPO**	[54], [163], [164], [165], [166]
	4VP + DVB	**TBPO**	[167], [168]
	4VP + EGDMA	**TBPO**	[76]
	mDEB	**TBPO**	[169]
	MAH + 4AS	**TBPO**	[157]
	MAH + DEGDVE	**TBPO**	[55], [170]
	MAH + S	**TBPO**	**[160]**, **[161]**, [162]
	MAH + DMAAm + DEGDVE	**TBPO**	[46], [47]
	MAH + VP + DEGDVE	**TBPO**	[170]
	NVCL	**TBPO**	[171]
	PhAc	**TBPO**	[172]
	VI	**TBPO**	**[173]**
	VI + VP	**TBPO**	[173]
	VP	**TBPO**	[173], **[174]**
	VP + EGDA	**TBPO**	**[174]**
	VP + VI	**TBPO**	[173]
	VP + MAH + DEGDVE	**TBPO**	[170]
Crosslinking	**DEGDVE** + MAH	**TBPO**	[55], [170]
	DEGDVE + nBA	**TBPO**	[55]
	DEGDVE + NIPAAm	**TBPO**	[56], [57], [58]
	DEGDVE + tBA	**TBPO**	[55], [88], [89]
	DEGDVE + DMAAm + MAH	**TBPO**	[46], [47]
	DEGDVE + MAH + VP	**TBPO**	[170]
	DVB	**TBPO**	[67], [167], [175]
	DVB + 4VP	**TBPO**	[167], [168]
	DVB + PFDA	**TBPO**	[72], [75], [77]
	EGDA	**TBPO**	[115], [137], [147], **[148]**, [153], **[175]**, **[176]**, [177], [178], [179]
	EGDA + DEAAm	**TBPO**	[41]
	EGDA + DMAEMA	**TBPO**	[96]
	EGDA + EGDMA	**TBPO**	[176]
	EGDA + GMA	**TBPO**	[121]
	EGDA + HEMA	**TBPO**	[89], **[129]**, [134], [135], [136], [137]
	EGDA + MAA	**TBPO**	[142]
	EGDA + neoPMA	**TBPO**	[147]

(continued overleaf)

Table 2.1 (Continued)

Polymer class	Monomer/precursor[b)]	Initiator[c)]	References[d), e), f)]
	EGDA + NIPAAm	**TBPO**	[41], [61]
	EGDA + PFDA	**TBPO**	[78], [79]
	EGDA + PFM	**TBPO**	[153]
	EGDA + PrgA	**TBPO**	[87]
	EGDA + VP	**TBPO**	**[174]**
	EGDA + AA + NIPAAm	**TBPO**	[41]
	EGDA + DMAAm + DEAAm	**TBPO**	[41]
	EGDA + DMAAm + NIPAAm	**TBPO**	[41]
	EGDA + HEMA + PFM	**TBPO**	[140]
	EGDMA	**TBPO**	[76], [176]
	EGDMA + 4VP	**TBPO**	[76]
	EGDMA + CHMA	**TAPO**	[90]
	—	**TBPO**	[95]
	EGDMA + DMAEMA	**TBPO**	[97], [98]
	EGDMA + EGDA	**TBPO**	[176]
	EGDMA + HEMA	**TBPO**	[138], [139]
	EGDMA + MAA	**TBPO**	[49], [50], [143]
	EGDMA + NIPAAm	**TBPO**	[62]
Ring opening	**EO**	**BF_3-DEE**	**[180]**
	IBF	—	[181]
	MDO	**TBPO**	[37]
	Trioxane	—	[182]

a) Only iCVD chemistries, that is, excludes piCVD, iPECVD, oCVD, and other polymer CVD variants.

b) 4AS = 4-amino styrene; 4VP = 4-vinyl pyridine; AA = acrylic acid; CHMA; D_3; D_4; DEAAm = diethyl acrylamide; DEAEA = diethylaminoethyl acrylate; DEGDVE = di(ethylene glycol) divinylether; DFHA = dodecafluoroheptyl acrylate; DMAAm = dimethyl acrylamide; DMAEMA = *N*,*N*-dimethylaminoethyl methacrylate; DMAMS = dimethylaminomethyl styrene; DPAEMA = 2-diisopropylaminoethyl methacrylate; DVB = divinyl benzene; EA; EGDA; EGDMA = ethylene glycol dimethacrylate; EO = ethylene oxide; FMA = furfuryl methacrylate; GMA; HEMA; HFDMA = heptadecafluorodecyl methacrylate; HFPO; HVDS; IBF = isobenzofuran; mDEB = *m*-diethynyl benzene; MAA = methacrylic acid; MAH = maleic anhydride; MCA = methyl α-chloroacrylate; MDO = 2-methylene-1,3-dioxepane; MMA = methyl methacrylate; nBA; nBMA = *n*-butyl methacrylate; nHA = *n*-hexyl acrylate; nPA = *n*-pentyl acrylate; nPrA = *n*-propyl acrylate; neoPMA = neopentyl methacrylate; NIPAAm = *N*-isopropyl acrylamide; NVCL = *N*-vinyl caprolactam; oNBMA = *o*-nitrobenzyl methacrylate; PFDA; PFEMA = perfluoroalkylethyl methacrylate; PFM = pentafluorophenyl methacrylate; PhAc = phenyl acetylene; PrgA = propargyl acrylate; PrgMA = propargyl methacrylate; S = styrene; tBA = *t*-butyl acrylate; V_3D_3; V_4D_4 = tetravinyltetramethylcyclotetrasiloxane.

c) BF_3 – DEE = boron trifluoride-diethyl etherate; PFBSF = perfluorobutanesulfonyl fluoride; PFOSF; TAPO; TBPO; TBPOB.

d) References exclude reviews.

e) References with comonomers are duplicated depending on which monomer in the comonomer series is cited first.

f) References in **bold** contain kinetic data.

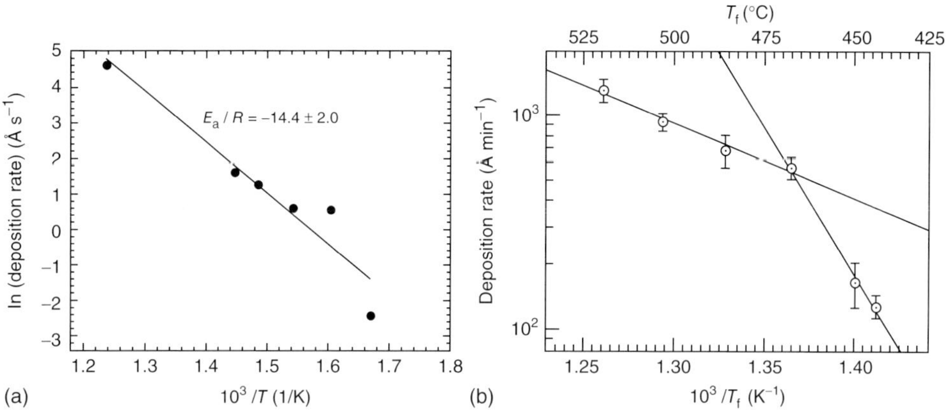

Figure 2.2 iCVD of polytetrafluoroethylene. (a) Arrhenius plot of deposition rate versus filament temperature from using self-initiated HFPO monomer [1]. Reprinted with permission from Ref. [1]. Copyright 1996, AIP Publishing LLC.) (b) Arrhenius plot of deposition rate versus filament temperature from using HFPO monomer and PFOSF initiator [19]. (Reprinted with permission from Ref. [19]. Copyright 2001, American Chemical Society.)

$CF_3CF(O)CF_2$) as the monomer precursor (see Table 2.1 and also Chapter 10). HFPO is a gas that is known to thermally decompose at temperatures as low as 150 °C to form difluorocarbene (CF_2) radicals with trifluoroacetyl fluoride ($CF_3C(=O)F$) being a relatively stable by-product gas [183–185]. These long-lived singlet radicals are then able to join together to form long chain $(CF_2)_n$ molecules of PTFE that is deposited on the substrate. In this way, HFPO behaves as an initiator as well, giving rise to self-initiated polymerization. Compared to tetrafluoroethylene as another potential monomer, the thermal decomposition of HFPO to form CF_2 is found to be much more favorable due to the strained epoxide ring versus the alkene double bond (ΔH_{rxn} of 23.6 vs 69.7 kcal mol^{-1}), [186] thereby allowing the significantly milder activation temperatures (>150 vs >800 °C) that makes iCVD unique.

Limb *et al.* studied the deposition kinetics of the self-initiated polymerization chemistry of HFPO to form PTFE [1]. Using a barrel-type reactor with a Nichrome filament wire, HFPO was thermally decomposed over a filament temperature range of 598–808 K. All other conditions were kept constant, with the reactor pressure at 1 Torr, substrate temperature at 293 K, and HFPO flow rate at 12.5 sccm. Fairly high deposition rates are obtained in this study, for example, a rate of 1.8 μm h^{-1} is achieved at a filament temperature of 691 K. As shown in Figure 2.2a, the measured deposition rate of the PTFE film as a function of filament temperature can be plotted in Arrhenius form to yield a linear relationship and from which an apparent activation energy (E_a) of 29 kcal mol^{-1} is derived. This value is close to the experimentally reported E_a of 31 kcal mol^{-1} for the thermal decomposition of HFPO [187]. As a result, it can be concluded that the thermolysis of HFPO to CF_2 is a key reaction step in the formation of PTFE.

In a density functional theory (DFT) study of the reaction pathways of HFPO and CF_2, Lau *et al.* found further support that there is a favorable route for the linking of CF_2 to form PTFE chains [186]. Based on the Becke and 3-parameter Lee-Yang-Parr hybrid functional (B3LYP) calculations with Gaussian basis sets, the addition of each CF_2 unit to an unterminated CF_2 chain radical is estimated to have an average reaction enthalpy (ΔH_{rxn}) of $-48.7\,kcal\,mol^{-1}$. On the other hand, the addition of a CF_2 unit to a perfluoroalkene double bond becomes increasingly endothermic. These results indicate that CF_2 is the active polymerization species, while the recombination of CF_2 to form tetrafluoroethylene (C_2F_4) represents essentially a dead-end to polymerization at least under typical iCVD operating conditions. This latter point has been verified with independent batch pyrolysis experiments [185] as well as with downstream gas phase FTIR analysis of the iCVD reactor effluent [2] that showed C_2F_4 to be a dominant gas phase by-product species of HFPO decomposition.

The ease with which CF_2 radicals can link up to form chains is further indicated by Cruden *et al.*, who performed several detailed UV-absorption studies to determine the concentration and fate of CF_2 during the pyrolysis of HFPO and the effect on the growth of PTFE [7, 8]. The deposition rate is found to be noticeable only when the CF_2 concentration reaches some threshold plateau. They suggest with the support of a kinetic model that this is due to the formation of oligomeric CF_2 species in the gas phase which consumes CF_2 and which are necessary to induce surface nucleation and growth, particularly under high CF_2 concentration, for example, high reactor pressure or high filament temperature. Similarly, Zhou *et al.* used a computational fluid dynamics (CFD) model to describe the deposition behavior of PTFE from HFPO pyrolysis and found that the significantly higher experimental growth rate at higher filament temperature, which deviated significantly from the model prediction, can be explained by the formation of CF_2 oligomers as growth species [9].

The choice of filament wire material can influence deposition rate, for example, the use of Nichrome gives a much higher rate compared to alumina (10–100 vs $<10\,nm\,min^{-1}$) [8]. A study of different filament materials by Yasuoka *et al.* showed that Ni has a catalyzing effect, although this catalytic activity can be poisoned by the carbonization and fluorination of the filament over time, particularly at high temperatures [10]. The filament can somewhat be regenerated by using an atomic H treatment.

The use of a separate initiator for the formation of PTFE using the HFPO chemistry is first reported by Pryce Lewis *et al.* [19]. As a result of the addition of a small amount of an initiator, perfluorooctanesulfonyl fluoride (PFOSF), a significant enhancement in deposition rate is observed. As shown in Figure 2.2b, an Arrhenius plot of the deposition rate with filament temperature again shows a linear trend; however, this time there are two rate-limiting regimes with a transition temperature at around 460 °C. Below 460 °C, the deposition rate increases strongly with higher filament temperature, yielding an apparent E_a of $263\,kJ\,mol^{-1}$ that is close to the C–S bond energy of $272\,kJ\,mol^{-1}$, [188] suggesting a reaction-limited regime in which the thermolysis of PFOSF is the rate determining step. Above 460 °C, the deposition rate increases much more slowly with filament temperature, yielding an apparent E_a of 68 kJ mol^{-1} that is attributed to a mass transfer-limited regime. These rate transitions are commonly observed in other CVD reaction systems, for example, in the thermal CVD of epitaxial silicon thin films from silane chemistries [189].

Pryce Lewis *et al.* proposed that the PFOSF initiator thermally decomposes in the vicinity of the heated filament wires to produce perfluoroalkyl radicals, which then act as initiation points for the addition of CF_2 units derived from the separate thermal decomposition of HFPO as the monomer precursor [19]. Owing to the larger perfluoroalkyl radicals compared to CF_2, they are believed to adsorb much more easily on the substrate surface, that is, a higher sticking coefficient, that can explain the faster observed polymer film growth. Interestingly, with both self-initiated [2] and PFSOF-initiated [19] polymerization, it is found that under low deposition rate and initiation-limited conditions, rod-like grains of PTFE are formed as evidenced by atomic force microscopy (AFM), while under higher deposition rate conditions in which polymer chain initiation or nucleation is sufficiently rapid and no longer limiting, spherical nodular grains are observed instead. This emphasizes that kinetics can have a great impact on the resulting film morphology.

In addition, by choosing different initiators to use with the HFPO chemistry, the deposition rate of PTFE can be tuned such that high growth rates >1 μm min^{-1} can be achieved [190]. Further, the choice of the initiator can influence end group functionality on the polymer chain, for example, using a methyl-terminated initiator results in PTFE chains having CH_3 end groups instead of CF_3 when using PFOSF [190].

2.3.2
Organosilicon Polymers

Another early polymer system adapted by iCVD is the general class of organosilicon polymers containing the siloxane Si–O–Si bond (see Table 2.1). Kwan *et al.* first used a cyclic siloxane ring monomer precursor, D_4 or octamethylcyclotetrasiloxane, that was self-initiated by a single heated tantalum wire to demonstrate the growth of organosilicon polymer [22]. Subsequently, Pryce-Lewis *et al.* relied on the self-initiated polymerization of D_4 and D_3, hexamethylcyclotrisiloxane, to form organosilicon polymers using a parallel array of filament wires that gives more uniform heating over a larger substrate area [21]. Based on spectroscopic evidence and the Arrhenius rate dependence on filament temperature, film growth is believed to occur through the incorporation of linear structures from ring opening (Si–O) reactions as well as cyclic structures from methyl abstraction (Si–C) of siloxane rings. With self-initiation, much higher filament temperatures are required (>500–800 °C) to dissociate the Si–O and Si–C bonds of the monomer precursor in order to derive faster polymer growth rates (>1 μm min^{-1}).

Murthy *et al.* first demonstrated the addition of a separate initiator, PFOSF, to enable the iCVD polymerization of another siloxane ring monomer, V_3D_3 or trivinyltrimethylcyclotrisiloxane using heated Nichrome wires [27]. Without

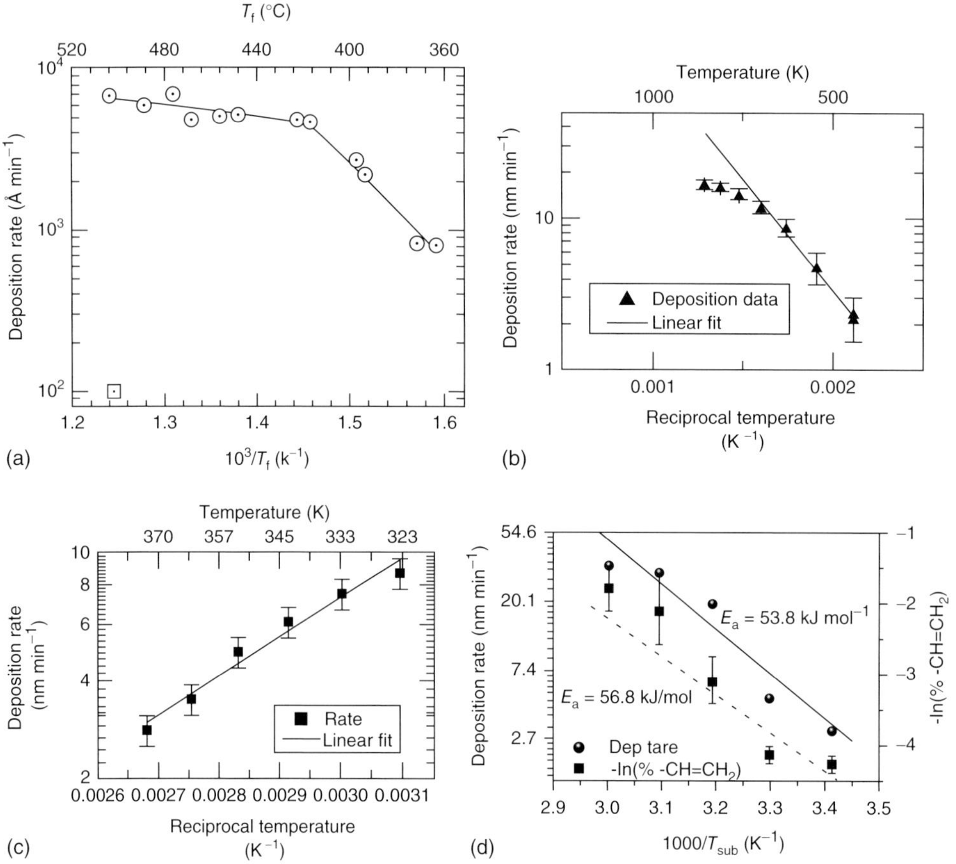

Figure 2.3 iCVD of organosilicon polymer. (a) Arrhenius plot of deposition rate versus filament temperature from using V_3D_3 monomer and PFOSF initiator [27]. (b) Arrhenius plot of deposition rate versus filament temperature from using V_3D_3 monomer and TBPO initiator [30]. (c) Arrhenius plot of deposition rate versus substrate temperature from using V_3D_3 monomer and TBPO initiator [30]. (d) Arrhenius plot of deposition rate and % of unreacted vinyl bonds versus substrate temperature from using HVDS monomer and TBPO initiator [23]. ((a) Reprinted with permission from Ref. [27]. Copyright 2002, American Chemical Society. (b,c) Reprinted with permission from Ref. [30]. Copyright 2006, American Chemical Society. (d) Reprinted with permission from Ref. [23]. Copyright 2009, American Chemical Society.)

the initiator, filament temperatures higher than 530 °C are needed to obtain deposition, and even then deposition is slow at around 10 nm min^{-1}. By adding PFOSF, much higher growth rates can be obtained (>100 nm min^{-1}) and at lower filament temperatures (<500 °C). As shown in Figure 2.3a, an Arrhenius plot of deposition rate with filament temperature yields two separate operating regimes. In the range of 350–420 °C, deposition rate is strongly dependent on filament temperature and growth is believed to be limited by the initiator although they did not report an activation energy. Above 420 °C, deposition rate dependence is much less sensitive and is attributed to mass transfer becoming rate-limiting.

Interestingly, they also used D_3 as a monomer for comparison and found that there is no enhancement in rate when PFOSF is used. Since the only difference between V_3D_3 and D_3 are the vinyl groups, this strongly indicates that vinyl groups are responsible for the enhanced growth when using the PFOSF initiator. This knowledge together with spectroscopic data suggest that polymerization occurs through the addition of vinyl bonds in a free radical mechanism that is initiated by the thermal decomposition of PFOSF as this provides the needed radicals to start chain propagation. Termination is believed to occur through perfluoroalkyl and SO_2F radicals capping the polymer radical chain ends as nuclear magnetic resonance (NMR) spectrosopy showed the presence of both these structural groups.

In another study, O'Shaughnessy *et al.* used a more common free radical initiator, *t*-butyl peroxide (TBPO), for the polymerization of V_3D_3 to form organosilicon polymer [30]. They looked at the effect of filament temperature as well as substrate temperature on deposition kinetics. Figure 2.3b shows that there is again two distinct regimes in the Arrhenius plot of deposition rate with filament temperature that is similar to the previous study. The high temperature region is attributed to mass transfer limitations given the low temperature sensitivity but with the low temperature region; since the calculated apparent E_a of 28.0 kJ mol^{-1} is much lower than the O–O bond energy, they think that the gas phase decomposition of TBPO is not likely to be rate-limiting and that the E_a value is a composite effect of competing processes. Instead, they believe that kinetics is more likely to be determined by the monomer propagation reaction given the large size and steric barriers presented by the bulky V_3D_3 monomer.

This seems to be supported by the effect of substrate temperature on growth rate, as seen in Figure 2.3c. The Arrhenius plot yields an apparent E_a of −23.2 kJ mol^{-1}. This negative value is atypical of normal rate behavior since this means that an increase in substrate temperature leads to a decrease in kinetics. This is explained by O'Shaughnessy *et al.* as a decrease in the equilibrium concentration of the monomer adsorbed on the substrate surface dominating over any increase in the propagation rate constant with increasing temperature.

To decouple the contrary effects of monomer adsorption and intrinsic kinetics as a result of substrate temperature changes, a separate work by Coclite *et al.* looked at the growth behavior for a fixed equilibrium concentration of the adsorbed monomer [23]. As will be discussed in more detail in the next section, this can be easily done by fixing the $P_M/P_{M,sat}$ ratio with various deposition runs. P_M is the partial pressure of the monomer vapor and $P_{M,sat}$ is the vapor pressure of the monomer evaluated at the substrate temperature so the ratio essentially gives the fractional saturation of the monomer at the substrate surface.

Using hexavinyldisiloxane (HVDS) as the monomer together with TBPO as the initiator, the effect of substrate temperature on deposition rate at a fixed $P_M/P_{M,sat}$ ratio of 0.30 is given in Figure 2.3d. With a constant concentration of adsorbed monomer, the rate behavior follows the normal Arrhenius behavior, giving an apparent E_a of 53.8 kJ mol^{-1}. Based on FTIR, the fraction of unreacted vinyl groups (relative to the monomer) in the resulting polymer film as a function of substrate temperature of deposition can also be plotted. This follows a similar trend as deposition rate, yielding an apparent E_a of 56.8 kJ mol^{-1}. The close agreement in E_a values implies that the dominant reaction pathway is surface propagation through the linkage of vinyl bonds.

For the case of V_3D_3 polymerization, using a thermal labile initiator like PFOSF and TBPO enables deposition at significantly faster kinetics and at a much milder activation temperature that minimizes monomer degradation and restricts reaction to chain addition through the vinyl bond. As a result, the organosilicon polymer from V_3D_3 consists of a carbon–carbon backbone and pendant siloxane rings [28]. Since all three vinyl groups in V_3D_3 can potentially undergo polymerization, there is the possibility of forming a crosslinked network [30]. This is supported by the observation that the polymer films are insoluble in common solvents like acetone, tetrahydrofuran, and dimethyl sulfoxide. At higher filament temperatures however, there is evidence to show the degradation of the V_3D_3 ring structure with vinyl abstraction and ring opening side reactions that eventually is believed to result in the formation of "T" groups consisting of silicon T-junctions connected to three other oxygens, that is, $(O)_3SiCH_3$ [28].

2.3.3
Acrylate and Methacrylate Polymers

Given the early success with synthesizing fluorocarbon and organosilicon polymers via a free radical initiator, acrylates and methacrylates present the next set of polymers made through the iCVD approach (see Table 2.1). Since these polymers are commonly polymerized in the liquid phase with a variety of different initiators including free radical initiators [191, 192], the iCVD technique naturally offers a viable means for creating them as thin films. It is therefore understandable that the number of acrylate and methacrylate polymers produced by iCVD continues to expand. These polymers are attractive as they offer a diverse range of chemical functionality and physical properties, and as a result a large number of applications based on iCVD processing have been demonstrated (refer to the chapters that follow).

The first polymer of this class produced by iCVD is polyglycidyl methacrylate (PGMA) from the work of Mao and Gleason [103]. Using TBPO as the initiator, deposition at a much higher rate and lower filament temperature is possible. For example, using a monomer flow rate of 2 sccm at a pressure of 0.5 Torr and substrate and filament temperatures of 25 and 220 °C, deposition rate is found to be only about 5 nm min^{-1} without TBPO and over 200 nm min^{-1} with 0.6 sccm of TBPO [193]. Even with filament temperatures above 220 °C, the deposition rate remained low at $<$10 nm min^{-1} without TBPO [103]. This work also reported for the first time molecular weight data of iCVD polymers with PGMA in the range of 16 000–33 000 g mol^{-1} that depended on deposition conditions.

In another study, Chan *et al.* demonstrated the use of triethylamine (TEA) as an alternative, non-peroxide free radical initiator to grow polymethyl methacrylate

(PMMA) [144]. Due to its stability, TEA required a higher filament temperature >450 °C to be thermally dissociated. At a filament temperature of 550 °C, the deposition rate is measured to be extremely slow at about 1 nm min^{-1} without any initiator. In contrast, with TEA, the rate increased to around 20 nm min^{-1}. The molecular weight is seen to increase from 500 to 3300 g mol^{-1} with an increase in monomer flow rate for a fixed initiator flow. The generally low polymer molecular weight appears to reflect the slow growth kinetics in this reaction chemistry.

One of the first detailed reports investigating iCVD kinetics comes from Lau and Gleason [48]. By using a homologous series of n-alkyl acrylate ($n = 2-6$) monomers and t-amyl peroxide (TAPO) as the initiator with a fixed monomer flow rate and other process conditions unchanged, a rapid rise in growth rate is seen with using an increasingly heavier monomer, showing a 300-fold rise from ethyl to hexyl acrylate. The number-average molecular weight (M_n) of the polymers correspondingly increased from 1000 to 61 000 g mol^{-1}. Since the only variable change here is the vapor pressure of the monomer at the substrate temperature, this gave one of the first indications that monomer adsorption is an important factor influencing deposition kinetics. This study also for the first time proposed the use of the saturation ratio, $P_M/P_{M,sat}$ (where P_M is the partial pressure of the monomer vapor and $P_{M,sat}$ is the vapor pressure of the monomer evaluated at the substrate temperature), as a suitable measure of the relative concentration of monomer adsorbed on the substrate surface. This parameter appears in numerous adsorption isotherms when relating to the amount of material adsorbed that becomes linear in the Henry's law limit [194].

Given this insight, another series of experiments were performed to study explicitly the effect of substrate temperature (17, 21, 25, and 29 °C), this time using ethyl acrylate (EA) as the monomer and all other conditions remaining constant (0.75 Torr pressure, 50.0 sccm monomer flow, 1.0 sccm initiator flow). Figure 2.4a shows that an Arrhenius plot of deposition rate with substrate temperature yields a positively sloped line, that is, an increase in rate with a decrease in temperature. This is true at three different filament temperatures (285, 310, and 360 °C) that were studied, and each of them yields similar slopes that averaged to an apparent E_a of −79.4 kJ mol^{-1}. When expressed as a function of the $P_M/P_{M,sat}$ ratio, the deposition rate shows a second-order dependence, at least in the low ratios (<0.03), that is, highly dilute conditions, studied. This dependence is actually seen in liquid phase free radical polymerization under monomer-starved conditions in which the concentration of activated initiator radicals (primary radicals) is high enough that they prematurely terminate growing polymer chain radicals, leading to a slower growth rate [191]. In the normal case of adequate monomer availability, the rate dependence is first order with respect to the monomer concentration.

To rationalize these results, a separate set of experiments were carried out to determine the adsorption behavior of EA monomer by using a quartz crystal microbalance attached to the iCVD reactor. The amount of adsorbed monomer (V_{ad}) was quantified as a function of $P_M/P_{M,sat}$ and fitted to a BET (Brunauer–Emmett–Teller) adsorption isotherm [195] as:

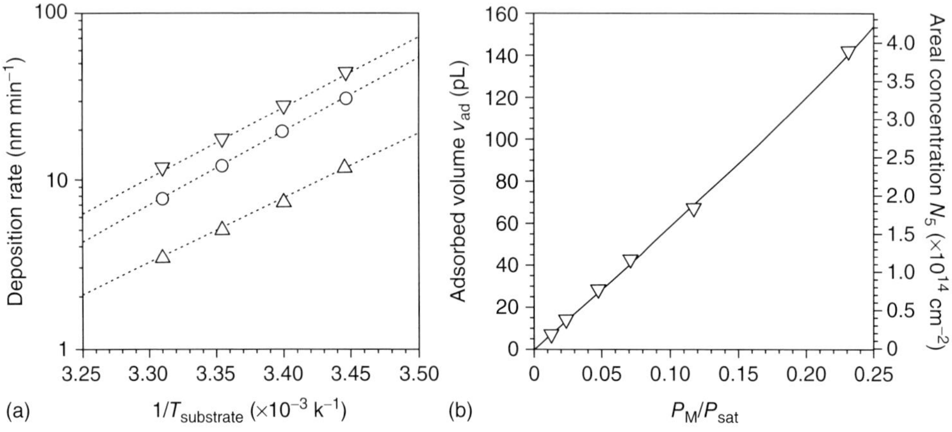

Figure 2.4 iCVD of polyethyl acrylate. (a) Arrhenius plot of deposition rate versus substrate temperature from using EA monomer and TAPO initiator [48]. (b) Adsorption data of EA at 27 °C and the fitted BET isotherm versus $P_M/P_{M,sat}$ [48]. ((a,b) Reprinted with permission from Ref. [48]. Copyright 2006, American Chemical Society.)

$$V_{ad}| = \frac{V_{ml}\, c\,(P_M/P_{M,sat})}{(1 - P_M/P_{M,sat})\,[1 - (1 - c)(P_M/P_{M,sat})]}$$

$$\text{where } c \approx \exp\left[\left(\Delta H_{des} - \Delta H_{vap}\right)/(RT)\right]$$

where ΔH_{des} is the heat of desorption, ΔH_{vap} is the heat of vaporization, and V_{ml} is the adsorbed volume of a monolayer. From Figure 2.4b, the fit allowed the values of V_{ml} and c to be derived as 292.6 pl and 1.950, respectively. Taking a literature value of ΔH_{vap} for EA, ΔH_{des} is calculated to be 40.9 kJ mol^{-1}, which is indicative of physisorption (20–80 kJ mol^{-1}) [196].

Since the surface conditions during deposition are dilute in monomer, the BET equation can first be simplified, which in the low concentration limit relates the surface monomer concentration [M] linearly to the $P_M/P_{M,sat}$ ratio. This can further be reduced by incorporating the Clausius–Clapeyron equation [197]. Finally, with the second order rate behavior, an activation energy can be written as:

$$\lim_{P_M/P_{M,sat}\to 0} V_{ad} = V_{ml}\, c\,(P_M/P_{M,sat}) \sim [\mathrm{M}]$$

$$P_{M,sat} = A \exp\left[-\Delta H_{vap}/(RT)\right]$$

$$[\mathrm{M}] \sim \exp\left[\Delta H_{des}/(RT)\right]$$

$$\text{rate} \sim [\mathrm{M}]^2 \sim \exp\left[2\Delta H_{des}/(RT)\right]$$

$$E_a = -2\Delta H_{des}$$

For EA, using the derived ΔH_{des} of 40.9 kJ mol^{-1}, this gives a prediction of E_a as −80.1 kJ mol^{-1} based on adsorption theory. The close agreement with the experimental value of −79.4 kJ mol^{-1} from kinetic data provides strong support that iCVD kinetics is adsorption-limited at least in the range of process conditions investigated.

Similar results have in fact been reproduced in other acrylate polymer systems, like perfluorodecyl acrylate (PFDA) [63] and ethylene glycol diacrylate (EGDA) [175]. For example, Gupta *et al.* found from kinetic data an apparent E_a of $-121.4\,\text{kJ mol}^{-1}$ for the growth rate of PFDA with substrate temperature [63]. Based on separate QCM adsorption experiments and fitting the BET equation, ΔH_{des} can be derived to be $60.4\,\text{kJ mol}^{-1}$, which when noting that the deposition rate is a second order dependence, a predicted E_a of $-120.9\,\text{kJ mol}^{-1}$ can be obtained, again showing remarkable agreement with the kinetic data.

Lau *et al.* followed up with another kinetic study, this time with the aim of developing a mechanistic model for describing the observed kinetic behavior [52]. A reaction mechanism for the iCVD process was proposed and shown in Figure 2.5a as a series of elementary steps involving gas phase initiator decomposition (step 1), monomer and primary radical adsorption (steps 2 and 3), surface polymerization through chain initiation, propagation, and termination reactions (steps 4–8). The surface reactions are analogous to the mechanism for free radical addition polymerization [191]. By making the pseudo steady state approximation on the active radical intermediates, the overall rate of polymerization (R_p) can be implicitly derived. Likewise, the kinetic chain length can also be derived and related to the molecular weight (M_n). The kinetic model focuses only on the surface reactions and lumps the initiator decomposition, precursor adsorption, and any other processes (e.g., loss to reactor walls or other sinks) into an "r" parameter that represents the net rate of "appearance" of primary radicals on the surface that ultimately take part in initiation, termination, or recombination events.

Using *n*-butyl acrylate (*n*BA) as the monomer together with TAPO initiator, experiments were conducted to obtain deposition rate and polymer molecular weight as a function of $P_M/P_{M,sat}$. However, instead of changing the substrate temperature, different monomer and initiator flow rates were used to alter the monomer partial pressure while other conditions remained fixed (1.0 Torr pressure, 23 °C substrate temperature, 260 °C filament temperature). A nitrogen patch flow was used to maintain a constant total flow rate (3.7 sccm) and therefore constant residence time. The ratio of initiator to monomer flow was also fixed at 0.233 to achieve the same relative gas concentration of the precursors. In this way, the experimental kinetic data is solely influenced by the amount of adsorbed monomer and not by other interfering factors such as temperature.

Deposition rate increased from 1.7 to $46.8\,\text{nm min}^{-1}$ as $P_M/P_{M,sat}$ increased from 0.03 to 0.17. Molecular weight (M_n) likewise increased from 2300 to $15\,200\,\text{g mol}^{-1}$. The polydispersity index (PDI) ranged between 2 and 3. The deposition rate was converted to a rate of polymerization (R_p in $\text{mol l}^{-1}\,\text{s}^{-1}$) by using QCM adsorption data together with fitting to the full BET equation that allowed the monolayer volume and thickness to be derived together with the assumption that reaction occurs within a monolayer thickness since the deposition conditions were at less than a full monolayer coverage (when $P_M/P_{M,sat}$ reaches 0.386). Likewise, the $P_M/P_{M,sat}$ ratio was converted to a surface monomer

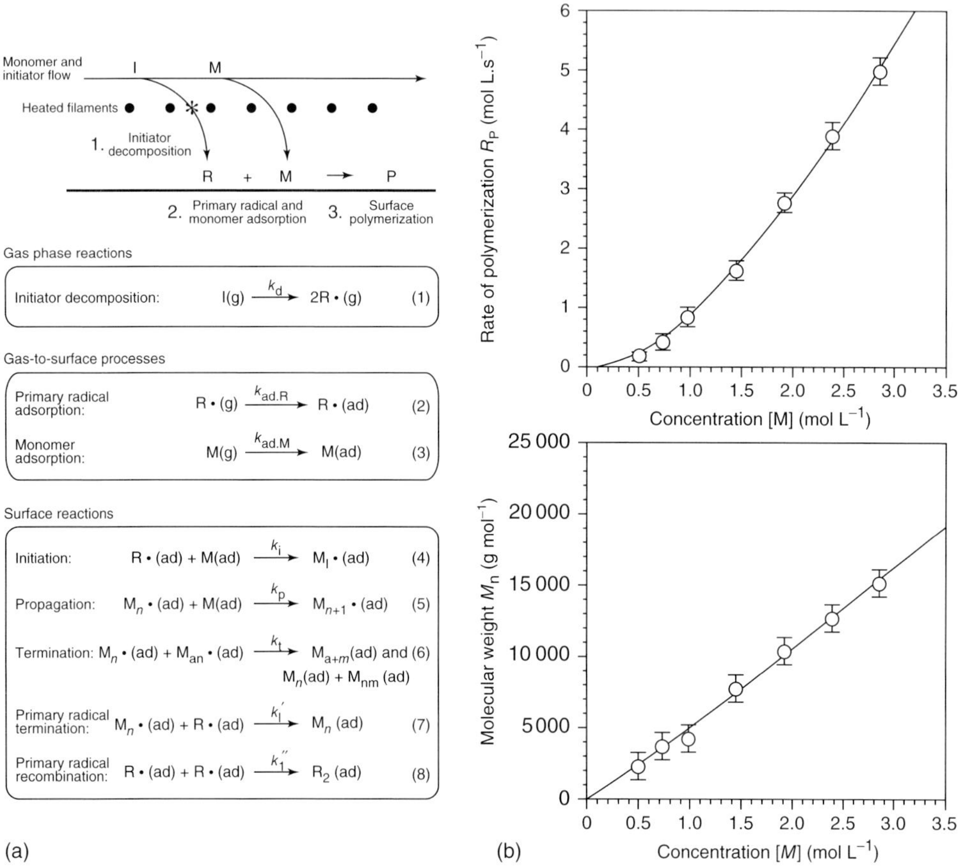

Figure 2.5 iCVD of poly-*n*-butyl acrylate. (a) Proposed free radical polymerization mechanism by iCVD [52]. (b) Plots of experimental R_p and M_n (discrete symbols) and the corresponding kinetic model fits (solid curves) versus surface monomer concentration $[M]$ from using nBA monomer and TAPO initiator [52]. ((a,b) Reprinted with permission from Ref. [52]. Copyright 2006, American Chemical Society.)

concentration $[M]$. In this way, the experimental data can be plotted as shown in Figure 2.5b as the discrete data points for both R_p and M_n as a function of $[M]$.

Using now the kinetic model, a multiresponse parameter estimation procedure was used to fit the model to experiment through a Levenberg–Marquardt minimization algorithm that minimized the error below a convergence criterion. The fitted curves as shown in Figure 2.5b match well with experimental data and the estimated parameters include the propagation rate constant (k_p), which came out to be 15 540 l mol^{-1} s^{-1}. This compares well with the value of 15 460 l mol^{-1} s^{-1} for liquid phase free radical polymerization of nBA, [198] providing clear support that iCVD polymerization kinetics closely follows that of their liquid phase counterpart.

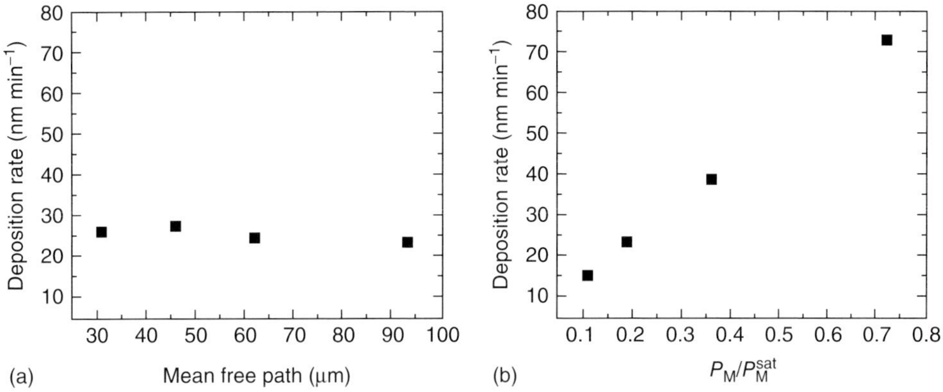

Figure 2.6 iCVD of polycyclohexyl methacrylate. (a) Plot of deposition rate versus mean free path of CHMA monomer and TBPO initiator molecules in the vapor phase [92]. (b) Plot of deposition rate versus $P_M/P_{M,sat}$ from using CHMA monomer and TBPO initiator [92]. ((a,b) Reprinted with permission from Ref. [92]. Copyright 2008, John Wiley & Sons, Inc.)

These trends again have been reproduced in other polymer systems. For example, in the iCVD of PFDA, parameter estimation from fitting the same kinetic model to experimental data gave the propagation rate constant (k_p) to be 11 418 l mol^{-1} s^{-1}, which again is similar to those of acrylate free radical polymerization in the liquid phase [63].

A more detailed understanding of gas phase effects started with the work of Baxamusa *et al.* They studied the iCVD polymerization and coating of polycyclohexyl methacrylate (PCHMA) using TBPO as the initiator [92]. The deposition rate was measured as a function of a change in total pressure (or mean free path) as well as a change in substrate temperature (or $P_M/P_{M,sat}$). Deposition rate is found to be almost independent of pressure or equivalently the mean free path (based on the kinetic theory of gases) of the vapor molecules, see Figure 2.6a. This means that homogenous gas phase events do not appear to have any significant contribution to growth. In contrast, deposition rate is seen to increase linearly with $P_M/P_{M,sat}$, see Figure 2.6b, which is typical of the first order dependence seen at sufficiently high monomer concentration in free radical polymerization [191]. This indicates that film formation depends more on the interaction of gas phase species with adsorbed monomer. Specifically, the activated initiator radicals are believed to chemisorb (rather than physisorb) and react directly from the gas phase with adsorbed monomer in an Eley–Rideal type heterogeneous reaction mechanism since these radicals have high volatility and short surface lifetime. Additional support comes from looking at the conformality of deposited films in microtrenches that is discussed in Chapter 5.

Ozaydin-Ince *et al.* followed up with further studies to understand the role of the initiator by looking at the effect of filament temperature on polymer growth from EGDA monomer and TBPO initiator [175]. An Arrhenius plot of deposition rate with filament temperature shows two operating regions with different slopes, as seen in Figure 2.7a. Conditions 1 and 2 correspond to a lower TBPO flow of 1 sccm and monomer flow of 0.4 and 1 sccm, respectively. Conditions 3 and 4 correspond to a higher TBPO flow of 3 sccm and monomer flow of 1 and 2 sccm, respectively. At lower temperatures, there is a strong rate dependence that is attributed to a reaction-limited regime in which the initiator decomposition reaction is believed to dominate. As such, the rate is not dependent on the gas phase concentration of monomer (initiator is constant) at least when there is no loading effect (conditions 3 and 4), yielding a single line that gives an apparent E_a of 93.2 kJ mol^{-1} (at conditions 1 and 2, there is a loading effect in which rates are artificially lowered due to insufficient vapor delivery). At higher temperatures, the rate dependence becomes less sensitive and this is believed to be a transition to a mass transfer-limited regime in which the transport of initiator radicals to the surface dominates. This leads to separate rate trends (instead of collapsing to a single relation) as the precursor flow rates are different for each case.

Because the E_a value in the reaction-limited regime does not correspond as expected to the reported E_a of 163.6 kJ mol^{-1} for the gas phase decomposition of TBPO [192], they rationalized this discrepancy to be due to the temperature gradient that exists between the filament and substrate, which essentially leads to a lower effective temperature. By re-graphing the Arrhenius plot against a logarithmic mean gas temperature based on the difference in temperatures at the filament and substrate, see Figure 2.7b, an apparent E_a of 166 kJ mol^{-1}, which is now close to the value for TBPO decomposition. They mention that previous reports [103, 145] on E_a of ~100 kJ mol^{-1} can also be adjusted to be closer to the value for TBPO decomposition based on the correction they adapted here.

Additional analysis was made to investigate the growth at different substrate temperatures, as given in Figure 2.7c. A maximum deposition rate of 17 nm min^{-1} is found at 33 °C, the lowest temperature used. Growth rate then drops sharply with an increase in temperature of up to 39 °C. In this range, a negative apparent E_a of −99.4 kJ mol^{-1} is obtained, which again indicates an adsorption-limited regime. Since the rate is again second order at the conditions tested, based on the relation of E_a with ΔH_{des} derived earlier, ΔH_{des} is estimated to be 49.7 kJ mol^{-1}, which is again comparable to those obtained for EA and PFDA and consistent with physisorption (20–80 kJ mol^{-1}) being the dominant adsorption mode for the monomer. Above 39 °C, deposition is observed to be more or less constant. This is hypothesized to be a transition to a regime in which surface reactions become important and there is a competing effect between these thermally activated processes and adsorption that somehow balances out.

Interestingly, Ozaydin-Ince *et al.* did a further study on EGDA-TBPO kinetics, this time looking at the effect of substrate temperature while maintaining a constant $P_M/P_{M,sat}$ of 0.35 and fixed monomer and initiator flow rates (1 and 2 sccm, respectively) [148]. In order to keep $P_M/P_{M,sat}$ constant, pressure needed to be varied as well to adjust P_M as $P_{M,sat}$ varied with substrate temperature. With $P_M/P_{M,sat}$ fixed, the adsorbed monomer concentration remains constant so that the effect of substrate temperature on surface reactions can be independently investigated. As seen in Figure 2.7d, the deposition rate does not appear to be

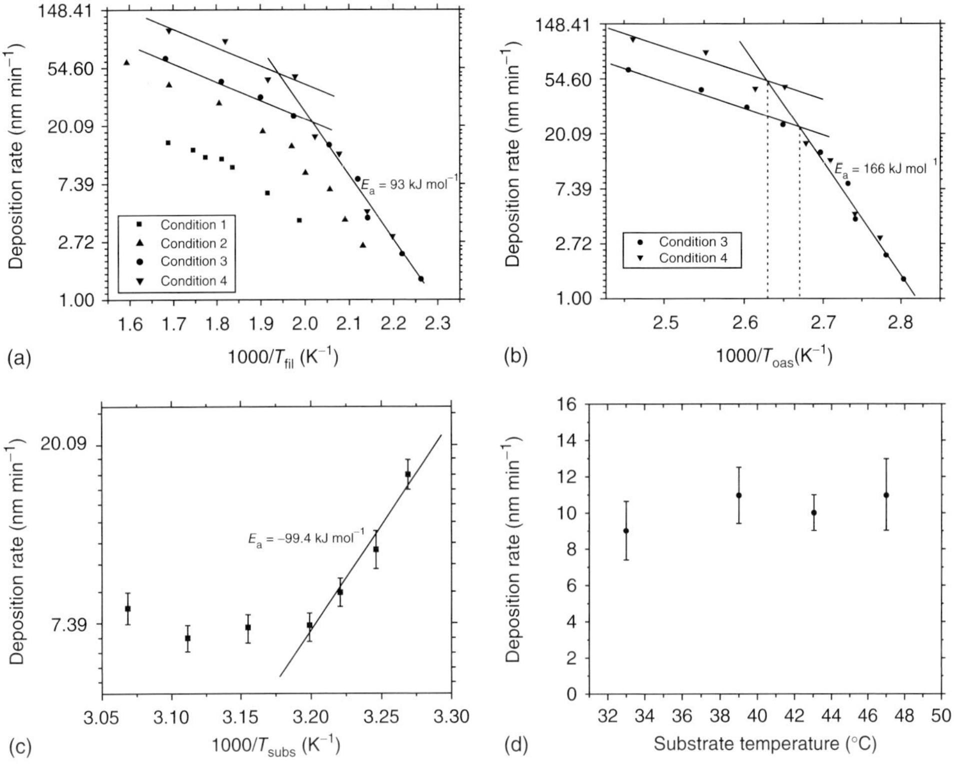

Figure 2.7 iCVD of polyethylene glycol diacrylate. (a) Arrhenius plot of deposition rate versus filament temperature from using EGDA monomer and TBPO initiator [175]. (b) Arrhenius plot of deposition rate versus logarithmic mean gas temperature from using EGDA monomer and TBPO initiator [175]. (c) Arrhenius plot of deposition rate versus substrate temperature from using EGDA monomer and TBPO initiator [175]. (d) Plot of deposition rate versus substrate temperature at constant $P_M/P_{M,sat}$ of 0.35 from using EGDA monomer and TBPO initiator [148]. ((a–c) Reprinted with permission from Ref. [175]. Copyright 2009, American Vacuum Society. (d) Adapted with permission from Ref. [148]. Copyright 2010, John Wiley & Sons, Inc.)

influenced strongly by temperature, leading to the conclusion that thermally activated surface reactions do not affect iCVD kinetics appreciably. Since pressure also changed with substrate temperature, the collision of vapor phase species does not seem like a significant factor as well.

Although most of the iCVD work on acrylate and methacrylate polymers constrain polymerization to $P_M/P_{M,sat} < 1$ and obtain reasonable growth conditions, there are reports of ways to further enhance deposition rate. By operating at $P_M/P_{M,sat} \simeq 1$, that is, at near or super-saturation conditions, growth typically becomes more rapid. For example, Bose *et al.* achieved high growth rate of >1.5 μm min^{-1} for the synthesis of polyhydroxyethyl methacrylate (PHEMA) from HEMA (2-hydroxyethyl methacrylate) and TAPO by operating at $P_M/P_{M,sat} \sim 1$, resulting in free-standing films of 100–200 μm [122, 123]. Interestingly, this produces ultrahigh molecular weight polymer, estimated to be about 820 000 g mol^{-1}, that gives PHEMA improved mechanical properties without the need for any crosslinking [123].

In another study, Tao *et al.* investigated $P_M/P_{M,sat} > 1$ conditions for the growth of PGMA using TBPO [115]. Significantly higher deposition rates can be obtained going up to about 600 nm min^{-1} at $P_M/P_{M,sat}$ of 2.18 and yielding a polymer with M_n of 420 000 g mol^{-1} and PDI of 2.4. They discovered that when depositing on a fluorinated surface under supersaturated conditions, the monomer vapor can undergo dropwise condensation, forming droplets that lead ultimately to film undulations. However, in certain cases where the surface is more wettable, can absorb the monomer, or has chemical groups to graft the polymer to the surface, interfacial surface energy is more favorable for forming a uniform film instead. For example, they found that by using a polymerized ethylene glycol diacrylate (PEGDA) layer below, the PGMA grown on top becomes a smooth film under supersaturated conditions, which they believe is due to the swelling of PEGDA by GMA (glycidyl methacrylate) that is absorbed into the layer as well as to unreacted acrylate groups on the PEGDA layer that react with GMA, thereby promoting wetting and preventing droplet nucleation. In contrast, for $P_M/P_{M,sat} < 1$, the type of substrate does not appear to influence growth as similar deposition rates are found for fluorinated and silicon substrates.

By changing the free radical initiator from TBPO to *t*-butyl peroxybenzoate (TBPOB), Xu *et al.* showed that the polymerization of CHMA (cyclohexyl methacrylate) is significantly faster using the less volatile TBPOB initiator (e.g., 5 vs 30 nm min^{-1} at a filament temperature of 270 °C), and the onset of appreciable deposition occurs at a lower filament temperature (150 vs 250 °C) [94]. At higher filament temperatures where the mass transport of initiated radicals is rate-limiting, the growth rate using TBPOB is roughly twice that with TBPO. This suggests that the heavier benzoate radical from TBPOB thermal decomposition compared to the *t*-butoxyl radical has a higher probability of sticking to the surface and react.

Bose *et al.* demonstrated that by repeatedly pulsing the filament heating on (10 min) and off (5 min), termed pulsed iCVD, growth conditions can be altered [90]. Compared to typical iCVD in which the filament is continuously heated during deposition, pulsed iCVD allows the filament to cool down during the off-period. This leads to substrate cooling as well since heat transfer from the filament source is lost in this off-time. As a result, there is a rapid acceleration of growth in this period with the lower substrate temperature (as the contact stage is at a fixed temperature), allowing for thicker films in a shorter total deposition time.

From all the kinetic studies performed on the iCVD of acrylate and methacrylate polymers, the current understanding is that the initiation reaction is the rate-determining step. The reaction rate can be expressed as $r_i = k_i[R\cdot]_g[M]_{ad}$ (where k_i is the initiation rate constant, $[R\cdot]_g$ is the gas phase concentration of primary initiator radicals, and $[M]_{ad}$ is the equilibrium adsorbed concentration of monomer), which highlights the fact that this is an Eley–Rideal type reaction mechanism. Under typical iCVD conditions in which filament temperature is generally high enough, there will be sufficient initiator radicals that this will not be limiting. The activation energy of the initiation rate constant is probably quite low with this chemisorption step that this will also not be a critical factor. Rather, the kinetics will be dominated by the surface availability of monomer given the strong dependence on the saturation ratio, $P_M/P_{M,sat}$. Surface polymerization follows the typical free radical polymerization mechanism. At sufficient monomer concentration, growth will be first order. This changes to second order when monomer concentration is lacking. The molecular weight generally increases with the growth rate as it is related to the kinetic chain length in the free radical mechanism. Polydispersity is also typical of free radical polymers.

2.3.4 Styrene and Other Vinyl Polymers

The number of kinetic studies on styrene and vinyl polymers are fewer given that the number of polymer systems studied are also fewer (see Table 2.1). In general, the kinetic behavior seems to follow trends similar to acrylate and methacrylate polymers. This is not too surprising since they all possess the C=C double bond through which free radical polymerization takes place. For example, Martin *et al.* found changes in operating regimes with filament temperature as well as a negative activation energy with substrate temperature in the deposition of polydimethylaminomethyl styrene (PDMAMS) using TAPO initiator [42].

One interesting report by Tenhaeff *et al.* demonstrated that appreciable growth is only possible with the copolymerization of styrene with maleic anhydride using TBPO initiator [160, 161]. Homopolymerization of either monomer is practically non-existent or slow (0–5 nm min^{-1}) and results in short oligomer chains. However, by introducing both monomers together, copolymers can be produced at substantially higher rate with a maximum observed rate of 75.4 nm min^{-1} and M_n in the range of 1400–4700 g mol^{-1}. As a result of the lack of reaction of the co-monomer with itself, perfectly alternating copolymers are made that is independent of the co-monomer vapor feed ratio. This copolymerization behavior is attributed to the favorable electronic interaction between the electron-donating styrene and electron-accepting maleic anhydride.

2.3.5 Ring Opening Polymers

In another proof that iCVD can adopt a wide range of conventional liquid phase polymerization chemistries, iCVD has been used in several ring opening polymerization depositions (see Table 2.1). Petruczok *et al.* showed the ring opening

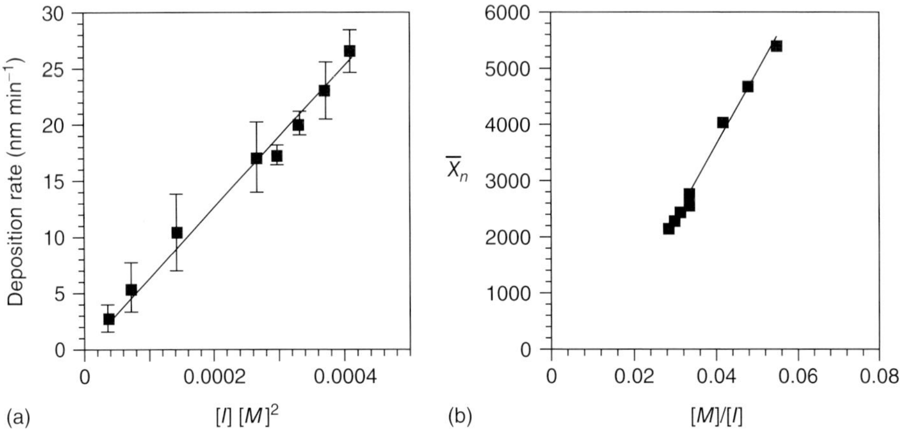

Figure 2.8 iCVD of polyethylene oxide. (a) Plot of deposition rate versus $[I][M]^2$ from using EO monomer and BF_3 initiator [180]. (b) Plot of kinetic chain length $\overline{X}_n$ versus $[M]/[I]$ from using EO monomer and BF_3 initiator [180]. ((a,b) Reprinted with permission from Ref. [180]. Copyright 2012, American Chemical Society.)

of 2-methylene-1,3-dioxepane (MDO) to form poly-ε-caprolactone (PCL) via a free radical ring opening polymerization using TBPO initiator in a closed batch iCVD process (instead of the typical continuous vapor flow operation) [37]. This approach is similar to the synthesis carried out in the liquid phase although no kinetic data is given [199].

Although growth data is scant for the iCVD ring opening process, kinetic behavior will most likely follow that done in the liquid phase. This is clearly demonstrated by Bose *et al.* who showed the first use of a cationic (rather than free radical) initiator, boron trifluoride (complexed with diethyl etherate), to enable the ring opening polymerization of ethylene oxide to form polyethylene oxide (PEO) [180]. Deposition kinetics and polymer molecular weight were measured for different partial pressures of monomer and initiator as well as substrate temperature to manipulate the P/P_{sat} ratio of both the monomer and initiator. Although the BF_3 Lewis acid typically uses water or some other proton-donating coinitiator to enable initiation in the liquid phase, [191] here in the iCVD environment, the lack of any significant water suggests a BF_3 self-initiation. This is part of a proposed reaction scheme, which under the pseudo steady state approximation, leads to the rate of polymerization R_p and kinetic chain length $\overline{X}_n$ being dependent on $[I][M]^2$ and $[M]/[I]$, respectively, where the concentrations refer to those at the surface, and therefore are directly related to their respective P/P_{sat} ratios. As shown in Figure 2.8, there is good agreement between the experimental kinetic data and model fits, indicating that the proposed cationic ring opening polymerization mechanism can potentially describe the rate behavior.

2.4 Summary

The iCVD approach provides a viable and valuable pathway for the synthesis and deposition of polymer thin films. Table 2.1 encompasses all the effort to date that has gone into the development and demonstration of iCVD as a polymer CVD process; in particular, the references in bold contain kinetic information. The thermal activation of a dedicated initiator species without degrading the monomer delivers a well-defined reaction pathway that leads to polymerization through initiator decomposition and monomer addition via chain initiation, propagation, and termination events. These elementary reactions define the iCVD polymerization mechanisms, whether free radical or ionic, that are remarkably similar to those found in liquid phase chemistries. A strong understanding of iCVD processing fundamentals is critical for achieving the desired polymer properties and film characteristics needed for effective integration in intended applications (see Chapters 5, 8, 10, 13–18). A good operating knowledge is also vital for the successful scale-up of iCVD reaction systems that can be aided by engineering analysis with dimensionless parameters (see Chapter 19). In studying iCVD kinetics, even though Arrhenius relationships of deposition rate are often formed with various iCVD parameters, careful consideration must be taken to understand clearly the underlying factors, whether rate coefficient, gas concentration, or surface concentration, that lead to the observed trends. Looking to the future, computational methods coupling fluid, heat, and mass transport, and reaction kinetics will be useful to further our understanding of deposition behavior.

References

1. Limb, S.J., Labelle, C.B., Gleason, K.K., Edell, D.J., and Gleason, E.F. (1996) *Appl. Phys. Lett.*, **68**, 2810.
2. Lau, K.K.S., Caulfield, J.A., and Gleason, K.K. (2000) *Chem. Mater.*, **12**, 3032.
3. Lau, K.K.S., Caulfield, J.A., and Gleason, K.K. (2000) *J. Vac. Sci. Technol., A*, **18**, 2404.
4. Lau, K.K.S. and Gleason, K.K. (2000) *J. Fluorine Chem.*, **104**, 119.
5. Lau, K.K.S. and Gleason, K.K. (2001) *J. Phys. Chem. B*, **105**, 2303.
6. Pryce Lewis, H.G., Weibel, G.L., Ober, C.K., and Gleason, K.K. (2001) *Chem. Vap. Deposition*, **7**, 195.
7. Cruden, B.A., Gleason, K.K., and Sawin, H.H. (2002) *J. Phys. D Appl. Phys.*, **35**, 480.
8. Cruden, B.A., Gleason, K.K., and Sawin, H.H. (2002) *J. Vac. Sci. Technol. B*, **20**, 690.
9. Zhou, J. and Wolden, C.A. (2003) *Thin Solid Films*, **430**, 28.
10. Yasuoka, H., Yoshida, M., Sugita, K., Ohdaira, K., Murata, H., and Matsumura, H. (2008) *Thin Solid Films*, **516**, 687.
11. Takachi, M., Yasuoka, H., Ohdaira, K., Shimoda, T., and Matsumura, H. (2009) *Thin Solid Films*, **517**, 3622.
12. Lau, K.K.S., Bico, J., Teo, K.B.K., Chhowalla, M., Amaratunga, G.A.J., Milne, W.I., McKinley, G.H., and Gleason, K.K. (2003) *Nano Lett.*, **3**, 1701.
13. Qi, H.J., Teo, K.B.K., Lau, K.K.S., Boyce, M.C., Milne, W.I., Robertson, J., and Gleason, K.K. (2003) *J. Mech. Phys. Solids*, **51**, 2213.

14. Lin, J., Murthy, S.K., Olsen, B.D., Gleason, K.K., and Klibanov, A.M. (2003) *Biotechnol. Lett.*, **25**, 1661.
15. Lau, K.K.S. and Gleason, K.K. (2007) *Surf. Coat. Technol.*, **201**, 9189.
16. Laird, E.D., Bose, R.K., Wang, W., Lau, K.K.S., and Li, C.Y. (2013) *Macromol. Rapid Commun.*, **34**, 251.
17. Laird, E.D., Bose, R.K., Qi, H., Lau, K.K.S., and Li, C.Y. (2013) *ACS Appl. Mater. Interfaces*, **5**, 12089.
18. Ölçeroğlu, E., Hsieh, C.-Y., Rahman, M.M., Lau, K.K.S., and McCarthy, M. (2014) *Langmuir*, **30**, 7556.
19. Pryce Lewis, H.G., Caulfield, J.A., and Gleason, K.K. (2001) *Langmuir*, **17**, 7652.
20. Murthy, S.K. and Gleason, K.K. (2002) *Macromolecules*, **35**, 1967.
21. Pryce Lewis, H.G., Casserly, T.B., and Gleason, K.K. (2001) *J. Electrochem. Soc.*, **148**, F212.
22. Kwan, M.C. and Gleason, K.K. (1997) *Chem. Vap. Deposition*, **3**, 299.
23. Coclite, A.M., Ozaydin-Ince, G., d'Agostino, R., and Gleason, K.K. (2009) *Macromolecules*, **42**, 8138.
24. Coclite, A.M., Ozaydin-Ince, G., Palumbo, F., Milella, A., and Gleason, K.K. (2010) *Plasma Processes Polym.*, **7**, 561.
25. Achyuta, A.K.H., White, A.J., Pryce Lewis, H.G., and Murthy, S.K. (2009) *Macromolecules*, **42**, 1970.
26. Achyuta, A.K.H., Polikov, V.S., White, A.J., Pryce Lewis, H.G., and Murthy, S.K. (2010) *Macromol. Biosci.*, **10**, 872.
27. Murthy, S.K., Olsen, B.D., and Gleason, K.K. (2002) *Langmuir*, **18**, 6424.
28. Murthy, S.K., Olsen, B.D., and Gleason, K.K. (2004) *J. Appl. Polym. Sci.*, **91**, 2176.
29. Murthy, S.K., Olsen, B.D., and Gleason, K.K. (2004) *Langmuir*, **20**, 4774.
30. O'Shaughnessy, W.S., Gao, M., and Gleason, K.K. (2006) *Langmuir*, **22**, 7021.
31. O'Shaughnessy, W.S., Murthy, S.K., Edell, D.J., and Gleason, K.K. (2007) *Biomacromolecules*, **8**, 2564.
32. O'Shaughnessy, W.S., Edell, D.J., and Gleason, K.K. (2008) *Thin Solid Films*, **516**, 684.
33. O'Shaughnessy, W.S., Edell, D.J., and Gleason, K.K. (2009) *Thin Solid Films*, **517**, 3612.
34. Sreenivasan, R., Bassett, E.K., Cervantes, T.M., Hoganson, D.M., Vacanti, J.P., and Gleason, K.K. (2012) *Microfluid. Nanofluid.*, **12**, 835.
35. Aresta, G., Palmans, J., van de Sanden, M.C.M., and Creatore, M. (2012) *J. Vac. Sci. Technol., A*, **30**, 041503.
36. Aresta, G., Palmans, J., van de Sanden, M.C.M., and Creatore, M. (2012) *Microporous Mesoporous Mater.*, **151**, 434.
37. Petruczok, C.D., Chen, N., and Gleason, K.K. (2014) *Langmuir*, **30**, 4830.
38. Trujillo, N.J., Baxamusa, S.H., and Gleason, K.K. (2009) *Chem. Mater.*, **21**, 742.
39. Trujillo, N.J., Wu, Q., and Gleason, K.K. (2010) *Adv. Funct. Mater.*, **20**, 607.
40. Yoo, Y., You, J.B., Choi, W., and Im, S.G. (2013) *Polym. Chem.*, **4**, 1664.
41. Pena-Francesch, A., Montero, L., and Borrós, S. (2014) *Langmuir*, **30**, 7162.
42. Martin, T.P. and Gleason, K.K. (2006) *Chem. Vap. Deposition*, **12**, 685.
43. Martin, T.P., Chan, K., and Gleason, K.K. (2008) *Thin Solid Films*, **516**, 681.
44. Mao, Y., Felix, N.M., Nguyen, P.T., Ober, C.K., and Gleason, K.K. (2006) *Chem. Vap. Deposition*, **12**, 259.
45. Mao, Y. and Gleason, K.K. (2006) *Macromolecules*, **39**, 3895.
46. Tenhaeff, W.E. and Gleason, K.K. (2009) *Thin Solid Films*, **517**, 3543.
47. Tenhaeff, W.E. and Gleason, K.K. (2009) *Chem. Mater.*, **21**, 4323.
48. Lau, K.K.S. and Gleason, K.K. (2006) *Macromolecules*, **39**, 3688.
49. Lau, K.K.S. and Gleason, K.K. (2007) *Macromol. Biosci.*, **7**, 429.
50. Lau, K.K.S. and Gleason, K.K. (2008) *Thin Solid Films*, **516**, 678–680.
51. Mao, Y. and Gleason, K.K. (2006) *Langmuir*, **22**, 1795.
52. Lau, K.K.S. and Gleason, K.K. (2006) *Macromolecules*, **39**, 3695.
53. Trujillo, N.J., Baxamusa, S.H., and Gleason, K.K. (2009) *Thin Solid Films*, **517**, 3615.
54. Haller, P.D., Bradley, L.C., and Gupta, M. (2013) *Langmuir*, **29**, 11640.

55. Sreenivasan, R., Bassett, E.K., Hoganson, D.M., Vacanti, J.P., and Gleason, K.K. (2011) *Biomaterials*, **32**, 3883.
56. Alf, M.E., Godfrin, P.D., Hatton, T.A., and Gleason, K.K. (2010) *Macromol. Rapid Commun.*, **31**, 2166.
57. Alf, M.E., Hatton, T.A., and Gleason, K.K. (2011) *Thin Solid Films*, **519**, 4412.
58. Alf, M.E., Hatton, T.A., and Gleason, K.K. (2011) *Langmuir*, **27**, 10691.
59. Frank-Finney, R.J., Haller, P.D., and Gupta, M. (2012) *Macromolecules*, **45**, 165.
60. Mehrabani, S., Kwong, P., Gupta, M., and Armani, A.M. (2013) *Appl. Phys. Lett.*, **102**, 241101.
61. Tekin, H., Ozaydin-Ince, G., Tsinman, T., Gleason, K.K., Langer, R., Khademhosseini, A., and Demirel, M.C. (2011) *Langmuir*, **27**, 5671.
62. Alf, M.E., Hatton, T.A., and Gleason, K.K. (2011) *Polymer*, **52**, 4429.
63. Gupta, M. and Gleason, K.K. (2006) *Langmuir*, **22**, 10047.
64. Gupta, M., Kapur, V., Pinkerton, N.M., and Gleason, K.K. (2008) *Chem. Mater.*, **20**, 1646.
65. Gupta, M. and Gleason, K.K. (2009) *Thin Solid Films*, **517**, 3547.
66. Baxamusa, S.H., Montero, L., Borrós, S., and Gleason, K.K. (2010) *Macromol. Rapid Commun.*, **31**, 735.
67. Asatekin, A. and Gleason, K.K. (2011) *Nano Lett.*, **11**, 677.
68. Haller, P.D., Frank-Finney, R.J., and Gupta, M. (2011) *Macromolecules*, **44**, 2653.
69. Kwong, P., Flowers, C.A., and Gupta, M. (2011) *Langmuir*, **27**, 10634.
70. Coclite, A.M., Shi, Y., and Gleason, K.K. (2012) *Adv. Funct. Mater.*, **22**, 2167.
71. Coclite, A.M., Shi, Y., and Gleason, K.K. (2012) *Adv. Mater.*, **24**, 4534.
72. Raghunathan, V., Yagüe, J.L., Xu, J., Michel, J., Gleason, K.K., and Kimerling, L.C. (2012) *Opt. Express*, **20**, 20808.
73. Karaman, M., Çabuk, N., Özyurt, D., and Köysüren, Ö. (2012) *Appl. Surf. Sci.*, **259**, 542.
74. Jeong, J.-M., Oh, M.S., Kim, B.J., Choi, C.-H., Lee, B., Lee, C.-S., and Im, S.G. (2013) *Langmuir*, **29**, 3474.
75. Yagüe, J.L. and Gleason, K.K. (2013) *Macromolecules*, **46**, 6548.
76. Chen, B., Kwong, P., and Gupta, M. (2013) *ACS Appl. Mater. Interfaces*, **5**, 12701.
77. Paxson, A.T., Yagüe, J.L., Gleason, K.K., and Varanasi, K.K. (2014) *Adv. Mater.*, **26**, 418.
78. Bradley, L.C. and Gupta, M. (2012) *Langmuir*, **28**, 10276.
79. Riche, C.T., Zhang, C., Gupta, M., and Malmstadt, N. (2014) *Lab Chip*, **14**, 1834.
80. Matin, A., Ozaydin-Ince, G., Khan, Z., Zaidi, S.M.J., Gleason, K., and Eggenspiler, D. (2011) *Desalin. Water Treat.*, **34**, 100.
81. Ozaydin-Ince, G., Matin, A., Khan, Z., Zaidi, S.M.J., and Gleason, K.K. (2013) *Thin Solid Films*, **539**, 181.
82. Matin, A., Khan, Z., Gleason, K.K., Khaled, M., Zaidi, S.M.J., Khalil, A., Moni, P., and Yang, R. (2014) *Sep. Purif. Technol.*, **124**, 117.
83. Matin, A., Shafi, H.Z., Khan, Z., Khaled, M., Yang, R., Gleason, K., and Rehman, F. (2014) *Desalination*, **343**, 128.
84. Amadei, C.A., Yang, R., Chiesa, M., Gleason, K.K., and Santos, S. (2014) *ACS Appl. Mater. Interfaces*, **6**, 4705.
85. Coclite, A.M., Lund, P., Di Mundo, R., and Palumbo, F. (2013) *Polymer*, **54**, 24.
86. Seidel, S. and Gupta, M. (2014) *J. Vac. Sci. Technol., A*, **32**, 041514.
87. Im, S.G., Kim, B.-S., Tenhaeff, W.E., Hammond, P.T., and Gleason, K.K. (2009) *Thin Solid Films*, **517**, 3606.
88. Kramer, N.J., Sachteleben, E., Ozaydin-Ince, G., van de Sanden, R., and Gleason, K.K. (2010) *Macromolecules*, **43**, 8344.
89. Ozaydin-Ince, G., Gleason, K.K., and Demirel, M.C. (2011) *Soft Matter*, **7**, 638.
90. Bose, R.K., Heming, A.M., and Lau, K.K.S. (2012) *Macromol. Rapid Commun.*, **33**, 1375.
91. Chan, K. and Gleason, K.K. (2006) *Macromolecules*, **39**, 3890.

92. Baxamusa, S.H. and Gleason, K.K. (2008) *Chem. Vap. Deposition*, **14**, 313.
93. Baxamusa, S.H. and Gleason, K.K. (2009) *Thin Solid Films*, **517**, 3536.
94. Xu, J. and Gleason, K.K. (2011) *ACS Appl. Mater. Interfaces*, **3**, 2410.
95. Chan, K. and Gleason, K.K. (2006) *J. Electrochem. Soc.*, **153**, C223.
96. Ye, Y. and Mao, Y. (2011) *J. Mater. Chem.*, **21**, 7946.
97. Yang, R., Xu, J., Ozaydin-Ince, G., Wong, S.Y., and Gleason, K.K. (2011) *Chem. Mater.*, **23**, 1263.
98. Yang, R. and Gleason, K.K. (2012) *Langmuir*, **28**, 12266.
99. Karaman, M. and Çabuk, N. (2012) *Thin Solid Films*, **520**, 6484.
100. Chen, G., Gupta, M., Chan, K., and Gleason, K.K. (2007) *Macromol. Rapid Commun.*, **28**, 2205.
101. Lau, K.K.S. and Gleason, K.K. (2006) *Adv. Mater.*, **18**, 1972.
102. Lau, K.K.S. and Gleason, K.K. (2008) *Thin Solid Films*, **516**, 674.
103. Mao, Y. and Gleason, K.K. (2004) *Langmuir*, **20**, 2484.
104. Mao, Y., Felix, N.M., Nguyen, P.T., Ober, C.K., and Gleason, K.K. (2004) *J. Vac. Sci. Technol. B*, **22**, 2473.
105. Gupta, M. and Gleason, K.K. (2006) *Thin Solid Films*, **515**, 1579.
106. Bakker, R., Verlaan, V., van der Werf, C.H.M., Rath, J.K., Gleasonb, K.K., and Schropp, R.E.I. (2007) *Surf. Coat. Technol.*, **201**, 9422.
107. Im, S.G., Bong, K.W., Lee, C.-H., Doyle, P.S., and Gleason, K.K. (2009) *Lab Chip*, **9**, 411.
108. Bakker, R., Verlaan, J., Verkerk, A.D., van der Werf, C.H.M., van Dijk, L., Rudolph, H., Rath, J.K., and Schropp, R.E.I. (2009) *Thin Solid Films*, **517**, 3555.
109. Ye, Y., Mao, Y., Wang, F., Lu, H., Qu, L., and Dai, L. (2011) *J. Mater. Chem.*, **21**, 837.
110. Spee, D.A., Bakker, R., van der Werf, C.H.M., van Steenbergen, M.J., Rath, J.K., and Schropp, R.E.I. (2011) *Thin Solid Films*, **519**, 4479.
111. Spee, D.A., van der Werf, C.H.M., Rath, J.K., and Schropp, R.E.I. (2011) *J. Nanosci. Nanotechnol.*, **11**, 8202.
112. Parker, T.C., Baechle, D., and Demaree, J.D. (2011) *Surf. Coat. Technol.*, **206**, 1680.
113. Yoshida, S., Kobayashi, T., Kumano, M., and Esashi, M. (2012) *J. Micro/Nanolithogr. MEMS MOEMS*, **11**, 023001.
114. Bong, K.W., Xu, J., Kim, J.-H., Chapin, S.C., Strano, M.S., Gleason, K.K., and Doyle, P.S. (2012) *Nat. Commun.*, **3**, 805.
115. Tao, R. and Anthamatten, M. (2012) *Langmuir*, **28**, 16580.
116. You, J.B., Min, K.-I., Lee, B., Kim, D.-P., and Im, S.G. (2013) *Lab Chip*, **13**, 1266.
117. Koysuren, O., Karaman, M., and Ozyurt, D. (2013) *J. Appl. Polym. Sci.*, **127**, 4557.
118. Spee, D.A., Schipper, M.R., van der Werf, C.H.M., Rath, J.K., and Schropp, R.E.I. (2013) *Thin Solid Films*, **532**, 84.
119. Kim, M.-J., Lee, B., Yang, K., Park, J., Jeon, S., Um, S.H., Kim, D.-I., Im, S.G., and Cho, S.-W. (2013) *Biomaterials*, **34**, 7236.
120. Kang, B.-J., Kim, H., Lee, S.K., Kim, J., Shen, Y., Jung, S., Kang, K.-S., Im, S.G., Lee, S.Y., Choi, M., Hwang, N.S., and Cho, J.-Y. (2014) *Acta Biomater.*, **10**, 3007.
121. Tao, R. and Anthamatten, M. (2013) *Macromol. Rapid Commun.*, **34**, 1755.
122. Bose, R.K. and Lau, K.K.S. (2009) *Chem. Vap. Deposition*, **15**, 150.
123. Bose, R.K. and Lau, K.K.S. (2010) *Biomacromolecules*, **11**, 2116.
124. Nejati, S. and Lau, K.K.S. (2011) *Nano Lett.*, **11**, 419.
125. Nejati, S. and Lau, K.K.S. (2011) *Thin Solid Films*, **519**, 4551–4554.
126. Jo, W., Freedman, K., Yi, D.K., Bose, R.K., Lau, K.K.S., Solomon, S.D., and Kim, M.J. (2011) *Biofabrication*, **3**, 015002.
127. Bose, R.K. and Lau, K.K.S. (2011) *Thin Solid Films*, **519**, 4415–4417.
128. Jo, W., Lee, J.H., and Kim, M.J. (2012) *J. Nanopart. Res.*, **14**, 699.
129. Chan, K. and Gleason, K.K. (2005) *Langmuir*, **21**, 8930.
130. Karaman, M., Kooi, S.E., and Gleason, K.K. (2008) *Chem. Mater.*, **20**, 2262.

131. Chen, B., Seidel, S., Hori, H., and Gupta, M. (2011) *ACS Appl. Mater. Interfaces*, **3**, 4201.
132. Montero, L., Gabriel, G., Guimera, A., Villa, R., Gleason, K.K., and Borrós, S. (2012) *Vacuum*, **86**, 2102.
133. Frank-Finney, R.J., Bradley, L.C., and Gupta, M. (2013) *Macromolecules*, **46**, 6852.
134. Ozaydin-Ince, G., Demirel, G., Gleason, K.K., and Demirel, M.C. (2010) *Soft Matter*, **6**, 1635.
135. Ozaydin-Ince, G., Dubach, J.M., Gleason, K.K., and Clark, H.A. (2011) *Proc. Natl. Acad. Sci. U.S.A.*, **108**, 2656.
136. Yagüe, J.L. and Gleason, K.K. (2012) *Soft Matter*, **8**, 2890.
137. Yin, J., Yagüe, J.L., Eggenspieler, D., Gleason, K.K., and Boyce, M.C. (2012) *Adv. Mater.*, **24**, 5441.
138. Ozaydin-Ince, G., Armagan, E., Erdogan, H., Buyukserin, F., Uzun, L., and Demirel, G. (2013) *ACS Appl. Mater. Interfaces*, **5**, 6447.
139. Kaya, A., Demiryurek, R., Armağan, E., Ozaydin-Ince, G., Sezen, M., and Koşar, A. (2013) *J. Micromech. Microeng.*, **23**, 115017.
140. Mari-Buye, N., O'Shaughnessy, S., Colominas, C., Semino, C.E., Gleason, K.K., and Borrós, S. (2009) *Adv. Funct. Mater.*, **19**, 1276.
141. You, J.B., Yoo, Y., Oh, M.S., and Im, S.G. (2014) *ACS Appl. Mater. Interfaces*, **6**, 4005.
142. Ye, Y., Mao, Y., Wang, H., and Ren, Z. (2012) *J. Mater. Chem.*, **22**, 2449.
143. McInnes, S.J.P., Szili, E.J., Al-Bataineh, S.A., Xu, J., Alf, M.E., Gleason, K.K., Short, R.D., and Voelcker, N.H. (2012) *ACS Appl. Mater. Interfaces*, **4**, 3566.
144. Chan, K. and Gleason, K.K. (2005) *Chem. Vap. Deposition*, **11**, 437.
145. Chen, X. and Anthamatten, M. (2008) *Polymer*, **49**, 1823.
146. Chen, X. and Anthamatten, M. (2009) *Langmuir*, **25**, 11555.
147. Lee, L.H. and Gleason, K.K. (2008) *J. Electrochem. Soc.*, **155**, G78.
148. Ozaydin-Ince, G. and Gleason, K.K. (2010) *Chem. Vap. Deposition*, **16**, 100.
149. Jeevendrakumar, V.J.B., Altemus, B.A., Gildea, A.J., and Bergkvist, M. (2013) *Thin Solid Films*, **542**, 81.
150. Haller, P.D., Flowers, C.A., and Gupta, M. (2011) *Soft Matter*, **7**, 2428.
151. Ma, M.L., Mao, Y., Gupta, M., Gleason, K.K., and Rutledge, G.C. (2005) *Macromolecules*, **38**, 9742.
152. Ma, M., Gupta, M., Li, Z., Zhai, L., Gleason, K.K., Cohen, R.E., Rubner, M.F., and Rutledge, G.C. (2007) *Adv. Mater.*, **19**, 255.
153. O'Shaughnessy, W.S., Mari-Buye, N., Borrós, S., and Gleason, K.K. (2007) *Macromol. Rapid Commun.*, **28**, 1877.
154. Im, S.G., Bong, K.W., Kim, B.-S., Baxamusa, S.H., Hammond, P.T., Doyle, P.S., and Gleason, K.K. (2008) *J. Am. Chem. Soc.*, **130**, 14424.
155. Im, S.G., Kim, B.-S., Lee, L.H., Tenhaeff, W.E., Hammond, P.T., and Gleason, K.K. (2008) *Macromol. Rapid Commun.*, **29**, 1648.
156. Xu, J. and Gleason, K.K. (2010) *Chem. Mater.*, **22**, 1732.
157. Xu, J., Asatekin, A., and Gleason, K.K. (2012) *Adv. Mater.*, **24**, 3692.
158. Martin, T.P., Kooi, S.E., Chang, S.H., Sedransk, K.L., and Gleason, K.K. (2007) *Biomaterials*, **28**, 909.
159. Yoo, G., Yoo, Y., Kwon, J.-H., Darpito, C., Mishra, S.K., Pak, K., Park, M.S., Im, S.G., and Yang, J.-W. (2014) *Green Chem.*, **16**, 312.
160. Tenhaeff, W.E. and Gleason, K.K. (2007) *Langmuir*, **23**, 6624.
161. Tenhaeff, W.E. and Gleason, K.K. (2007) *Surf. Coat. Technol.*, **201**, 9417–9421.
162. Lee, C.-H., Tenhaeff, W., and Gleason, K.K. (2009) *Thin Solid Films*, **517**, 3619.
163. Tenhaeff, W.E., McIntosh, L.D., and Gleason, K.K. (2010) *Adv. Funct. Mater.*, **20**, 1144.
164. Petruczok, C.D. and Gleason, K.K. (2012) *Adv. Mater.*, **24**, 6445.
165. Petruczok, C.D., Choi, H.J., Yang, S.Y., Asatekin, A., Gleason, K.K., and Barbastathis, G. (2013) *J. Microelectromech. Syst.*, **22**, 54.
166. You, J.B., Kim, S.Y., Park, Y.J., Ko, Y.G., and Im, S.G. (2014) *Langmuir*, **30**, 916.
167. Petruczok, C.D., Yang, R., and Gleason, K.K. (2013) *Macromolecules*, **46**, 1832.
168. Yang, R., Jang, H., Stocker, R., and Gleason, K.K. (2014) *Adv. Mater.*, **26**, 1711.

169. Bakker, R., Weijers, P., Spee, D.A., van Steenbergen, M.J., van der Werf, C.H.M., Rath, J.K., and Schropp, R.E.I. (2011) *Thin Solid Films*, **519**, 4418.
170. Arora, W.J., Tenhaeff, W.E., Gleason, K.K., and Barbastathis, G. (2009) *J. Microelectromech. Syst.*, **18**, 97.
171. Lee, B., Jiao, A., Yu, S., You, J.B., Kim, D.-H., and Im, S.G. (2013) *Acta Biomater.*, **9**, 7691.
172. Bakker, R., Weijers, P., van der Werf, C.H.M., Rath, J.K., and Schropp, R.E.I. (2010) *Phys. Status Solidi A*, **207**, 647.
173. Chen, G., Lau, K.K.S., and Gleason, K.K. (2009) *Thin Solid Films*, **517**, 3539.
174. Chan, K., Kostun, L.E., Tenhaeff, W.E., and Gleason, K.K. (2006) *Polymer*, **47**, 6941.
175. Ozaydin-Ince, G. and Gleason, K.K. (2009) *J. Vac. Sci. Technol., A*, **27**, 1135.
176. Ozaydin-Ince, G. and Gleason, K.K. (2010) *J. Electrochem. Soc.*, **157**, D41.
177. Lee, E., Faguet, J., Brcka, J., Akiyama, O., Liu, J., and Toma, D. (2011) *Thin Solid Films*, **519**, 4571.
178. Yang, R., Buonassisi, T., and Gleason, K.K. (2013) *Adv. Mater.*, **25**, 2078.
179. Yin, J., Yagüe, J.J.L., Boyce, M.C., and Gleason, K.K. (2014) *ACS Appl. Mater. Interfaces*, **6**, 2857.
180. Bose, R.K., Nejati, S., Stufflet, D.R., and Lau, K.K.S. (2012) *Macromolecules*, **45**, 6915.
181. Choi, H.-G., Amara, J.P., Martin, T.P., Gleason, K.K., Swager, T.M., and Jensen, K.F. (2006) *Chem. Mater.*, **18**, 6339.
182. Loo, L.S. and Gleason, K.K. (2001) *Electrochem. Solid-State Lett.*, **4**, G81.
183. Sargeant, P.B. (1970) *J. Org. Chem.*, **35**, 678.
184. Mahler, W. and Resnick, P.R. (1974) *J. Fluorine Chem.*, **3**, 451.
185. Kennedy, R.C. and Levy, J.B. (1976) *J. Fluorine Chem.*, **7**, 101.
186. Lau, K.K.S., Gleason, K.K., and Trout, B.L. (2000) *J. Chem. Phys.*, **113**, 4103.
187. Knickelbein, M.B., Webb, D.A., and Grant, E.R. (1984) *Mater. Res. Soc. Proc.*, **38**, 23.
188. Sanderson, R.T. (1976) *Chemical Bonds and Bond Energy*, 2nd edn, Academic Press, New York.
189. Eversteyn, F.C. (1974) *Philips Res. Rep.*, **29**, 45.
190. Lau, K.K.S., Murthy, S.K., Pryce Lewis, H.G., Caulfield, J.A., and Gleason, K.K. (2003) *J. Fluorine Chem.*, **122**, 93.
191. Odian, G. (2004) *Principles of Polymerization*, 4th edn, John Wiley & Sons, Inc., Hoboken, NJ.
192. Brandrup, J., Immergut, E.H., and Grulke, E.A. (eds) (2003) *Polymer Handbook*, 4th edn, John Wiley & Sons, Inc., Hoboken, NJ.
193. Lau, K.K.S., Mao, Y., Pryce Lewis, H.G., Murthy, S.K., Olsen, B.D., Loo, L.S., and Gleason, K.K. (2006) *Thin Solid Films*, **501**, 211.
194. Flory, P.J. (1953) *Principles of Polymer Chemistry*, Cornell University Press, Ithaca, NY.
195. Adamson, A.W. and Gast, A.P. (1997) *Physical Chemistry of Surfaces*, 6th edn, John Wiley & Sons, Inc, New York.
196. Hayward, D.O. and Trapnell, B.M.W. (1964) *Chemisorption*, 2nd edn, Butterworth, London.
197. Atkins, P. and de Paula, J. (2014) *Atkins' Physical Chemistry*, 10th edn, Oxford University Press, Oxford.
198. Beuermann, S., Paquet, D.A., McMinn, J.H., and Hutchinson, R.A. (1996) *Macromolecules*, **29**, 4206.
199. Bailey, W.J., Ni, Z., and Wu, S.-R. (1982) *J. Polym. Sci. Polym. Chem. Ed.*, **20**, 3021.

3
Copolymerization and Crosslinking

Yu Mao

3.1
Introduction

Copolymerization offers a simple and unique method for introducing desirable structure and properties into polymer thin films synthesized by initiated chemical vapor deposition (iCVD). iCVD copolymerization does not involve monomer fragmentation as in the process of plasma copolymerization, resulting in a well-designed polymer structure with regular repeating units and pedant functional groups. iCVD copolymerization creates opportunities in synthesizing copolymers that would have monomer miscibility issues in solution phase. More importantly, the monomer type and flow ratio can be varied during the process of film growth, providing the capability to independently control and optimize specific copolymer film properties.

Understanding iCVD copolymerization fundamentals enables better control of the copolymer composition and structure. Copolymer reactivity ratios are convenient for characterizing copolymerization behaviors. Similar to that of liquid copolymerization, the iCVD copolymerization kinetics are governed by the concentrations of monomers, the monomer reactivity ratios, and the copolymerization temperature. Since most iCVD copolymerization processes involve diffusion of monomer vapors through gas phase and heterogeneous reaction on solid surface, several factors are to be considered in studying the fundamentals: (i) determination of monomer surface concentration; (ii) modification of reactivity ratios due to the heterogeneous reaction; and (iii) surface temperature.

Analysis of the copolymer composition and structure is of significance in understanding the copolymerization process and controlling the copolymer properties. Infrared (IR) spectroscopy and other thin- film characterization methods have been applied in the analysis of chemical composition and film structure. Relating copolymer composition and structure to film properties establishes the processing–structure–property relationship in studying iCVD copolymers. Together, the understanding of this relationship enables us to design and synthesize iCVD copolymers with the desired properties.

CVD Polymers: Fabrication of Organic Surfaces and Devices, First Edition. Edited by Karen K. Gleason.

This chapter will focus on the fundamentals of copolymerization kinetics, analysis of iCVD copolymer composition and structure, and understanding of factors in tuning the copolymer properties. Applications of iCVD copolymers will be discussed in subsequent chapters.

3.2
Copolymer Composition and Structure

3.2.1
Confirmation of iCVD Copolymerization

Monomers can simultaneously participate in polymer chain propagation when mixed under appropriate reaction conditions in solution phase. During the heterogeneous reactions of iCVD, copolymerization has been verified by studying the sequence distribution in the synthesized polymers. In the iCVD of maleic anhydride (Ma) and styrene (S), the chemical shift of quaternary C3 in the ^{13}C nuclear magnetic resonance (NMR) was used to probe the sequence information (Figure 3.1 a and b) [1]. The signal between 136 and 140 ppm was assigned to the alternating Ma–S–Ma triad. In addition, the absence of signal between 141.5 and 146 and 145–147.5 ppm, corresponding to the semi-alternating

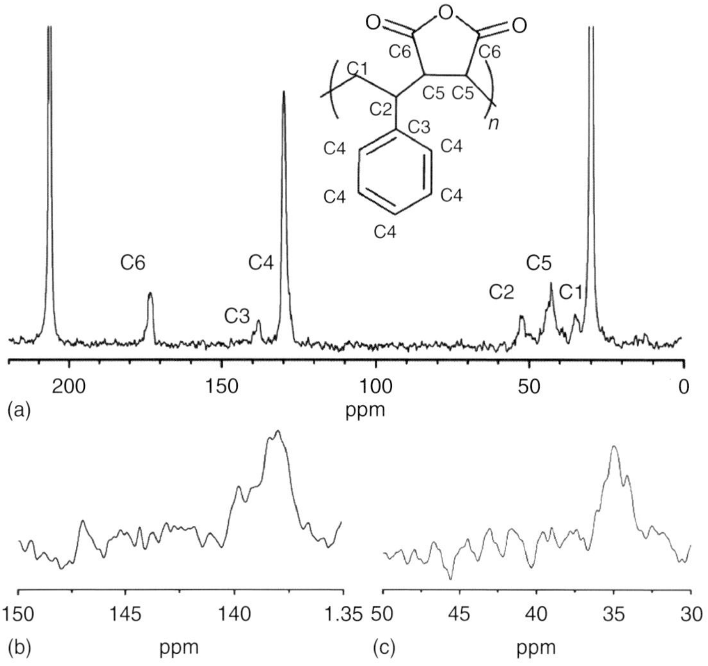

Figure 3.1 ^{13}C NMR spectrum of iCVD polymer of maleic anhydride and styrene (a), along with (b) the expanded region of quaternary carbon C3 and (c) the DEPT CH_2 subspectrum. (Adapted with permission from Ref. [1]. Copyright 2007 American Chemical Society.)

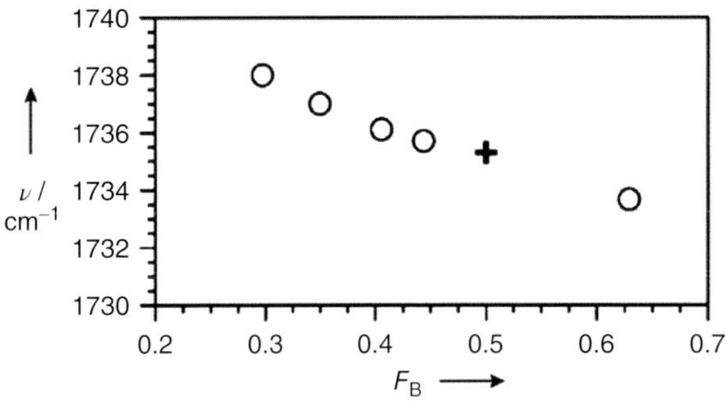

Figure 3.2 Systematic shift in the EA carbonyl peak position as a function of the EA fraction in the iCVD polymer of MA and EA. (Adapted with permission from Ref. [2]. Copyright 2007 Wiley-VCH.)

(S–S–Ma and Ma–S–S) and nonalternating (S–S–S) triads, respectively, confirms the copolymerization of Ma and S. Further verification is found in the distortionless enhancement polarization transfer (DEPT) CH_2 subspectrum for the C1 methylene carbon of styrene (Figure 3.1c). The signals at 33–37, 37–42, and 42–47 ppm correspond to the alternating, semi-alternating, and nonalternating triads, respectively. Only the signal at 33–37 ppm is present, while the other two are absent, confirming the formation of alternating Ma–S–Ma triads.

In the series of iCVD polymers synthesized by varying the monomer ratio of methacrylic acid (MA) and ethyl acrylate (EA) [2], the Fourier transform infrared (FTIR) absorption of the carbonyl group in EA demonstrated a systematic peak shift from 1733 to 1738 cm^{-1}. This peak shift increased with the decreasing fraction of EA (Figure 3.2). This correlation excluded the origin of hydrogen bonds formed between side groups (i.e., the MA carboxylic acid and the EA carbonyl group) in causing the shift, because the hydrogen bonds would weaken the oscillation of the C=O bond, resulting in a decrease in the wavenumber. Therefore, the systematic peak shift indicated copolymerization instead of homopolymerization of two different polymers in iCVD, because only copolymerization can change the electron cloud distribution along the polymer backbone and result in the correlation between carbonyl peak shift and the EA fraction.

3.2.2 Analysis of Copolymer Composition

The copolymer composition can be quantitatively calculated using the spectrum analysis in FTIR. Copolymer films can be directly deposited on IR transparent substrates with high quality and uniform thickness, which makes samples ready for accurate FTIR analysis. On the other hand, FTIR serves as a nondestructive characterization method for iCVD copolymers because IR has minimal impact on the film composition and structure. According to the Beer–Lambert Law, at the same film thickness, the absorbance of a mode is proportional to the moiety concentration and the absorption coefficient.

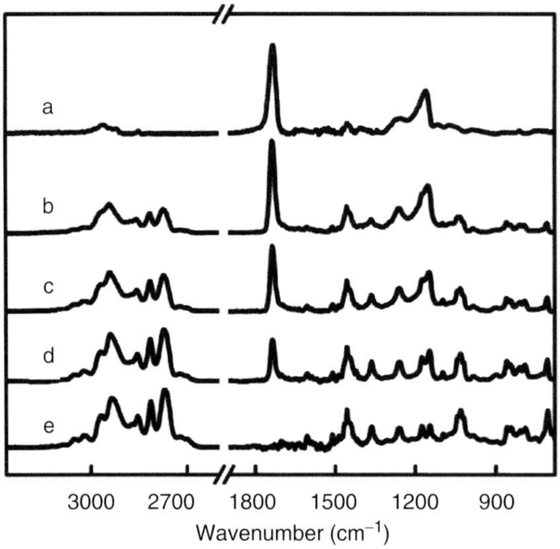

Figure 3.3 FTIR spectra of poly(ethylene glycol diacrylate) (a), poly(dimethylaminomethylstyrene-*co*-ethylene glycol diacrylate) with varied composition (b–d), and poly(dimethylaminomethylstyrene) (e) [3]. (Reproduced by permission of the Royal Society of Chemistry.)

In copolymers where distinct characteristic absorption peaks can be identified for each component, the ratio of components can be calculated by comparing the copolymer absorption with the homopolymer absorption. The underlying assumption for this simple calculation is that the absorption coefficient of a particular mode in the copolymers is the same as that in the homopolymer. Figure 3.3 shows FTIR spectra of poly(dimethylaminomethylstyrene-*co*-ethylene glycol diacrylate) films, compared with those of poly(dimethylaminomethylstyrene) and poly(ethylene glycol diacrylate) [3]. The absorption peaks at 2768 and 2814 cm^{-1} were assigned to the C-H stretching in the tertiary amine group in dimethylaminomethylstyrene (DMAMS), and the absorption peak at 1735 cm^{-1} was assigned to the C=O stretching in ethylene glycol diacrylate (EGDA). The molar ratio of DMAMS to EGDA in the copolymer films, $n_{\mathrm{DMAMS}}/n_{\mathrm{EGDA}}$, can be calculated using the equation:

$$n_{\mathrm{DMAMS}}/n_{\mathrm{EGDA}} = \frac{A_{\mathrm{N-C-H}}A*_{\mathrm{C=O}}M_{\mathrm{EGDA}}}{A_{\mathrm{C=O}}A*_{\mathrm{N-C-H}}M_{\mathrm{DMAMS}}}$$

where $A_{\mathrm{N-C-H}}$ and $A^{*}_{\mathrm{N-C-H}}$ are the peak areas of the N-C-H absorption in the spectra of copolymer and homopolymer, respectively, $A_{\mathrm{C=O}}$ and $A^{*}_{\mathrm{C=O}}$ are the peak areas of the C=O absorption in the spectra of copolymer and homopolymer, respectively, and M_{DMAMS} and M_{EGDA} are the molecular weights of the DMAMS and EGDA repeating units, respectively.

For copolymer spectra where components shared the same or overlapping characteristic peaks such as the C=O stretching at 1700–1740 cm^{-1}, the net contribution from one component can be identified either by resolving the peak into two separate peaks using the peak decoupling programs [2] or by referencing to a different characteristic peak through a sensitivity ratio *r*. In the analysis of copolymer poly(2-hydroxyethyl methacrylate-*co*-ethylene glycol diacrylate) [P(HEMA-*co*-EGDA)] [4], *r* is calculated using the quotient of C=O peak area and O-H region area in PHEMA. The net contribution from the C=O in the HEMA component to the total $A_{C=O}$ in the copolymer is rA_{O-H}, where A_{O-H} is the peak area of O-H absorption in copolymer. Therefore, the molar ratio of EGDA to HEMA, n_{EGDA}/n_{HEMA}, can be calculated as

$$n_{EGDA}/n_{HEMA} = \frac{(A_{C=O} - rA_{O-H})/2}{rA_{O-H}}$$

Note that there is a factor of 2 because each EGDA unit has two C=O bonds. This calculation requires the assumption that the carbonyl absorption coefficient is equivalent in HEMA units as in EGDA units. X-ray photoelectron spectroscopy (XPS) high resolution scan in the oxygen region was used to verify the calculation of the molar ratio n_{EGDA}/n_{HEMA}. Figure 3.4 shows the n_{EGDA}/n_{HEMA} ratios obtained from FTIR and XPS analysis of P(HEMA–EGDA) copolymer deposited at different EGDA partial pressures. The good agreement between the two sets of results indicates the validity of the assumption. The correlation of the n_{EGDA}/n_{HEMA} ratio with the EGDA partial pressure provides a guideline for controlling the composition of the copolymers.

It is noted that the carbonyl absorption coefficient is not equivalent for some polyacrylic copolymer components. For the copolymer of poly(vinylpyrrolidone-*co*-ethylene glycol diacrylate), the FTIR spectra analysis has to be corrected using factors obtained from XPS analysis before the copolymer composition can be determined [5]. It is estimated that the C=O absorption coefficient in the

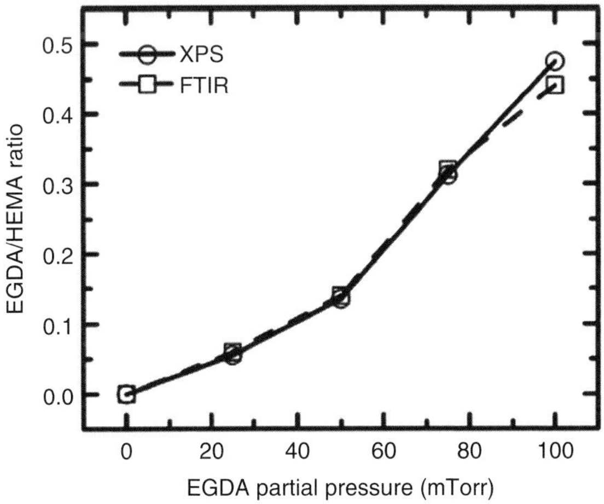

Figure 3.4 EGDA/HEMA ratio in the P(HEMA–EGDA) films as a function of the partial pressure of EGDA during iCVD. (Reproduced with permission from Ref. [4]. Copyright 2005 American Chemical Society.)

vinylpyrrolidone component is approximately two times of that in the ethylene glycol diacrylate component.

3.2.3 Compositional Gradient

The iCVD copolymer films can have either a homogeneous composition or a compositional gradient, as shown in Figure 3.5. Compositional gradient is a common strategy for forming integral films when specific surface properties, different from those in the bulk, are required. In addition, the gradual composition change along the film thickness can introduce interesting physical, chemical, or biological properties that are desired in many applications.

While traditional fabrication of compositionally graded films usually involves meticulously designed fabrication with long duration, iCVD provides a simple and fast method for synthesizing compositionally graded copolymer films. The feed ratio of monomers and other deposition conditions can be adjusted *in situ*, enabling control of the film composition along with the vertical growth of the film, resulting in a graded structure. In the synthesis of graded poly(2-hydroxyethyl methacrylate-*co*-pentafluorophenyl methacrylate) films [6], 2-hydroxyethyl methacrylate (HEMA) was initially introduced to form a hydrogel layer, followed by the introduction of pentafluorophenyl methacrylate (PFM). As PFM molecules gradually adsorb on the hydrogel surface, they were copolymerized into the hydrogel. This order of monomer introduction resulted in a copolymer gradient, with HEMA enriched at the bottom and PFM enriched at the top.

Figure 3.6 shows the distinct FTIR difference between three homogeneous copolymers with varying PFM compositions and three graded copolymers with the same near-surface feed compositions. The intensity of the fluorinated phenyl ring absorption at 1522 cm^{-1} indicates the relative amount of PFM component in both copolymer films. At the same 71% feed composition, the peak intensity in the homogeneous copolymer is much stronger than that in the graded copolymer, because PFM is present across the entire homogeneous copolymer, while the graded copolymer only has PFM incorporated at the surface layer.

Time-of-flight secondary ion mass spectrometry (ToF-SIMS) was used to probe the depth profile of PFM in the graded copolymer. Figure 3.7 shows the ion intensity associated with the charged PFM monomer fragment as a function of the film depth [6]. The homogeneous copolymer shows a relatively constant PFM content across the film depth. The graded copolymer shows comparable ion intensity in the first 20 nm of the film, but the intensity gradually decays at a depth below that

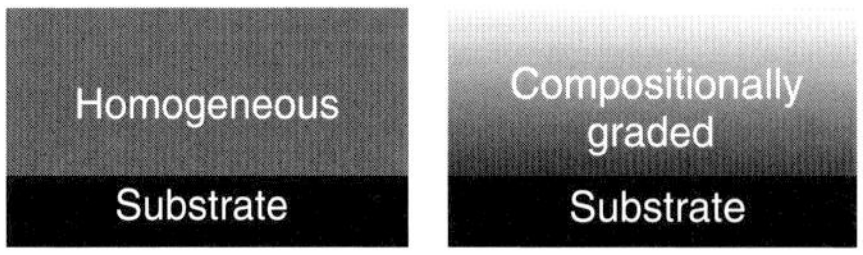

Figure 3.5 Schematics of the homogeneous (a) and compositionally graded (b) copolymer films.

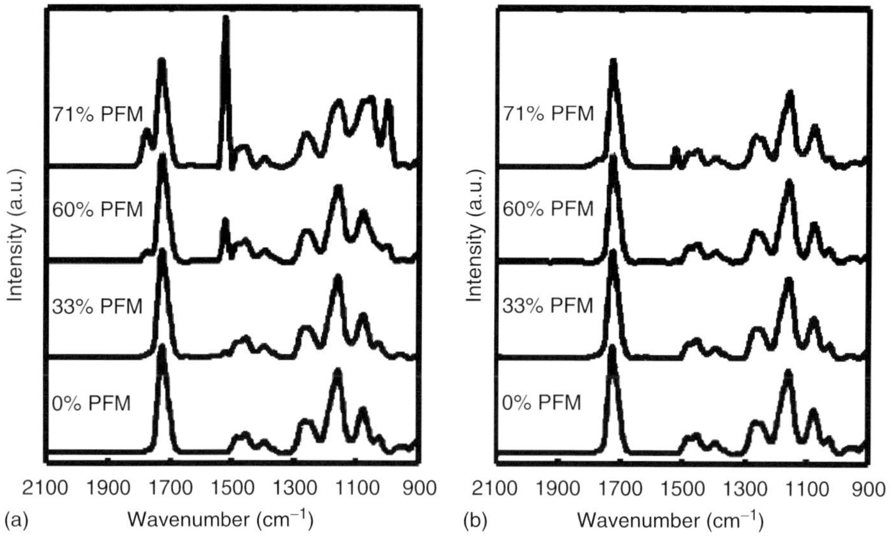

Figure 3.6 FTIR spectra of (a) homogeneous copolymers and (b) graded copolymers, along with the spectra of poly(2-hydroxyethyl methacrylate). The PFM composition refers to gas phase molar percentage during iCVD. (Reproduced with permission from Ref. [6]. Copyright 2009 American Chemical Society.)

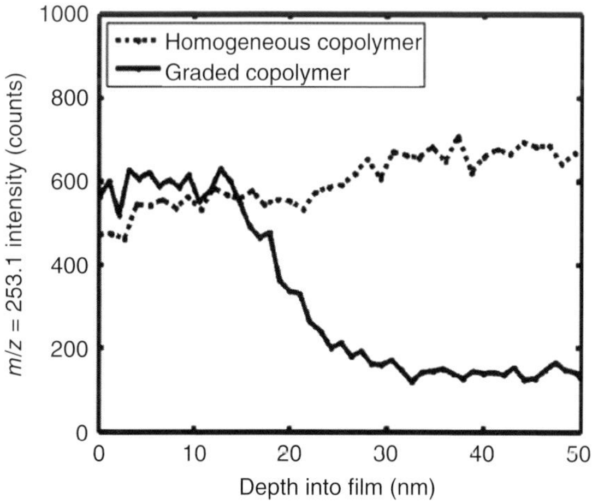

Figure 3.7 ToF-SIMS depth profile showing signal from $m/z = 253.1$, corresponding to a positively charged PFM fragment, as a function of film depth for both the homogeneous and graded copolymers. (Reproduced with permission from Ref. [6]. Copyright 2009 American Chemical Society.)

point. The depth profile is consistent with the fact that PFM was introduced in the final stage of iCVD copolymerization. It also provides direct confirmation for the PFM compositional gradient throughout the thickness of the graded copolymer.

The narrow confinement of the PFM component at the surface allows immobilization of biomolecules while retaining the swelling property of the bulk hydrogel. The PFM moiety provides excellent reactive sites for single-step reactions with ligands bearing amine groups at mild conditions. Meanwhile, the swollen water content of the graded copolymer can be maintained high, because the bulk of the film is composed of poly(2-hydroxyethyl methacrylate) hydrogel. The compositional gradient decouples the underlying physical properties of the hydrogel from the chemical properties of the surface activity, allowing two properties to be combined together in one copolymer film.

3.3 Copolymerization Kinetics

3.3.1 Copolymerization Equation and Reactivity Ratio

Systematic control of the copolymer composition was demonstrated in the synthesis of many iCVD copolymers, such as poly(methacrylic acid-*co*-methyl α-chloroacrylate) and poly(methacrylic acid-*co*-ethyl acrylate), by controlling the feed fractions of precursor monomers [4, 7]. The mole fractions of copolymer components were quantitatively analyzed using FTIR or XPS [3, 4], as discussed in Section 3.2.2. Because CVD copolymerization occurs on the surface instead of in the vapor phase, the monomer feed fractions were actually determined by calculating the ratio of monomer concentrations on the reaction surface. The monomer surface concentration can be derived using Henry's law limit by relating with the saturation ratio, which is equal to the ratio of monomer partial vapor pressure to the monomer saturation vapor pressure [2].

The Fineman–Ross copolymerization equation was commonly used to relate the iCVD copolymer composition with the monomer feed fraction [8]:

$$\frac{f_A(1-2F_A)}{F_A(1-f_A)} = r_B + \left[\frac{f_A^2\left(F_A-1\right)}{F_A(1-f_A)^2}\right] r_A$$

where F_A is the mole fraction of component A in the copolymer, f_A is the feed fraction of monomer A, r_A and r_B are the reactivity ratios of monomer A and B, respectively. By plotting $\frac{f_A(1-2F_A)}{F_A(1-f_A)}$ versus $\frac{f_A^2(F_A-1)}{F_A(1-f_A)^2}$, the monomer reactivity ratios can be calculated from the slope and intercept of the regression line. In Figure 3.8a, the fraction of MA (F_A) in the copolymer P(MAA-*co*-EA) is plotted as a function of the MA feed fraction (f_A). The increase in f_A results in a systematic increase in F_A. Figure 3.8b shows the plot of the Fineman–Ross copolymerization equation,

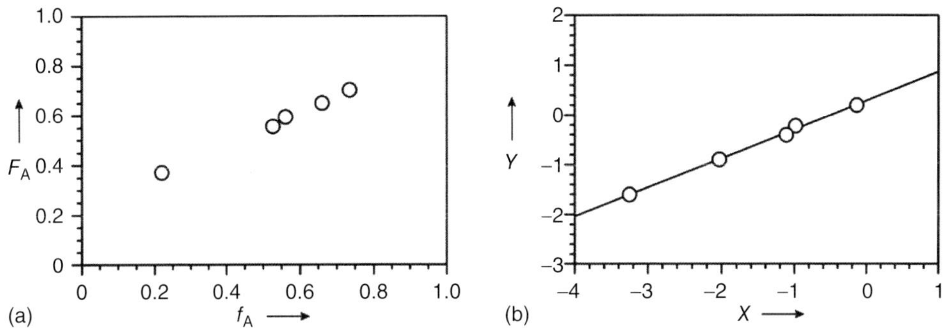

Figure 3.8 (a) An increase in methacrylic acid feed fraction on the surface (f_A) results in a systematic increase in the fraction of methacrylic acid (F_A) in the copolymer of P(MAA-*co*-EA). (b) Plotting in the form of the Fineman–Ross copolymerization equation. (Reproduced with permission from Ref. [2] Copyright 2007 Wiley-VCH.)

where Y is equal to $\frac{f_A(1-2F_A)}{F_A(1-f_A)}$ and X is equal to $\frac{f_A^2(F_A-1)}{F_A(1-f_A)^2}$. From the plot, the reactivity ratios of MA and EA were determined to be 0.58 and 0.29, respectively.

The Kelen–Tudos plot [9] is another linear graphical method for evaluating monomer reactivity ratios. By following the same notation as in the Fineman–Ross copolymerization equation, the copolymer composition can relate to the monomer feed fraction through the equation

$$\eta = r_A\xi - \frac{r_B}{\alpha}(1-\xi)$$

where

$$\eta = \frac{f_A(1-f_A)(2F_A-1)}{(1-f_A)^2F_A\alpha + f_A{}^2(1-F_A)}$$

and

$$\xi = \frac{f_A{}^2(1-F_A)}{(1-f_A)^2F_A\alpha + f_A{}^2(1-F_A)}$$

The parameter α was introduced to symmetrically distribute data on the plot. α is defined as

$$\alpha = \sqrt{F_{min} \times F_{max}}$$

and

$$F = \left(\frac{f_A}{1-f_A}\right)^2 \times \frac{(1-F_A)}{F_A}$$

Figure 3.9 shows the Kelen–Tudos plots for three copolymerization studies: (i) Ma with S, (ii) MAA with EA, and (iii) 2,2,3,3,4,4,5,5,6,6,7,7-dodecafluoroheptyl acrylate (DFHA) with glycidyl methacrylate (GMA). The fitted regression line for the Ma/S copolymerization showed a slope of zero, indicating zero reactivity ratios for both Ma and S monomers. For the MAA/EA copolymerization, the reactivity ratios of MAA and EA were derived from the Kelen–Tudos plot to be 0.56 and

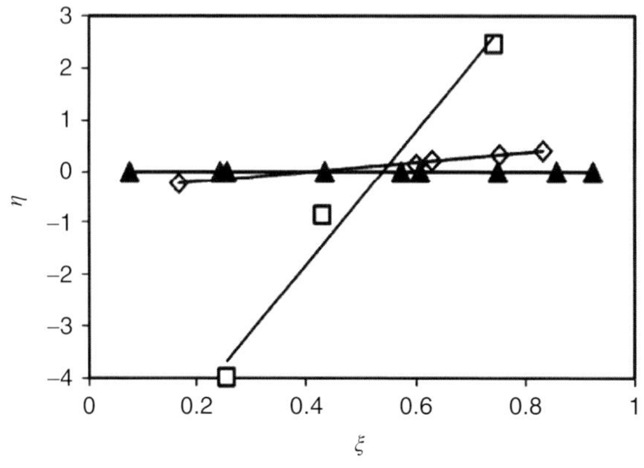

Figure 3.9 Kelen–Tudos plot for the Ma/S (triangle), MAA/EA (diamond), and DFHA/GMA (square) copolymer systems. (Reproduced with permission from Ref. [1]. Copyright 2007 American Chemical Society.)

0.28, respectively. The obtained results were close to the values determined using the Fineman–Ross method. The DFHA/GMA copolymerization was found to bear reactivity ratios of 5.91 and 0.09 for DFHA and GMA, respectively. The much higher reactivity ratio of DFHA when compared to that of GMA indicates the preferential addition of DFHA to all propagating radicals, which explains the enrichment of DFHA in the copolymers even at small DFHA feed fractions [10].

It is noted that during iCVD copolymerization, divinyl monomers such as EGDA are commonly used to crosslink the polymer films *in situ*. Under the assumption that all the double bonds in the crosslinking agents reacted during iCVD, which has been verified in FTIR and XPS studies [3, 4], the divinyl monomer can be viewed as two independent monomers whose overall structure is similar to that of the divinyl monomer. For example, one EGDA molecule can be considered to be the addition of two methyl acrylates molecules. In this case, the reactivity ratios can be fitted using the copolymerization equations discussed above. The only modification for the equations would be to double the divinyl monomer concentration to get the concentration of the "imaginary" monomer. In the copolymerization of HEMA with EGDA, the reactivity ratios were determined from the modified Fineman–Ross plot to be 0.22 and 2.11 for HEMA and MA, respectively [11].

Since the reactivity ratio indicates the tendency of a propagating radical for self-propagation (adding the same monomer as the ending unit) over cross-propagation (adding a monomer that is different from the ending unit), understanding the reactivity ratios obtained from iCVD is essential for guiding the choices of co-monomers and deposition conditions. In the iCVD copolymerization of monomer pairs such as styrene and maleic anhydride, the obtained reactivity ratios are remarkably close to the values obtained in solution phase, indicating a similar environment for these monomers under iCVD and solution copolymerization conditions.

For reactivity ratios not reported in literature, they can be empirically estimated using the $Q-e$ scheme, which correlates monomer reactivity ratio with monomer structure by separating the monomer reactivity into a resonance factor Q and a polar factor e. In the iCVD copolymerization of MAA and EA, the reactivity ratio of EA is comparable to the value obtained using the $Q-e$ scheme; however, the reactivity ratio of MAA is much lower than the $Q-e$ value [2]. Similar trends were observed in the pairs of HEMA/EGDA [4] and MAA/methyl α-chloroacrylate [11], where monomers with higher polarity were found to have lower reactivity ratios than the $Q-e$ values. One cause could be the lack of solvent effect during iCVD, which inhibits the formation of hydrogen bonds and the orientation of these monomers with the same monomer units in the propagating radicals. As a result, self-propagation during vapor deposition is not as favored as in the solution phase. The origin of errors could also be defects in $Q-e$ estimation and nonlinear reaction kinetics. The nonlinearity is caused by the termination of propagating chain radicals with initiator radicals during iCVD polymerization [12]. Consequently, the copolymerization equation will have to be evaluated adequately in a more complex way.

Overall, the study on the copolymerization equations and reactivity ratios of iCVD systems provides guidance in understanding the relative activities of monomers and the sequence behavior during iCVD. More importantly, it allows the synthesis of copolymer films with the desired composition by controlling deposition conditions that affect the monomer feed fractions. Recognizing the similarity of iCVD copolymerizations to those in solution phase, the monomer reactivity ratios obtained experimentally or empirically can serve as guidelines for directing iCVD copolymerizations.

3.3.2
Types of iCVD Copolymerization

Two different types of copolymerization behavior have been observed in iCVD: random copolymerization and alternating copolymerization. In random copolymerization, both monomers enter the propagating chains in random placement, although the monomer with a high reactivity ratio takes a larger portion in the overall copolymer composition. In alternating copolymerization, each propagating chain preferentially adds the monomer that is different from the ending unit, resulting in an alternating arrangement of monomers and an equal amount of each monomer in the overall copolymer composition.

As indicated in the study of reactivity ratios, random copolymerization is common in iCVD [2, 11, 7]. Alternating copolymerization has been observed in the depositions of poly(styrene-*alt*-maleic anhydride) [1] and poly(4-aminostyrene-*alt*-maleic anhydride) [13]. In comparison, iCVD of the corresponding homopolymers, including polystyrene, poly(4-aminostyrene), and poly(maleic anhydride), was either not observed or quite slow, while the simultaneous introduction of monomer pairs resulted in a significant increase in copolymerization rates (Figure 3.10). The deposition rate and molecular weight

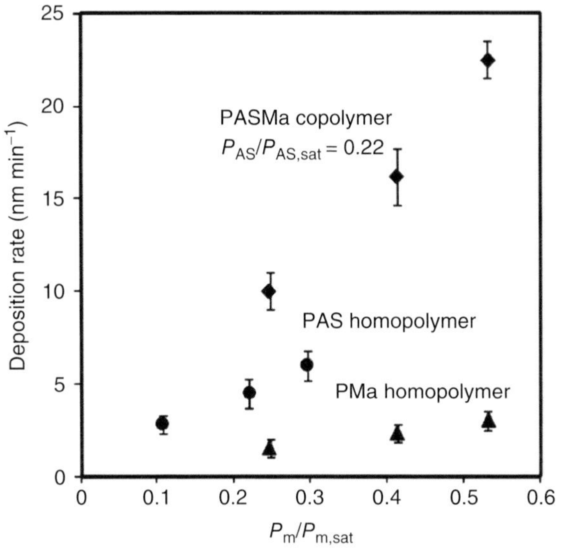

Figure 3.10 Increase in deposition rates of poly(4-aminostyrene-*alt*-maleic anhydride), poly(4-aminostyrene), and poly(maleic anhydride) with the saturation ratio of maleic anhydride. (Reproduced with permission from Ref. [13]. Copyright 2012 Wiley-VCH.)

further increase with the saturation ratio of monomers, consistent with the surface polymerization scheme during CVD.

The composition of iCVD alternating copolymers was found to be nonrelated with the monomer feed ratio. In poly(styrene-*alt*-maleic anhydride), the maleic anhydride mole fraction was calculated from XPS analysis to be $49.3 \pm 1.6\%$, even though the mole feed fraction of maleic anhydride varied from 9 to 80% [1]. It is speculated that styrene and maleic anhydride molecules form pairs similar to what was described in the complex participation model in solution phase [14]. During iCVD, the surface concentrates the molecules within a liquid-like thin layer. When the molecules adsorb to the substrate surface, favorable electronic interactions could occur between the electron-donating styrene and the electron-accepting maleic anhydride. In this scenario of surface adsorption, pairs can be formed without solvent coordination.

3.4 Tunable Properties of iCVD Copolymers

3.4.1 Mechanical Properties

The mechanical properties of iCVD copolymers can be tuned by incorporating a crosslinking component into the copolymers. By increasing the crosslinker

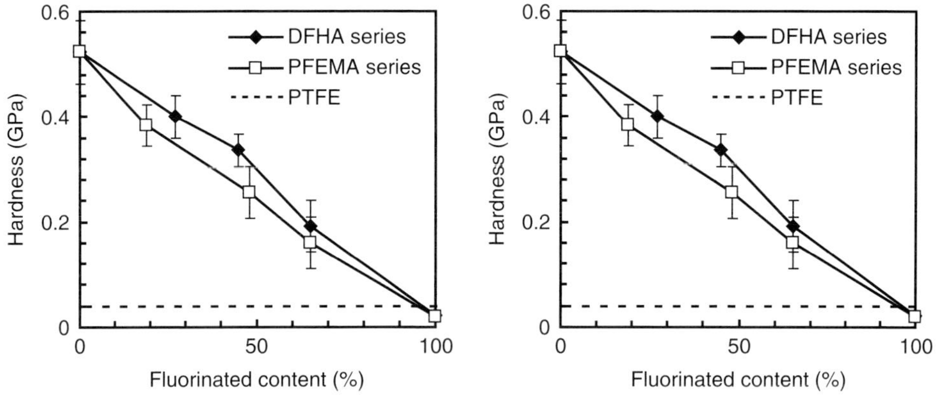

Figure 3.11 Mechanical properties of glycidyl fluorinated copolymers with varied fluorinated content after annealing 14 h. Both the (perfluoroalkyl)ethyl methacrylate (PFEMA) and the DFHA copolymer series demonstrated enhancement in hardness and modulus compared with poly(tetrafluoroethylene) (PTFE). (Reproduced with permission from Ref. [10]. Copyright 2006 American Chemical Society.)

amount in the feed, the crosslinking degree can be raised, resulting in enhanced mechanical properties. In poly(4-vinylpyridine-*co*-divinylbenzene) copolymer films, the Young's modulus can be tuned between 3.4 and 5.8 GPa by adjusting the flow rates of 4-vinylpyridine and divinylbenzene monomers [15].

Another strategy for improving the mechanical properties is to introduce self-crosslinkable components. In the series of fluorinated copolymers, GMA was copolymerized to introduce epoxy groups [10]. Post-deposition annealing of the copolymers induced ring opening reactions between epoxy pairs, resulting in a crosslinked structure. The post annealing does not require any initiators or catalysts, and the crosslinking degree can be systematically varied by controlling the annealing time. Because of the weak intermolecular cohesion between fluorinated chains, the fluorinated homopolymers (at fluorinated content 100%) have very low hardness and modulus (Figure 3.11). With the decrease of the fluorinated content (thus the increase of GMA content), both the hardness and modulus of the glycidyl fluorinated copolymers were significantly improved.

Highly crosslinked copolymer films with elastic modulus exceeding 20 GPa were reported in the alternating copolymer of 4-aminostyrene (4-AS) and Ma after annealing [13]. The reaction between neighboring amino and anhydride pedant groups creates a massively crosslinked network, demonstrating elastic modulus far greater than typical polymers (0.5–5 GPa). It is interesting that these strong films maintain high flexibility with no cracks being formed even after 200 cycles of repeated folding. The combination of high modulus and flexibility is ascribed to the unique molecular architecture in the annealed copolymer of 4-As and Ma, shown in Figure 3.12. The average connectivity number, *r*, of the annealed copolymer film was calculated to be 2.385, which is below the rigidity percolation threshold value of 2.4, suggesting a highly crosslinked, yet nonrigid structure. These

(a)

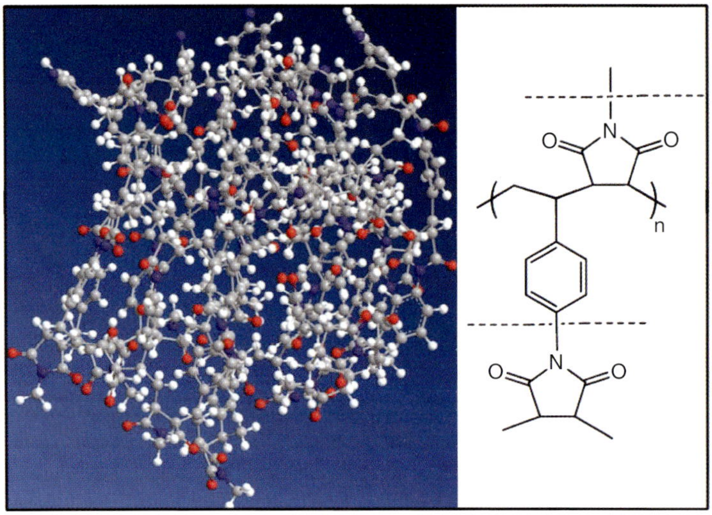

(b)

Figure 3.12 Structures of (a) as-deposited and (b) annealed copolymer of 4-AS and Ma. The 3-D molecular structures were obtained from molecular dynamics simulation using Chem3D software. (Reproduced with permission from Ref. [13]. Copyright 2012 Wiley-VCH.)

results demonstrated significant achievements in controlling the iCVD copolymer mechanical properties and fabricating copolymer thin films that are both strong and tough.

3.4.2 Swelling

Hydrogels synthesized from iCVD copolymerization present swelling behavior. The degree of swelling closely relates with the crosslinking degree, which can be systematically controlled by varying the feed ratio of crosslinker monomers. There are various methods to measure the degree of swelling. One simple approach is to measure the weight of water uptake relative to the dry-state weight of the hydrogel [16]. Another way is to calculate water uptake content through ellipsometry analysis [4]. The analysis involves modeling the swollen hydrogel

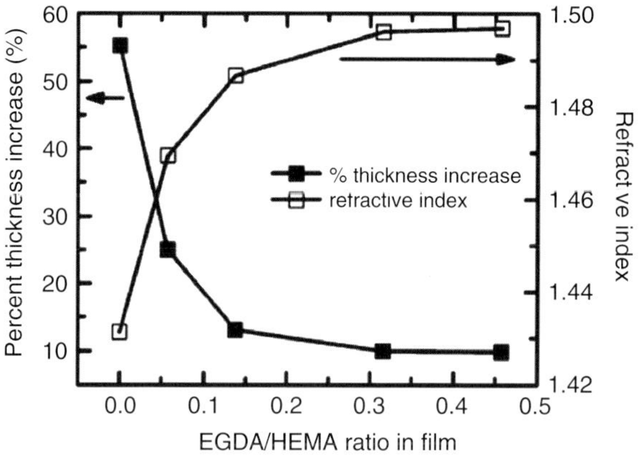

Figure 3.13 Changes in film thickness and refractive index of swollen film as functions of the EGDA/HEMA ratio in the film. (Reproduced with permission from Ref. [4]. Copyright 2005 American Chemical Society.)

as a composite material with the polymer as the matrix and water as the filler. One typical method for modeling composite materials is the effective medium approximation. Figure 3.13 shows the thickness and the refractive index of the swollen P(HEMA-*co*-EGDA) films with varied EGDA/HEMA [4]. With the increase of EGDA/HEMA ratio, the hydrogel swells less, because the polymer chains are restricted more tightly as the crosslinking degree increases. The linear PHEMA has a thickness increase of 55%, while the most crosslinked copolymer (EGDA/HEMA = 0.45) has an increase of only 10%. The effect of EGDA/HEMA ratio on the hydrogel swelling is also observed in the refractive index measurements. With the ratio increase, the refractive index increases, because the EGDA crosslinking hinders the hydrogel to uptake water, and water has a lower refractive index ($\eta_{633} = 1.33$) than the polymer matrix ($\eta_{633} = 1.49 - 1.51$). The decrease of hydrogel swelling with the increase of crosslinker content has been observed in other iCVD deposited hydrogels as well [16].

The hydrogel swelling also depends on the film composition. At the same temperature, poly-(*N*,*N*-dimethylaminoethyl methacrylate-*co*-ethylene glycol diacrylate) P(DMAEMA-*co*-EGDA) hydrogels demonstrate a much higher swelling degree than P(HEMA–EGDA) hydrogels with the same EGDA content. It is noted that hydrogels synthesized by CVD are usually attached to solid substrates. At the same composition, the swelling of hydrogels attached on substrates is lower than the swelling of unconstrained bulk hydrogels because of the substrate effect. On comparison, the volumetric swelling degrees between surface attached hydrogel and unconstrained hydrogel showed a square root relationship between the former and the latter [17]. In addition, the swelling of hydrogels confined in nanopores is further suppressed compared with the swelling of hydrogels attached to planar surfaces due to the space confinement imposed by the pores [16].

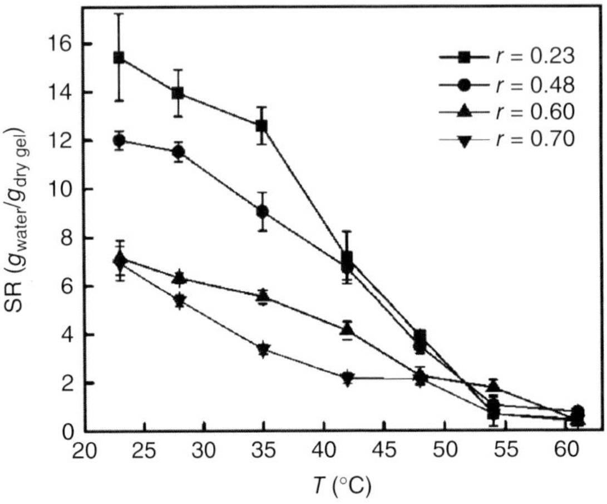

Figure 3.14 The temperature dependence of the swelling ratio (SR) of P(DMAEMA-*co*-EGDA) hydrogel with varied EGDA/DEMAEMA molar ratio *r* [16]. (Reproduced by permission of The Royal Society of Chemistry.)

3.4.3
Thermal Properties

iCVD offers the capability to control the copolymer composition and structure for tuning of thermal properties. The effect of film composition on the thermal responsiveness of P(DMAEMA-*co*-EGDA) hydrogels was studied. The swelling ratios of hydrogels with varied EGDA/DEMAEMA molar ratio, *r*, was measured as a function of the temperature (Figure 3.14). At 23 °C, a swelling ratio of 15.4 was observed for the film with $r = 0.23$. As the temperature increased, the swelling ratio decreased significantly. At 54 °C, the film only had a swelling ratio of approximately 0.5. This significant change is attributed to the change from the intermolecular hydrogel–water hydrogen bonding to the intra-molecular hydrogen bonding within the hydrogel at elevated temperatures. The change in the swelling ratio was observed to decrease with the increase of EGDA/DMAEMA ratio, suggesting that the increasing degree of crosslinking prevents the hydrogel network from responding to heat. As a result, hydrogels with lower EGDA/DMAEMA ratio showed much steeper temperature response, and the response became linear as the EGDA/DMAEMA ratio increased.

The dynamic response of hydrogel films to temperature change was measured using quartz crystal microbalance with dissipation monitoring. Poly(*N*-isopropylacrylamide) hydrogel crosslinked with ethylene glycol dimethacrylate showed a transition temperature of 15 °C [18], while crosslinking with di(ethylene glycol) divinyl ether resulted in a transition temperature of 28 °C [19]. The hydrogel structure plays an important role in determining the transition kinetics in response to heat. The time constant associated with the response of the graded poly(*N*-isopropylacrylamide-*co*-ethylene glycol dimethacrylate) film was reported to be significantly lower (<143 s) than that of the homogeneous copolymer counterpart (1000–8000 s). These results suggest that the iCVD hydrogel thermal response can be optimized for desired transition kinetics.

3.4.4 Surface Properties

Hydrophilic surfaces can be created by surface modification with iCVD hydrogels. P(HEMA-*co*-EGDA) hydrogels demonstrate a receding contact angle of 17° [4]. An even lower contact angle of 11° was reported in poly(vinylpyrrolidone-*co*-ethylene glycol diacrylate) hydrogels [20]. Systematic control over hydrogel composition is essential for tailoring the surface hydrophilicity. In P(HEMA-*co*-EGDA) hydrogels, both the advancing and receding contact angles increase with increasing EGDA ratio. The increase in the EGDA ratio results in a decrease in hydrophilicity simply because the hydroxyl content is decreased. The EGDA ratio in the hydrogel could be tuned by varying the partial pressure of the monomer.

There are multiple factors to be understood in tuning the surface hydrophobicity of fluorinated copolymers. Figure 3.15 compares the critical surface tension of the glycidyl fluorinated copolymers with that of poly(tetrafluoroethylene) (PTFE) [10]. The desired low surface energy can be obtained in the copolymers with a small content of the fluorinated components. This is attributed to the segregation of the lower-surface-energy fluorinated segment to the air–polymer surface, as observed in the angle-resolved X-ray spectroscopy study. The critical surface tension varied with copolymer composition, but the values only had limited dependence on the content of the fluorinated components because of the fluorine surface enrichment. With the same amount of fluorinated component, the critical surface tension of (perfluoroalkyl)ethyl methacrylate (PFEMA) copolymers was

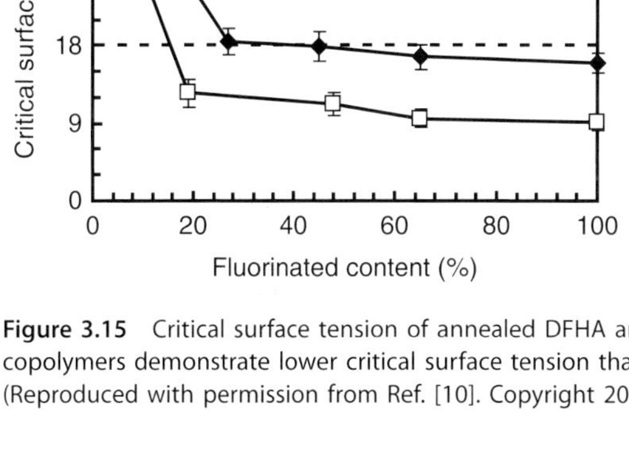

Figure 3.15 Critical surface tension of annealed DFHA and PFEMA copolymer series. Both copolymers demonstrate lower critical surface tension than that of the benchmark PTFE. (Reproduced with permission from Ref. [10]. Copyright 2006 American Chemical Society.)

observed to be much lower than that of the DFHA copolymers, indicating the significantly different contribution of CF_3 and CF_2H end groups in lowering the surface energy of copolymer films.

3.5 Conclusions

Copolymerization via iCVD proceeds with similar kinetics as that in solution phase copolymerization. Reactivity ratios can be determined using the Fineman–Ross method and the Kelen–Tudos method. Deviation of reactivity ratios from estimated values in solution phase can be possibly attributed to the lack of solvent effect and the nonlinear termination kinetics during iCVD polymerization. Both random and alternating copolymerizations were observed. The copolymerization process conditions such as monomer flow sequence and feed ratio can be systematically adjusted for the control of copolymer composition and structure, resulting in the tuning of copolymer properties, including, but not limited to, mechanical properties, swelling, thermal properties, and surface properties.

References

1. Tenhaeff, W.E. and Gleason, K.K. (2007) Initiated chemical vapor deposition of alternating copolymers of styrene and maleic anhydride. *Langmuir*, **23** (12), 6624–6630.
2. Lau, K.K.S. and Gleason, K.K. (2007) All-dry synthesis and coating of methacrylic acid copolymers for controlled release. *Macromol. Biosci.*, **7** (4), 429–434.
3. Ye, Y.M., Song, Q., and Mao, Y. (2011) Single-step fabrication of non-leaching antibacterial surfaces using vapor crosslinking. *J. Mater. Chem.*, **21** (1), 257–262.
4. Chan, K. and Gleason, K.K. (2005) Initiated chemical vapor deposition of linear and cross-linked poly(2-hydroxyethyl methacrylate) for use as thin-film hydrogels. *Langmuir*, **21** (19), 8930–8939.
5. Chan, K. *et al.* (2006) Initiated chemical vapor deposition of polyvinylpyrrolidone-based thin films. *Polymer*, **47** (20), 6941–6947.
6. Montero, L. *et al.* (2009) Thin hydrogel films with nanoconfined surface reactivity by photoinitiated chemical vapor deposition. *Chem. Mater.*, **21** (2), 399–403.
7. Mao, Y. and Gleason, K.K. (2006) Positive-tone nanopatterning of chemical vapor deposited polyacrylic thin films. *Langmuir*, **22** (4), 1795–1799.
8. Odian, G. (1981) *Principles of Polymerization*, Wiley-Interscience, New York.
9. Kelen, T. and Tudos, F. (1975) Analysis of linear methods for determining copolymerization reactivity ratios. 1. New improved linear graphic method. *J. Macromol. Sci. -Chem. A*, **9** (1), 1–27.
10. Mao, Y. and Gleason, K.K. (2006) Vapor-deposited fluorinated glycidyl copolymer thin films with low surface energy and improved mechanical properties. *Macromolecules*, **39** (11), 3895–3900.
11. Yague, J.L. and Gleason, K.K. (2012) Systematic control of mesh size in hydrogels by initiated chemical vapor deposition. *Soft Matter*, **8** (10), 2890–2894.
12. Lau, K.K.S. and Gleason, K.K. (2006) Initiated chemical vapor deposition (iCVD) of poly(alkyl acrylates): an

experimental study. *Macromolecules*, **39** (10), 3688–3694.
13. Xu, J.J., Asatekin, A., and Gleason, K.K. (2012) The design and synthesis of hard and impermeable, yet flexible, conformal organic coatings. *Adv. Mater.*, **24** (27), 3692–3696.
14. Ha, N.T.H. and Fujimori, K. (1998) Theoretical study of the copolymerization of styrene and maleic anhydride prepared in carbon tetrachloride and in N,N-dimethylformamide. *Acta Polym.*, **49** (8), 404–410.
15. Petruczok, C.D., Yang, R., and Gleason, K.K. (2013) Controllable cross-linking of vapor-deposited polymer thin films and impact on material properties. *Macromolecules*, **46** (5), 1832–1840.
16. Ye, Y.M. and Mao, Y. (2011) Vapor-based synthesis of ultrathin hydrogel coatings for thermo-responsive nanovalves. *J. Mater. Chem.*, **21** (22), 7946–7952.
17. Toomey, R., Freidank, D., and Ruhe, J. (2004) Swelling behavior of thin, surface-attached polymer networks. *Macromolecules*, **37** (3), 882–887.
18. Alf, M.E., Hatton, T.A., and Gleason, K.K. (2011) Novel N-isopropylacrylamide based polymer architecture for faster LCST transition kinetics. *Polymer*, **52** (20), 4429–4434.
19. Alf, M.E. *et al.* (2010) Sharp hydrophilicity switching and conformality on nanostructured surfaces prepared via initiated chemical vapor deposition (iCVD) of a novel thermally responsive copolymer. *Macromol. Rapid Commun.*, **31** (24), 2166–2172.
20. Tenhaeff, W.E. and Gleason, K.K. (2008) Initiated and oxidative chemical vapor deposition of polymeric thin films: iCVD and oCVD. *Adv. Funct. Mater.*, **18** (7), 979–992.

4
Non-Thermal Initiation Strategies and Grafting

Daniel D. Burkey

4.1
Introduction

Much of this text focuses on chemical vapor deposition that is initiated via thermal processes, either via direct activation of a target monomer or by activation of an initiator. In some cases, thermal processes may not be desirable, for example, in the case of a thermally sensitive substrate. Even under vacuum, the use of heated filaments in close proximity to the substrate results in some convective and especially radiative loading. Since iCVD and related processes are typically absorption driven, this thermal load generally requires an active cooling strategy to maintain the substrate at the desired temperature. This can be challenging if the substrate is complex or a poor thermal conductor. To address some of these issues, attention is now turned toward several alternative schemes for initiation, as well as to strategies for improving adhesion between a chemically vapor deposited thin film and a variety of substrates that might be of interest.

4.2
Initiation Strategies

4.2.1
Plasma Initiation Strategies

4.2.1.1 Plasma Enhanced Chemical Vapor Deposition (PECVD)

Plasma enhanced chemical vapor deposition (PECVD) is a well-developed and widely used methodology for fabrication of thin films from a variety of precursors, both organic and inorganic [1]. Plasmas can be generated in a variety of ways, but generally involve subjecting a monomer in the gas phase to an energetic discharge through application of electrical energy. Typical methods include radio frequency plasmas (RF, 13.56 MHz commonly), atmospheric plasmas (APs), and microwave plasmas (MW, 2.45 GHz, commonly). As indicated by d'Agostino, reactor geometry can also influence the discharge type [2]. Examples include the following:

CVD Polymers: Fabrication of Organic Surfaces and Devices, First Edition. Edited by Karen K. Gleason.

- Electrodeless systems, such as MW or high-frequency RF with resonant cavity coupling.
- External electrodes, where the electrodes are exterior to the reactor and are capacitively coupled (i.e., the plasma is formed between a charged electrode and a ground or biased electrode) or inductively coupled (i.e., the plasma is formed by electromagnetic induction, often via an energized coil).
- Internal electrodes, where the electrodes are interior to the reactor and are capacitively or DC coupled.

In terms of deposition of thin polymer films from organic precursors, careful selection of the plasma conditions is required in order to retain the desired molecular features in the monomer. At moderate to high plasma powers, organic molecules typically undergo numerous fragmentation pathways, and the resulting thin films are generally amorphous in structure and do not bear a strong resemblance to traditional polymers from the same starting monomer [3]. Thus, it can be challenging to use PECVD to deposit thin organic films with a well-defined structure. Palumbo *et al.* quantify this in the plasma deposition of poly(ethylene oxide)-like (PEO) thin films from 2,5,8,11-tetraoxa-dodecane (Triglyme) [4]. X-ray photoelectron spectroscopy (XPS) peaks in the polymer films corresponding to C-O (PEO-like) structures decreased steadily with increasing plasma power, whereas C-H (amorphous hydrocarbon) peaks increased steadily over the same range.

Similarly, Zhao *et al.* report on the plasma polymerization of conjugated polynitrile thin films [5]. For deposition of 4-biphenylcarbonitrile, deposition at 25 W resulted in thin films that showed strong evidence of conjugation and aromatic group retention in the Fourier Transform Infrared Spectroscopy (FTIR) spectrum. Conversely, thin films deposited at 55 W show considerably diminished aromatic content and more aliphatic carbon absorptions, indicating a breakup of the aromatic ring. Additionally, UV–vis absorption spectra for the two films show a marked change in both the intensity and the position of the absorption band, from ~370 nm in the 25 W sample to <360 nm in the 55 W sample. As Zhao *et al.* indicated, the shift to shorter wavelengths for the absorption band is typically indicative of reduced conjugation length, which is consistent with the destruction of the aromatic groups at higher deposition powers.

As it is clear that power plays a critical role in functional group retention in organic thin films, Yasuda characterized the effect of plasma power and monomer flowrate on deposition by using the composite parameter W/FM, where W is the discharge power of the plasma, in Watts, F is the monomer flowrate in cubic centimeters per minute, and M is the molecular weight of the monomer. This is essentially a measure of the energy input per unit mass of the monomer, and some general trends in the structure and properties of plasma deposited polymers can be observed by plotting various quantities versus W/FM as a parameter. For example, Morosoff noted that a plasma polymerized polytetrafluoroethylene (PTFE) was fluorine deficient when W/F was high, which corresponded to either high power or low flow conditions. Conversely, a fluorine-rich polymer was obtained when W/F was low [6]. It has been argued that plasma polymers of similar structure and properties should be obtained if the W/FM ratio is held constant, despite varying powers and flowrates. For structural retention of specific chemical functionalities, a low power-per-unit-mass of monomer is required, so as to prevent significant fragmentation of the monomer. Thus, operation at either low powers or high flowrates would be recommended. For a more complete discussion of the W/FM parameter and its limitations, as well as a discussion of the plasma physics, the reader is referred to Chapter 6.

4.2.1.2 Pulsed-Plasma Enhanced Chemical Vapor Deposition (PPECVD)

Some success in retaining desired molecular functionality in plasma-deposited organic thin films has been achieved by moving to a pulsed-plasma enhanced chemical vapor deposition approach (PPECVD). While, in PECVD, energy is supplied continuously to the monomer, leading to multiple fragmentation paths, in PPECVD, energy is applied in short bursts, allowing the molecules exposed to the electric field to return to a ground state in between pulses. Förch *et al.* characterize the mechanism of pulsed-plasma deposition in terms of the plasma on-time, t_{on}, the plasma off-time, t_{off}, and the input power, P_{peak} [7]. The duty cycle, DC, of the pulsed plasma is given as

$$DC = \frac{t_{on}}{(t_{on} + t_{off})} \tag{4.1}$$

which is the ratio of plasma on-time to the total cycle time. Förch *et al.* also define the "equivalent power" that a substrate and growing film are exposed to during a pulsed-plasma deposition. Since the plasma is only on for a fraction of the total deposition time, the average power experienced is given as

$$P_{eq} = DC * P_{peak} \tag{4.2}$$

In this way, a high peak power can be maintained, allowing easy generation of the plasma, but the monomer exposure to the plasma is limited, potentially preserving desired functional groups. As the plasma off-time approaches zero, the plasma becomes continuous, the duty cycle becomes one, and the equivalent power is simply equal to the peak input power. In terms of Yasuda's W/FM, low-duty cycles have the effect of keeping the W value low, which promotes structure retention.

In the plasma on-time, conditions are akin to what they would be under continuous deposition, with a variety of activated species in the gas phase and also in contact with the substrate. During t_{on}, there is thus a competition between thin film deposition and ablation. During t_{off}, many of the activated species, such as ions and electrons, are rapidly relaxed back to a ground state and are no longer reactive. Comparatively, organic radicals can be substantially longer lived, and as such, contribute greatly to thin film growth during the plasma off-time. The pulsing of the energy changes the relative distribution of activated species in the plasma as compared to the continuous mode, and can result in the selective activation of some molecules. This in turn can result in the retention of selected monomer structural

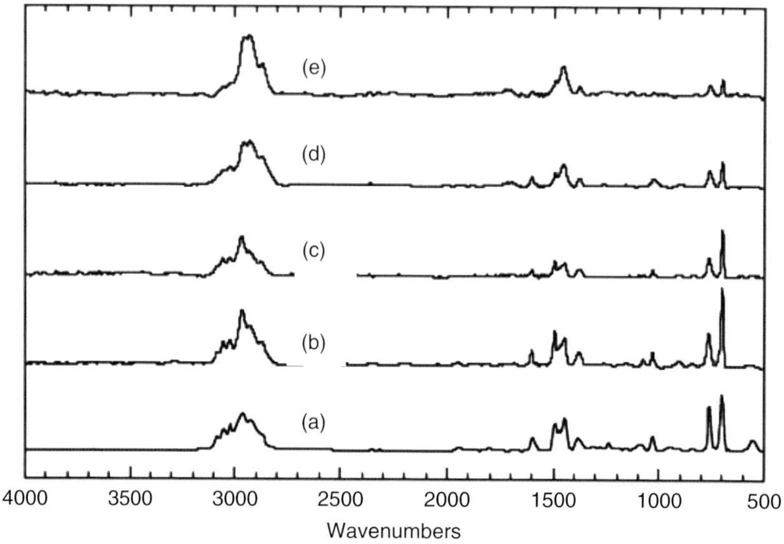

Figure 4.1 FTIR spectra of (a) low-molecular weight (LMW) PaMS standard, (b) 10/100 PPECVD sample, (c) 40 W continuous PECVD sample, (d) 10/20 PPECVD sample, and (e) 200 W continuous PECVD sample. (Reprinted with permission from Ref. [8]. Copyright 2003 Wiley-VCH.)

elements in the thin film. In general, however, reducing the duty cycle or equivalent power to improve structure retention comes at the expense of the ability to sustain the plasma. As Förch *et al.* describe, for duty cycles less than 1/100, the decay of charge carriers during the plasma off-time means that significant energy is required to reignite the plasma. This results in very slow deposition rates, as well as conditions similar to continuous plasma, since high powers are required to reignite the plasma even for a short duration.

As an example, consider the deposition of poly(α-methylstyrene) from α-methylstyrene (aMS) monomer under a variety of plasma conditions, as shown in Figure 4.1 [8]. As seen in Figure 4.1a, conventionally polymerized a MS retains numerous indicators of functional group retention. Most specifically, the two aryl C-H bending modes at 700 and 760 cm^{-1}, as well as the >3000 cm^{-1} C-H stretching in substituted aryl rings. In Figure 4.1b, a 10/100 pulsed plasma thin film is shown. As indicated above, the 10/100 duty cycle refers to the plasma duty cycle, which in this case is in milliseconds. As shown, the FTIR spectrum of the pulsed-plasma film and the standard are very similar, indicating a high degree of structural similarity between the two. Contrast this with Figure 4.1e, where a 200 W continuous plasma deposition is shown. The FTIR features previously apparent that correspond to specific aryl vibration modes are largely diminished or absent, indicating that the aryl functionality has been lost. Indeed, the most prominent absorptions in that spectrum are general C-H stretching, indicating a largely amorphous organic thin film.

4.2.1.3 Microwave Plasmas

MW plasmas, as compared to RF plasmas, typically operate at higher electron and ion densities, but at comparatively lower energies for both [9]. As such, neutral radical species can play a larger role in thin film deposition, which for radical-driven deposition processes can lead to a greater degree of monomer structure retention, or at least a significantly different structure from lower frequency plasmas. Labelle and Gleason demonstrated that thin films deposited via 1,1,2,2-tetrafluoroethane in both a MW plasma system and a RF plasma system had significantly different chemical structures, as determined by C1s XPS and optical emission spectroscopy (OES) [10]. The MW plasma samples had significantly higher CF_2 fractions than the samples produced via RF plasma, indicating differential activation of the monomer dependent upon the plasma source.

In general, however, thin films that are produced via direct plasma activation of the target monomer tend to lack the structural fidelity of traditionally polymerized materials or those produced via iCVD-type processes, due to the numerous and often nonspecific activation pathways that are available during a plasma process.

4.2.1.4 Initiated Plasma Enhanced Chemical Vapor Deposition (iPECVD)

In a manner similar to thermally initiated CVD processes, it is possible to use a plasma to activate an initiator molecule to begin a polymerization reaction, rather than using the plasma to activate the monomer directly. This presents similar advantages to thermal CVD processes using an initiator, in that a low-energy-activated initiator molecule can be used, which promotes the preservation of the desired monomer structure. Coclite and Gleason demonstrate the deposition of the organosilicon monomer 1,3,5-trivinyl-1,1,3,5,5-pentamethyltrisiloxane (TVTSO) using a conventional PECVD process, a thermally activated iCVD, and an initiated plasma process they call iPECVD, for initiated PECVD [11]. In the iPECVD process, an MW plasma (2.45 GHz) at 50 W was used. Additionally, the electrode area was relatively large, resulting in a plasma density of only $0.07\,W\,cm^{-2}$. Conventional RF plasma discharges often have plasma densities of $0.2\,W\,cm^{-2}$, about three times higher. As MW plasmas have been shown to have ion energies that are half that of RF generated plasmas, they are especially well suited to low-power applications where selective activation of a molecule is desired. While some deposition does occur for the monomer alone at very low power densities, the addition of a labile initiator, *tert*-butyl peroxide (TBPO), results in an almost 10-fold increase in deposition rate. The authors hypothesize that similarly to the use of TBPO in thermal iCVD, only a low-power plasma is required to break the weak oxygen–oxygen bond in the peroxide, leading to radical formation that can then initiate polymerization of the monomer vinyl groups as expected. This mechanism is supported by FTIR data shown in Figure 4.2, which compares the TVTSO polymer films grown via the processes described above. As shown, the low-power (50 W) iPECVD sample shows the closest structural resemblance to the thermal iCVD sample. Increasing

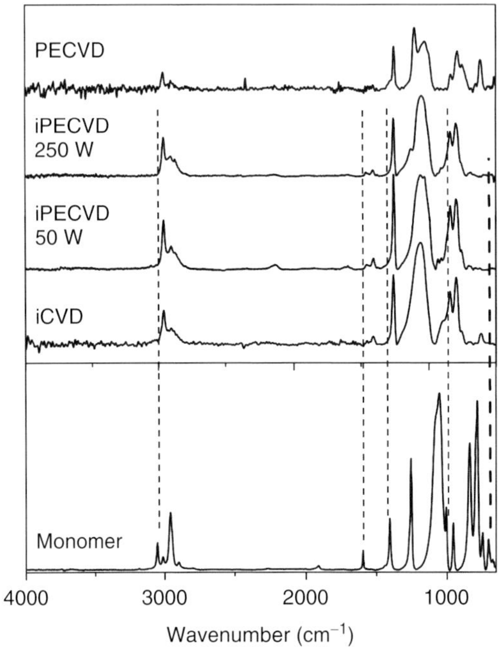

Figure 4.2 FTIR spectra of TVTSO monomer and polymers deposited by plasma-free iCVD, iPECVD at 50 and 250 W, and conventional PECVD without initiator. The dashed lines indicate the vinyl group's absorptions. The polymer spectra are normalized by the measured thickness. (Reprinted with permission from Ref. [11]. Copyright 2012 Wiley-VCH.)

the plasma power to 250 W, and then transitioning to straight PECVD shows increasing structural changes away from the monomer base structure, likely due to increased fragmentation of the monomer itself at higher plasma powers. The lack of vinyl moieties in the polymerized film, seen in the monomer at ~3150 and ~1600 cm^{-1}, suggest that radical polymerization through the vinyl groups is a dominant mechanism in all of the depositions.

4.2.1.5 Plasma Initiation Summary

In summary, in all plasma CVD processes, increasing plasma power and increasing monomer exposure to the plasma tend to increase fragmentation of the original structure and lead to more amorphous final products. Preserving monomer structure or desirable functional groups in the plasma setting requires a careful control of the monomer's exposure to the plasma. As discussed, this can be achieved through the use of low-power-density plasmas or pulsed plasmas, or through the use of initiator molecules that fragment at energies significantly lower than that of the monomer itself.

4.2.2 Photoinitiation Strategies

4.2.2.1 Photoactive Initiator Molecules

Schemes involving photoactive initiator molecules are similar to their analogous thermally activated pathways, differing only in the mechanism used to generate radicals. Using a photoactive initiator species may in some cases present advantages over performing the same polymerization using thermal initiation, the largest being the reduction of thermal loading on the substrate. Even in a vacuum, the convective and radiative loading from an array of hot filaments can be significant, and without adequate temperature control, can result in elevated substrate temperatures and corresponding changes in deposition rate due to changes in reactant absorption. Concepts such as the ratio of monomer partial pressure (P_M) or the initiator partial pressure (P_I) to their saturation pressures (P_{SAT}) at the reaction conditions are as important in piCVD systems as they are in thermal iCVD, as the monomer and initiator surface adsorption are critical in controlling aspects of the deposition such as rate and conformality [12].

Depending on the initiator structure and photoactivity, photoinitiator molecules can be classified broadly into two main categories: gas-phase photoinitators, which will radicalize and initiate in the gas phase, and surface-phase photoinitiators, which require the photoinitiator to be adsorbed on the desired substrate and activated prior to the introduction of the monomer species of interest.

Gas Phase Photoinitiators Gas phase photoinitiator molecules are generally Type-I photoinitiators, meaning they undergo a unimolecular decomposition to yield radicals, which then go on to initiate polymerization. A critical consideration when choosing a photoinitiator for a piCVD application is the volatility – the initiator should have sufficient vapor pressure such that it remains in the vapor phase at the conditions in the CVD reactor. Additional considerations are the photolytic efficiency – how many radicals can be generated per photon of incident light, as well as the wavelength and intensity of the light. For a photoinitiator to be effective, its absorption spectrum should be closely matched to the emission spectrum of the light source. As such, molecules that contain well-defined absorptions at commonly accessible wavelengths are potential photoinitiators. For example, azo-containing compounds have been shown to decompose under UV irradiation, and Chan *et al.* used 2,2′-azobis(2-methylpropane) (ABMP) to generate *tert*-butyl radicals and nitrogen via UV irradiation in the 250–350 nm range, as shown in Equation 4.3 [13].

$$(CH_3)_3CN = NC(CH_3)_3 + h\nu \rightarrow 2(CH_3)_3C\bullet + N_2(g) \quad (4.3)$$

When combined with a monomer capable of undergoing radical-initiated polymerization, such as glycidal methacrylate (GMA), a poly(glycidal methacrylate) (PGMA) thin film was formed, as expected. Once a radical species has been generated, the polymerization reaction for photoinitiated systems proceeds similarly

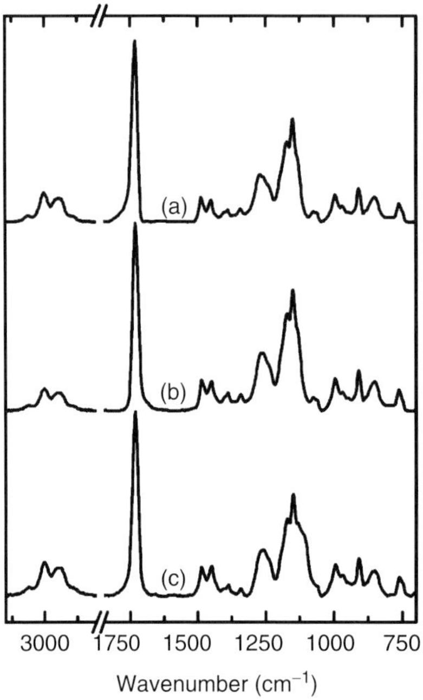

Figure 4.3 FTIR spectra of (a) piCVD PGMA film, (b) iCVD PGMA film, and (c) spin-cast PGMA obtained from polymer source. The absorption peaks at 901, 848, and 760 cm^{-1} are assigned to the characteristic bands of the epoxide group. (Reprinted with permission from Ref. [13]. Copyright 2005 American Chemical Society.)

to those that have been initiated by the thermal radical generation. In Figure 4.3, Chan *et al.* compared the FTIR spectra of the PGMA thin films fabricated using the photointiated CVD process (a), a thermal iCVD process (b), and traditional spin-casting (c). As can be seen, the three spectra are virtually identical, indicating that the structure of the three polymers is consistent.

Other small photoactive molecules, such as peroxides, are also viable photo initiators under the right conditions. TBPO, widely used as a thermal iCVD initiator, also decomposes under UV irradiation, but typically only at efficient yields at lower UV wavelengths (<300 nm) where the energies involved can be damaging to both the desired monomer and the growing polymer film.

Surface Photoinitiators In contrast to a gas-phase reaction using a Type I photoinitiator, a surface-driven model using a Type II photoinitiator may also be considered. Type II photoinitiators rely on a bimolecular process for radical generation, which in this case is the reaction between a photoinitiator and the surface to be deposited on. Some of the most common Type II photoinitiators are in the benzophenone family. As described by Martin *et al.*, methods for initiating polymerization using Type II photoinitiators can be categorized into all-solution techniques, hybrid techniques, and all-vapor phase techniques [14]. In the all-solution implementation, both the photoinitiator and the monomer are combined in a common solvent with the object to be coated and then exposed to the correct wavelength of light. In the hybrid approach, the photoinitiator alone is solubilized, and then applied to the desired substrate and the solvent evaporated, leaving only the photoinitiator behind. This surface is then exposed to both irradiation and the monomer in the vapor phase, thereby enabling polymerization. In the all-vapor phase model, both the photoinitiator and the monomer are in the vapor phase, and irradiation in the presence of the desired substrate leads to polymer deposition. The all-vapor phase method has the most applicability to chemically vapor deposited systems, and is the only method applicable when the substrate to be coated is not compatible with liquid-phase approaches.

Using benzophenone as an example, the reaction begins with the absorption of light and the transition to an activated singlet state [15]. This singlet state then transitions to a triplet state, and it is here that the bimolecular nature of the Type II photoinitiator comes into play, as the triplet state must interact with a second molecule to generate a radical. In this case, the surface to be coated acts in this capacity, and the activated benzophenone abstracts a hydrogen atom from the surface, generating a surface radical and a benzopinacol radical (BP-OH•). In the presence of a suitable monomer species, the surface radical is then free to initiate the polymerization reaction. Termination of this reaction can occur when two growing chains interact, when the growing chain connects with the benzopinacol radical, or when two benzopinacol radicals effectively dimerize. As Deng *et al.* point out, the benzopinacol radicals are typically not as reactive as the surface-generated radicals, and their primary role comes in termination reactions. Figure 4.4 shows the general reaction pathway for the benzophenone system. Since this method results in a polymer thin film that is tethered to the substrate, the use of Type II photoinitiaors is also a viable strategy for grafting chemically vapor deposited polymers to surfaces. This will be discussed later in this chapter.

4.2.2.2 Photoactive Monomer Species

In some cases, monomer species can be inherently photoactive, and as such can autopolymerize without the need for an additional photoinitiator. Wang and Brown reported on the self-initiated polymerization of a variety of acrylates and methacrylates, including acrylic acid, methacrylic acid, glycidal acrylate, 2-hydroxyethyl acrylate, 2-hydroxyethyl methacrylate (HEMA), *N*-(hydroxymethyl acrylamide, and poly(ethylene glycol) methacrylate, all in the neat liquid phase [17]. It is noted that many acrylic and styrenic monomers are inherently photoactive, likely due to the extended π-bonding network; carbonyl-alkenyl in acrylates and aryl-alkenyl in styrenes. However, at ambient conditions, the autopolymerization rates can be very slow, generally making the addition of a photoinitiator desirable. The above monomers have been demonstrated to be the exception to this rule, generally polymerizing extensively within seconds of irradiation. In the gas phase, several acrylic monomers have been shown to also be photoactive, with thin film deposition occurring upon exposure of the monomer vapor to UV irradiation. For piCVD of inherently photoactive monomers, the potential advantages are numerous. Without the need for thermal or plasma activation, the deposition conditions are mild, and tight control over substrate temperature is possible, allowing for the coating of otherwise delicate substrates. Additionally,

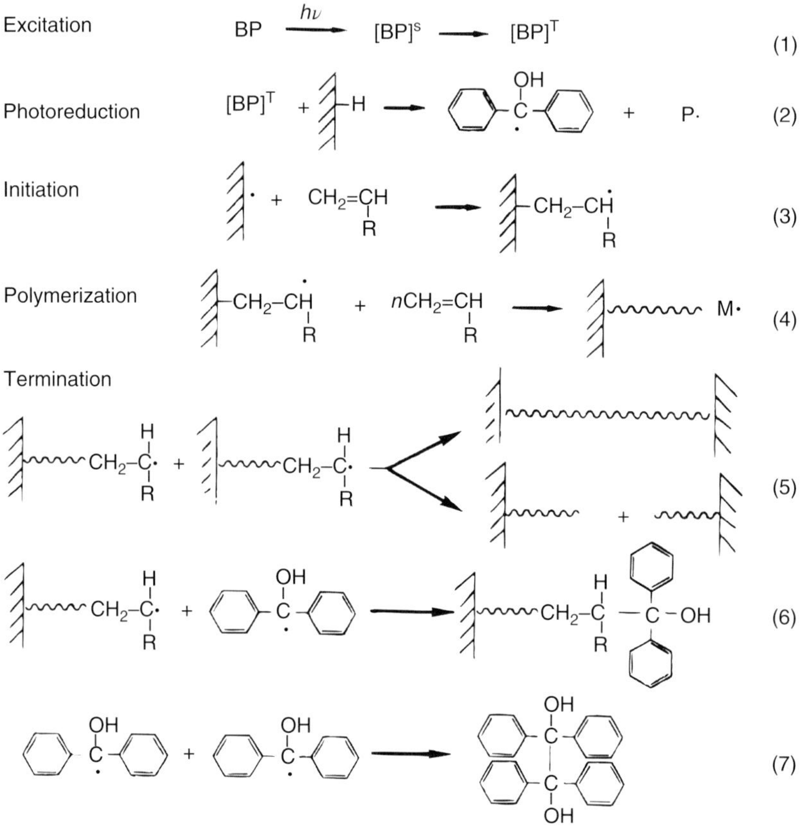

Figure 4.4 Excitation, photoreduction, initiation, polymerization, and termination steps for a benzophenone-initiated photopolymerization reaction. (Reprinted with permission from Ref. [16]. Copyright 1996 American Chemical Society.)

the reactor scheme is typically simplified, because for a homopolymer only a single reagent is needed, and no hardware other than a light source is required for activation.

Baxamusa *et al.* demonstrated the deposition of poly(2-hydroxyethyl methacrylate) (pHEMA) using HEMA alone as a monomer under 254 nm UV light [18]. The general reaction scheme for HEMA deposition is shown in Figure 4.5, where the HEMA monomer, either in the gas phase or at the surface, is activated by UV light, and self-polymerizes. Suresh *et al.* demonstrated a similar result for hexyl acrylate (HA) [19].

Mechanistically, Abrahamson *et al.* observed the photodecomposition of various aldehydes and ketones into radical species and carbon oxides [20]. Morita and Sadakiyo have specifically posited a mechanism for UV fragmentation of acrylates, based on a two-photon pathway, shown in Figure 4.6 [21]. In this pathway, an acrylate monomer fragments upon UV irradiation, eventually yielding

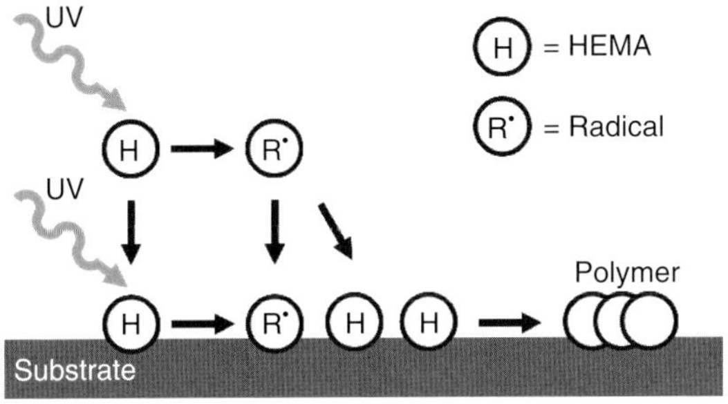

Figure 4.5 Proposed piCVD reaction mechanism. Ultraviolet light decomposes HEMA monomer into radical species either at the substrate or in the vapor phase. These radicals initiate polymerization by reacting with surface adsorbed monomer. Reprinted with permission from Ref. [18]. Copyright 2008 American Chemical Society.)

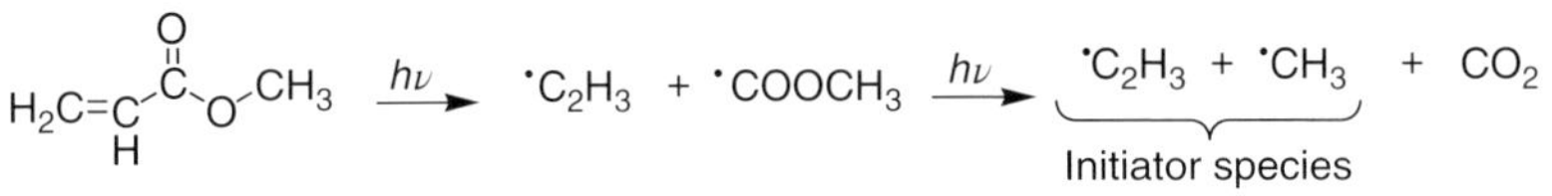

Figure 4.6 Proposed decomposition reaction for acrylates under ultraviolet irradiation. (Adapted from Ref. [21]. Copyright 1995 Elsevier Science S.A.)

multiple alkyl radicals that are capable of initiating polymerization, as well as CO_2. Rosenfeld and Weiner extensively studied the photodecomposition of acrylic and methacrylic acid under Ultraviolet-B (UVB) (249 nm) and Ultraviolet-C (UVC) (193 nm) irradiation in the gas phase, and determined that decarboxylation was the primary method of decomposition [22]. An interesting note here is that Rosenfeld and Weiner were also able to comment on the differences observed between the photoactive behavior of the acrylic and methacrylic acid in solution versus that in the gas phase. As noted by Wang and Brown, in the liquid phase, the acrylic acids underwent autopolymerization, whereas in the gas phase, UV irradiation resulted in decarboxylation [17]. It is postulated that under UV irradiation, a triplet state is formed, and the interactions of that triplet state with other ground-state acrylic acid molecules allow for the initiation of polymerization in the solution phase. Other triplet-state molecules can undergo relaxation to the ground state via vibrational interactions with a solvent. As a result, these relaxed states no longer have sufficient energy to undergo the decarboxylation mechanism, and polymerization is preferred. In contrast, in the gas phase, the frequency of these relaxational collisions is reduced due to the orders-of-magnitude density differences between the liquid and gas phases. This results in the triplet states retaining enough energy to undergo unimolecular relaxation mechanisms resulting in decarboxylation. These mechanistic differences can also partially explain the differences in observed piCVD behavior of some acrylic monomers – that is, only a subset of the acrylate monomers that have been studied in piCVD experiments have been demonstrated to auto-polymerize. This could be due to the rapid decarboxylation of certain monomer species under UV irradiation in the gas phase, while other, more stable monomers are long-lived enough in the triplet-state to initiate polymerization. Because the growing polymer film also contains elements of the original photoactive monomer species, active sites may also be generated in the polymer, which can significantly increase the deposition rate once an initial polymer layer has been formed. This behavior is observed specifically in the HA system, where the apparent deposition rate increases with time and the thickness of polymer deposited [19].

4.2.2.3 Photoinitiation Summary

In summary, piCVD is an attractive alternative to thermally activated iCVD processes in that it eliminates any thermal loading on the substrate, and can thus be used for the most delicate samples, at the same time producing polymers that are nearly identical to their iCVD and traditionally polymerized counterparts. Additionally, the ability to move the excitation source external to the reactor and the lack of the need for heated filaments or plasma source greatly simplifies the reactor setup needed. Lastly, the use of UV photoinitiators often can lead to direct grafting of the polymer to the desired substrate due to the mechanism of activation. Disadvantages, however, include the fact that a general, labile, UV initiator with a high photolytic efficiency at reasonable intensities and wavelengths has not been demonstrated, and only a small subset of desired monomers are UV active on their own. This is in contrast to thermal iCVD, in which organic peroxides, such as TBPO or *tert*-amyl peroxide have proven to be thermally labile under most iCVD conditions and have thus enjoyed wide use as initiators in iCVD systems.

4.3 Grafting

While one of the most powerful aspects of the CVD technique is the ability to coat a wide variety of surfaces, this can also present challenges. Differences in surface energies, mechanical stresses, thermal stresses, or chemical incompatibilities can result in the films delaminating from the substrate, which typically eliminates the desirable properties of the coating. One of the most common ways to increase adhesion between a CVD thin film and its substrate is through the use of grafting, that is, covalently bonding the desired polymer film to its substrate. Improving the adhesion of thin films to their substrates has numerous benefits. In tribological applications, strong bonds between the coating and substrate lead to increased abrasion resistance and prolong the coating's useful life [23], and in electronic or conductive thin films, such as those produced by oxidative chemical vapor deposition (oCVD), grafting to the surface may enhance the electronic properties of the thin film, as charge carriers may also now be conducted through an extended network that includes the substrate. We can broadly categorize grafting into two main approaches: chemical modification of the substrate itself to improve bonding, and the application of an additional molecule, an adhesion promoter, to facilitate bonding between the polymer and the substrate.

4.3.1 Surface Modification of Organic Substrates

As many of the CVD processes of interest follow radical-based mechanisms, one straightforward grafting strategy is to generate radicals on the substrate. In this way, the desired monomer is activated by its interaction with the surface radical, and the polymer growth originates from the surface. An inherent limitation in this approach is the finite number of surface sites that are available for initiation, which can limit the polymer growth rate. For organic substrates, radical generation at the surface can be accomplished by exposing the substrate to a plasma discharge prior to thin film deposition. The bombardment of the surface by ions, electrons, and activated neutrals can generate reactive radical sites that can then be used as attachment points for the desired monomer species. As Chu *et al.* describe, upon exposure to a plasma discharge, hydrogen abstraction from the organic substrate is the primary method by which radicals are created [23]. If the plasma exposure is brief enough and the substrate is not re-exposed to ambient conditions prior to the deposition step, then these radicals can serve as initiation points for the deposition of the desired polymer thin films from the gas phase. Since only the substrate surface is exposed to the plasma, the activation sites are confined to a narrow region at the surface, and this is often sufficient to significantly alter the surface chemistry of the substrate as compared to the bulk. For example, the use of an oxygen plasma can render hydrophobic polymers more hydrophilic, while fluorine-containing monomers, such as SF_6, CF_4, or C_2F_6 can impart hydrophobic properties to surfaces. Alternatively, plasma treatment of a surface can prime it for activation upon exposure to atmosphere. Heinz *et al.* describe the growth of polymer thin films on plasma-activated PEO thin films. The PEO substrates are exposed to a plasma discharge, and upon exposure to the atmosphere, radical sites react with atmospheric oxygen to generate peroxides that are then reactive toward radical polymerizations upon thermal activation of the peroxides [24]. Alternatively, the plasma-treated surface can be exposed to an adhesion promoter, such as a functionalized silane, which then reacts with the surface radicals to generate a functional surface. Yin *et al.* used this method to functionalize a stretchable PDMS surface using trichlorovinylsilane (TCVS) [25]. This TCVS-treated stretched surface was then used as the substrate to deposit various combinations of acrylate copolymers. Upon relaxation of the stretching, highly ordered, wrinkled microtopologies were observed. Without grafting, delamination of the coating occurs during the stretching/unstretching of the flexible substrate.

Formation of a densely crosslinked adhesion promotion layer is also possible using an all-CVD approach. Ye *et al.* formed a highly crosslinked polymer base layer from poly(dimethylaminomethyl styrene) (PDMAMS) and ethylene glycol diacrylate (EGDA) [26]. The PDMAMS-*co*-EGDA base layer was deposited onto silicon wafer, and Ye *et al.* showed that the crosslinked copolymer was highly stable to ultrasonication, whereas the PDMAMS homopolymer was quickly destroyed under the same conditions. Since the copolymer film contains many of the same chemical functionalities as the homopolymer, it was used as a substrate to graft a top layer of the PDMAMS homopolymer for use as an antimicrobial coating. The PDMAMS-*co*-EGDA copolymer showed reduced antimicrobial activity, so the presence of the homopolymer as the top layer was necessary for the desired functionality. In these cases, adhesion promotion of the desired polymer thin film is accomplished on an organic substrate via a modification of the original surface.

As noted above in the photoinitiation section, application of a Type-II photoinitiator such as benzophenone can also lead to grafting, as the photoactivated benzophenone generates surface radical sites via hydrogen abstraction from an organic substrate. Martin *et al.* used this technique to produce surface-grafted thin films on nylon fibers from both styrene- and acrylate-based monomers [14]. As compared to non-grafted samples from the same monomers produced via traditional iCVD, the grafted samples retained over 80% of the original coating mass after a series of durability tests. The non-grafted samples retained approximately 25% of the coating mass, and the coating under SEM was visibly degraded.

For thin films produced via oCVD, the proposed mechanism for these polymerizations lends itself to relatively easy grafting strategies. As related by Im *et al.*, oCVD depositions rely on radical cations formed using Friedel–Craft catalysts (e.g., $FeCl_3$, $AlCl_3$, and $CuCl_2$) to produce conjugated polymers, such as poly(3,4-ethylenedioxythiophene) (PEDOT), polypyrrole, and polythiophene. As such, aromatic groups present at the substrate surface also will be subject to activation by the catalyst, allowing activated monomers to attach covalently. Thus, common plastic substrates containing aromatic groups, such as polystyrenes, polycarbonates, and poly(terephthalates) are all able to support direct grafting of conjugated polymers via this mechanism, shown in Figure 4.7 [27].

4.3.2
Surface Modification of Inorganic Substrates

For inorganic substrates, generation of radical sites is generally more difficult, and thus grafting to these substrates requires the formation of an adhesion promotion layer, which is bonded to the substrate and provides a surface amenable to polymer deposition.

Yang *et al.* recently described a novel approach to fabrication of an adhesion promotion layer using a variation of traditional iCVD [28]. In this modified process, a silicon substrate is first treated with HF to generate Si-H surface sites. TBPO is then introduced to the iCVD reactor, with filament temperatures greater than 300 °C. Yang *et al.* showed that above this temperature, methyl radicals ($CH_3\bullet$) are favored over *t*-butoxy radicals, and their greater reactivity results in the generation of surface radical sites via hydrogen abstraction. By alternating the substrate exposure to TBPO and a functionalized dialkenyl monomer (in this example, decadiene) at these conditions, a functionalized surface can be created. Yang *et al.* followed this functionalization step with traditional iCVD using TBPO and EGDA, forming a Poly(ethylene glycol diacrylate) (PEGDA) thin film that was effectively tethered to the silicon substrate. Nanoscratch testing and SEM imagery of the non-grafted films showed clear delamination of the PEGDA layer, while substrates that

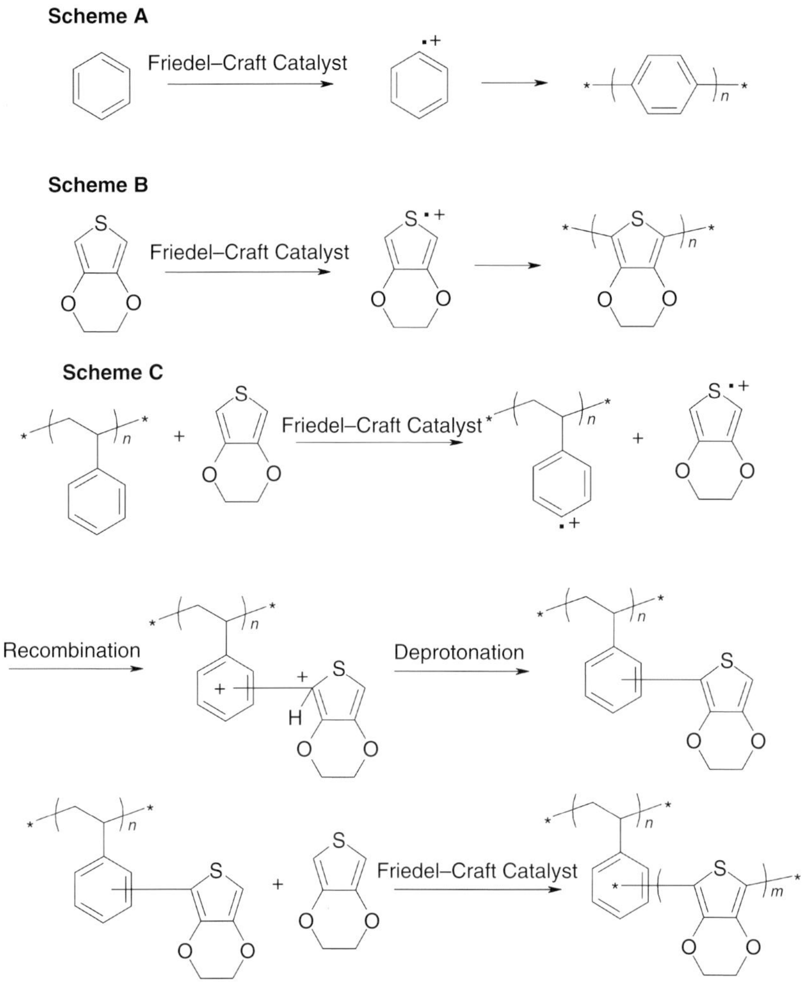

Figure 4.7 A schematic representation of the reaction mechanism involved in grafting PEDOT films onto aromatic polymeric substrates. (Reprinted with permission from Ref. [27]. Copyright 2007 Wiley-VCH.)

had been functionalized according to the above protocol showed no such delamination.

Another common way of preparing inorganic substrates for polymer deposition is through the use of silane coupling agents [29]. These coupling agents are generally multifunctional silanes, with one functional group that can be readily hydrolyzed by the presence of hydroxyl groups on the surface, which bonds the silane to the substrate. By careful choice of the second functionality of the silane, appropriate surfaces for a variety of polymerization reactions can be fabricated. The first step, then, is the creation of surface hydroxyl groups on the substrate to prepare it for silane exposure. Both liquid-phase and vapor-phase mechanisms

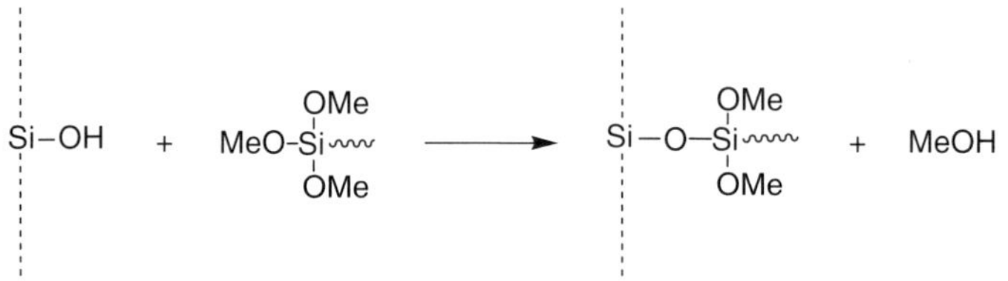

Figure 4.8 General silane coupling reaction on a hydroxylated surface by a functionalized alkoxysilane. The alkoxysilane undergoes a condensation reaction with the surface, yielding a primary alcohol (methanol in this case), and a grafted silane.

exist to accomplish this. If the substrate is compatible, exposing it to a Pirhana solution (H_2SO_4 + H_2O_2) is an effective way to generate surface hydroxyl groups [30]. The combination of the acid and the peroxide yield a powerful oxidizing agent that converts surface sites to a hydroxyl moiety that can then be further reacted. For surfaces that are incompatible with a Piranha treatment, exposing the substrate to an oxygen plasma can achieve surface hydroxylation as well. As Choi *et al.* point out in their example silicon substrate, the oxygen radicals that are generated in the plasma will break Si-Si bonds as well as Si-O bonds at the surface, resulting in the formation of active dangling bonds and hydroxyl groups [31]. Once the surface hydroxyl groups have been created, then the substrate can be brought into contact with the silane coupling agent. Reactive silanes typically have one or more hydrolyzable substituents, as well as at least one reactive substituent that is chosen so as to optimize the polymerization of the final thin film. Functionalized alkoxysilanes are some of the most common silane coupling agents, as they yield primary alcohols when undergoing a condensation reaction with surface hydroxyl groups. Figure 4.8 details the reaction mechanism for a generic alkoxysilane with a hydroxylated surface. For iCVD, piCVD, or other radical-based polymerizations, the selection of the silane functional group will be one that can participate in the chain growth mechanism common to those polymerizations. For example, a vinyl, allyl, or acryl functionality on the silane will allow either an initiator molecule or an activated monomer unit to react with the silane agent, providing a graft point on the surface for continued polymerization. Since the functional group on the silane is indistinguishable from the functional group in the target monomer, the addition of an additional initiator during the deposition can result in high deposition rates, because the silane functionality is simply incorporated into the growing polymer chain.

Silane coupling agents of this type are generally small molecules with sufficiently high vapor pressure to allow them to be volatilized at moderate vacuum and temperatures. Duchet *et al.* describe the vapor-phase silanization process used to functionalize glass, silicon wafers, and silica particles [32]. Following cleaning and hydroxylation, the surface to be treated is placed in a vacuum chamber and dried at 140 °C for 2 h at 0.2 Pa to remove physiosorbed water. Removal of any physisorbed water is crucial; if the coupling agent is introduced without proper drying, then it will react with the adsorbed water rather than the immobilized silanols created via hydroxylation, and the silane will not be bonded to the surface. For small alkyl silanes (C4–C8), temperatures of ~70 °C were used, with higher temperatures required for increasing molecular weights. Sufficient vacuum is then formed to volatilize the silane, which adsorbs onto the hydroxylated surface. Depending upon the specific conditions the silane reacts, forming a functionalized surface now suitable for polymerization. Duchet *et al.* used a reaction time of 48 h, and after the reaction is complete, excess silane is removed from the surface via subsequent washes with tetrahydrofuran (THF) and xylenes, and then drying in nitrogen. Im *et al.* used this general method, depositing TCVS on silicon wafer, glass, quartz, and PDMS to serve as an adhesion promoter for poly(glycidyl methacrylate) used in a microfluidic device [33]. Paxson *et al.* used both TCVS as well as trichloro(1*H*,1*H*,2*H*,2*H*-perfluorooctyl) silane as adhesion promotion layers in the copolymerization of (1*H*,1*H*,2*H*,2*H*)- perfluorodecyl acrylate (PFDA) with divinyl benzene (DVB) on aluminum substrates in a heat-transfer application [34]. The PFDA-*co*-DVB polymer film imparted sufficient hydrophobicity to facilitate dropwise condensation of steam, which resulted in significantly higher heat transfer coefficients as compared to film condensation. The grafting of the copolymer to the aluminum surface was critical to the coating surviving for extended periods at elevated temperatures.

Additionally, Trujillo *et al.* used the TCVS as an adhesion promoter in the fabrication of nano-patterned surfaces using iCVD [35]. Without the addition of the adhesion promotion layer, the iCVD polymer would have been quickly removed during the aggressive sonication in THF required during processing. The importance of the adhesion promotion layer in this application is readily apparent in Figure 4.9. In Figure 4.9a in the film deposited with the TCVS pre-treatment, the pattern formed during the processing step is clearly visible, whereas the same pattern vanishes almost instantaneously (Figure 4.9b, no TCVS pre-treatment) after a 5 s rinse in THF.

Grafting of conjugated polymers via oCVD is also possible using inorganic substrates by applying the silane coupling mechanism. In this application, an aromatically functionalized silane, such as phenyltrichlorosilane (PTCS) is attached to the substrate, which then allows Freidel–Craft activation of the phenyl group and subsequent grafting. Im *et al.* was able to show considerably improved adhesion to substrates using both inherently aromatic substrates as well as substrates with aromatic adhesion promoters attached. Figure 4.10 demonstrates this visually [25]. As can be seen, in Figure 4.10a, PEDOT deposited on an aromatic substrate such as poly(ethylene terephthalate) (PET) (left) shows no delamination after a standard ASTM tape test as well as 3 h of ultrasonication, whereas the sample prepared on polypropylene (right) shows almost complete delamination. Similar results are seen in Figure 4.10b, in this case on a silica glass substrate. The slide on the left of Figure 4.10b, treated with PTCS, shows no delamination after 1 h of ultrasonication, whereas the sample on the right on untreated glass shows significant degradation after only 5 min at the same conditions. As oCVD thin films produced in this way are often rinsed in solvent to remove spent catalyst, ensuring that the film does not delaminate during this processing step is an essential quality control issue.

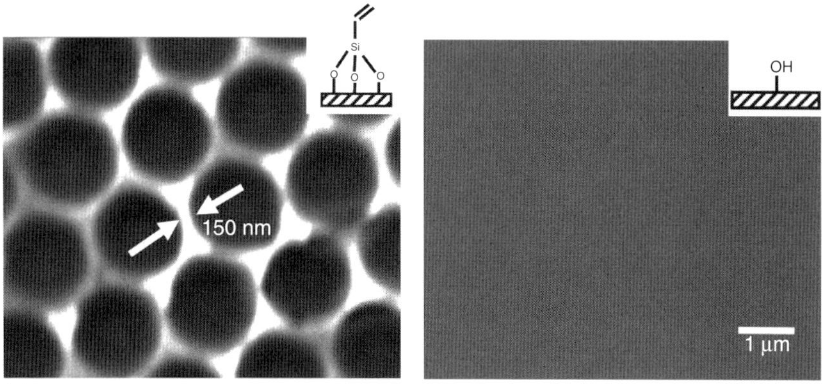

(a) (b)

Figure 4.9 Grafted polymer patterns produced from 1 µm particles (SEM image for poly(butyl acrylate) shown) typically generate features 150 nm in width (a). Because of the vinyl pretreatment, these patterns are robust and withstand several hours in THF. The silylation step is necessary to covalently attach the iCVD polymer to the substrate. Without this step the polymer is easily washed away during the template lift-off, leaving behind a clean substrate (SEM image (b)). The insets depict the corresponding substrate functionality prior to patterning. (Reprinted with permission from Ref. [35]. Copyright 2009 American Chemical Society.)

4.3.3 Grafting Summary

Grafting provides a reliable way to improve adhesion of CVD thin films to their desired substrates. A variety of techniques exist for both organic/polymer surfaces as well as inorganic surfaces. In many cases, direct modification of a substrate is possible to promote adhesion. In cases where direct modification is not possible, the use of an adhesion promotion layer that facilitates bonding between the desired substrate and the polymer thin film can be employed.

4.4 Summary

As shown in this chapter, as well as the remainder of this text, there is a wide diversity in the techniques that are available in the vapor deposition of polymeric thin films. While plasma techniques have been established for decades, the challenge in selective activation of monomers remains an obstacle in using them to fabricate well-defined organic films with retention of desired functionality. However, as shown, extension of the chemical initiator concept to PECVD systems has improved PECVD's utility in this area.

Photoinitiated chemical vapor deposition systems provide an alternative to thermal iCVD techniques, without typically losing the advantageous preservation of desired chemical functionality and similarity to traditionally polymerized

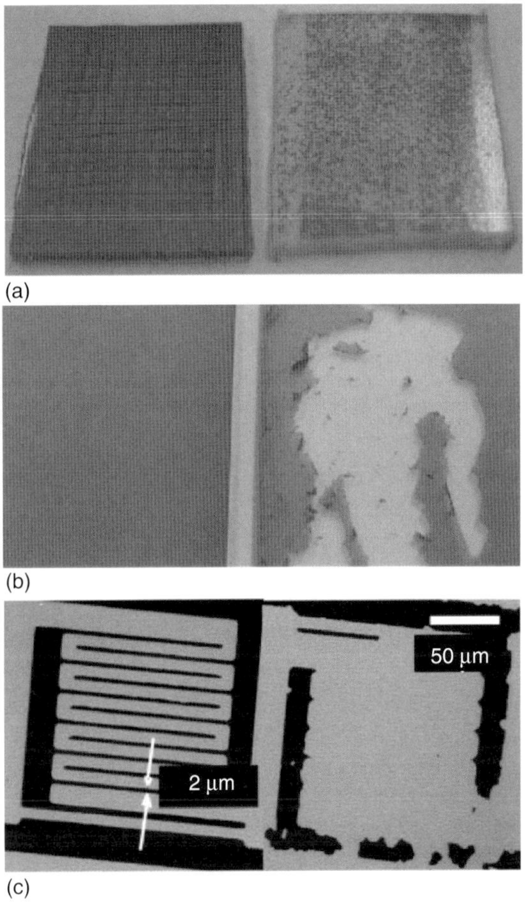

Figure 4.10 (a) PEDOT grafted on PET (left) and PEDOT on PP, ungrafted (right) after tape testing; (b) PEDOT grafted (left) and ungrafted (right) on glass after ultrasonication for 5 min; and (c) conventional lift-off photolithography pattern on Si wafers with grafted PEDOT (left) and with ungrafted PEDOT (right). The scale bars represent a length of 50 µm. (Reprinted with permission from Ref. [27]. Copyright 2007 Wiley-VCH.)

materials that are the hallmarks of the iCVD technique. piCVD allows for simplified reactor geometries, as well as precise thermal control of the system and substrate, but the lack of a universal, easily activated photoinitiator with wide applicability similar to the thermally activated peroxides used in iCVD has limited its general extensibility.

Lastly, grafting techniques take advantage of the radical-initiated nature of most of the CVD processes described, in that they either generate radicals at surfaces which then serve as initiation points for the growing polymer, or, they rely on attachment of chemical functionalities to the substrate that can be directly incorporated into the backbone of the growing polymer. In either case, the covalent attachment of the polymer to its substrate significantly improves the durability and longevity of coating – an advantage in the commercialization and wide adoption of organic polymer thin films fabricated via CVD.

References

1. Biederman, H. and Slavınská, D. (2000) Plasma polymer films and their future prospects. *Surf. Coat. Technol.*, **125**, 371–376.
2. d'Agostino, R., Cramarossa, F., Fracassi, F., and Illuzzi, F. (1990) in *Plasma Deposition, Treatment, and Etching of Polymers* (ed. R. d'Agostino), Academic Press, San Diego, CA, pp. 95–163.
3. Jacob, M.V., Easton, C.D., Woods, G.S., and Berndt, C.C. (2008) Fabrication of a novel organic polymer thin film. *Thin Solid Films*, **516**, 3884–3887.
4. Palumbo, F., Favia, P., Vulpio, M., and D'Agostino, R. (2001) RF plasma deposition of PEO-like films: diagnostics and process control. *Plasmas Polym.*, **6**, 163–174.
5. Zhao, X.-Y., Wang, M., Wang, Z., and Zhang, B. (2008) Structural and dielectric properties of conjugated polynitrile thin films deposited by plasma polymerization. *Thin Solid Films*, **516**, 8272–8277.
6. Morosoff, N. (1990) in *Plasma Deposition, Treatment, and Etching of Polymers* (ed. R. d'Agostino), Academic Press, San Diego, CA, pp. 1–94.
7. Förch, R., Zhang, Z., and Knoll, W. (2005) Soft plasma treated surfaces: tailoring of structure and properties for biomaterial applications. *Plasma Processes Polym.*, **2**, 351–372.
8. Burkey, D.D. and Gleason, K.K. (2003) Structure and thermal properties of thin film poly (alpha-methylstyrene) deposited via plasma-enhanced CVD. *Chem. Vap. Deposition*, **9**, 65–71.
9. Finke, B., Schröder, K., and Ohl, A. (2009) Structure retention and water stability of microwave plasma polymerized films from allylamine and acrylic acid. *Plasma Processes Polym.*, **6**, S70–S74.
10. Labelle, C.B. and Gleason, K.K. (2001) Pulsed plasma deposition from 1,1,2,2-tetrafluoroethane by electron cyclotron resonance and conventional plasma enhanced chemical vapor deposition. *J. Appl. Polym. Sci.*, **80**, 2084–2092.
11. Coclite, A.M. and Gleason, K.K. (2012) Initiated PECVD of organosilicon coatings: a new strategy to enhance monomer structure retention. *Plasma Processes Polym.*, **9**, 425–434.
12. Baxamusa, S.H. and Gleason, K.K. (2008) Thin polymer films with high step coverage in microtrenches by initiated CVD. *Chem. Vap. Deposition*, **14** (9-10), 313–318.
13. Chan, K. and Gleason, K.K. (2005) Photoinitiated chemical vapor deposition of polymeric thin films using a volatile photoinitiator. *Langmuir*, **21**, 11773–11779.
14. Martin, T.P., Sedransk, K.L., Chan, K., Baxamusa, S.H., and Gleason, K.K. (2007) Solventless surface photoinitiated polymerization: grafting Chemical Vapor Deposition (gCVD). *Macromolecules*, **40**, 4586–4591.
15. Deng, J., Wang, L., Liu, L., and Yang, W. (2009) Developments and new applications of UV-induced surface graft polymerizations. *Prog. Polym. Sci.*, **34**, 156–193.
16. Yang, W. and Rånby, B. (1996) Radical living graft polymerization on the surface of polymeric materials. *Macromolecules*, **29**, 3308–3310.
17. Wang, H. and Brown, H.R. (2004) Self-initiated photopolymerization and photografting of acrylic monomers. *Macromol. Rapid Commun.*, **25**, 1095–1099.
18. Baxamusa, S.H., Montero, L., Dubach, J.M., Clark, H.A., Borros, S., and Gleason, K.K. (2008) Protection of sensors for biological applications by photoinitiated chemical vapor deposition of hydrogel thin films. *Biomacromolecules*, **9**, 2857–2862.

19. Suresh, A., Anastasio, D., and Burkey, D.D. (2014) Potential of hexyl acrylate monomer as an initiator in photo-initiated CVD. *Chem. Vap. Deposition*, **20**, 5–7.
20. Abrahamson, E.W., Littler, J.G.F., and Vo, K.-P. (1966) Spectroscopic basis of carbonyl photochemistry. I. The role of excited-state geometry in the photodecomposition of formaldehyde. *J. Chem. Phys.*, **44**, 4082–4086.
21. Morita, H. and Sadakiyo, T. (1995) Laser-induced polymeric film formation from gaseous methyl acrylate. *J. Photochem. Photobiol., A Chem*, **87**, 163–167.
22. Rosenfeld, R.N. and Weiner, B.R. (1983) Photofragmentation of acrylic acid and methacrylic acid in the gas phase. *J. Am. Chem. Soc.*, **105**, 6233–6236.
23. Chu, P., Chen, J., Wang, L., and Huang, N. (2002) Plasma-surface modification of biomaterials. *Mater. Sci. Eng., R*, **36**, 143–206.
24. Heinz, P., Brétagnol, F., Mannelli, I., Sirghi, L., Valsesia, A., Ceccone, G., Gilliland, D., Landfester, K., Rauscher, H., and Rossi, F. (2008) Poly(N-isopropylacrylamide) grafted on plasma-activated poly(ethylene oxide): thermal response and interaction with proteins. *Langmuir*, **24**, 6166–6175.
25. Yin, J., Yagüe, J.L., Eggenspieler, D., Gleason, K.K., and Boyce, M.C. (2012) Deterministic order in surface micro-topologies through sequential wrinkling. *Adv. Mater.*, **24**, 5441–5446.
26. Ye, Y., Song, Q., and Mao, Y. (2011) Solventless hybrid grafting of antimicrobial polymers for self-sterilizing surfaces. *J. Mater. Chem.*, **21**, 13188–13194.
27. Im, S.G., Yoo, P.J., Hammond, P.T., and Gleason, K.K. (2007) Grafted conducting polymer films for nano-patterning onto various organic and inorganic substrates by oxidative chemical vapor deposition. *Adv. Mater.*, **19**, 2863–2867.
28. Yang, R., Buonassisi, T., and Gleason, K.K. (2013) Organic vapor passivation of silicon at room temperature. *Adv. Mater.*, **25**, 2078–2083.
29. Plueddemann, E.P. (1991) *Silane Coupling Agents*, Plenum Press, New York.
30. Rye, R.R., Nelson, G.C., and Dugger, M.T. (1997) Mechanistic aspects of alkylchlorosilane coupling reactions. *Langmuir*, **13**, 2965–2972.
31. Choi, S.-W., Choi, W.-B., Lee, Y.-H., Ju, B.-K., Sung, M.-Y., and Kim, B.-H. (2002) The analysis of oxygen plasma pretreatment for improving anodic bonding. *J. Electrochem. Soc.*, **149** (1), G8–G11.
32. Duchet, J., Gerard, J.-F., Chapel, J.-P., and Chabert, B. (2000) Grafting of alkylchlorosilanes onto silica from solution for adhesion enhancement. *J. Adhes. Sci. Technol.*, **14**, 691–718.
33. Im, S.G., Bong, K.W., Lee, C.-H., Doyle, P.S., and Gleason, K.K. (2009) A conformal nano-adhesive via initiated chemical vapor deposition for microfluidic devices. *Lab Chip*, **9**, 411–416.
34. Paxson, A.T., Yagüe, J.L., Gleason, K.K., and Varanasi, K.K. (2014) Stable dropwise condensation for enhancing heat transfer via the initiated chemical vapor deposition (iCVD) of grafted polymer films. *Adv. Mater.*, **26**, 418–423.
35. Trujillo, N.J., Baxamusa, S.H., and Gleason, K.K. (2009) Grafted functional polymer nanostructures patterned bottom-up by colloidal lithography and initiated chemical vapor deposition (iCVD). *Chem. Mater.*, **21**, 742–750.

5
Conformal Polymer CVD

Salmaan Baxamusa

5.1
Introduction

One of the distinguishing features of vapor-phase deposition is its ability to conformally coat topographically complex substrates. Historically, this has been problematic for polymer coatings because polymer research has focused almost exclusively on liquid-phase synthesis and casting. Polymer vapor deposition technologies such as plasma polymerization, parylene, molecular layer deposition (MLD), oxidative chemical vapor deposition (oCVD), and initiated chemical vapor deposition (iCVD) allow for conformal polymer coatings that are not accessible through liquid means. This chapter will discuss the fundamental physical mechanisms that govern the conformality of polymer chemical vapor deposition (CVD) coatings, review the relevant experimental evidence, and finally discuss several examples of new applications enabled by conformal CVD polymer coatings.

5.2
Vapor Phase Transport

In the vapor phase, reactants diffuse via intermolecular collisions into very small geometries such as pores, trenches, and even the spaces between woven fibers. The mean free path, which is the average distance between intermolecular collisions in the vapor phase, can be engineered over a wide range of length scales because it varies as the inverse of pressure. At atmospheric pressure, the mean free path is only dozens of nanometers, but at typical deposition pressures ($\sim$100 mTorr) the mean free path can be nearly a millimeter and at high vacuum the mean free path exceeds a meter. The ratio of the mean free path to the characteristic geometric length scale is known as the Knudsen number (Kn) and is a key parameter in vacuum deposition. When $\mathrm{Kn} < 1$, molecular movement is governed by classical diffusion that results from the intermolecular collisions. This diffusion allows reactive species to impinge evenly at all surfaces,

CVD Polymers: Fabrication of Organic Surfaces and Devices, First Edition. Edited by Karen K. Gleason.

resulting in conformal deposition. When $Kn > 1$, the flow regime is known as molecular flow. Intermolecular collisions are infrequent in molecular flow and the trajectory of reactive species from their source to the deposition substrate is based on line-of-sight, resulting in nonconformal deposition. This is a common limitation for high-vacuum physical vapor deposition (PVD) techniques such as low-pressure sputtering or electron beam PVD. Even at the higher pressures used for CVD, the local Knudsen number can easily exceed unity when coating complex topographies because the relevant length scale can shrink to sub-micron levels.

In practice, this line-of-sight limitation is mitigated because reactant molecules that impinge on the substrate may either stick via irreversible reaction or reflect back into the gas phase, a concept quantified by a parameter called the "sticking probability." When $Kn > 1$, CVD techniques can conformally coat topographically complex substrates if the sticking probability is low enough. An additional factor unique to polymer CVD is that the growing polymer chain is both reactive and has negligible vapor pressure due to its molecular weight. Therefore, reactions between surface-bound molecules are important in polymer vapor deposition, and must be considered in conjunction with gas phase transport when analyzing the factors that govern conformal coverage of the polymer coating.

5.3 Conformal Polymer Coating Applications

The ability to grow conformal polymer coatings directly from vapor-phase reactants make polymer CVD ideal for depositing organic coatings in the interior of high-aspect ratio features. Such features include trenches and overhangs that may be found in integrated circuitry, the narrow and long pores of filtration membranes or microfluidic devices, and complicated topographies of microelectromechanical (MEMS) devices. Lithography or surface patterning also requires conformal deposition within structures defined by the interface between contact masks and the substrate.

Conformal coatings may also be desired for exterior coatings around particles or fibers. Emerging advanced manufacturing techniques such as 3D-printing can be used to fabricate meta-materials with complicated geometries, and polymer coating technologies will eventually be needed for these structures as well. Conceptually, conformal exterior coatings are governed by the same physics and chemistry as interior coatings.

Some applications may require only *qualitatively* conformal coatings; that is, a polymer coating that is continuous around all surfaces but not necessarily of uniform thickness. Examples of applications requiring qualitatively conformal coatings include surface modification to control surface energy [1], protective coatings for biosensors or membranes [2, 3], or the introduction of surface chemical moieties for subsequent functionalization [4]. In other applications, it may be important to have *quantitatively* conformal coatings; that is, coatings that must be of uniform thickness on all surfaces. This is particularly important in optical coatings such as dielectric mirrors [5], but is also critical for gas permeable or diffusion barrier films and drug-release devices. Because optimizing the conformality requires tradeoffs with other deposition properties – most importantly deposition rate – understanding whether an application requires qualitatively or quantitatively conformal coatings is an important step in designing the deposition process.

5.4 Conformal Polymer Coating Technologies

In liquid-phase coating technologies such as spin or dip coating, surface tension effects cause severe coating nonuniformities when the capillary length of the solvent is larger than the length scale of the geometry to be coated [6]. Capillary length is a material property that is largely invariant from one solvent to another and is typically on the order of 1 mm. This renders liquid-phase coating unable to evenly coat the ever-shrinking feature sizes of electronic, biomedical, optical, or mechanical devices. Vapor-phase deposition removes the constraints imposed by solvent surface tension.

The deposition of poly(*p*-xylylene) (known by its trade name parylene) is one of the oldest vapor-phase polymerization techniques, having been commercialized by Union Carbide in the 1960s. In parylene deposition, a paracyclophane monomer decomposes into *p*-xylylene monomer units at high temperatures (~550 °C); these monomer units react at the substrate to form the polymer. Parylene coatings are in general found to be conformal [7]. An excellent monograph on the mechanism of parylene deposition is available [8].

Another major vapor-phase polymerization technique is plasma CVD, first popularized in the 1970s. In plasma CVD, a glow discharge decomposes the gaseous monomer into reactive fragments. These reactive fragments recombine at a surface, often through complex and unexpected chemical pathways, to form a polymer film. The presence of nonuniform electric fields and charged species can lead to coating nonuniformities, especially in corners and at edges [9]. The deposition rate of plasma polymers is determined by a balance between concurrent etching and deposition, so any place on the substrate where these relative rates are altered will have nonuniform and nonconformal deposition [10]. Conformal deposition is possible to some degree [11], although systematic studies to understand the conformality of plasma CVD polymers are limited and this aspect of plasma polymerization remains poorly understood. Conformality of inorganic films deposited via plasma CVD has been considered [12].

The gold standard in conformal coating is atomic layer deposition (ALD), in which bifunctional precursors are sequentially introduced into a vacuum chamber to form inorganic metal and ceramic coatings. ALD chemistry is essentially a condensation reaction between two precursors, which bears strong resemblance to reaction pathways for step growth polymerization. Indeed, a variant of ALD, termed molecular layer deposition was first described in the 1990s and uses organic precursors to grow polymers or inorganic/organic polymer hybrids [13–15]. Because each precursor is reactive only with the other, the reaction is self-limiting and can be used to grow films approximately one molecular layer at a time. MLD is therefore quantitatively conformal in theory, although incomplete exposure of the substrate to the reactant will lead to nonuniformity in film thickness. Models have been developed to determine the effect of mass transport limitations on the conformality of ALD coatings; these models also apply to MLD [16, 17]. One of the limitations of MLD is that deposition rates are slow, on the order of nm/min, but the available material set is growing.

oCVD is a technique for depositing conducting polymers. Step growth of monomers is achieved with the use of an oxidant, such as an evaporated metal halogen salt or halogen gas. Like other vapor phase processes, oCVD films are conformal over micron scale features [18]. The physical processes governing the conformality of oCVD films have not been studied in detail, but the use of a halogen gas as opposed to an evaporated metal halogen salt has been shown to result in more conformal films [19].

Finally, iCVD, a vapor deposition analog of free-radical polymerization, has been used extensively to create conformal coatings [20, 21]. The balance of this chapter will focus heavily on iCVD films, as their conformality has been given considerable analytical attention. Many of the concepts discussed in this chapter may be adapted to other polymer CVD mechanisms, and the qualitative conclusions of iCVD studies are broadly applicable to vapor phase polymer deposition. Parallels to parylene deposition, whose kinetics have been studied extensively, will be highlighted in the development of the reaction-diffusion model in Section 5.6.

Some examples of conformal polymer deposition are shown in Figure 5.1. Figure 5.1a shows iCVD, spin-coated, and plasma polymer films inside of trench structures. The iCVD film is conformal around the trench. The nonuniform spin-coated film structure reflects the effect of surface tension in the confined geometry. The plasma polymer is somewhat continuous although the coating is depleted around the exterior corners and sidewall coverage is not uniform. Figure 5.1b is an X-ray photoelectron spectroscopy (XPS) map demonstrating parylene deposition within the serpentine channels of a microfluidic device. A conformal MLD coating around a nanoparticle is shown in Figure 5.1c.

5.5
Gas and Surface Reactions

Polymer CVD involves two key steps: transport of reactants from the vapor phase to the substrate, and polymerization which lead to film formation. In general, polymerization occurs at the surface because high molecular weight growing chains have effectively zero vapor pressure. For example, iCVD is a free-radical polymerization process whose rate is determined by the reaction between growing polymer chains and monomer molecules. The rate expression for this process is [25]

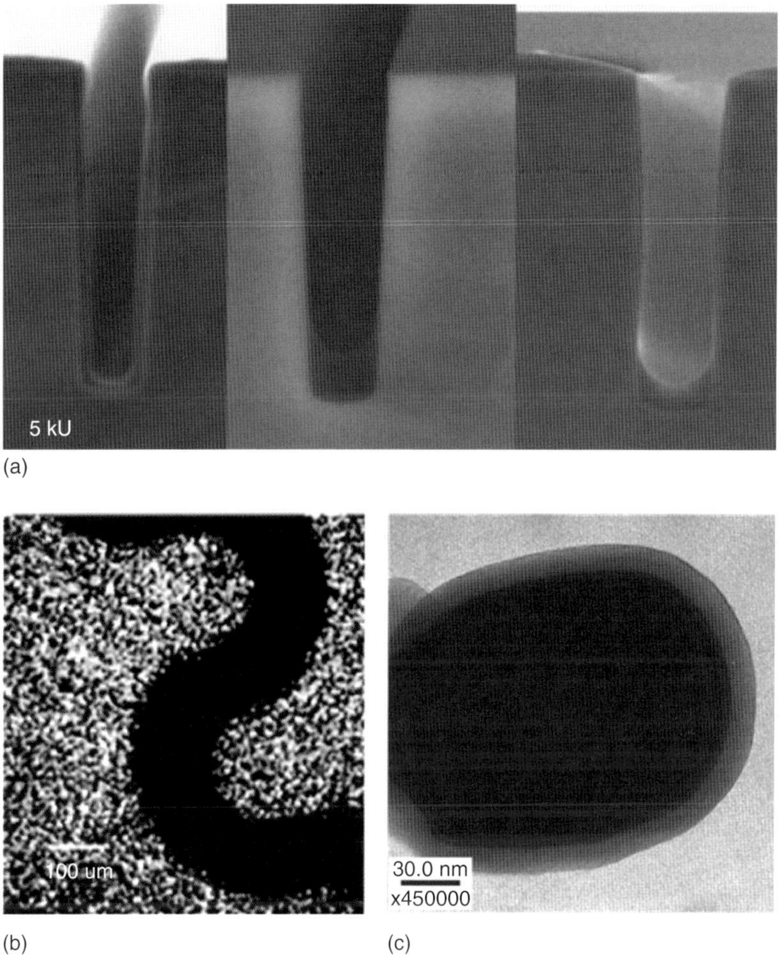

Figure 5.1 (a) Polymer coating in trench structures (approximately 2 μm wide by 7 μm deep) deposited via (left to right) iCVD, spin-coating, and plasma polymerization. (b) XPS map of Si showing no silicon within a 100 μm width serpentine microfluidic channel coated with parylene. (c) MLD coating conformal around a nanoparticle. (Reprinted with permission from (a) [22], copyright 2008 Wiley (b) [23], copyright 2005 American Chemical Society (c) [24], copyright 2008 American Chemical Society.)

$$\text{rate} = k_p C_M \left(\frac{f k_d}{k_t} C_I \right)^{1/2} \tag{5.1}$$

where k_p, k_d, and k_t are kinetic rate constants for propagation, initiator dissociation, and termination respectively; C_M is the monomer concentration; C_I is the initiator concentration; and f is the initiator efficiency. A detailed kinetic and mechanistic study of the iCVD of butyl acrylate shows rate constants that are

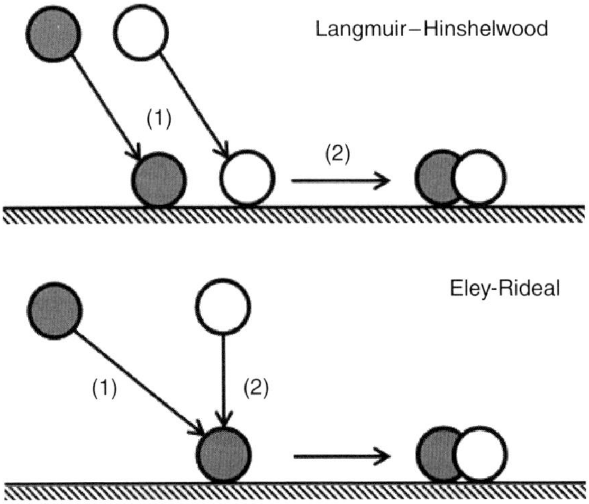

Figure 5.2 Two modes of bimolecular surface reactions. In the Langmuir–Hinshelwood mechanism, the reactants separately adsorb to the surface (1) followed by a reaction between the adsorbed species (2). In the Eley–Rideal mechanism, the first reactant adsorbs to the surface (1), followed by a reaction between the gaseous second species and the adsorbed first species (2).

nearly equal to those for liquid-phase polymerization; growing chain reacts primarily with the liquid-like surface-adsorbed monomer [26]. Parylene deposition proceeds by the reaction of a *p*-xylylene unit with a surface-adsorbed growing chain [8].

Surface reactions can occur by one of two mechanisms: the *Langmuir–Hinshelwood* mechanism, where two surface-bound molecules react, or the *Eley–Rideal* mechanism, where a gas phase molecule collides and reacts with a surface-bound molecule. These mechanisms are illustrated in Figure 5.2. Eley–Rideal is the most likely mechanism for chain initiation, a hypothesis supported by the fact that the time scale for a surface reaction is about a factor of 10 longer than the surface lifetime of an adsorbed reactant. Instead, gaseous species are likely to react directly with a surface-bound molecule, as in the Eley–Rideal mechanism. When this irreversible chemisorption occurs, the reactant is said to "stick" to the surface. This has been observed for both iCVD and parylene deposition. The so-called sticking probability is an important parameter that will be explored in detail in Section 5.6.

While gas phase transport does not appear explicitly in polymer CVD rate expressions, the supply of reactants to the surface depends wholly on adsorption from the vapor phase: monomer units must adsorb to the surface at the same rate that they are consumed by polymerization. In the case of iCVD, primary radicals must also appear at the surface to maintain a steady-state of growing polymer chains. Primary radicals are formed by homolysis of the initiator, either thermolytically or photolytically, in the gas phase. In the photolytic case, there is some evidence that radicals may also be formed at the substrate [27].

Based on the importance of supplying reactants to the surface, it is not surprising that deposition conformality is governed in part by vapor phase processes. In the following section, we will discuss models to understand how to predict and control the conformal deposition of CVD polymers. The interplay between vapor transport and surface reaction is best understood as a balance between supply and consumption. Where the supply of reactants to the surface is abundant, polymerization will take place equally on all surfaces. Where the supply is depleted by the consumption of the reactants, the deposition rate will be locally lower.

5.6 The Reaction-Diffusion Model

5.6.1 Reaction and Diffusion in a Pore

A simplified steady-state 1-D reaction-diffusion model of gaseous molecules traveling through a straight circular pore of radius r and length L can be used to understand deposition conformality. In this section, we modify the methods of Asatekin *et al.* and Komiyama *et al.* to develop a generalized set of governing equations [2, 28]. In the absence of gas-phase reactions, the concentration C_i of the ith species can be expressed as

$$\frac{Dr}{2}\frac{\mathrm{d}^2 C_i}{\mathrm{d}x^2} = \upsilon_i \tag{5.2}$$

with boundary conditions:

$$C_i|_{x=0} = C_{i,o} \tag{5.3}$$

$$D_i \left.\frac{\mathrm{d}C_i}{\mathrm{d}x}\right|_{x=L} = \upsilon_i|_{x=L} \tag{5.4}$$

where x is the spatial dimension, D_i is the diffusivity, υ_i is the local surface flux to the wall of the pore due to reaction, and $C_{i,o}$ is the concentration at the pore entrance. The geometry of the pore, governing equations, and boundary conditions are shown in Figure 5.3.

Equation 5.3 specifies the concentration at the entrance of the pore and Equation 5.4 states that the flux of the ith species at the bottom of the pore is equal to the surface flux due to reaction. Note that the classical diffusivity does not apply in this analysis because $\mathrm{Kn} > 1$ for CVD in a narrow pore. For molecular flow in a pore, the diffusivity based on kinetic theory is

$$D_i = \frac{2r}{3}\overline{V}_i \tag{5.5}$$

where $\overline{V}_i$ is the mean gas velocity. Analogous expressions valid in different geometries (trenches, square pores, etc.) are available [29].

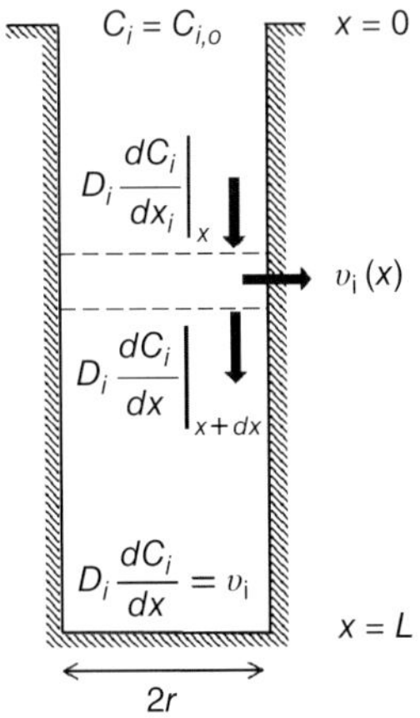

Figure 5.3 Schematic for reaction and diffusion in a pore of radius r and depth L. The boundary conditions at the top (Equation 5.3) and bottom (Equation 5.4) of the pore are shown, and the governing equation (Equation 5.2) is based on a species balance on the differential slice.

We will consider these governing equations in the context of iCVD because there are several experimental results that provide a basis for comparison. Analogous arguments apply for other polymer CVD systems. In iCVD, there are two species of interest in the gas phase: the monomer M and the initiating radical I (the terms initiator and primary radical are also used interchangeably to refer to this species).

The steady-state surface flux υ_i has a physical interpretation for the monomer: it is simply the deposition rate because the net amount of monomer that leaves the gas phase incorporates into the growing film. No such interpretation can be made for the initiating radical. Therefore, we adopt the concept of a sticking probability Γ_i, which is the probability that an impinging gas phase molecule irreversibly sticks to the surface. Therefore, the steady-state surface flux is the product of the sticking probability Γ_i with the surface-incident flux N_i.

$$\upsilon_i = \Gamma_i N_i = \frac{1}{4}\Gamma_i C_i \overline{V}_i \tag{5.6}$$

where we have evaluated N_i using ideal kinetic theory.

Equation 5.2 can therefore be rewritten as:

$$\frac{d^2\psi_i}{d\lambda^2} = \Phi_i^2 \psi_i \tag{5.7}$$

where we have introduced the nondimensional variables $\lambda = x/L$ and $\psi_i = C_i/C_{i,o}$, and grouped parameters into a so-called Thiele modulus Φ_i

$$\Phi_i^2 = \frac{3\Gamma_i}{4}\left(\frac{L}{r}\right)^2 \tag{5.8}$$

The Thiele modulus frequently appears in reaction-diffusion problems and is often assumed to be constant. However, the presence of the sticking probability in the Thiele modulus warrants caution in the analysis of both the monomer and the initiator concentration profiles. The sticking probability of the monomer has a clear physical relationship to the surface reaction rate given by Equation 5.1:

$$\frac{1}{4}\Gamma_{M}C_{M}\overline{V}_{M} = h_{ml}k_{p}KC_{M}\left(\frac{fk_{d}}{k_{t}}C_{I}\right)^{1/2} \tag{5.9}$$

where we have used the subscripts M and I to denote the monomer and the initiator respectively; the monolayer height h_{ml} is used to convert the volumetric rate expression to a surface rate expression; and the adsorption constant K relates the gas phase monomer concentration to the surface concentration [26]. Any appropriate rate expression can be used here to extend this analysis to other systems or polymer deposition reaction mechanisms. Equation 5.9 states that the net impingement of the monomer is equal to the deposition rate, which itself depends upon the concentration of the initiator. For the primary radical, the sticking probability has been experimentally verified to depend on the monomer surface concentration [22]. Therefore, Equation 5.7 must be solved simultaneously for both the monomer and the initiator. When combined with the mixed boundary condition Equation 5.4, it becomes clear that this boundary value problem is not amenable to an analytical solution. However, some general observations can be made based on the form of Equation 5.7. Equation 5.7 does have an analytical solution when simplifying assumptions are made; this solution is discussed in Section 5.6.4.

The Thiele modulus represents the ratio of reaction rate to diffusion rate (for the case of the monomer, this can be seen by substituting Equations 5.9 and 5.5 into Equation 5.8). Systems with a large Thiele modulus are diffusion-limited and those with a small Thiele modulus are reaction-limited. In the limit of small Thiele modulus (i.e., at slow reaction rates), Equation 5.7 has the solution $\psi_i = 1$; that is, the concentration does not vary down the length of the pore. This leads to an important general conclusion for the conformality of polymer CVD films: *conformal deposition is preferred when the surface reaction rate (equivalently, the rate of irreversible adsorption onto the surface) is lower than the gas diffusion rate.*

The form of the Thiele modulus also suggests two corollaries: (i) *conformal deposition is preferred when the sticking probability is low and* (ii) *conformal deposition is hindered when the aspect ratio of the feature is large.* The Thiele modulus captures this competition between sticking probability and aspect ratio.

A common method for evaluating the conformality of films deposited in pores is to measure (e.g., via electron microscopy) the step coverage S, defined as the ratio of the film thickness at the bottom of a feature to the thickness at the top. This is equivalent to the ratio of the deposition rates at the top and the bottom. Based on Equation 5.1, the step coverage for a free-radical iCVD system can be written in terms of the nondimensionalized reactant concentrations

$$S = \frac{r_b}{r_t} = \left(\psi_I\right)^{1/2}\psi_M\Big|_{\lambda=1} \tag{5.10}$$

Again note that any appropriate rate expression can be used here, allowing for adaptability to other kinetic models.

Two limiting cases for the step coverage can be readily identified.

Case 1: $\Phi_M \ll \Phi_I$.

When the monomer concentration profile is nearly uniform throughout the pore ($\psi_M \sim 1$) relative to the initiator concentration profile, the step coverage is determined entirely by the initiator. The step coverage is thus $S = (\psi_I)^{1/2}|_{\lambda=1}$. This condition occurs when the sticking probability for the monomer is small. Equation 5.9 shows that this corresponds to a slow polymerization rate (low rate constant k_p, a low rate of initiator homolysis k_d, a low initiator efficiency f, or high termination rate k_t) or when the initiator sticking probability Γ_I is much larger than the monomer sticking probability Γ_M.

Case 2: $\Phi_I \ll \Phi_M$

When the initiator concentration profile is nearly uniform throughout the pore ($\psi_I \sim 1$) relative to the monomer concentration profile, the step coverage is determined entirely by the monomer. The step coverage is thus $S = \psi_M|_{\lambda=1}$. As opposed to Case 1, this condition is achieved when the polymerization rate is high or when the initiator sticking probability Γ_I is much smaller than the monomer sticking probability Γ_M.

Figure 5.4 shows numerical solutions to Equation 5.7 solved simultaneously for both the monomer and initiator. To capture the coupled nature of the concentration profiles, these solutions assume that the monomer sticking probability varies with the square root of the initiator concentration (as in Equation 5.9) and that the initiator sticking probability varies linearly with the monomer concentration. The ratio of sticking probabilities is Γ_M/Γ_I =0.1, 1, and 10 in the three solutions, corresponding to case 1, the transition between case 1 and case 2, and case 2, respectively. These solutions highlight the importance of the relative magnitudes of the sticking probabilities in determining the step coverage.

Figure 5.5 shows the step coverage, as defined by Equation 5.10, for a variety of sticking probabilities and aspect ratios. In these solutions, the ratio Γ_M/Γ_I was varied. The step coverage decreases with increasing aspect ratio and with increasing initiator sticking probability; the step coverage becomes insensitive to the initiator sticking probability as its value drops below the monomer sticking probability. This represents the transition from case 1 to case 2 described above; when $\Gamma_I < \Gamma_M$, the step coverage is determined entirely by the monomer.

The following sections review several iCVD studies on deposition conformality in the context of the models developed in this section.

5.6.2
Initiator Controlled Consumption

The first systematic experiments on the conformality of iCVD coatings studied step coverage inside of trenches [22]. Step coverage was not dependent on gas mean free path, consistent with the fact that Kn > 1 inside the trench structures. An analytical model was proposed to connect the number of wall-molecule collisions with the deposition conformality. While this model lacks the rigor of the transport model developed in Section 5.6.1, it has intuitive appeal because it

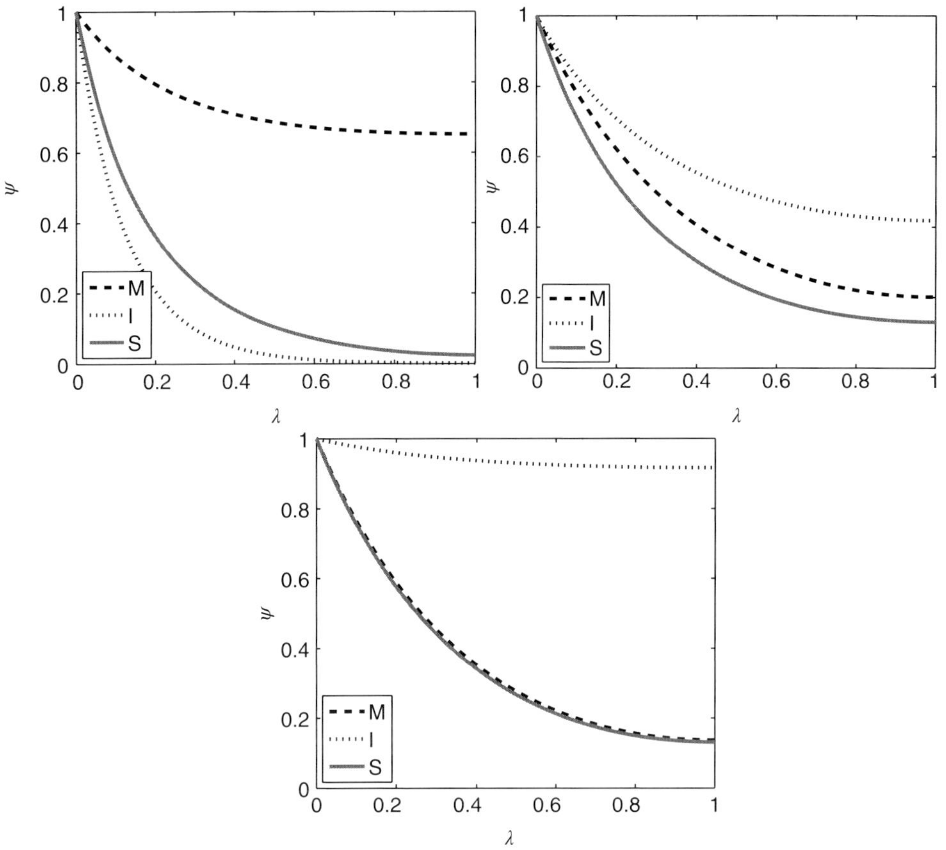

Figure 5.4 Numerical solutions to Equation 5.7 in a pore with aspect ratio $L/r = 100$ and a ratio of sticking probabilities (a) $\Gamma_M/\Gamma_I = 0.1$, (b) $\Gamma_M/\Gamma_I = 1$, and (c) $\Gamma_M/\Gamma_I = 10$.

is based on a physical mechanism in which reactive species transport down the length of the pore via a wall-collision process.

Consider a reactive molecule that travels down the depth L of the trench with opening width $2r$. It will collide with the wall n times as it travels to the bottom of the trench. If the molecule reacts with the trench wall with sticking probability Γ then the step coverage will be

$$S = (1 - \Gamma)^n \tag{5.11}$$

In the original analysis, it was assumed that the deposition rate was proportional to the flux of the reactive molecule. In the case of the initiating radical, the deposition rate is proportional to the square root of its flux (see Equation 5.1), so the exponent n can be replaced by $n/2$. This convention will be adopted here.

The number of collisions n was estimated to be the product of the collision frequency $v_{gas}/(2r)$ and the diffusive time scale L^2/D, where D is the Knudsen

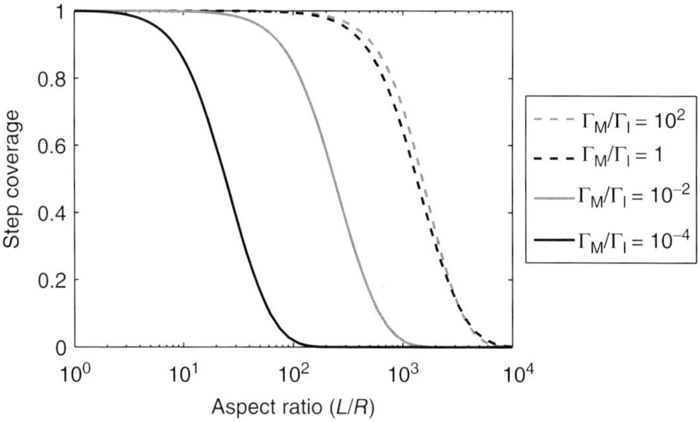

Figure 5.5 Step coverage S as a function of pore aspect ratio L/r, based on numerical solutions to Equation 5.7.

diffusivity inside the trench:

$$n = \frac{3}{4}\left(\frac{L}{r}\right)^2 \tag{5.12}$$

The prefactor will vary based on the geometry of the trench. Here, we have assumed a straight, cylindrical pore for consistency with Section 5.6.1. This differs from the prefactor used in most experimental studies, which is based on Knudsen diffusion in rectangular trenches of infinite length.

Substituting Equation 5.12 into Equation 5.11 and assuming $\Gamma \ll 1$ yields

$$\ln(S) = -\frac{3}{8}\Gamma\left(\frac{L}{r}\right)^2 = -\frac{1}{2}\Phi^2 \tag{5.13}$$

As in Equation 5.7, the Thiele modulus appears as the factor that governs step coverage. Thus, even for a simple approach based on physical intuition, the dependence of the step coverage on the aspect ratio and sticking probability can still be derived.

A number of studies have measured sticking probabilities using Equation 5.13 and consistently find $\Gamma = 10^{-2}$ to 10^{-1} for a variety of deposition conditions and monomer-initiator pairs [22, 30, 31]. The magnitude of the sticking probability yields a critical insight into the deposition mechanism for iCVD. For typical deposition monomer partial pressures, the sticking probability of the monomer would have to be on the order of 10^{-6} in order to be consistent with the deposition rate. Therefore, experiments that analyze step coverage based on this model fall into Case 1 identified in Section 5.6.1: the conformality is initiator consumption controlled. The measured sticking probabilities are thus the sticking probability of the initiator.

The simpler Equation 5.13 is in general consistent with the numerical solutions of Equation 5.7 in commonly encountered experimental regimes. For example, the initiator sticking probability in a trench with aspect ratio $L/r = 10$ and step coverage of 0.86 can be calculated as 0.4×10^{-2} using Equation 5.13 and 1.0×10^{-2}

using Equation 5.7. The agreement is remarkable considering that Equation 5.13 was derived using scaling arguments to determine collision frequency and diffusive time scales. The discrepancy may be resolved through a more accurate estimate of n by, for example, accounting for the reflection mechanism (e.g., specular, diffuse, cosine) of the impinging primary radical. Nevertheless, sticking probabilities determined by Equation 5.13 should be considered accurate to within experimental uncertainty.

5.6.3 Factors Affecting the Initiator Sticking Probability

Several iCVD studies have reported the effect of deposition conditions on the initiator sticking probability. Such studies are best understood within the context of the Eley–Rideal mechanism: the sticking probability of the initiating radical increases as the probability of irreversible reaction with a surface-bound monomer increases.

Classical adsorption theory (as might be used in catalysis) predicts that the sticking probability of a species will decrease as the surface becomes more heavily populated by adsorbed molecules. This is due to the Langmuir-type adsorption occupying the surface sites that accommodate the impinging species. In iCVD higher monomer surface coverage actually promotes irreversible adsorption of the primary radical because it increases the probability that an impinging primary radical strikes and reacts with a monomer [22]. This deviation from the classical adsorption theory makes it clear that the monomer serves as the adsorption site for the primary radical.

From a practical perspective, monomer surface coverage is easily controlled by changing the partial pressure of the monomer gas during deposition through a combination of deposition pressure, substrate temperature, and monomer flowrate. The fractional saturation of the monomer gas P/P^{sat} is the relevant normalized parameter and is the primary process variable used to tune film conformality. A low initiator sticking probability, and therefore a high degree of conformality, can be achieved by operating under conditions where $P/P^{sat} < 0.1$. Figure 5.6 demonstrates the control over step coverage within trenches by varying P/P^{sat}.

This can be generalized to other polymer CVD systems: *whenever a reactive molecule adsorbs by irreversible chemical reaction with a surface species, a low coverage of the surface species will enhance deposition conformality.* The chief drawback of operating under these conditions is that the deposition rate is slower.

The reactivity of the monomer affects the initiator sticking probability as well. For similar values of P/P^{sat}, a divinyl acrylate monomer was found to have a greater sticking probability than a monovinyl methacrylate monomer, suggesting that the increased vinyl concentration or the increased steric accessibility of the acrylate (or both) is responsible for the enhanced sticking of the primary radical. Additional evidence for the Eley–Rideal mechanism in iCVD was provided in that study by showing that the substrate temperature, when holding the fractional

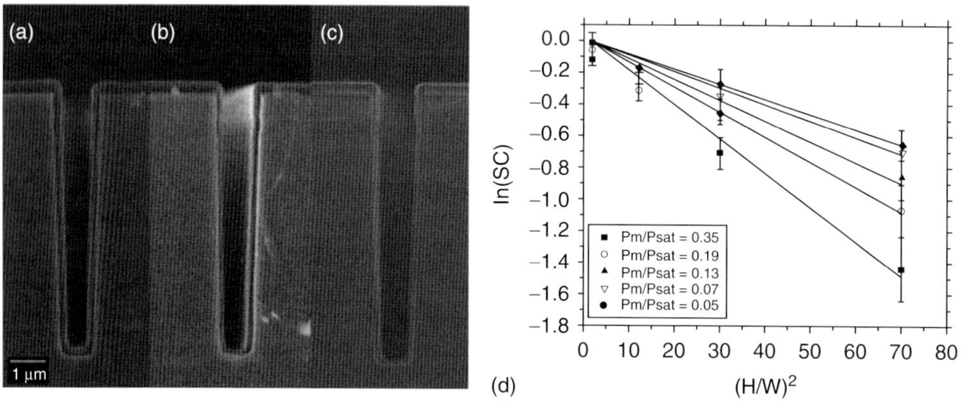

Figure 5.6 Films deposited in trenches at P/P^{sat} = (a) 0.05, (b) 0.13, (c) 0.35. Step coverage decreases as P/P^{sat} increases. (d) Fits to Equation 5.13; the slopes are used to compute the initiator sticking probability between 0.018 (at $P/P^{sat} = 0.05$) and 0.043 (at $P/P^{sat} = 0.35$). (Reprinted with permission from Ref. [30]. Copyright 2010 Wiley.)

saturation constant, does not affect the sticking probability. Thermally activated surface diffusion or reaction therefore does not play an important role in the deposition mechanism [30].

An analogy can be drawn between the parylene monomer and the initiating radical in iCVD, as both serve as the reactive impinging species in their respective depositions. In parylene deposition, an impinging reactant will only stick to the surface if it impinges directly on and reacts with a reactive chain end [8]. The sticking probability has been reported to vary between 10^{-5} and 10^{-3} at high- and low-substrate temperatures, respectively. If thermally activated surface processes (such as reaction) were important, the sticking probability would increase, and not decrease, with temperature [7].

The most common initiating radical for iCVD is a *tert*-butoxy radical, and the studies cited in this section invariably utilized this initiator. When the initiating radical was instead a mixture of oxybenzoate and *tert*-butoxy radicals, the sticking probability was found to increase by a factor of 3. One possibility is that the sticking probability of a primary radical is affected by vapor pressure; oxybenzoate is considerably heavier than *tert*-butoxy. It is also larger, and so its physical cross-section – the surface area sampled by the primary radical – during impingement may also play a role. Yet another possibility is that the oxybenzoate is simply more reactive and is therefore more likely to initiate a growing polymer chain upon impingement [31].

5.6.4
Monomer Controlled Consumption

Most iCVD sticking probability measurements have been performed on trench structures with aspect ratios less than 10. These studies have invariably been performed in systems where the conformality is controlled by initiator consumption, that is, the sticking probability of the initiator is much larger than that of the monomer. There are several reports of iCVD inside of pores with aspect ratios of over 100, which appear to be monomer-consumption controlled [2, 32, 33]. Asatekin *et al.* developed a reaction-diffusion transport model for the monomer similar to the one developed in Section 5.6.1. By neglecting the coupled behavior with the initiator and assuming zero flux of monomer at the bottom of the pore, an analytical solution for the monomer concentration profile was obtained:

$$\psi_{\mathrm{M}} = \frac{e^{\Phi_{\mathrm{M}}\lambda} + e^{\Phi_{\mathrm{M}}(2-\lambda)}}{1 + e^{2\Phi_{\mathrm{M}}}} \tag{5.14}$$

The zero-flux boundary condition is equivalent to setting the right-hand side of Equation 5.4 to zero and is accurate when the sticking probability is small, as is the case for the monomer.

In these systems, the Thiele modulus for the monomer was larger, by about a factor of 10, than that of the initiator [2]. Equation 5.8 can be used to estimate the monomer sticking probability as approximately 10^{-6}, in line with estimates in other studies. However, the initiator sticking probability was approximately 10^{-7}, far lower than those measured in low aspect ratio trench structures for the same monomer-initiator system ($\sim 10^{-2}$) [30]. Therefore, there are systems in which simultaneous monomer and initiator consumption are important. The reason for the dramatic decrease in initiator sticking probability is unclear. One explanation is that the initiator sticking probability depends upon the aspect ratio of the pore, although there is no known physical mechanism linking the two. Figure 5.7, from a different study on a similar system, shows electron microprobe analysis of the fluorine content obtained through the cross-section of a membrane coated with fluoropolymer. The presence of fluorine at both the top and bottom of the membrane pore shows successful deposition within these high aspect ratio features.

5.6.5 Other Polymer CVD Systems

Conformal films are the result of processing conditions where the surface reaction, rather than vapor phase transport, is the rate-limiting step. For iCVD and parylene deposition, the key is to operate under conditions where the sticking probabilities of the reactants are low. Other polymer CVD systems can be understood in the same way.

The conformality of conducting polymers deposited by oCVD may be controlled by the sticking probability of the oxidant. Bromine gas oxidant results in more conformal films than evaporated $FeCl_3$ oxidant. The sticking probability of bromine gas is likely to be much lower than the evaporated metal salt due to its considerably higher volatility [19], consistent with iCVD studies showing decreased conformality when the volatility of the initiating radical is lowered [31].

The general conclusion that conformality is enhanced when vapor phase transport is not a limiting process is also observed in MLD. A useful chemical processing analogy between CVD and MLD is that the former resembles a continuous

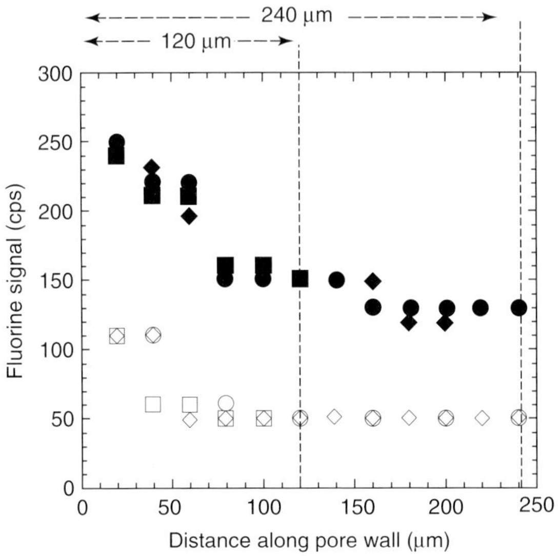

Figure 5.7 Electron microprobe fluorine signal within a membrane pore coated with a fluoropolymer. Unfilled data points were obtained after 2 min of coating and filled data points were obtained after 5 min of coating. Squares are membrane pores 120 μm deep; diamonds and circles are membrane pores 240 μm deep. (Reprinted with permission from Ref. [32]. Copyright 2008 American Chemical Society.)

process where *rates* are important and the latter resembles a batch process where *durations* are important. Consistent with this analogy, ALD models predict that step coverage is a strong function of *total* exposure time of reactants due to the time required to diffuse reactants into small pores [16]. This has been observed experimentally in MLD systems [34].

Conformality of plasma CVD is not well understood, but the species formed by fragmentation of organic gases in plasma CVD are thought to be highly reactive. Within the context of sticking probability, these molecular fragments are likely to have very high sticking probabilities and therefore conformal plasma polymer films are difficult to achieve.

A summary of experimental results for a variety of systems is presented in Table 5.1.

5.7 Applications

Conformal polymer CVD is a key part of a number of applications reported in the literature. This section is not an exhaustive list of such applications, but rather highlights the growing number of fields in which the conformality of polymer CVD films contributes to or even enables applications. Several of these applications are the subject of more detailed discussion in Part III of this book.

Table 5.1 Summary of experimental results on factors affecting polymer CVD conformality.

System	Parameter	Effect	References
iCVD	P/P^{sat} of monomer near substrate	Increasing P/P^{sat} decreases sticking probability	[22, 30]
iCVD	Monomer reactivity	Increased reactivity decreases sticking probability	[30]
iCVD	Mean free path	No effect	[22]
iCVD	Substrate temperature	No effect (when P/P^{sat} held constant)	[30]
Parylene	Substrate temperature	Increased substrate temperature decreases sticking probability. Fractional saturation not controlled	[7]
oCVD, iCVD	Reactant volatility	Increased volatility of initiator (iCVD) or oxidant (oCVD) decreases sticking probability	[19]
MLD	Exposure (product of pressure and time)	Step coverage increases with increasing exposure during reaction cycles	[34]

The quantitative understanding of polymer CVD conformality is based on deposition inside of well-defined trenches and pores. Direct application of these concepts can be made to patterning of surfaces and deposition inside of membranes.

Contact masks used for surface patterning resemble low aspect ratio trenches. For example, dual-polymer patterned surfaces can be fabricated via iCVD using commercial TEM grids as a contact mask. Because the mask has a finite thickness, the unmasked regions reside at the bottom of a trench whose width is the opening of the mask (7.5–15 μm) and whose depth is the thickness of the mask (~15 μm) [35]. Colloids can also be used as contact masks for patterning; the interstitial spaces between colloidal particles as small as 1 μm serve as openings through which reactants must diffuse to reach the surface [36]. The film thickness in the unmasked regions cannot be measured *in situ*, so a witness substrate must be used to monitor film thickness. By operating under conditions that yield high step coverage, the film thickness on the witness serves as a suitable surrogate for the thickness within the mask. Surface patterns can thus be fabricated with *in situ* thickness control. These techniques can be easily adapted to macroscopically curved substrates. Polymer patterns with feature sizes as small as 15 μm have been demonstrated on 3-mm diameter rods [37].

Membranes play a central role in technologies for water purification and gas separations. Conformal polymer CVD makes it possible to modify membrane surfaces throughout pores, which can have openings <100 nm and aspect ratios exceeding 1000 : 1. Fluorinated iCVD coatings deposited on membranes with an 80 : 1 aspect ratio showed step coverage of approximately 50% even without optimizing the fractional saturation of the monomer [32]. The detailed reaction-diffusion model from Section 5.6.4 was used to analyze coverage in 1000 : 1 aspect ratio pores in track-etched polycarbonate membranes. Conformal hydrophobic coatings in these membranes were shown to enhance the selective diffusion of a hydrophobic molecule relative to a hydrophilic analog [2].

Conformal deposition to completely fill the inside of high-aspect ratio pores has also been used to template the fabrication of cylindrical polymer microstructures. If the coating preferentially deposits at the mouth of the pore, then the opening will eventually pinch off and block the diffusion of reactants into the pore, limiting the growth of the polymer cylinder. Therefore, these templated structures are only accessible when deposition conditions result in conformal coatings [33]. Within mesoporous materials with irregular pore structure, MLD is an attractive technique for depositing polymers. Oxidation of highly conformal MLD films deposited within mesoporous zeolite forms zeolite-supported nanoporous networks that enhance the selectivity of gas separation of H_2 from N_2 by a factor of 100 as compared to uncoated zeolite [38].

In addition to templating structures, pore-filling via polymer CVD is an important end to itself. Ceramic moisture barrier layers frequently fail due to water permeation through pinholes with diameters ranging between 1 and 1000 nm. Adsorption isotherms show that cyclic siloxane monomers can adsorb within these pinholes, suggesting that iCVD may be an effective means for healing defects in ceramic water-barrier layers [39]. Defect planarization, which is closely related to pore-filling, has been demonstrated with cyclic siloxane iCVD polymers deposited using a plasma-assisted process [40]. Gap-filling and planarization of micron-sized trench structures for electrical applications is also possible using parylene deposition [41].

Polymer deposition on nonplanar substrates has found application in miniature devices requiring surface modification. Figure 5.8a shows a cantilever structure coated with a reactive iCVD polymer that was used to create a chemical sensor: the film stress induced by the reaction between the polymer coating and an analyte transduced the chemical signal into a mechanical signal via bending of the cantilever [42].

Several studies have used polymer CVD to modify microfluidic devices. Parylene derivatives were used to modify assembled devices with a channel opening of 75 μm × 100 μm and channel lengths of up to 2.8 mm. In analogy with step coverage, the ratio of the film thickness within the microfluidic channel to the film thickness outside the channel ranged from ~1% to 89%. Better coverage was observed for parylene derivatized with lighter side groups, suggesting that the sticking probability increased when the vapor pressure of the precursor was low, or, equivalently, when the fractional saturation was high. These results are consistent with the systematic studies discussed in Section 5.6 of this chapter. Biotin could be subsequently immobilized on the reactive parylene coatings [23].

Coating microfluidic devices prior to assembly has enabled new device-bonding techniques. The standard technique is to activate poly(dimethylsiloxane) (PDMS) and glass surfaces using oxygen plasma and then to immediately place the surfaces in contact. The PDMS surface is typically patterned with the microfluidic channel topography while the glass surface is flat. By coating the PDMS with an epoxy-containing polymer and coating the glass with an amine-containing

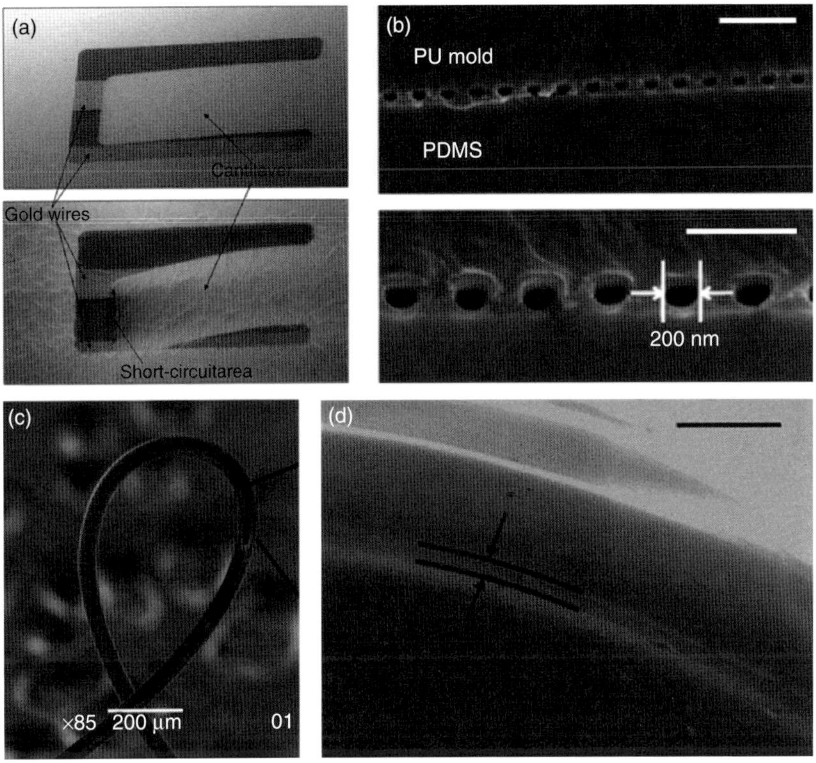

Figure 5.8 (a) 50 μm × 20 μm × 0.1 μm microcantilever coated with reactive polymer before (top) and after (bottom) reaction with analyte. (b) Microfluidic device with 200 nm wide channels bonded by reaction between coatings on the PU and PDMS pieces. Scale bars represent 1 μm (top) and 500 nm (bottom). (c) Wire coated with flexible, insulating polymer. (d) 195 nm thick anti-microbial coating (between arrows) around a nylon fiber. A layer of carbon is deposited over the nylon-coating assembly for imaging purposes; scale bar represents 1 μm. (Reprinted/adapted with permission from (a) Ref. [42], copyright 2009 IEEE (b) Ref. [43], copyright 2008 Royal Society of Chemistry (c) Ref. [44], copyright 2007 American Chemical Society, and (d) Ref. [45], copyright 2007 Elsevier.)

polymer, the surface of each piece contains one part of a two-part glue. The bonding proceeds by the assembly and subsequent reaction between epoxy and amine. Conformal deposition on the patterned substrate is important for producing a continuous bond between the two pieces and for maintaining the fidelity of the channel geometry. In Figure 5.8b, such a scheme has been used to assemble devices with channels as thin as 200 nm [43]. Utilizing coatings as opposed to traditional plasma-based bonding also allows assembly of non-PDMS microfluidic devices. Such bonds display good hydrolytic stability and can withstand over 150 psi without failing [46, 47].

While most analytical work on polymer CVD has been performed on interior coatings in trenches and pores, exterior coatings around masses of particles and fibers are a key area of interest. Powder beds and fiber mats can be thought of as a collection of non-planar substrates with tortuous channels defined by the interstices between the individual particles and fibers; therefore, deposition conditions which yield conformal films are required to coat such substrates.

For particles, agitation is often required because of contact shadowing at the point where particles touch each other or the deposition stage. Rotating iCVD reactors have been designed to aid the agitation process [48]. Microspheres (1 – 100 μm) have been coated via iCVD [3, 48] including the encapsulation of 25 μm drug particles with enteric coatings [49]. Smaller nanoparticles, including carbon nanotubes, have been coated via iCVD as well [48]. MLD can coat nanoparticles smaller than 100 nm, including particles as small as 7 nm [34, 50].

Plasma polymerization has been used to coat individual fibers with acetaldehyde-derived polymer films for cell-adhesion purposes [51]. Similar results were achieved using iCVD to coat wires with an electrically passivating polymer as shown in Figure 5.8c [44]. However, fibers wound into tows were not evenly coated in a plasma polymerization process due to shadowing effects [52]. This is consistent with the observed conformality plasma polymers (see, for example, Figure 5.1a). Wound or woven fiber mats are an important class of substrate because the ability to coat commercial textiles enables applications in self-cleaning and sterile consumer products. Polymer-textile systems exhibiting superhydrophobicity and anti-microbial activity have been demonstrated by iCVD [45, 53]. Figure 5.8d shows a cross-section of a fiber from a woven fabric coated with an anti-microbial polymer.

5.8
Conclusion

Polymer CVD technologies enable conformal coatings of complex topographies that have feature sizes ranging from nanometers to millimeters. Liquid-phase coatings are unable to produce conformal coatings on these size scales due to the effects of solvent surface tension. In polymer CVD, conformal coatings result when vapor-phase reactants transport evenly to all surfaces of the substrate. This transport occurs in competition with the surface reaction and conformality is best achieved when the surface reaction rate is lower than the gas diffusion rate, a concept embodied by the Thiele modulus. For reaction and diffusion in a pore, the Thiele modulus depends on the sticking probability of the reactant and the aspect ratio of the pore. Systems with a small Thiele modulus – that is, systems with low sticking probabilities and small aspect ratios – result in more conformal coatings. A number of chemical and physical factors have been found to control the sticking probability, which can be quantitatively estimated based on reaction-diffusion models. Conformal polymer coatings via plasma polymerization, parylene deposition, iCVD, oCVD, and MLD are central to a number of applications demonstrated in the literature.

Acknowledgment

This chapter was prepared by Lawrence Livermore National Laboratory under contract DE-AC52-107NA27344.

References

1. Chu, K.-H., Xiao, R., and Wang, E.N. (2010) Uni-directional liquid spreading on asymmetric nanostructured surfaces. *Nat. Mater.*, **9**, 413.
2. Asatekin, A. and Gleason, K.K. (2011) Polymeric nanopore membranes for hydrophobicity-based separations by conformal initiated chemical vapor deposition. *Nano Lett.*, **11**, 677.
3. Baxamusa, S.H., Montero, L., Dubach, J.M., Clark, H.A., Borros, S., and Gleason, K.K. (2008) Protection of sensors for biological applications by photoinitiated chemical vapor deposition of hydrogel thin films. *Biomacromolecules*, **9**, 2857.
4. Montero, L., Baxamusa, S.H., Borros, S., and Gleason, K.K. (2008) Thin hydrogel films with nanoconfined surface reactivity by photoinitiated chemical vapor deposition. *Chem. Mater.*, **21**, 399.
5. Karaman, M., Kooi, S.E., and Gleason, K.K. (2008) Vapor deposition of hybrid organic–inorganic dielectric bragg mirrors having rapid and reversibly tunable optical reflectance. *Chem. Mater.*, **20**, 2262.
6. Stillwagon, L.E., Larson, R.G., and Taylor, G.N. (1987) Planarization of substrate topography by spin coating. *J. Electrochem. Soc.*, **134**, 2030.
7. Fortin, J.B. and Lu, T.M. (2002) A model for the chemical vapor deposition of poly(para-xylylene) (parylene) thin films. *Chem. Mater.*, **14**, 1945.
8. Fortin, J.B. and Lu, T.M. (2003) *Chemical Vapor Deposition Polymerization: The Growth and Properties of Parylene Thin Films*, Springer.
9. Macak, E.B., Münz, W.-D., and Rodenburg, J.M. (2003) Plasma–surface interaction at sharp edges and corners during ion-assisted physical vapor deposition. Part I: edge-related effects and their influence on coating morphology and composition. *J. Appl. Phys.*, **94**, 2829.
10. Yasuda, H. (1985) *Plasma Polymerization*, Academic Press.
11. Yeo, L.P., Yan, Y.H., Lam, Y.C., and Chan-Park, M.B. (2006) Design of experiment for optimization of plasma-polymerized octafluorocyclobutane coating on very high aspect ratio silicon molds. *Langmuir*, **22**, 10196.
12. Fridman, A. (2008) *Plasma Chemistry*, Cambridge University Press.
13. Du, Y. and George, S.M. (2007) Molecular layer deposition of nylon 66 films examined using in situ FTIR spectroscopy. *J. Phys. Chem. C*, **111**, 8509.
14. Yoshimura, T., Tatsuura, S., and Sotoyama, W. (1991) Polymer-films formed with monolayer growth steps by molecular layer deposition. *Appl. Phys. Lett.*, **59**, 482.
15. Zhou, H. and Bent, S.F. (2013) Fabrication of organic interfacial layers by molecular layer deposition: present status and future opportunities. *J. Vac. Sci. Technol., A*, **31**, 040801.
16. Gordon, R.G., Hausmann, D., Kim, E., and Shepard, J. (2003) A kinetic model for step coverage by atomic layer deposition in narrow holes or trenches. *Chem. Vap. Deposition*, **9**, 73.
17. Ja-Yong, K., Ji-Hoon, A., Sang-Won, K., and Jin-Hyock, K. (2007) Step coverage modeling of thin films in atomic layer deposition. *J. Appl. Phys.*, **101**, 73502.
18. Im, S.G., Kusters, D., Choi, W., Baxamusa, S.H., van de Sanden, M.C.M., and Gleason, K.K. (2008) Conformal coverage of poly(3,4-ethylenedioxythiophene) films with tunable nanoporosity via oxidative chemical vapor deposition. *ACS Nano*, **2**, 1959.
19. Chelawat, H., Vaddiraju, S., and Gleason, K. (2010) Conformal, conducting

poly(3,4-ethylenedioxythiophene) thin films deposited using bromine as the oxidant in a completely dry oxidative chemical vapor deposition process. *Chem. Mater.*, **22**, 2864.

20. Alf, M.E., Asatekin, A., Barr, M.C., Baxamusa, S.H., Chelawat, H., Ozaydin-Ince, G., Petruczok, C.D., Sreenivasan, R., Tenhaeff, W.E., Trujillo, N.J., Vaddiraju, S., Xu, J., and Gleason, K.K. (2010) Chemical vapor deposition of conformal, functional, and responsive polymer films. *Adv. Mater.*, **22**, 1993.
21. Baxamusa, S.H., Im, S.G., and Gleason, K.K. (2009) Initiated and oxidative chemical vapor deposition: a scalable method for conformal and functional polymer films on real substrates. *Phys. Chem. Chem. Phys.*, **11**, 5227.
22. Baxamusa, S.H. and Gleason, K.K. (2008) Thin polymer films with high step coverage in microtrenches by initiated CVD. *Chem. Vap. Deposition*, **14**, 313.
23. Chen, H.-Y., Elkasabi, Y., and Lahann, J. (2005) Surface modification of confined microgeometries via vapor-deposited polymer coatings. *J. Am. Chem. Soc.*, **128**, 374.
24. Dameron, A.A., Seghete, D., Burton, B.B., Davidson, S.D., Cavanagh, A.S., Bertrand, J.A., and George, S.M. (2008) Molecular layer deposition of alucone polymer films using trimethylaluminum and ethylene glycol. *Chem. Mater.*, **20**, 3315.
25. Odian, G. (2004) *Principles of Polymerization*, John Wiley & Sons, Inc.
26. Lau, K.K.S. and Gleason, K.K. (2006) Initiated chemical vapor deposition (iCVD) of poly(alkyl acrylates): a kinetic model. *Macromolecules*, **39**, 3695.
27. Suresh, A., Anastasio, D., and Burkey, D.D. (2014) Potential of hexyl acrylate monomer as an initiator in photo-initiated CVD. *Chem. Vap. Deposition*, **20**, 5.
28. Komiyama, H., Shimogaki, Y., and Egashira, Y. (1999) Chemical reaction engineering in the design of CVD reactors. *Chem. Eng. Sci.*, **54**, 1941.
29. Ohring, M. (2001) *Materials Science of Thin Films*, Elsevier Science.
30. Ozaydin-Ince, G. and Gleason, K.K. (2010) Tunable conformality of polymer coatings on high aspect ratio features. *Chem. Vap. Deposition*, **16**, 100.
31. Xu, J. and Gleason, K.K. (2011) Conformal polymeric thin films by low-temperature rapid initiated chemical vapor deposition (iCVD) using tert-butyl peroxybenzoate as an initiator. *ACS Appl. Mater. Interfaces*, **3**, 2410.
32. Gupta, M., Kapur, V., Pinkerton, N.M., and Gleason, K.K. (2008) Initiated chemical vapor deposition (iCVD) of conformal polymeric nanocoatings for the surface modification of high-aspect-ratio pores. *Chem. Mater.*, **20**, 1646.
33. Ozaydin-Ince, G., Dubach, J.M., Gleason, K.K., and Clark, H.A. (2011) Microworm optode sensors limit particle diffusion to enable in vivo measurements. *Proc. Natl. Acad. Sci. U.S.A.*, **108**, 2656.
34. Loscutoff, P.W., Lee, H.-B.-R., and Bent, S.F. (2010) Deposition of ultrathin polythiourea films by molecular layer deposition. *Chem. Mater.*, **22**, 5563.
35. Baxamusa, S.H., Montero, L., Borrós, S., and Gleason, K.K. (2010) Self-aligned micropatterns of bifunctional polymer surfaces with independent chemical and topographical contrast. *Macromol. Rapid Commun.*, **31**, 735.
36. Trujillo, N.J., Baxamusa, S.H., and Gleason, K.K. (2009) Grafted functional polymer nanostructures patterned bottom-up by colloidal lithography and initiated chemical vapor deposition (iCVD). *Chem. Mater.*, **21**, 742.
37. Petruczok, C.D. and Gleason, K.K. (2012) Initiated chemical vapor deposition-based method for patterning polymer and metal microstructures on curved substrates. *Adv. Mater.*, **24**, 6445.
38. Yu, M., Funke, H.H., Noble, R.D., and Falconer, J.L. (2011) H_2 separation using defect-free, inorganic composite membranes. *J. Am. Chem. Soc.*, **133**, 1748.
39. Aresta, G., Palmans, J., van de Sanden, M.C.M., and Creatore, M. (2012) Evidence of the filling of nano-porosity in SiO_2-like layers by an initiated-CVD monomer. *Microporous Mesoporous Mater.*, **151**, 434.

40. Coclite, A.M. and Gleason, K.K. (2012) Global and local planarization of surface roughness by chemical vapor deposition of organosilicon polymer for barrier applications. *J. Appl. Phys.*, **111**, 073516.
41. Zhang, X., Dabral, S., Howard, B.J., Bica, J., Chiang, C., Ochoa, V., Ficalora, P.J., Steinbruchel, C.O., Bakhru, H., Lu, T.-M., and McDonald, J.F. (1993) Parylene as a conformal insulator for submicron multilayer interconnection. *Proc. SPIE*, **1805** (30).
42. Arora, W.J., Tenhaeff, W.E., Gleason, K.K., and Barbastathis, G. (2009) Integration of reactive polymeric nanofilms into a low-power electromechanical switch for selective chemical sensing. *J. Microelectromech. Syst.*, **18**, 97.
43. Im, S.G., Bong, K.W., Lee, C.-H., Doyle, P.S., and Gleason, K.K. (2009) A conformal nano-adhesive via initiated chemical vapor deposition for microfluidic devices. *Lab Chip*, **9**, 411.
44. O'Shaughnessy, W.S., Murthy, S.K., Edell, D.J., and Gleason, K.K. (2007) Stable biopassive insulation synthesized by initiated chemical vapor deposition of poly(1,3,5-trivinyltrimethylcyclotrisiloxane). *Biomacromolecules*, **8**, 2564.
45. Martin, T.P., Kooi, S.E., Chang, S.H., Sedransk, K.L., and Gleason, K.K. (2007) Initiated chemical vapor deposition of antimicrobial polymer coatings. *Biomaterials*, **28**, 909.
46. Bong, K.W., Xu, J., Kim, J.-H., Chapin, S.C., Strano, M.S., Gleason, K.K., and Doyle, P.S. (2012) Non-polydimethylsiloxane devices for oxygen-free flow lithography. *Nat. Commun.*, **3**, 805.
47. Xu, J. and Gleason, K.K. (2010) Conformal, amine-functionalized thin films by initiated chemical vapor deposition (iCVD) for hydrolytically stable microfluidic devices. *Chem. Mater.*, **22**, 1732.
48. Lau, K.K.S. and Gleason, K.K. (2006) Particle surface design using an all-dry encapsulation method. *Adv. Mater.*, **18**, 1972.
49. Lau, K.K.S. and Gleason, K.K. (2007) All-dry synthesis and coating of methacrylic acid copolymers for controlled release. *Macromol. Biosci.*, **7**, 429.
50. Bent, S., Loscutoff, P.W., and Clendenning, S. (2010) Fabrication of organic thin films for copper diffusion barrier layers using molecular layer deposition. *MRS Online Proceedings Lib.*, **1249**, F02.
51. Hadjizadeh, A. (2010) Acetaldehyde plasma polymer-coated PET fibers for endothelial cell patterning: chemical, topographical, and biological analysis. *J. Biomed. Mater. Res. B Appl. Biomater.*, **94B**, 11.
52. Pittman, C.U. Jr.,, Jiang, W., He, G.R., and Gardner, S.D. (1998) Oxygen plasma and isobutylene plasma treatments of carbon fibers: determination of surface functionality and effects on composite properties. *Carbon*, **36**, 25.
53. Ma, M., Gupta, M., Li, Z., Zhai, L., Gleason, K.K., Cohen, R.E., Rubner, M.F., and Rutledge, G.C. (2007) Decorated electrospun fibers exhibiting superhydrophobicity. *Adv. Mater.*, **19**, 255.

6
Plasma Enhanced-Chemical Vapor Deposited Polymers: Plasma Phase Reactions, Plasma–Surface Interactions, and Film Properties

Mariadriana Creatore and Alberto Perrotta

6.1
Introduction: Chemical Vapor Deposition Methods, Advantages, and Challenges

The tailoring of the surface energy of a selected substrate in order to meet specific requirements represents an ever-green and extensive field of research and application. The surface energy, and, in general, surface properties, can be affected either directly, that is, by means of a surface treatment, or by the deposition of a thin (from few to several hundred nanometers) layer. In addition to the traditional wet chemistry-based approach (e.g., self-assembled monolayers SAMs and sol–gel), gas phase-based methods are also widely available. They are acknowledged as very efficient (and often cost-effective) methods to modify the surface properties of a substrate toward hydrophilicity/hydrophobicity/oleophobicity [1–3], to improve the adhesion [4] or to generate a selected surface chemistry for biomaterial applications [5–7].

A well-known gas-phase approach is chemical vapor deposition (CVD) in which one or more volatile molecular (either organic or inorganic) precursors decompose either in the gas phase or at the substrate. CVD processes are categorized on the basis of the activation step (or energy source) selected for the decomposition of the precursor(s) into active species (e.g., radicals). In thermal CVD, energy in the form of heat is supplied to the substrate to activate the precursor molecules and promote surface reactions. In plasma enhanced-chemical vapor deposition (PE-CVD, Figure 6.1a,b) the high reactivity of plasma species (generally, electrons and ions) is responsible for the decomposition of the molecules into depositing radicals, and this occurs already in the gas phase. Hot-wire and initiated CVD (i-CVD) (Figure 6.1c), instead, adopt a hot filament/grid to decompose the precursor molecule (hot-wire CVD) into radicals or homolytically cleave an initiator molecule (i-CVD). This latter then promotes the polymerization of an unsaturated monomer adsorbed at the substrate surface according to the classical radical polymerization steps of initiation–propagation–termination. The advantage of PE-CVD and hot-wire/i-CVD methods with respect to the traditional CVD approach is their compatibility with thermosensitive substrates (e.g., polymers) due to the high reactivity at low thermal budget of the plasma

CVD Polymers: Fabrication of Organic Surfaces and Devices, First Edition. Edited by Karen K. Gleason.

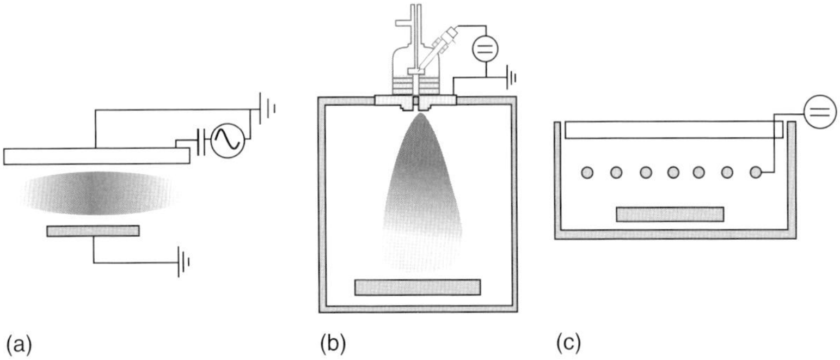

Figure 6.1 Schematics of (a) radio-frequency (rf) capacitively coupled PE-CVD; (b) direct current (dc) cascaded arc generated-expanding thermal plasma CVD; and (c) hot wire/initiated CVD setup.

medium and the independent control of the grid and substrate temperatures in hot-wire/i-CVD.

Although several and diverse gas-phase approaches are known to promote thin film polymer deposition, including sputtering [8], laser ablation, and more recently molecular layer deposition [9], the field of PE-CVD (and i-CVD) of polymers remains fascinating because it succeeds in polymerizing monomer units also at low pressure (<1 Torr), that is, under conditions where conventional (chain-growth) polymerization would not lead to film deposition, because of the low ceiling temperature at that low pressure. Furthermore, the majority of organic molecules characterized by saturated bonds, which would not be considered for conventional polymerization, is yet polymerized by plasmas. These observations point out that the plasma polymer growth mechanisms are very different from the conventional polymerization routes.

At the same time, the above-mentioned advantages are counter-balanced by the well-established observation that plasma polymerization, and, in general, plasma deposition, are highly system-dependent, that is, results in terms of polymer properties are strongly dependent on the chosen plasma setup (e.g., plasma source, electrode configuration) and process parameters (e.g., power density, gas flow rates, and pressure) windows.

No matter their variability, plasma polymerization processes are widely adopted in many fields, ranging, for example, from electronics and photovoltaics to composites, microfluidics, and cell biology. The modification of the surface chemistry (and other properties) reaches out the engineering of patterned substrates [10] (Figure 6.2a) and nanofabrication [11–13]. At the same time, the variability in plasma processes hampers the translation of scientific results and applications from one setup to another, as well as it challenges potential process scale-up. Such difficulties arise mainly from: (i) the control on the process parameters is by far not sufficient to provide knowledge on the internal parameters of a plasma process and only the application of diagnostics can provide correlations among different plasma systems and (ii) a complete description of the plasma polymerization/deposition mechanisms is complex in its derivation (Figure 6.2b) [14, 15].

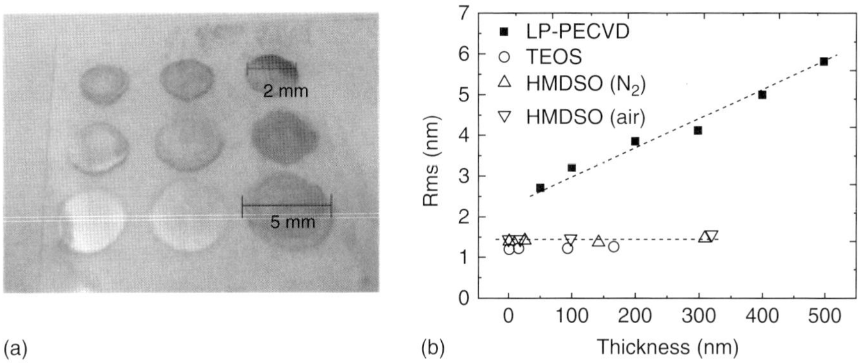

Figure 6.2 (a) Nickel pads, developed upon electrodeless plating on a fluorinated ethylene–propylene co-polymer functionalized with amino-groups. A μPlasmaPrint setup – a multi pin-to-plate dielectric barrier discharge – fed with 3-aminopropyl trimethoxysilane has been adopted for the functionalization [10]. (b) Surface roughness evolution as function of the film thickness of SiO_2-like layers deposited in a low-pressure PE-CVD setup [14] and in an atmospheric pressure dielectric barrier discharge (DBD) [15]. *Why is there no surface roughness development in the atmospheric pressure DBD deposited layers?*

While providing the reader with references to excellent books and reviews in the field of plasma physics [16, 17] and plasma processing of materials and, applications [18–21], we dedicate the present chapter to:

- An overview of the production routes of the active species in low-pressure plasmas,[1)] highlighting the role of selected plasma diagnostics for radical (and their surface reactivity) detection.
- A discussion of the role of ions in plasma polymerization, supported by an overview of the phenomena occurring in the plasma sheath region.
- A critical overview of the approaches reported in literature toward the description of plasma polymerization through a macroscopic approach.
- Selected examples of plasma polymerization processes focused on precursor structure retention/fragmentation, densification via ion bombardment, and topography evolution with film thickness.

Although this chapter addresses PE-CVD processes, selected plasma polymer properties will be compared with those from i-CVD processes. The i-CVD

1) Although atmospheric pressure (direct) plasmas for thin film deposition/polymerization are widely investigated and appreciated also in the industrial environment, this chapter will not cover their plasma chemistry because their complexity in terms of discharge regime and energetic species could not be summarized in few lines. Instead, they have been very recently addressed in an excellent review from Massines *et al.* [22]. In Section 6.5.3 an example of atmospheric pressure dielectric barrier discharge deposited layer will be discussed to highlight the role of plasma–surface interactions in a plasma where the ratio surface-to-volume is much higher than in low pressure PE-CVD.

approach [23, 24], extensively discussed throughout this book, is in fact acknowledged as the ultimate gas-phase approach mimicking the conventional polymerization.

6.2
Plasma Parameters, Plasma Phase Reactions, and the Role of Diagnostics

PE-CVD processes are characterized by gas phase as well as (heterogeneous) surface reactions. In the gas phase the kinetics of each reaction is controlled by the density of the participating species and their temperature(s). If we consider the classification of plasmas [16] (Figure 6.3) on the basis of the electron density (equal to the ion density) n_e and the ionization degree, material processing is generally carried out under low- density plasma (e.g., an RF driven parallel plate configuration) conditions where n_e is in the range of 10^9-10^{10} cm^{-3} and the ionization degree is in the range of 10^{-6} to 10^{-3} and high-density plasma conditions (e.g., inductively coupled and electron cyclotron resonance plasmas) where the electron density is as high as $10^{10}-10^{13}$ cm^{-3} and the ionization degree is in the range 10^{-4} to 10^{-1}. Processing of materials and, specifically, PE-CVD of polymers, occurs in the so-called cold plasmas, that is, where a non-equilibrium exists among the electron and ion/gas temperatures. Briefly, the coupling of the input power with electrons is strong, whereas the energy coupling between electrons and ions and between neutrals and surfaces is weak. Therefore, electrons have a temperature as high as $5\times10^3-10^5$ K (i.e., from few tenths of electronvolts to few electronvolts) whereas the ion/gas temperature (as well as the wall temperature with which the neutrals and ions have a strong energy coupling) is in the range of 300–500 K. This temperature gap allows classifying plasmas as low thermal budget systems, where reactivity in the gas phase is controlled by the electron temperature. The energy

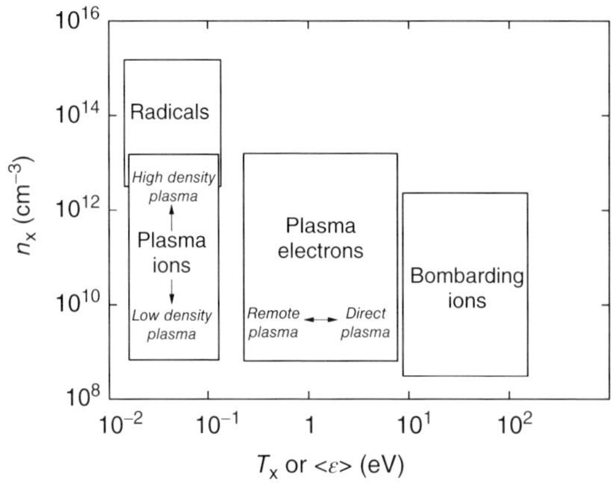

Figure 6.3 Classification of plasmas on the basis of the species density n_x and temperature T_x. The ranges of density and temperature for the ions and radicals as well as of the ions accelerating in the plasma sheath are also indicated.

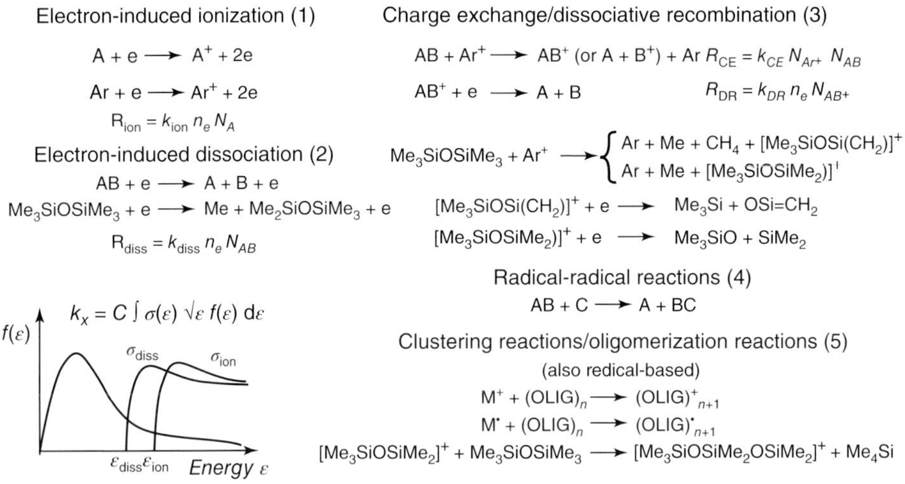

Figure 6.4 Schematics of selected chemical reactions in the plasma phase: electron impact-induced (paths 1 and 2) and (Ar^+,e) induced charge exchange (CE)/dissociative recombination (DR) (path 3). Radical–radical (path 4) and clustering reactions (path 5) are also reported. The chosen model molecule is HMDSO.

coupled into the gas (mixture) either via a DC or an AC electric field promotes the acceleration of electrons, causing ionization and leading to a specific electron density and electron energy distribution function. Depending on the pressure in the system, electrons may thermalize, leading to a Maxwell energy distribution,[2] which allows defining the electron temperature of the plasma system.

The electron energy is then transferred either through elastic collisions, or via ionization and dissociation reactions (paths (1) and (2) in Figure 6.4) [16]. These latter are responsible for the generation of reactive species in the plasma phase, according to a reaction constant as indicated in Figure 6.4. Furthermore, selected plasma systems have a remote character, that is, there is no power being coupled directly into the gas mixture. Because low electron temperatures develop (see Figure 6.3), electrons cannot be held responsible for ionization and dissociation processes. For example, in the case of the expanding thermal plasma, ETP, (Figure 6.1b), the (Ar^+,e) emanating from the DC cascade arc plasma source into the downstream region (characterized by a T_e of 0.3 eV) are held responsible for molecule activation [25, 26] (path (3), Figure 6.4). Ar ions promote charge exchange (CE) reactions, through which neutral molecules become activated. Radicals are then formed via dissociative recombination between the low-energy electron and the previously formed molecular ion.

2) Under low pressure conditions, however, this is a rather strong approximation, since electron collisions are rarely occurring and, therefore, thermalization is impeded. Furthermore, specific loss channels determine the preferential depletion of high energy electrons through excitation, ionization and dissociation processes.

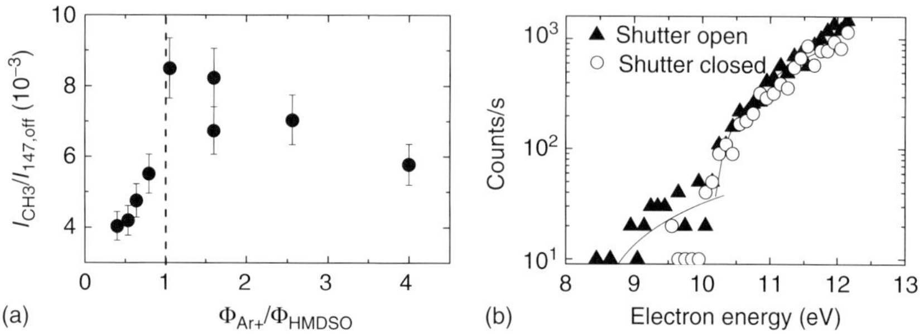

Figure 6.5 (a) CH_3 (15 amu) mass spectrometry signal normalized to the HMDSO parent ion signal (147 amu) as function of the Ar ions available per HMDSO molecule. The Ar ion flow rate has been determined by means of Langmuir probe measurements under the assumption that the directed particle velocity along the expansion is constant. (b) Electron energy scan for the 58 amu-mass peak ($SiMe_2$ and $OSi{=}CH_2$) in Ar/HMDSO plasma in the presence/absence of a shutter blocking the plasma beam. Experimental conditions reported in [27].

One valid example is represented by the dissociation paths (path (3), Figure 6.4) which the hexamethyldisiloxane (HMDSO) molecule undergoes when injected in the downstream chamber of an ETP system for the deposition of organosilicon polymers [27]. The CE reaction constant has been found equal to $4 \times 10^{-16}\,m^3\,s^{-1}$. In comparison, the electron induced dissociation rate constant for $T_e = 0.3\,eV$ is equal to $8 \times 10^{-25}\,m^3\,s^{-1}$. Three-stage molecular beam threshold ionization-mass spectrometry (TI-MS) has allowed the identification of the reaction products: the CE process has a dissociative character, leading to CH_3 abstraction (Figure 6.5a). Under the experimental conditions investigated (n_e/[HMDSO] $> 10^{-3}$), the ions produced via CE reactions lead to negligible ion-induced oligomerization. Instead, they undergo recombination reactions with electrons, leading to Si–O bond dissociation. The overall HMDSO dissociation path is hypothesized to be responsible for the low carbon concentration ($SiCO_{0.8}$) in films deposited under the experimental conditions reported in [28]: while $SiMe_3$ radicals most certainly do not contribute to film growth but recombine at the surface to $SiMe_4$ and $SiMe_3H$, $SiMe_2$ and $OSiCH_2$ are identified as the most probable deposition precursors. In Figure 6.5b, two contributions are in fact observed: one with an appearance energy (AE) of 8.3 eV, which is not present when a shutter prevents the plasma beam from reaching the MS housing, and the other with an AE of 10.2 eV, whose intensity is not affected by the shutter insertion. This latter is attributed to Me_2Si originating from the dissociative ionization of $SiMe_3H$ and $SiMe_2H_2$ in the MS. The earlier AE is associated to $SiMe_2$ and $OSi{=}CH_2$ and the signal disappearance in the presence of the shutter points out the species reactivity at the surface. Another valid example of plasma diagnostics allowing to infer information on the species reactivity is the IRIS (imaging of radicals interacting with surfaces) technique introduced by Fisher *et al.* [29].

The role of plasma diagnostics is fundamental to defining the plasma parameters and identifying the dissociation paths of (plasma deposition) precursor molecules:

selected examples are optical emission spectroscopy [30], gas/plasma phase FT-IR spectroscopy [31], and other mass spectrometry-based approaches, such as gas chromatography MS [32], selected ion flow tube MS and proton transfer reaction MS [33], just to cite a selected number of examples focusing on the siloxane chemistry. Particularly, positive ion MS data have often highlighted an extensive molecular oligomerization [34] induced by ion-molecule reaction taking place in the plasma phase when operating at low power density (path (5), Figure 6.4).

6.3 Plasma Polymerization: Is It Just Chemistry? The Role of Ions in Film Growth

The characteristics of a plasma polymer may not always reflect the composition of the plasma chemistry, instead it can strongly be affected by what is generally addressed as *plasma–surface interactions.* The latter includes not only the adsorption processes dependent on the specific sticking coefficient of a radical species, but also the desorption (reactive) steps, insertion reactions of radicals into strained bonds and processes either assisted or controlled by (positive) ionic species. As highlighted earlier, measurements in the plasma bulk demonstrate the existence of ion-molecule reactions and therefore of ionic species which could contribute to film growth. As shown in Figure 6.6, while the plasma bulk

$$n_e(x) = n_s e^{\frac{eV(x)}{kT_e}} \qquad \bar{u}_e(x) = \sqrt{\frac{8kT_e}{\pi m_e}} \qquad j_e = \frac{1}{4} e n_s \bar{u}_e e^{\frac{eV_{wall}}{kT_e}}$$

$\boldsymbol{n_e}(x)$ Boltzmann distribution of electrons, $\boldsymbol{u_e}$ average electron velocity in the plasma, $\boldsymbol{n_s}$ electron density at the sheath edge, $\boldsymbol{V}(x)$ potential (<0) in the (pre)sheath region with respect to the plasma potential, $\boldsymbol{V_{wall}}$ potential (<0) at the wall, $\boldsymbol{j_e}$ electron current from the plasma

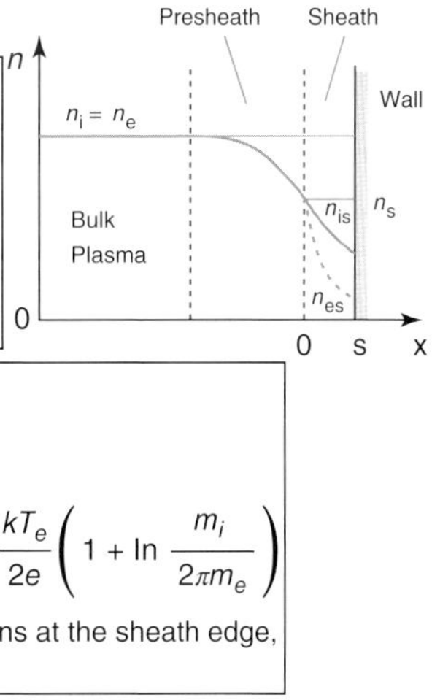

$$u_{is} \geq u_{Bohm} = \sqrt{\frac{kT_e}{m_i}} \qquad n_{is} = n_{es} = n_p e^{\frac{eV_p}{kT_e}} = e^{-1/2} n_p$$

$$j_i = Qen_{is}u_{is} = e^{-1/2} n_p \sqrt{\frac{kT_e}{m_i}} \qquad V_{float} = V_{plasma} - V_{wall} = \frac{kT_e}{2e}\left(1 + \ln\frac{m_i}{2\pi m_e}\right)$$

$\boldsymbol{j_i}$ ion current in the sheath, $\boldsymbol{Q}$ average ion charge state, $\boldsymbol{u_{is}}$ velocity of ions at the sheath edge, $\boldsymbol{n_{is}}$ ion density in the sheath, $\boldsymbol{V_{float}}$ floating potential

$$j_i = \frac{4}{9e}\varepsilon_0 \frac{V_0^{3/2}}{s^2}\sqrt{\frac{2e}{m_i}} \qquad s_{Child} = \frac{\sqrt{2}}{3}\lambda_D\left(\frac{2eV_0}{kT_e}\right)^{3/4} \qquad \lambda_D = \sqrt{\frac{\varepsilon_0 kT_e}{n_i e^2}}$$

$\boldsymbol{j_i}$ ion current in the sheath in case of a negative sheath potential $\boldsymbol{V_0}$ (Child law for space chargelimited ion current), $\boldsymbol{l_D}$ electron Debye length, $\boldsymbol{s_{Child}}$ sheath thickness

Figure 6.6 Selected equations to be applied in the analysis of the electron/ion densities and currents in the sheath region for a floating and a biased wall [16, 18, 35].

is considered (quasi-)neutral, any time a plasma meets a solid (the chamber wall, the electrode, the substrate, or a probe), a boundary layer or sheath develops [16, 35]. If the surface is floating, the net current must be zero and the space charge will develop in order to balance the flow of electrons and ions through the sheath. The velocity of ions at the sheath edge u_{is} obeys the Bohm sheath criterion and results higher than the random (thermal) ion velocity. From $j_e = j_i$, the potential at the wall V_{wall} can be calculated, and by knowing the plasma potential V_p ($=kT_e/2e$), the floating potential V_{float} is determined. The ions reaching the wall will therefore gain an energy equal to eV_{float}. In case a negative bias V_0 is applied to the wall (e.g., an externally biased substrate), the Child law for space-charge limited ion current applies. The sheath becomes a Child law sheath with a thickness s_{Child} larger than the Debye length λ_D if $kT_e << eV_0$. The ions will then gain an energy $e(V_p - V_0)$.

The physical contribution of ions (in terms of energy and flux) to film growth, with respect to the chemical process of (radical) polymerization can be quantified in: effective participation to film growth, creation of active sites via surface atom displacement (thereby enhancing the probability for radicals to stick to the surface of the growing film) up to ion damage phenomena, such as bulk atom displacement and etching/sputtering [18]. Such physical contribution to film growth may be attenuated to a limited or major extent if the sheath is not collisionless [36], that is, if the ion-neutral mean free path λ_i (practical formula: $\lambda_i[\text{cm}] = 1/(330\ p[\text{Torr}])$ where p is the pressure in the system) is no longer larger than the sheath thickness. An increase in pressure promotes, therefore, ion-neutral collisions and can nullify any surface activation by ions during film growth.

6.4 Considerations on the Macroscopic Kinetics Approach to Plasma Polymerization

As earlier mentioned in Section 6.1, plasma polymerization processes need to be translated from one setup to another as well as scaled up to industrial level, where often complex diagnostic tools are not available or the plasma chamber is not accessible for inspection. Both cases would virtually require a correlation of specific external process parameters with the plasma polymer properties.

Several studies have aimed to understand the scaling properties of external process parameters: Yasuda proposed to adopt the W/FM parameter, where W is the power, F is the monomer flow rate, and M is its molecular weight, as this scaling parameter allows the comparison among different plasma polymerization processes and among input parameters [37, 38].

More recently, it has been shown through a round-robin study on a model system, that is, plasma polymerization of acrylate monomer carried out on several reactors in different laboratories, that high levels of variability in terms of polymer properties are present, even though (nominally) identical W/F conditions are adopted on the several reactors [39]. The inherent deficiency of the parameter may generate from differences in power coupling efficiency to the systems as well

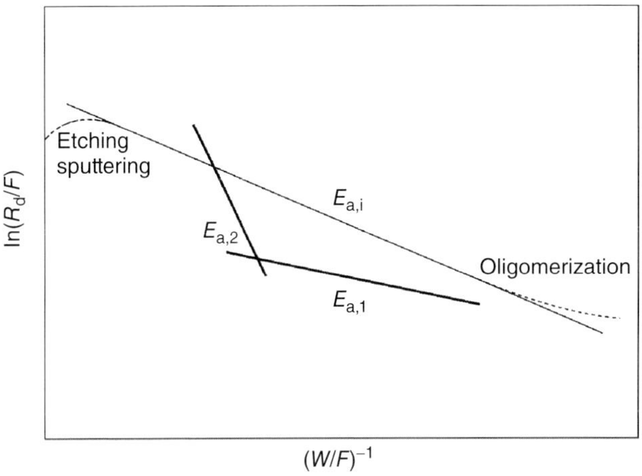

Figure 6.7 General Arrhenius-type plot for plasma polymerization processes, according to the macroscopic approach described in [40]. Ideally, the process would be described by the activation energy $E_{a,i}$. However, deviations from linearity can occur at high/low power input. Also for intermediate power input the experimental data can be described by more than one activation energy, that is, $E_{a,1}$ and $E_{a,2}$, depending on the evolution of the plasma chemistry with the power input.

as different reactor geometries affecting the polymerization process (e.g., in terms of dimensions, injection of gases with respect to the active plasma zone).

The quest for a macroscopic kinetics approach, that is, where the kinetics of the plasma (polymerization) process are controlled by a macroscopic parameter, has led to the definition of an equation where the reaction parameter W/F (J cm^{-3}) controls the mass deposition rate (R_m) per gas flow F, associating an activation energy to the polymerization process (Figure 6.7) [40]

$$\frac{R_m}{F} = Ge^{\left(\frac{-E_a}{(W/F)}\right)} \tag{6.1}$$

Equation 6.1 intuitively states that W/F, that is, the energy invested per particle within the plasma zone, determines R_m. G is a constant dependent on the chosen reactor and process and E_a represents the (overall) specific energy required to initiate chemical reactions leading to film deposition. The equation has been shown to hold for several monomers (in the same plasma reactor) and gas mixtures (after the addition of an appropriate flow factor weighting the contribution of the different monomers to the polymerization). This approach assumes that the polymerization is limited by the activation of the monomer in the plasma phase (i.e., the deposition occurs in a reaction kinetics-limited regime) and not limited instead by mass transport to the substrate. Furthermore, no interaction between the plasma phase and the surface are taken into consideration, that is, a potential role of ion bombardment (see Section 6.3) or the surface itself being a source of active species.

This macroscopic kinetics approach has been widely applied to the plasma polymerization of acrylic acid: several growth regimes develop as function of the energy input, that is, whether the deposition is controlled by radical polymerization, by oligomerization or etching phenomena leading to a decrease in R_m. For example, in the case of radical polymerization, the Arrhenius-type plot showed a linear range under a moderate energy input, with an E_a of 7 eV (per molecule), corresponding to the dissociation energy of the unsaturated bond in acrylic acid. Progressively higher energy inputs have led to the identification of other linear ranges where monomer fragmentation may be favored at other chemical bonds and where the total depletion of acrylic acid and conversion in mixtures of CO_2/C_2H_4 lead to different activation energy values [41].

Since the activation energy concept is linked to the active plasma zone, a modification of Equation 6.1 has been also proposed where a correction for the plasma active zone allows comparing different plasma reactor geometries [41].

Because of the complexity of the plasma polymerization mechanism [42–44] and the need for a universal macroscopic approach, which includes also the role of plasma–surface interactions, initial studies [45] have been recently reported in literature aimed to disclose the elements affecting the pre-factor *G* in the Arrhenius-like equation. *G* is dependent on the energy invested per monomer molecule, the rate for the chemical reaction, the efficiency of power coupling into the gas, the monomer gas flow rate, the molecular masses of the deposition precursors, and of the abstracted radicals/species and the sticking probability.

This macroscopic approach has been an extensive subject of discussion in literature [46]. Although it is generally recognized that the macroscopic scaling is useful in obtaining empirical insights into the kinetics of a plasma process, the limitations of the approach have been highlighted as well. For example, the growth rate R_m is an ill-defined parameter; instead, the growth flux should be adopted. Equation 6.1 cannot be considered to be universal because modifications are necessary in order to adopt the approach also in other plasma systems. Finally, it holds as long as the investigated external parameters affect only the chemical reactions in the plasma phase. Any deviation from the so-called chemical quasi-equilibrium would definitely require the application of diagnostics to learn about the internal parameters of the plasma (electron density, electron energy distribution function, self-bias voltage, ion mean free path), the growth, and the ion fluxes and ion energy in the sheath.

6.5 Polymer Film Characteristics

6.5.1 Plasma Polymer Chemistry: From Precursor Fragmentation to Retention

A characteristic of plasma deposition is the synthesis of thin films showing (opto-chemical) properties other than the pristine precursor structure. One of the most

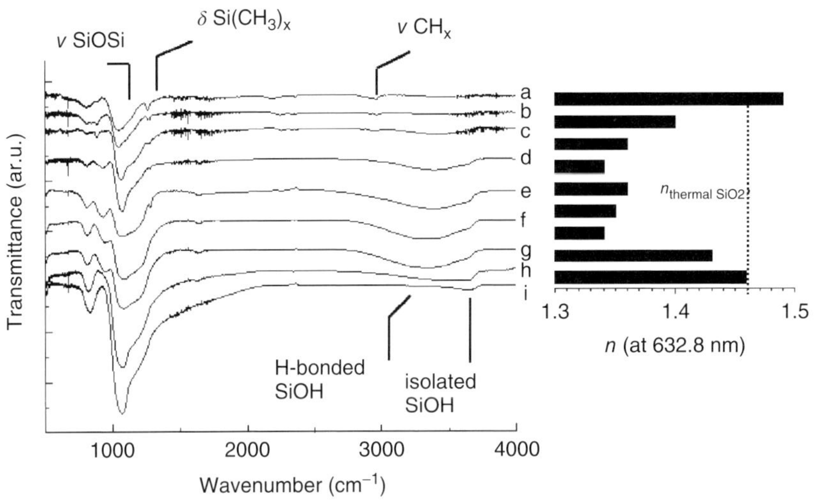

Figure 6.8 FTIR spectra and *in situ* refractive index values of ETP-deposited SiO_2-like films. The experimental conditions are reported in [51].

investigated deposition process is the plasma synthesis of SiO_2-like[3)] layers from organosilicon precursors admixed to oxidant species (e.g., O_2, air, N_2O) [47–50].

In Figure 6.8, the FTIR spectra and refractive index values are reported for ETP-CVD deposited organosilicon- and SiO_2-like films [51]. In this study, a plasma chemistry-controlled process (i.e., occurring in the absence of ion bombardment, $V_{float} = 2-3$ eV because $T_e = 0.3$ eV) has led to the tailoring of the film chemistry from silicone-like to silica-like films. The ETP set-up allows for the independent control of the plasma parameters (e.g., T_e, $\Phi_{Ar^+,e}$), the gas phase chemistry (e.g., O_2 and HMDSO flow rates) and substrate conditioning (e.g., temperature). At constant plasma source parameters (Ar flow rate Φ_{Ar}, arc current I_{arc}), the $\Phi_{Ar^+,e}$ emanating from the arc is constant and equal to 1.25 sccs (i.e., higher than Φ_{HMDSO}), according to Langmuir probe measurements [27]. The activation of the HMDSO molecule occurs via the paths described in Figure 6.4; furthermore the CE reaction constant of Ar^+ with HMDSO (4×10^{-16} m^3 s^{-1}) [27] has been found to be higher than the one with O_2 (6×10^{-17} m^3 s^{-1}) [52]. Therefore, a decrease in carbon content is observed in the spectra a–d and supported by the decrease in refractive index n, according to an increase in O_2 flow rate, which progressively oxidizes either in the plasma phase or at the surface the carbon-containing functionalities. The IR spectrum in trace e shows a higher absorption of the $Si(CH_3)_x$-related band when compared to trace d: although the HMDSO dilution in O_2 is kept constant, the increase in the pressure enhances the growth flux (also witnessed by an increase in film porosity, that is, decrease of the n value and increase in area of

3) With SiO_2-like (often reported also as SiO_x in literature) we refer to a class of silica layer films not strictly being characterized by the same opto-chemical properties of thermal silica, but instead possessing either a residual carbon and/or silanol (SiOH) content.

the SiOH-related bands) and this presumably limits the competing heterogeneous reactions of O_2 with the surface of the growing layer. Only an increase in Ar flow leads to the deposition of carbon-free SiO_2-like layers, because of the enhanced O_2 activation via CE reaction. The porosity of the layer is then decreased, as supported by the increase in n, by an increase in substrate temperature (traces f–i): this latter leads to an enhanced diffusion of active species on the surface of the growing film and promotes condensation reactions of SiOH groups to SiOSi bridges.

Although inorganic thin films, such as SiO_2, are widely acknowledged for their mechanical, optical, and (O_2 and H_2O vapor) barrier properties, it is often desired to retain specific functional groups present in the pristine precursor chemical structure for diverse applications, for example, adhesion improvement, immobilization of biomolecules, and enhanced permeability properties. Ngamou *et al.* [53] have reported on the ETP deposition of perm-selective hybrid silica layers from an organically bridged monomer, 1,2-bis(triethoxysilyl)-ethane (BTESE). The replacement of the hydrolysable Si-O-Si network in the silica matrix by the Si–$(CH_2)_2$-bridge is known to improve the hydrothermal stability of the membrane and, therefore, the retention of the monomer bridge is investigated. The key to the monomer (bridge) retention is the control on $\Phi_{Ar^+,e}$ with respect to Φ_{BTESE}: because of the CE and dissociative recombination processes described in Figure 6.4, it is expected that a limited activation/dissociation of the monomer occurs for $\Phi_{Ar^+,e}/\Phi_{BTESE} \leq 1$. In Figure 6.9, the FTIR spectra and monomer retention percentage are presented as function of Φ_{BTESE} and confirm that the largest bridge retention is achieved for $\Phi_{Ar^+,e}/\Phi_{BTESE} \leq 1$.

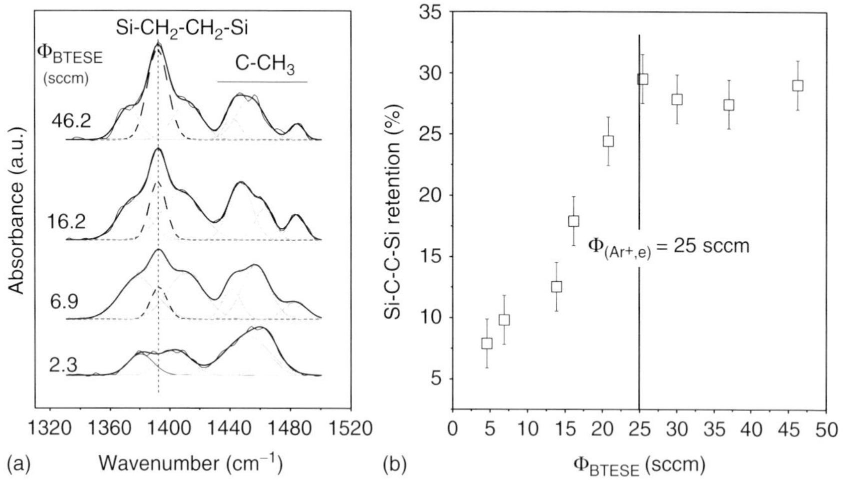

Figure 6.9 (a) FTIR absorption spectra of plasma-polymerized BTESE layers. (b) Retention of the Si-CH_2-CH_2-Si bridge quantified by the absorption area of the band at 1391 cm^{-1} as function of Φ_{BTESE}. The $\Phi_{Ar^+,e}$ is also reported [53].

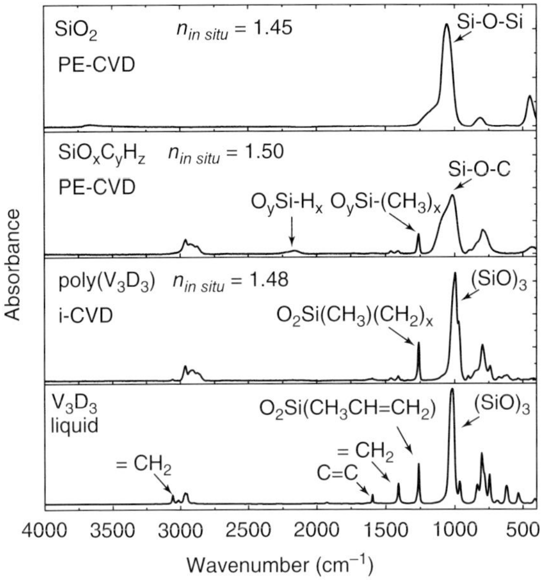

Figure 6.10 FT-IR spectra of the i-CVD poly(V_3D_3), PE-CVD $SiO_xC_yH_z$, and PE-CVD SiO_2 layers: the absorption bands are identified. The spectrum of the liquid phase V_3D_3 monomer is included. The *in situ* refractive index values are also reported [56].

In addition to the example of monomer retention in remote plasmas, selected plasma processes in direct plasmas, also aimed to retain specific functionalities are present in literature. An intuitive approach is to limit the available power density per molecule in order to control/limit the chemical bond dissociation in the monomer [54, 55]. In Ref. [54] the control on the power density allows to retain the surface concentration of carboxylic groups up to 10^{-6} mol cm^{-2} in plasma polymers deposited from ethylene/acrylic acid mixtures. Another approach is to pulse the plasma power according to the duty cycle parameter (DC, i.e., $t_{on}/(t_{on}+t_{off})$) where t_{on} and t_{off} are the intervals during which the plasma is on and off, respectively.

The total retention of the monomer structure can be instead achieved with initiated-CVD, briefly addressed in Section 6.1. A comparison between plasma deposited and i-CVD polymers from the same monomer 1,3,5-trivinyl-1,3,5-trimethyl cyclotrisiloxane (V_3D_3) is presented in Figure 6.10, together with a PE-CVD SiO_2-like layer and the liquid monomer spectra. The i-CVD technique is able to retain the entire monomer structure while providing a monomer-to-polymer conversion degree up to 87%, as pointed out from the massive decrease of the C=C bond-related absorption band (at 3017 and 3058 cm^{-1}) [56]. Several examples are presented from the group of Gleason for i-CVD deposited acrylate- and organosilicon-polymers [57].

6.5.2
Densification of the Film Micro-structure

As earlier observed in Figure 6.8, an approach toward plasma deposited (polymer) layer densification is to increase the temperature [51, 58]. Often, low-temperature plasma processes are needed, because polymer substrates are of interest, for example, for (gas, moisture) barrier applications. An alternative route is the application of an external bias voltage to the substrate, which aims to provide (and control) the energy of the bombarding ions and/or increase their flux toward their substrate. In Section 6.3, the roles of ion (bombardment) toward the surface of the growing film have been reported. In the present section two experimental routes toward ion-assisted remote plasma deposition of SiO_2-like layers are discussed. The influence of ions during film growth in terms of ion-to-growth flux and ion energy, $\Gamma_{ion}/\Gamma_{growth}$ and E_{ion}, [59–61] on the microstructure of the SiO_2-layers is investigated. Figure 6.11a shows the ETP setup, coupled with an rf bias system and a pulse-shaped (p-s) system [62]. Although simple in its implementation, the rf bias solution induces the ignition of an additional plasma in front of the substrate, that is, ions are not only accelerated but also generated. Furthermore, because the ion sheath transit time is usually shorter than the rf period and ions respond to the instantaneous sheath voltage, the final energy of

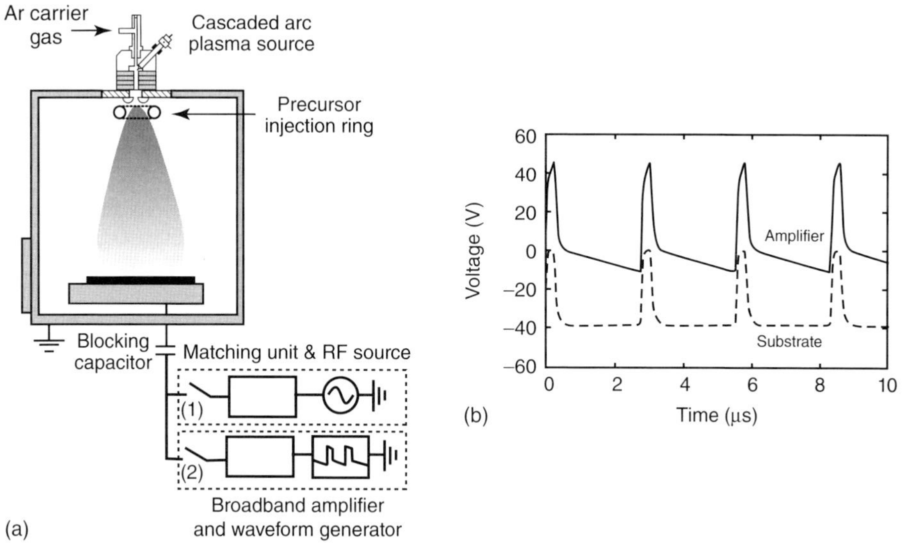

Figure 6.11 (a) Schematics of the ETP setup coupled with an rf and a pulse-shaped substrate biasing system. In the p-s biasing mode the output of the amplifier is coupled to the substrate through a ceramic capacitor. (b) Potential for the amplifier and the substrate: the one on the amplifier output decreases linearly during two pulses in order to keep the potential on the substrate constant, since the capacitor is charged by the positive ion current. From the potential difference, the potential on the coupling capacitor is determined and the ion current is calculated as $I_p = C_c\ (dV/dt)$, where C_c is the capacitance of the capacitor.

the ions depends on the phase of the rf cycle when they enter the sheath and this leads to the development of a bi-modal ion energy distribution function (IED) [63]. This would eventually subject the growing layers to bombarding ions having either too low energy to induce densification or too high energy, inducing material damage. Instead, the p-s bias substrate bias developed by Wang and Wendt [64] leads to a narrow, almost mono-energetic IED because of the mostly constant substrate potential that accelerates the ions toward the substrate. The waveform applied to the substrate consists of a negative voltage peak combined with a short voltage spike toward floating potential (Figure 6.11b). The ion energy is tunable by varying the applied bias voltage and it is possible to derive the ion flux to the substrate directly from the measured, time-dependent potentials. Also, because it operates at a much lower frequency than rf (i.e., 1–215 kHz), there is virtually no extra plasma ignition, that is, the ion flux is constant.

The effect of the ion bombardment during SiO_2 growth, as monitored by the change in the refractive index of the layer (by means of spectroscopic ellipsometry [65]) as well as in the residual porosity (as quantified by ellipsometric porosimetry[4] [66, 67]), is reported in Figure 6.12. Both biasing techniques lead to film densification (where the refractive index of thermal silica-1.46-is taken as reference). However, the rf bias induces densification under a higher G_{ion}/G_{growth} than the p-s bias (Figure 6.12c) [68] because of the ignition of a plasma in front of the substrate. Interestingly, the $\Gamma_{ion}/\Gamma_{growth}$ value of 0.37 for the p-s bias is in agreement with simulation and experimental studies on SiO_2 densification reported in literature [59, 60]. Although the critical energy per SiO_2 unit[5] $E_{u,c}$ is found to be in the same range when operating with the two bias techniques (Figure 6.12c), it can be concluded from Figure 6.12b that the $\Gamma_{ion}/\Gamma_{growth}$ and E_i parameters are not interchangeable in terms of densification: while a high $\Gamma_{ion}/\Gamma_{growth}$ promotes a fast removal of meso-porosity (in the range 0–15 V), higher energy values are necessary to remove residual micro-porosity. In the case of the p-s bias, the decrease of meso- and micro-porosity is gradual and occurs simultaneously because of the limited $\Gamma_{ion}/\Gamma_{growth}$ value.

A similar study is presented by Hegemann *et al.* [69] where plasma polymers are deposited from C_2H_4-, CO_2/C_2H_4-, and NH_3/C_2H_4-fed mixtures in a

4) Ellipsometric porosimetry is a technique combining spectroscopic ellipsometry with adsorption/desorption studies, with the purpose of quantifying the fraction of open porosity, that is, accessible to the ambient. In these studies, the uptake of a probe molecule within the open porosity of the layer is followed *in situ* by monitoring the changes in the refractive index which the layer undergoes. From the shape of the isotherm it is possible to discern between micro-(<2 nm in pore diameter) and meso-(>2 nm in pore diameter) porosity.

5) The parameters in Figure 6.12c have been calculated as follows: $\Gamma_{growth} = (\rho_{SiO_2} \cdot N_A \cdot R_d)/M_{SiO_2}$ where ρ_{SiO_2} is the layer density determined by Rutherford Back-Scattering Spectroscopy, N_A is the Avogadro's number and R_d the deposition rate determined by *in situ* Spectroscopic Ellipsometry. Γ_{ion}, in the case of the p-s bias, is calculated as $I_{ion}/(e{\cdot}A)$ where I_{ion} is directly measured from the slope of the potential of the capacitor, e is the electron charge and A the effective biased area of the substrate. In the case of the rf bias, Γ_{ion} is equal to $P_{rf}/(e{\cdot}V_{dc}{\cdot}A)$ with P_{rf} the rf power provided to the system and V_{dc} the measured potential at the substrate.

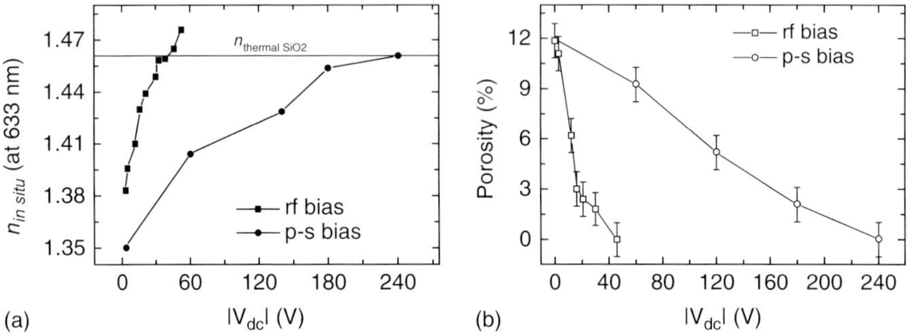

(c)

Parameter	rf bias	p-s bias
$(\Gamma_{ion}/\Gamma_{growth})_c$	2.7	0.37
$E_{i,c}$ (eV)	40	240
$E_{u,c}$ (eV)	108	89

Figure 6.12 Effect of the rf and p-s voltage on SiO_2-like layers: (a) *in situ* refractive index, (b) porosity relative content as determined by means of ellipsometric porosimetry, and (c) critical values of the densification parameters [68].

symmetric, capacitively coupled rf reactor. An energy per deposited carbon atom is found in the range of 5–7 eV, above which highly cross-linked polymers are deposited. When the energy overcomes 25 eV, ion-assisted etching processes are observed.

When referring to the i-CVD case, Aresta *et al.* [56] reported on the valuable application of *in situ* spectroscopic ellipsometry during the growth of organosilicon polymers. Upon chamber degassing, a thickness loss is observed, which is attributed to the presence of unreacted monomer units and oligomers out-diffusing from the polymer bulk. In Figure 6.13a, the thickness loss (Δt) is reported as function of the substrate temperature (i.e., as function of the P_M/P_{sat}[6)] parameter) together with the corresponding adsorbed monomer thickness prior to deposition. As P_M/P_{sat} decreases, the thickness loss becomes less pronounced and then negligible at a value of P_M/P_{sat} below 0.1. This range defines the i-CVD process window under which stable polymers are deposited. The *in situ* and *ex situ* refractive index values have been also followed as function of the substrate temperature (Figure 6.13b). Upon thickness loss, the *n* value after chamber degassing is found to increase with respect to the as-deposited value due to a collapse of the layer thickness upon out-diffusion of the unreacted monomer. As the process window is approached, that is, the thickness loss becomes negligible, the *in situ* refractive index increases pointing out a densification of

6) The i-CVD monomer surface concentration can be related to a non-dimensional parameter given by the ratio of the monomer partial pressure (P_M) and its saturated vapor pressure (P_{sat}).

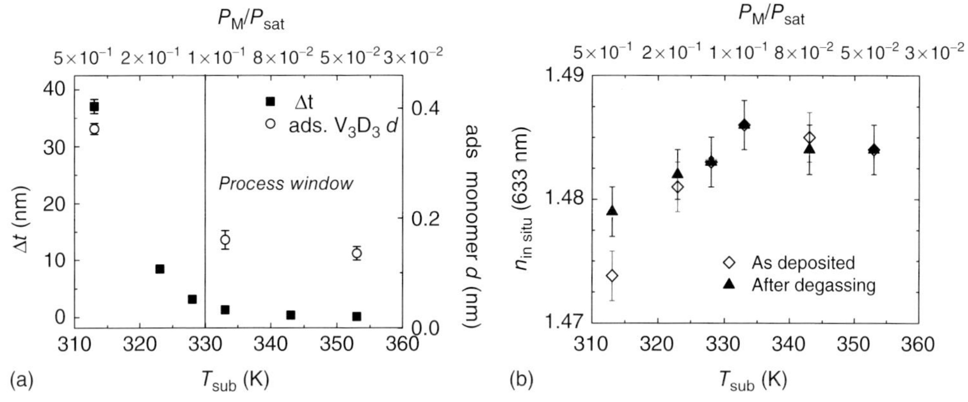

Figure 6.13 V_3D_3 i-CVD deposition: (a) absolute thickness loss and adsorbed monomer thickness as function of the substrate temperature and (b) as-deposited and after degassing refractive index values as function of the substrate temperature [56].

the layer and stays constant upon chamber degassing. Gleason *et al.* reported on the mechanistic study of the deposition of poly-acrylates via i-CVD. Studies related to the densification/cross-linking of the i-CVD polymers are also carried out with more conventional diagnostics, such as gel permeation chromatography [23].

6.5.3
Plasma Polymer Topography

Surface roughness controls many important physical and chemical properties of thin films. Depending on the application, either its minimization or maximization is desired, as for example, in the case of hydrophobicity and hydrophilicity. One of the most known examples is the plasma deposition of fluorocarbon polymers showing a range of morphologies as function of the plasma parameters [70].

The development of morphology and roughness during deposition of thin films is a direct result of the growth processes on the substrate surface. As the atoms are deposited on a surface, they do not arrive at the same time uniformly across the surface. This random fluctuation, which is inherent to the deposition process, may create a surface growth front roughness. This latter competes with surface smoothing processes, such as surface diffusion, to form a rough morphology if the experiment is performed at either a sufficiently low temperature and/or at high growth rate. In addition, the growth front roughness can also be enhanced by processes such as geometrical shadowing and step-edge barriers [71].

The topography of a thin film can be described in terms of standard deviation of the surface heights, referred as the interface width or root-mean-square (rms) roughness. The characteristic behavior of the interface width forms the basis of the dynamic scaling theory (DST). The dynamic scaling hypothesis was formulated to provide a quantitative methodology for describing the self-affine surfaces. A self-affine surface is one that is invariant under anisotropic transformation, that is, when the structure is rescaled with different scaling factors in lateral and vertical directions, the new surface is statistically similar to the original surface. From the rms surface roughness development, $w(L,t)$, the growth exponent β can be determined according to

$$w(L,t) \sim t^{\beta} \tag{6.2}$$

where L is the system size and t is the time. Growth exponent $\beta = 0$ depicts a perfect epitaxial growth, $\beta = 0.25$ indicates that surface diffusion governs the roughness evolution, and $\beta = 0.5$ denotes a random process. β values larger than 0.5 signifies anomalous scaling due to the nonlocal effects such as shadowing, which produce a rapid increase in surface roughness during film growth [71, 72]. This power-law increase in roughness with time or thickness can be observed until a saturation regime is reached where the rms roughness is constant after a certain crossover time t_x. The values of w_{sat} and t_x depend on the system size L observed according to

$$w_{\text{sat}}(L) \sim L^{\alpha} \tag{6.3}$$

$$t_x \sim L^{z} \tag{6.4}$$

where α is called the roughness exponent and z is the dynamic exponent. For the systems where the growth follows the principles of dynamic scaling [73] it can be shown that

$$z = \frac{\alpha}{\beta} \tag{6.5}$$

Such analysis allows to identify the different surface processes that control the thin film growth (diffusion, condensation, surface reactivity, lateral growth, shadowing effects, etc.).

Although thin film surfaces of different nature have been studied extensively for scaling concepts [74, 75], PE-CVD layers have received little attention within the concepts of DST. In Ref. [76] we have therefore applied DST to understand a rather unique result already presented in Figure 6.2b. The atmospheric pressure (AP) DBD SiO_2 layer roughness has been found to remain close to that of the pristine polymer substrate, no matter which organosilicon precursor has been used or whether the discharge was operating in pure N_2 or air. This indicates that the factor which governs such an exceptional phenomena is more of a physical than of a chemical origin occurring in the uniform glow-like discharge regime used in [15]. The lack of roughness evolution is rather unique because low-pressure PE-CVD layers are generally characterized by an increase in surface roughness [11]. The DST analysis has shown that α was close to 1 for the whole thickness series and polymer substrates, suggesting a self-similar growth of SiO_2 on polymers. A deeper analysis reported in [76] shows that $\beta = 0$, while a high dynamic exponent ($z > 4$) is observed, in contradiction to Equation 6.5. A small increase in roughness and extreme low β values are typical of molecular beam epitaxial growth systems. Similarly, during the growth of hydrogenated amorphous carbon in electron cyclotron wave resonance and amorphous Si films in magnetron sputtering systems [77], a value of β equal to 0 is reported. This ultra-smoothness is attributed to the downhill mass transport process leading to surface leveling, induced by the negative substrate biasing. However, in all these cases, a very low value (0–0.1) of α is observed, in disagreement with our results. This clearly indicates that we deviate from other studies reported so far in literature. A value of β equal to 0 suggests that other non-local factors in addition to diffusion contribute to smooth films. A layer-by-layer growth [78] aided by physical means could be hypothesized, because the smoothing phenomena are found to be independent of precursor or reactive gas plasma chemistry. The deposition may be promoted by the surface charge distribution of electrons, which acts as a specific site for mild fragmented or activated film precursors. Another hypothesis is that the atmospheric pressure dielectric barrier discharge (AP-DBD) layers grow similarly to the conformal and layer-by-layer like mode of atomic layer deposited (ALD) oxides because the scaling exponents show strong similarities. Similar observations were drawn also in the case of organosilicon-like layers deposited by AP PE-CVD [79].

This analysis, reporting on the uniqueness of SiO_2 (and organosilicon) film growth by AP-DBD in terms of scaling exponents, together with the conclusion that the film growth does not fall in any of the universality classes reported so far represents just one of the several examples where a clear picture of the plasma–surface interaction and, in general, polymerization mechanism is still lacking. Although the development in the field of plasma diagnostics (devoted to the measurement of internal plasma parameters and identification of chemical reactions) and (*in situ*, real time) thin film diagnostics has strongly supported the understanding of plasma processes, the plasma (polymer and thin film) community has the necessity to develop and adopt diagnostic methods for gaining insight into the processes of plasma–surface interaction.

Acknowledgments

The authors acknowledge Prof. M.C.M. van de Sanden (TU/e and DIFFER, the Netherlands) for the valuable discussions on the results presented in Section 6.5. The colleagues at InnoPhysics B.V., FujiFilm Europe B.V., Energy Research Centre of the Netherlands (ECN) and Holst Centre are acknowledged for the cooperation and scientific discussions. M2i (Material to Innovation Institute), DPI (Dutch Polymer Institute), the Marie Curie fellowship (5th Framework EC programme), the Dutch Technology Foundation (STW) and the Netherlands Organization for Scientific Research (NOW)-VENI and Aspasia programs are acknowledged for the financial support.

References

1. Milella, A., Di Mundo, R., Palumbo, F., Favia, P., Fracassi, F., and d'Agostino, R. (2009) *Plasma Processes Polym.*, **6** (6-7), 460.
2. Ruiz, A., Valsesia, A., Ceccone, G., Gilliland, D., Colpo, P., and Rossi, F. (2007) *Langmuir*, **23** (26), 12984.
3. Chase, J.E. and Boerio, F.J. (2003) *J. Vac. Sci. Technol., A*, **21** (3), 607.
4. Ardelean, H., Petit, S., Laurens, P., Marcus, P., and Arefi-Khonsari, F. (2005) *Appl. Surf. Sci.*, **243**, 304.
5. Favia, P., Sardella, E., Lopez, L.C., Laera, S., Milella, A., Pistillo, B., Intranuovo, F., Nardulli, M., Gristina, R., and d'Agostino, R. (2008) *Plasma Assisted Decontamination of Biological and Chemical Agents*, NATO Science for Peace and Security Series A-Chemistry and Biology, Springer, Dordrecht.
6. Ratner, B.D. and Bryant, S.J. (2004) *Annu. Rev. Biomed. Eng.*, **6**, 41.
7. Girard-Lauriault, P.L., Mwale, F., Iordanova, M., Demers, C., Desjardins, P., and Wertheimer, M.R. (2005) *Plasma Processes Polym.*, **2** (3), 263.
8. Biederman, H. (2000) *Vacuum*, **59**, 594.
9. George, S.M., Yoon, B., and Dameron, A.A. (2009) *Acc. Chem. Res.*, **42** (4), 498.
10. Schalken, J.R.G., Creatore, M., Verhoeven, P., and Stevens, A. (2015) *Nanosci. Nanotechnol. Lett.*, **7**, 62.
11. Kylián, O., Choukourov, A., and Biederman, H. (2013) *Thin Solid Films*, **548**, 1.
12. Mariotti, D., and Sankaran, R. M. (2011) *J. Phys. D: Appl. Phys.*, **44**, 174023.
13. Levchenko, I., and Ostrikov, K., (2007) *J. Phys. D: Appl. Phys.*, **40**, 2308.
14. Wuu, D.S., Lo, W.C., Chiang, C.C., Lin, H.B., Chang, L.S., Horng, R.H., Huang, C.L., and Gao, Y.J. (2005) *Surf. Coat. Tech.*, **197** (2–3), 253.
15. Premkumar, P.A., Starostin, S.A., Creatore, M., de Vries, H., Paffen, R.M.J., Koenraad, P.M., and van de Sanden, M.C.M. (2010) *Plasma Proces. Polym.*, **7**, 635.
16. Lieberman, M.A. and Lichtenberg, A.J. (eds) (1994) *Principles of Plasma Discharges for Materials Processing*, Wiley Interscience, New York.
17. Franklin, R.N. (ed.) (1977) *Plasma Phenomena in Gas Discharges*, Oxford University Press, Oxford.
18. Anders, A. (ed.) (2000) *Handbook of Plasma Immersion Ion Implantation and Deposition*, John Wiley & Sons, Inc., New York.
19. Bunshah, R.F., McGuire, G.E., and Rossnagel, S.M. (eds) (1994) *Handbook of Deposition Technologies for Films and Coatings*, Elsevier, Oxford.
20. d'Agostino, R., Favia, P., Oehr, C., and Wertheimer, M.R. (eds) (2005) *Plasma Processes and Polymers*, Wiley-VCH Verlag GmbH & Co. KGaA, Weinheim.
21. d'Agostino, R., Favia, P., and Fracassi, F. (eds) (1997) *Plasma Processing of Polymers*, Springer.
22. Massines, F., Sarra-Bournet, C., Fanelli, F., Naudé, N., and Gherardi, N. (2012) *Plasma Processes Polym.*, **9**, 1041.
23. Lau, K.K.S. and Gleason, K.K. (2006) *Macromolecules*, **39** (10), 3688.
24. Lau, K.K.S. and Gleason, K.K. (2006) *Macromolecules*, **39** (10), 3695.
25. van de Sanden, M.C.M., de Regt, J.M., and Schram, D.C. (1994) *Plasma Sources Sci. Technol.*, **3**, 501.
26. van de Sanden, M.C.M., Severens, R.J., Kessels, W.M.M., Meulenbroeks, R.F.G., and Schram, D.C. (1998) *J. Appl. Phys.*, **84**, 2426.
27. Creatore, M., Barrell, Y., Benedikt, J., and van de Sanden, M.C.M. (2006) *Plasma Sources Sci. Technol.*, **15**, 421.
28. Creatore, M., Kessels, W.M.M., Barrell, Y., van de Benedikt, J., and Sanden, M.C.M. (2004) *Mater. Sci. Semicond. Process.*, **7**, 283.
29. Liu, D.P., Martin, I.T., Zhou, J., and Fisher, E.R. (2006) *Pure Appl. Chem.*, **78**, 1187.
30. Favia, P., Creatore, M., Palumbo, F., Colaprico, V., and d'Agostino, R. (2001) *Surf. Coat. Technol.*, **142-144**, 1.
31. Magni, D., Descheneaux, C., Hollenstein, C., Creatore, M., and Fayet, P. (2001) *J. Phys. D Appl. Phys.*, **34** (1), 87.

32. Fanelli, F., d'Agostino, R., and Fracassi, F. (2011) *Plasma Processes Polym.*, **8**, 932.
33. Steele, D.A., Short, R.D., Brown, P., and Mayhew, C. (2011) *Plasma Processes Polym.*, **8** (4), 287.
34. Alexander, M.R., Jones, F.R., and Short, R.D. (1997) *J. Phys. Chem. B*, **101** (18), 3614.
35. Anders, A. (2004) *Surf. Coat. Technol.*, **183**, 301–311.
36. Benilov, M.S. (2009) *Plasma Sources Sci. Technol.*, **18**, 014005.
37. Yasuda, H.K. (2005) *Plasma Processes Polym.*, **2**, 293.
38. Yasuda, H. and Wang, C.R. (1985) *J. Polym. Sci.: Polym. Chem. Ed.*, **23**, 87.
39. Whittle, J.D., Short, R.D., Steele, D.A., Bradley, J.W., Bryant, P.M., Jan, F., Biederman, H., Serov, A.A., Chokurov, A., Hook, A.L., Ciridon, W.A., Ceccone, G., Hegemann, D., Körner, E., and Michelmore, A. (2013) *Plasma Processes Polym.*, **10**, 767.
40. Hegemann, D., Hossain, M.M., Körner, E., and Balazs, D.J. (2007) *Plasma Processes Polym.*, **4**, 229.
41. Hegemann, D., Körner, E., and Guimond, S. (2009) *Plasma Processes Polym.*, **6**, 246.
42. Short, R.D. and Steele, D.A. (2010) *Plasma Processes Polym.*, **7**, 366.
43. Hegemann, D., Körner, E., and Guimond, S. (2010) *Plasma Processes Polym.*, **7**, 371.
44. Friedrich, J. (2011) *Plasma Processes Polym.*, **8**, 783.
45. Hegemann, D. (2013) *J. Phys. D Appl. Phys.*, **46**, 205204.
46. (a) von Keudell, A. and Benedikt, J. (2010) *Plasma Processes Polym.*, **7**, 376; (b) van de Sanden, M.C.M. (2010) *Plasma Processes Polym.*, **7**, 887; (c) Whittle, J.D., Steele, D.A., and Short, R.D. (2012) *Plasma Processes Polym.*, **9**, 840; (d) d'Agostino, R. and Palumbo, F. (2012) *Plasma Processes Polym.*, **9**, 844.
47. Bousquet, A., Goullet, A., Leteinturier, C., Coulon, N., and Granier, A. (2008) *Eur. Phys. J. Appl. Phys.*, **42**, 3.
48. Vallee, C., Goullet, A., Granier, A., van der Lee, A., Durand, J., and Marliere, C. (2000) *J. Non Cryst. Solids*, **272** (2-3), 163.
49. Reuter, R., Rügner, K., Ellerweg, D., de los Arcos, T., von Keudell, A., and Benedikt, J. (2012) *Plasma Processes Polym.*, **9**, 1116.
50. Pulpytel, J., Kumar, V., Peng, P., Micheli, V., Laidani, N., and Arefi-Khonsari, F. (2011) *Plasma Processes Polym.*, **8** (7), 664.
51. Creatore, M., Kilic, M., O'Brien, K., Groenen, R., and van de Sanden, M.C.M. (2003) *Thin Solid Films*, **427**, 137.
52. Gaucherel, P. and Rowe, B. (1977) *Int. J. Mass Spectrom. Ion Phys.*, **25**, 211.
53. Ngamou, P.H.T., Overbeek, J.P., Kreiter, R., van Veen, H.M., Vente, J.F., Wienk, I.M., Cuperus, P.F., and Creatore, M. (2013) *J. Mater. Chem. A*, **1**, 5567.
54. Pistillo, B.R., Perrotta, A., Gristina, R., Ceccone, G., Nardulli, M., d'Agostino, R., and Favia, P. (2011) *Surf. Coat. Technol.*, **205**, S534.
55. Palumbo, F., Favia, P., Vulpio, M., and d'Agostino, R. (2001) *Plasmas Polym.*, **6** (3), 163.
56. Aresta, G., Palmans, J., van de Sanden, M.C.M., and Creatore, M. (2012) *J. Vac. Sci. Technol., A*, **30** (4), 041503.
57. Tenhaeff, W.E. and Gleason, K.K. (2008) *Adv. Funct. Mater.*, **18** (7), 979.
58. Creatore, M., Cigal, J.-C., Kroensen, G.M.W., and van de Sanden, M.C.M. (2005) *Thin Solid Films*, **484**, 104.
59. Lefèvre, A., Lewis, L.J., Martinu, L., and Wertheimer, M.R. (2001) *Phys. Rev. B*, **64**, 115429.
60. Martinu, L., Klemberg-Sapieha, J.E., Küttel, O.M., Raveh, A., and Wertheimer, M.R. (1994) *J. Vac. Sci. Technol., A*, **12**, 1360.
61. Petrov, I., Adibi, F., Greene, J.E., Hultman, L., and Sundgren, J.–.E. (1993) *Appl. Phys. Lett.*, **63**, 36.
62. Kudlacek, P., Rumphorst, R.F., and van de Sanden, M.C.M. (2009) *J. Appl. Phys.*, **106**, 073303.
63. Kawamura, E., Vahedi, V., Lieberman, M.A., and Birdsall, C.K. (1999) *Plasma Sources Sci. Technol.*, **8**, R45.
64. Wang, S. B. and Wendt, A.E. (2000) *J. Appl. Phys.*, **88**, 643.
65. Fujiwara, H. (ed.) (2007) *Spectroscopic Ellipsometry: Principles and Applications*, John Wiley & Sons, Ltd, Chichester.

66. Baklanov, M.R., Mogilnikov, K.P., Polovinkin, V.G., and Dultsev, F.N. (2000) *J. Vac. Sci. Technol. B*, **18**, 1385.
67. Perrotta, A., van Beekum, E.R.J., Aresta, G., Jagia, A., Keuning, W., van de Sanden, M.C.M., Kessels, W.M.M., and Creatore, M. (2014) *Microporous Mesoporous Mater.*, **188**, 163.
68. Terlinden, N.M., van de Sanden, M.C.M., and Creatore, M. (2008) *Microstructural Characterization of Plasma-Deposited SiO_2-Like Thin Films*, Eindhoven University of Technology.
69. Hegemann, D., Körner, E., Blanchard, N., Drabik, M., and Guimond, S. (2012) *Appl. Phys. Lett.*, **101**, 211603.
70. Milella, A. (2005) Plasma processing of polymers, in *Plasma Processes and Polymers* (eds P. Favia, C. Oehr, and R.M. Wertheimer), Wiley-VCH Verlag GmbH, Weinheim.
71. Pelliccione, M., Karabacak, T., Gaire, C., Wang, G.-C., and Lu, T.-M. (2006) *Phys. Rev. B*, **74**, 125420.
72. Baloniak, T. and von Keudell, A. (2008) *Plasma Processes Polym.*, **5**, 653.
73. Barbasi, A.L. and Stanley, H.E. (1995) *Fractal Concepts in Surface Growth*, Cambridge University Press, Cambridge.
74. Karabacak, T., Zhao, Y.P., Wang, G.C., and Lu, T.M. (2001) *Phys. Rev. B*, **64**, 085323.
75. Borras, A., Gill, A.Y., Barranco, A., Cotrino, J., and Elipe, A.R.G. (2007) *Phys. Rev. B*, **76**, 235303.
76. Premkumar, A.P., Starostin, S.A., de Vries, H., Creatore, M., Koenraad, P.M., MacDonald, W.A., and van de Sanden, M.C.M. (2012) *Plasma Processes Polym.*, **9**, 1194.
77. Cubero, A.R., Gaco, R., and Vazquez, L. (2011) *Appl. Phys. Lett.*, **98**, 011904.
78. Dennler, G., Houdayer, A., Segui, Y., and Wertheimer, M.R. (2001) *J. Vac. Sci. Technol., A*, **19**, 2320.
79. Premkumar, P.A., Starostin, S.A., de Vries, H., Creatore, M., Koenraad, P.M., and van de Sanden, M.C.M. (2013) *Plasma Processes Polym.*, **10**, 313.

7
Fabrication of Organic Interfacial Layers by Molecular Layer Deposition: Present Status and Future Opportunities[1)]

Han Zhou and Stacey F. Bent

7.1
Introduction

Polymeric thin films have attracted a great deal of attention because of their widespread application in many fields, such as microelectronics [2], biomedical materials [3], and energy related devices [4]. As nanotechnology has undergone tremendous development, high quality polymer coatings of nanometer dimensions are increasingly needed to enable miniaturization of a variety of devices. However, achieving a high level of control over thin film properties at the nanoscale is difficult and presents many challenges for conventional polymeric film coating methods such as spin-coating. For example, the miniaturization of semiconductor devices often requires continuous pinhole-free coatings [5]. But pinholes are frequently found in spin-coated films [6]. Furthermore, conformal coatings can be difficult to achieve by conventional methods on substrates with high-aspect-ratio features [7, 8]. In addition to thickness conformality, compositional uniformity is another challenge for the spin-coating method. Traditionally, polymers need to be synthesized in the bulk phase before coating, leading to additional difficulties in obtaining uniform chemical compositions of polymer films at very small thicknesses. To overcome the difficulties associated with spin-coating and other wet processes, a number of dry processes have been developed for depositing polymeric thin films. One such technique, chemical vapor deposition (CVD), has been used to make various polymer films, including poly(*p*-xylene), poly(glycidyl methacrylate), and poly(styrene) [2, 9].

A related dry process known as molecular layer deposition (MLD) can provide exquisite control over film thickness, composition, and conformality at the molecular level. MLD utilizes sequential, self-limiting surface reactions to build up a thin film on a substrate one molecular layer at a time. Figure 7.1 illustrates a generic MLD scheme using sequential and self-limiting surface reactions. A typical MLD cycle includes two precursor doses with a purging step immediately

1) This chapter was reprinted with permission from J. Vac. Sci. Technol. A 31, 040801 (2013), Copyright 2013, American Vacuum Society [1].

CVD Polymers: Fabrication of Organic Surfaces and Devices, First Edition. Edited by Karen K. Gleason.

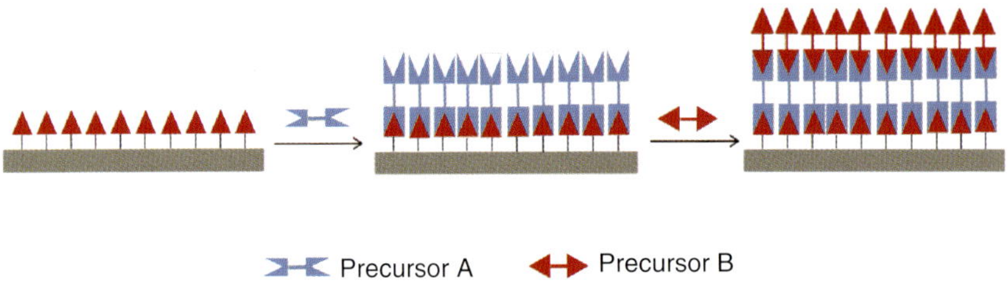

Figure 7.1 Schematic of the MLD process based on self-limiting surface reactions.

following each dose to remove excess precursor molecules from the reaction system. During each dose step, bifunctional precursor molecules react with surface reactive groups via the corresponding linking chemistry, such as amide coupling or urea coupling, add a molecular layer of precursor on the surface, and leave the surface terminated with new reactive sites for further film growth. The self-limiting nature of the surface reactions results in several characteristic features of MLD film growth, such as a linear growth rate and saturation behavior with respect to the precursor doses. Here we illustrate these properties using polyurea MLD films deposited using 1,4-phenylene diisocyanate (PDIC) and ethylenediamine (ED) as an example. Figure 7.2a shows that the film thickness increases linearly with increasing number of MLD cycles. Moreover, the film growth rate saturates with increasing precursor doses as demonstrated in Figure 7.2b,c [10].

MLD has many advantages over other film deposition methods. First, its self-limiting nature ensures excellent coating conformality, even for high-aspect-ratio features. Secondly, its layer-by-layer growth behavior enables fine tuning of chemical composition as it allows for the embedding of desired chemical functionalities at controlled positions in the deposited thin film by using well designed precursors. Hence, MLD can serve as a powerful deposition technique for thin film polymers and thereby address the challenges that nanotechnology brings to polymeric coating.

Before the development of the MLD technique, a number of other dry processes for depositing polymeric thin films were explored, such as initiated chemical vapor deposition (iCVD) [9, 11]. The CVD methods often require free radicals to grow the polymeric film. The free radicals are typically generated by plasma or photons, which may cause the loss of the precursors' functionalities, although good results have been obtained by some radical-based CVD methods. Vapor deposition polymerization (VDP) is a special type of CVD technique using milder growth conditions [12]. During a typical VDP process, a simultaneous flow containing two kinds of bifunctional precursors, for example, dicarboxylic acid dichlorides and diamines, is introduced into the chamber, and polymerization products are deposited onto the substrate. Some polymeric films like polyimide and polyamide have been achieved using this method [13]. However, the co-condensation of the precursor molecules may interfere with the ability to achieve a high degree of control over some film properties, like molecular orientation. Unlike VDP, in which the precursors are delivered simultaneously, in MLD the

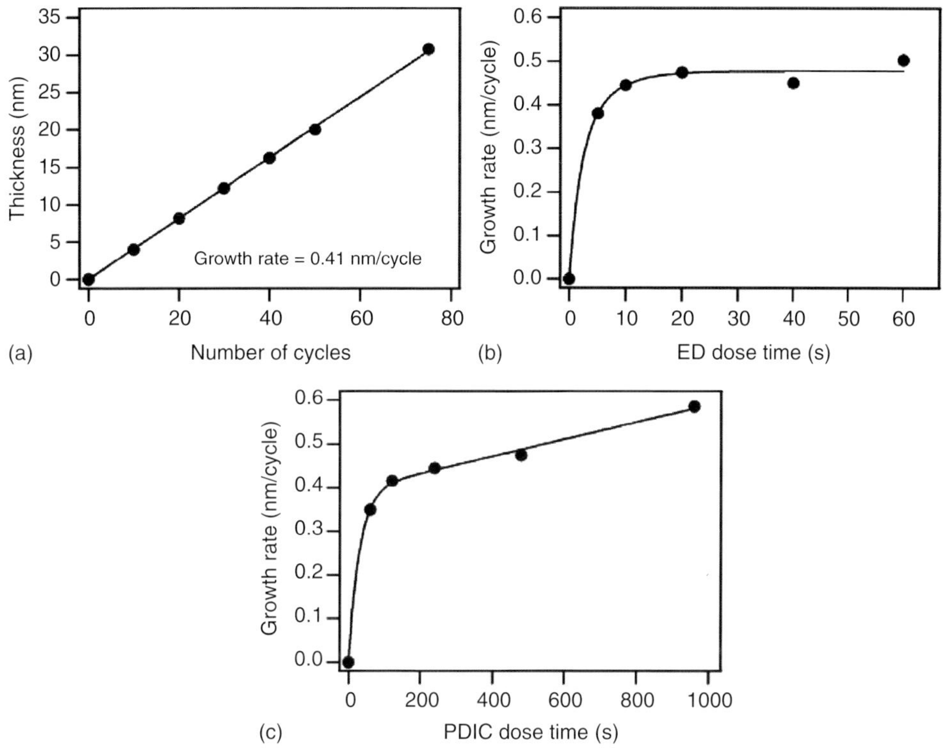

Figure 7.2 (a) Plot of polyurea MLD film thickness as a function of number of cycles. Saturation curves showing growth rate as a function of (b) ED dose time, and (c) PDIC dose time while keeping the dose time of the other reactant constant. The deviation from true saturation in (c) is likely due to a CVD component of the reaction when the purge time was not sufficient at the highest PDIC doses. (Reprinted with permission from Loscutoff et al., ACS Nano 4, 331 (2010). Copyright (2010) American Chemical Society.)

two precursors are introduced into the reactor separately. By this means, more precise control over film growth can be expected using MLD.

In practice, MLD is an analog of atomic layer deposition (ALD), which utilizes sequential, self-limiting reactions to grow inorganic thin films, including many metal oxides and metals. ALD has been successfully used in semiconductor fabrication processes to deposit important inorganic materials, such as high-κ gate oxide and copper diffusion barriers [14], as well as in a number of energy-related devices [15]. From a processing aspect, MLD is very similar to ALD since they both use a sequence of dosing and purging steps. Hence MLD can be integrated smoothly into manufacturing, taking advantage of ALD apparatuses. On the other hand, various thin films have been successfully synthesized using MLD, including pure organic polymeric thin films and novel organic-inorganic hybrid thin films. As a result, it is foreseen that the MLD technique will gain increasing interest and wider application in the near future.

7.2
MLD Coupling Chemistry

7.2.1
Pure Organic MLD

Alternating VDP, in which precursors were introduced in a stepwise fashion, can be viewed as the prototype of MLD [13, 16–18]. More systematic MLD work was developed beginning in the mid-2000s, centered on depositing pure

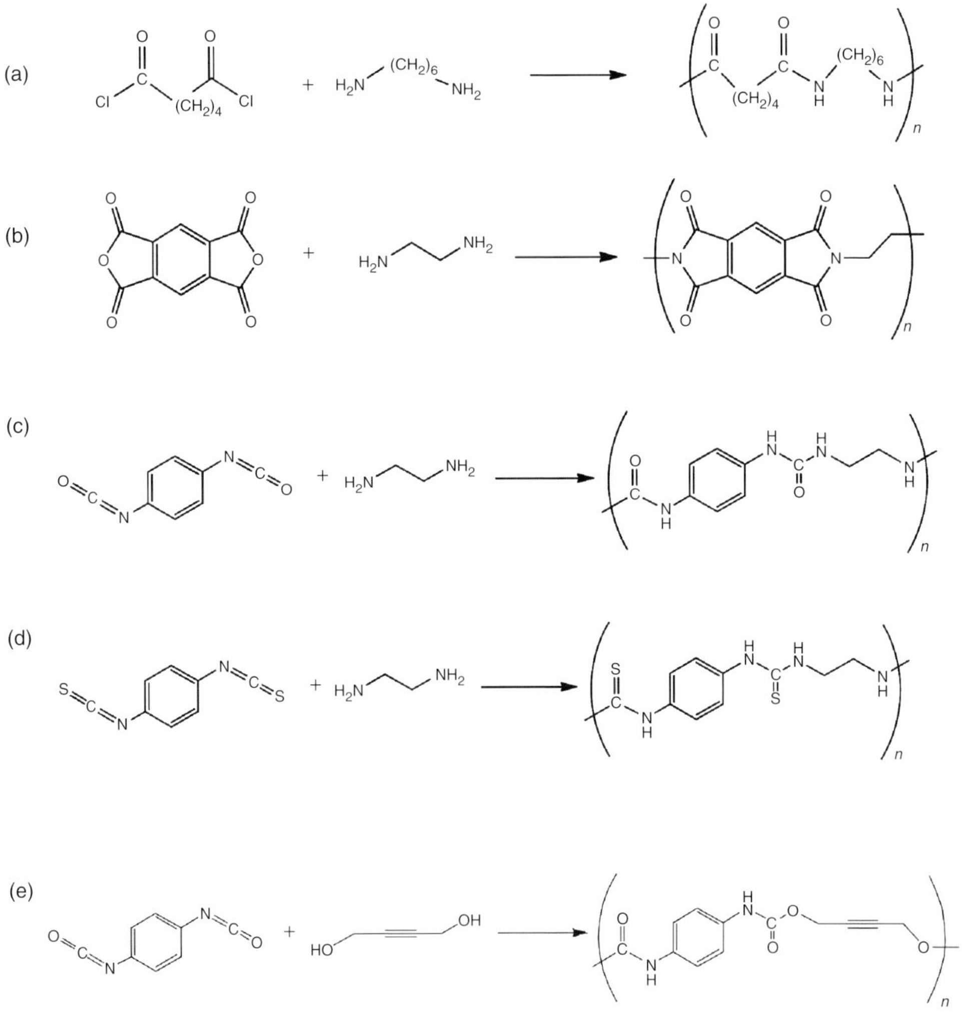

Figure 7.3 MLD reactions for (a) polyamide, (b) polyimide, (c) polyurea, (d) polythiourea, (e) polyurethane, (f) polyazomethine, (g) polyester, (h) aluminum alkoxide, (i) zinc alkoxide, and (j) titanium alkoxide based films.

(f)

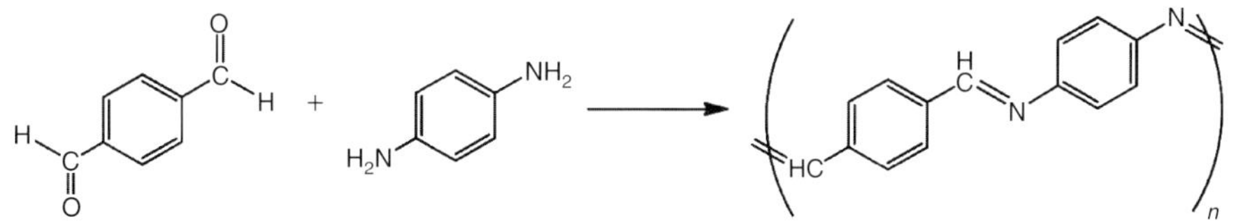

(g)

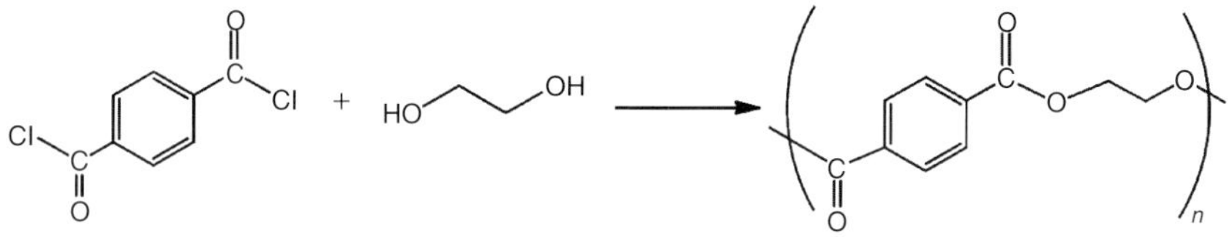

(h)

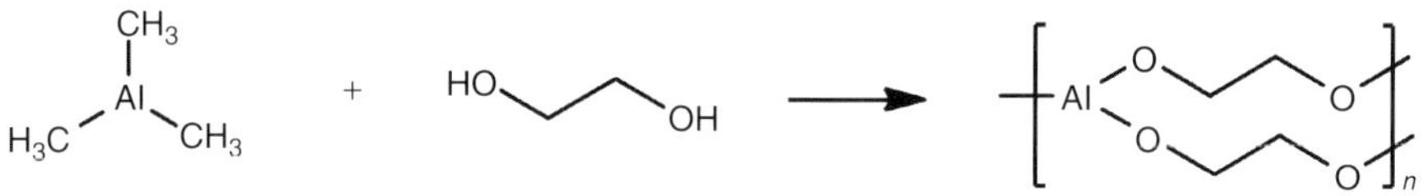

(i)

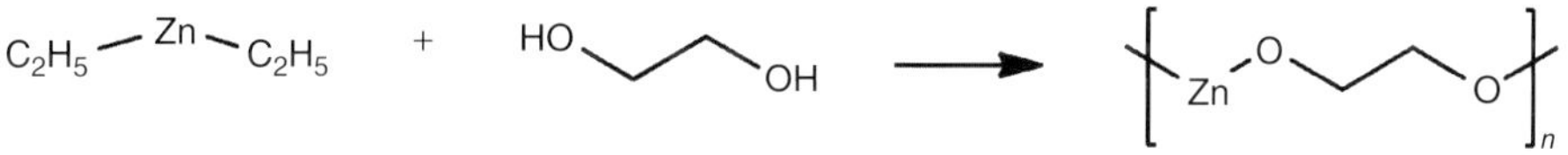

(j)

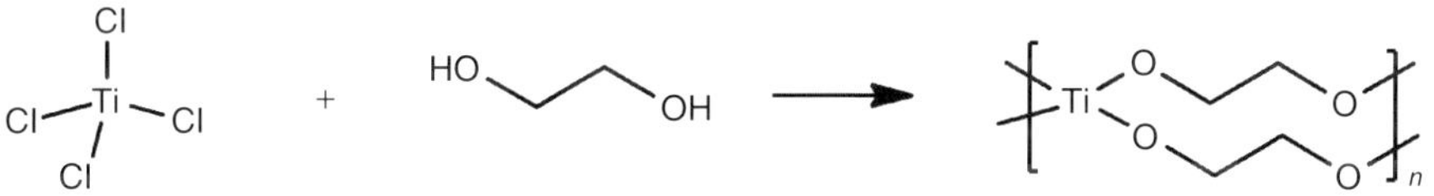

Figure 7.3 (*Continued*)

organic films, including polyurea [10, 19], polyamide [20, 21], and polyimide [22]. Other organic MLD films have also been investigated, like polythiourea [23], polyurethane [24], polyazomethine [25], and polyester [26] films. Each MLD organic film was deposited via the denoted chemical linkage, similar to traditional solution-phase polymer synthesis. Although in this review we will refer to the MLD films by their polymer analog, in many cases the films may be short oligomeric chains bound to the substrate surface.

7.2.1.1 **Polyamide**

Polyamide MLD films were deposited via the amide linking reaction between diacyl chloride and diamine [18, 20, 21, 27]. For example, nylon-66 films were made using adipoyl chloride and 1,6-hexanediamine [21]. As another example, poly(*p*-phenylene terephthalamide) MLD films were made using terephthaloyl chloride and *p*-phenylenediamine (PPDA) [20]. Figure 7.3a shows the MLD coupling reaction used to deposit the nylon-66 polyamide MLD film. As shown in Figure 7.4a,b, the presence of characteristic polyamide IR absorption peaks, namely the amide I mode around 1650 cm^{-1} and amide II mode around 1540 cm^{-1}, confirmed that the MLD films were deposited via amide linkages. Moreover, the increasing peak intensity with the increasing number of MLD cycles demonstrated the growth of the polyamide films. Growth saturation behavior during the precursor pulse was also observed, confirming that self-limiting surface reactions took place during the MLD process.

A growth rate of 17 Å/cycle was measured for nylon-66 MLD, which was very close to the length of the molecular repeating unit [21]. In contrast, the poly(*p*-phenylene terephthalamide) film had a maximum growth rate of only 4.0 Å/cycle, much less than the expected length of a molecular unit [20]. This low growth rate can be attributed to a few factors, including tilting of the polymer chain toward the surface and "dual reaction" of both functional groups of the precursor, causing a capping of surface reactive sites. For both polyamide MLD films, the growth rate decreased with increasing deposition temperature, suggesting that precursor residence time on the surface is a determinant factor. At a high deposition temperature, precursor molecules desorb quickly from the surface before the coupling reaction takes place, leading to a slow growth rate.

7.2.1.2 Polyimide

Polyimide MLD films have been deposited using dicarboxylic anhydride and diamine precursors via imide linkages [16, 22]. Figure 7.3b shows one such MLD system, the chemical reaction between 1,2,3,5-benzenetetracarboxylic anhydride (pyromellitic dianhydride, PMDA) and ethylenediamine (EDA). Putkonen *et al.* deposited a number of polyimide MLD films using PMDA and different diamine precursors, including ethylenediamine (EDA), 1,6-diaminohexane (DAH), 1,4-phenylenediamine (PDA), and 4,4′-oxydianiline (ODA) [22]. According to thickness and Fourier transform infrared spectroscopic (FTIR) measurements on deposited MLD thin films, successful polyimide film deposition was achieved using all four diamine precursors. As shown in Figure 7.5a, characteristic imine C-N stretching absorption peaks around 1380 cm^{-1} were present in the FTIR spectra of deposited films. In addition, the imidization reaction between dicarboxylic anhydride and diamine precursors went to completion, as measured by the absence of any characteristic IR absorption peaks from reaction intermediates like isoimide or amide [22]. According to polyimide film growth curves shown in Figure 7.5b, the polyimide film growth was highly linear with the number of MLD cycles, suggesting a layer-by-layer growth behavior. In addition to the linear growth rate, self-limiting growth was also verified by the observation of the saturation of the film growth with increasing precursor pulsing times. Similar to polyamide MLD films [20, 21], from Figure 7.5c it can be seen that the growth rates of polyimide films also decreased at higher deposition temperature, again implying that the precursor molecule residence time on the growth surface is crucial for the growth rate. The polyimide films grown on a planar substrate had

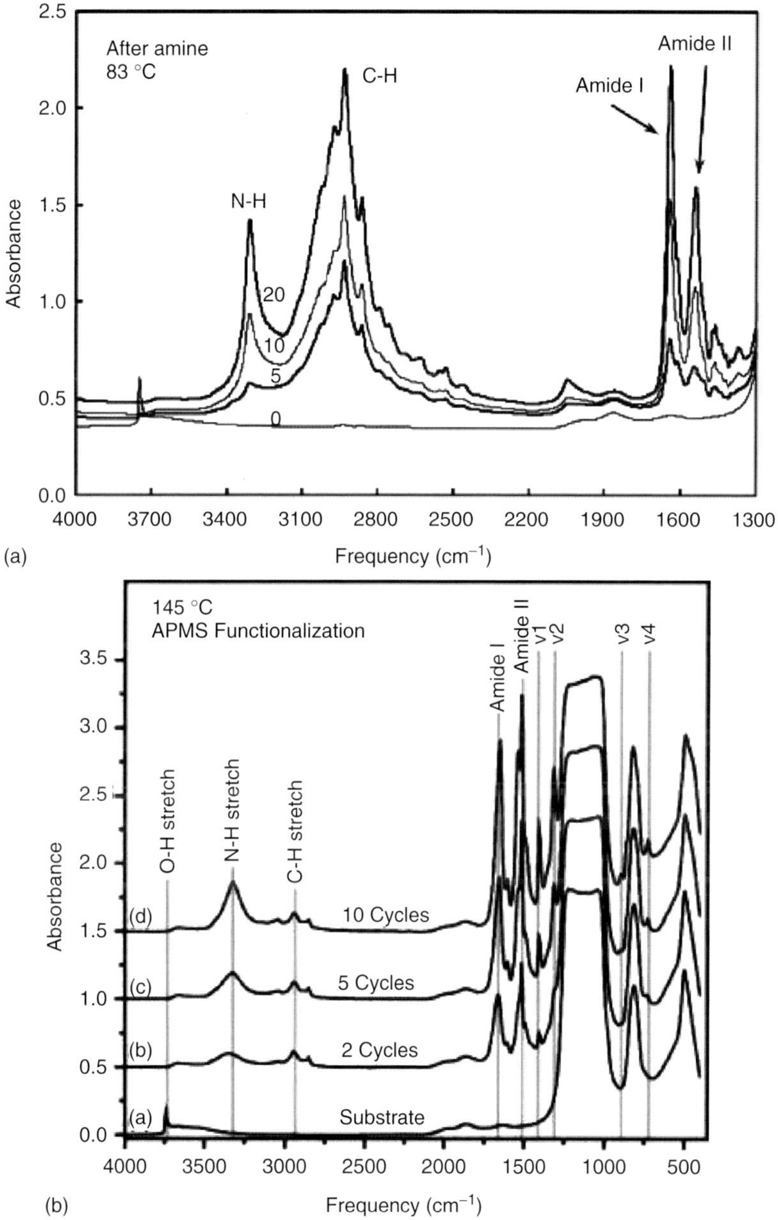

Figure 7.4 (a) FTIR spectra of nylon-66 MLD films after 0, 5, 10, and 20 MLD cycles. (Reprinted with permission from Y. Du and S. M. George, J. Phys. Chem. C 111, 8509 (2007). Copyright (2007) American Chemical Society.) (b) FTIR spectra of poly(*p*-phenylene terephthalamide) MLD films after varying numbers of MLD cycles. (Reprinted with permission from Adamczyk et al., Langmuir 24, 2081 (2008). Copyright (2008) American Chemical Society.)

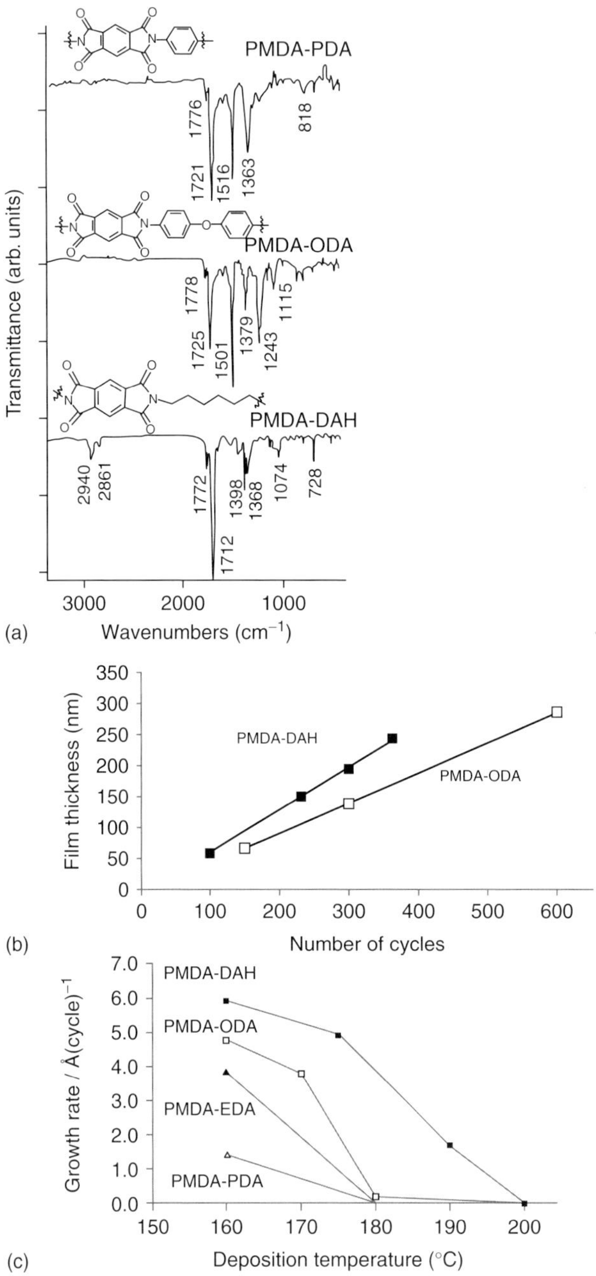

Figure 7.5 (a) FTIR spectra of polyimide MLD films, (b) growth curves of polyimide MLD films, and (c) growth rate of polyimide MLD films as a function of deposition temperature. (Reprinted with permission from Putkonen et al., J. Mater. Chem. 17, 664 (2007). Copyright (2007) Royal Society of Chemistry.)

very smooth surfaces, as the roughness measured by atomic force microscopy (AFM) was less than 1% of the film thickness [22].

7.2.1.3 Polyurea

Polyurea MLD was investigated using diisocyanate and diamine precursors on a silicon substrate [10]. The MLD reaction is shown in Figure 7.3c. Previously, polyurea ultrathin films with up to four molecular layers were first deposited on a germanium surface under ultra-high-vacuum condition [19]. Compared to polyamide, polyimide, or polyazomethine, an advantage of polyurea and polyurethane chemistry is that no by-products are generated in the MLD process and thus no by-products are present to affect subsequent deposition. Using PDIC and ethylenediamine (ED), polyurea thin films were successfully grown and characterized. A linear growth rate of 4.1 Å/cycle was obtained at room temperature, whereas a theoretical chain length for one polyurea repeating unit was calculated to be 12.6 Å [10]. This discrepancy can be explained by a tilted polymer chain orientation and/or by subsaturation coverage for each cycle. Self-limiting growth characteristics were also observed for both PDIC and ED precursor doses. FTIR spectra confirmed the formation of urea linkages by the observation of characteristic IR bands due to amide I and amide II modes, as shown in Figure 7.6a. Hydrogen bonding between adjacent polyurea polymer chains was also evident spectroscopically, since the amide I mode showed a hydrogen bonding-induced red shift compared to the theoretically calculated spectrum. The chemical composition was studied by X-ray photoelectron spectroscopy (XPS) and a C/N/O ratio of 4.9/1.5/1 was measured, while the theoretical ratio was 5/2/1. The good agreement between the experimental and theoretical composition demonstrated the capability of MLD to deposit stoichiometric polymer films. Figure 7.6b shows a transmission electron microscopy (TEM) image of a polyurea MLD thin film deposited on a silicon wafer. It can be seen that the polyurea coating is highly uniform with little thickness variation on the planar substrate.

The MLD process allows for good design flexibility to achieve certain desired properties of the deposited films, for instance, by embedding desired chemical functionalities into the backbone of precursor molecules while keeping the end groups unchanged so that the reactivity toward the counter precursor is retained. For example, a sulfur-containing polyurea MLD film was deposited by using 2,2′-thiobis(ethylamine)ethane (TBEA) in place of ethylenediamine [10]. The incorporation of sulfur was confirmed by the presence of sulfur signal as well as the stoichiometric composition measured by XPS. Moreover, the MLD technique can provide unprecedented control over film composition at a molecular level, which was demonstrated in this study as well [10]. By alternating cycles to form blocks of PDIC/ED and PDIC/TBEA polyurea layers, nanolaminates were formed. Figure 7.6c shows the depth profiles of a nanolaminate with a PDIC/TBEA block sandwiched by PDIC/ED blocks. As expected, sulfur from the PDIC/TBEA block was only detected in the middle portion of the nanolaminate, showcasing the ability of MLD to achieve well controlled film composition in the vertical direction.

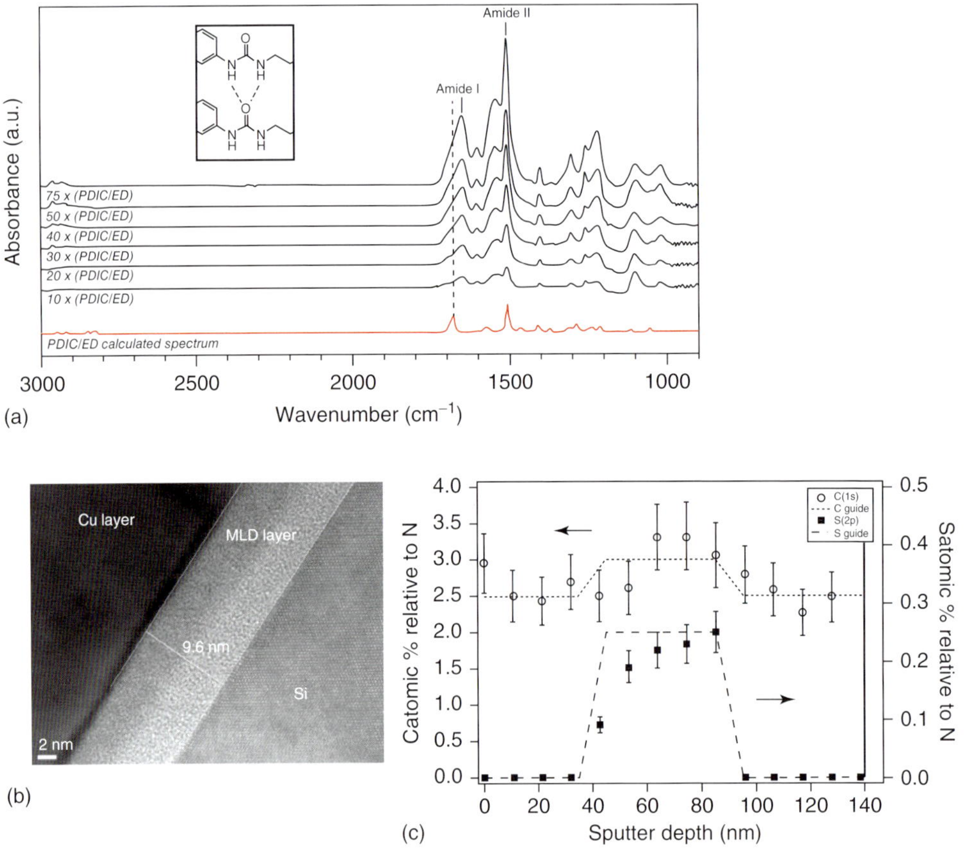

Figure 7.6 (a) FTIR spectra of polyurea films after increasing MLD cycles and a simulated IR spectrum from density functional theory (DFT) calculations. The inset illustrates hydrogen bonding between adjacent polymer chains, consistent with the broadening of the amide I peak. (b) TEM image of polyurea MLD film with copper evaporated on top. (c) Depth profile of a nanolaminate with the PDIC/TBEA block in middle. (Reprinted with permission from Loscutoff et al., ACS Nano 4, 331 (2010). Copyright (2010) American Chemical Society.)

7.2.1.4 **Polythiourea**

Diisothiocyanate and diamine were used to deposit polythiourea MLD films at room temperature [23]. The MLD reaction using phenylene diisothiocyanate (PDITC) and ethylenediamine is shown in Figure 7.3d. The polythiourea film exhibited a constant growth rate of 1.9 Å/cycle and self-saturating growth behavior was observed for both precursors. The IR spectra of deposited films agreed well with expected vibrational bands of polythiourea polymer. Moreover, XPS measurement revealed that the films were deposited with stoichiometric composition [23]. Thus the polythiourea MLD process exhibited all important MLD growth characteristics. The conformality of MLD films was demonstrated using polythiourea-coated silica nanoparticles, as shown by the TEM image in

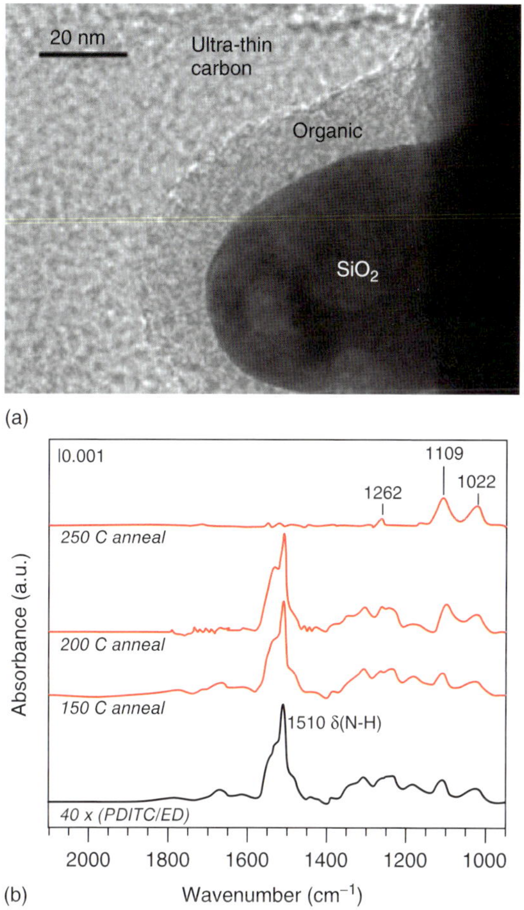

Figure 7.7 (a) TEM image of polythiourea MLD film on a silica nanoparticle on a TEM grid coated with ultrathin carbon and (b) FTIR spectra of polythiourea MLD films after annealing at various temperatures. (Reprinted with permission from Loscutoff et al., Chem. Mater. 22, 5563 (2010). Copyright (2010) American Chemical Society.)

Figure 7.7a. It is evident that the MLD film can form a highly conformal coating over the nanoparticle substrate.

Additionally, the thermal stability of the polythiourea films was examined by a series of annealing experiments [23]. As shown in Figure 7.7b, the film was stable to at least 200 °C under vacuum with a nitrogen flow, with complete film desorption or decomposition observed at 250 °C.

7.2.1.5 **Polyurethane**

Polyurethane MLD was carried out using diisocyanate (1,4-phenylene diisocyanate) and diol (2-butyne-1,4-diol) precursors, with the MLD reaction

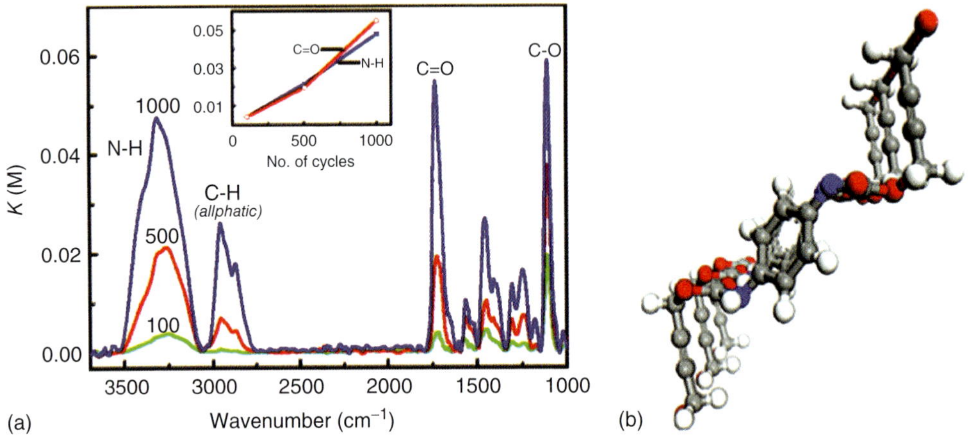

Figure 7.8 (a) FTIR spectra of polyurethane films after 100, 500, and 1000 MLD cycles. The inset shows intensities of C=O and N-H stretching bands. (b) Structure of polyurethane MLD chains from a horizontal view. (Reprinted with permission from Lee et al., Science 301, 818 (2003). Copyright (2003) The American Association for the Advancement of Science.)

illustrated in Figure 7.3e [24]. Film thicknesses were measured by cross-sectional scanning electron microscopy (SEM) and a linear growth rate of approximately 6.5 Å/cycles was determined on silicon or glass substrate at a deposition temperature of 120 °C. The successful urethane linkage formation of the MLD film was verified by FTIR spectra, which revealed the characteristic N-H, C-H, C=O, and C-O stretching bands, with intensities increasing as the number of MLD cycles increased, as shown in Figure 7.8a. In addition, polarized grazing angle IR study indicated that the N-H, C-H, C=O, and C-O bonds were parallel to the substrate surface. By comparing the IR bandwidths of MLD-grown and spin-on polyurethane films, it was found that the MLD films had wider N-H and C=O stretching peaks, implying the formation of hydrogen bonding between these chemical moieties. It was concluded that the polymer chains in the MLD film could achieve a more ordered structure resulting from the hydrogen bonding. Based on these findings, together with theoretical modeling, a molecular structure of the polyurethane MLD chains was proposed as shown in Figure 7.8b.

7.2.1.6 **Polyazomethine**

Using dialdehyde and diamine precursors, polyazomethine MLD films were deposited on gold substrates via imine linkages by Yoshimura *et al.* [25]. The chemical reaction for a conjugated polyazomethine MLD with terephthalaldehyde (TPA) and PPDA precursors is shown in Figure 7.3f. The film growth was characterized by FTIR spectroscopy. The IR bands attributed to benzene ring mode vibrations were detected in sample films after deposition. Moreover, it was found in this study that initial functionalization of the gold substrate to form an amine-terminated self-assembled monolayer (SAM) prior to MLD was crucial to achieving a continuous film growth, which can be explained by a higher density of surface reactive groups resulting from the SAM modification compared to a bare gold surface [25]. We note that a SAM promoting layer had a much less pronounced effect when using an SiO_2 substrate for the polyamide or polyimide films as discussed in previous sections [21, 22], likely because the bare SiO_2 surface already has sufficient surface OH groups to initiate the MLD reaction.

7.2.1.7 Polyester

Ivanova *et al.* [26] recently reported a polyester MLD system with diacyl chloride and diol precursors. More specifically, polyethylene terephthalate thin films were deposited using terephthaloyl chloride and ethylene glycol (EG) at 145 °C [26]. Figure 7.3g displays the MLD reaction. The characteristic aromatic ester C=O vibrational band was observed in the FTIR spectrum of the deposited film, confirming the formation of an ester linkage. The polyester films exhibited a maximum growth rate of 3.3 Å/cycle at 145 °C [26]. The growth rate decreased quickly as deposition temperature increased, which is consistent with other MLD systems. Interestingly, in this study it was shown that the initial surface functionalization of the silicon substrate had a strong facilitating effect on film growth. While the growth rate obtained on a silicon substrate pretreated with an amine-terminated SAM was 3.3 Å/cycle, the MLD films directly grown on a silicon substrate showed a growth rate of only 0.5 Å/cycle [26].

7.2.2 Organic–Inorganic Hybrid MLD

ALD has been widely demonstrated as a powerful technique for the deposition of inorganic thin films. Examples are ALD of Al_2O_3 [28] and HfO_2 [29] by using metalorganic precursors, such as trimethylaluminum (TMA) and hafnium tetrachloride, and inorganic counter-reactants, for example, O_2 and H_2O [28]. By replacing the inorganic counter-reactants with appropriate organic precursors, MLD can be blended with ALD to produce organic-inorganic hybrid materials. Although not purely organic MLD films, these hybrid systems are reviewed here because they engender many of the best properties of both ALD and MLD films.

The family of organic–inorganic hybrid MLD films has been produced using a combination of metalorganic precursors and bifunctional organic counter-reactants in a sequential, step-wise deposition fashion [30–32]. Moreover, some applications of these novel materials have been developed [33, 34].

7.2.2.1 Organic–Inorganic Hybrid MLD with Homo-Bifunctional Precursors

The majority of organic-inorganic hybrid MLD films have been deposited with homo-bifunctional organic precursors, for example, diol, dicarboxylic acid, or diamine. These systems are reviewed here.

Aluminum Alkoxide Aluminum alkoxides or "alucones," are the most extensively studied organic-inorganic hybrid films deposited by MLD. TMA was used as the metalorganic precursor and EG [35] and dicarboxylic acids [36, 37] were exploited as the organic counter-reactant, respectively. The range of organic reactants demonstrates the tunability of MLD films achievable by varying the backbone of the precursor molecules.

Taking aluminum alkoxide MLD using TMA and EG as an example [35], the overall reaction is shown in Figure 7.3h. Self-saturating growth behavior during both TMA and EG pulses was observed by quartz crystal microbalance (QCM) measurement. Similar to pure organic MLD, the growth rate of TMA/EG aluminum alkoxide MLD is also strongly temperature dependent. The growth rate decreased monotonically from 4 Å/cycle at 85 °C to 0.4 Å/cycle at 175 °C, as can be seen from Figure 7.9a. In contrast, the film density remained a constant of $1.5\,g\,cm^{-3}$ regardless of the deposition temperature. This film density is much lower than that of Al_2O_3, which is $2.9\,g\,cm^{-3}$, due to the incorporation of organic components. FTIR measurement also confirmed the incorporation of the $-O-CH_2-CH_2-O-$ backbone in the deposited films, as the CH_2, C–C, and C–O stretching bands were present in their IR spectra [35].

In situ FTIR studies performed during film growth revealed a deviation of the growth mechanism from the ideal model in which the film surface after the EG pulse is terminated with reactive hydroxyl groups for further reaction with TMA, because no OH absorption was observed in the IR spectra after the EG dose. Moreover, it was also evident that TMA continued to be absorbed into the deposited polymer film [35]. As a result, the authors proposed a diffusion-facilitated growth model, as displayed in Figure 7.9b. It was suggested that TMA can diffuse into the film and form new reactive sites, leading the reaction between TMA and EG into a more CVD-likely mechanism.

Interestingly, aluminum alkoxides formed with TMA and EG degraded over time in ambient air [35]. A 20% film thickness loss was observed by 200 h after film deposition, possibly due to moisture uptake and subsequent reaction between water and the deposited aluminum alkoxide film.

Besides diols like EG, other homo-bifunctional organic precursors have also been used to react with TMA, forming various aluminum alkoxide MLD thin films. Klepper *et al.* used a number of dicarboxylic acids, including both saturated linear molecules, for example, glutaric acid and pimelic acid [37], and unsaturated linear molecules, for example, maleic acid and *trans, trans*-muconic acid [36]. For all dicarboxylic acid precursors studied, self-saturating growth behavior was observed for the hybrid MLD process. Like aluminum alkoxide grown with EG, film growth rates using dicarboxylic acids also decreased with increasing deposition temperature. Film densities ranged from 1.4 to $2.9\,g\,cm^{-3}$, which is less than the density of Al_2O_3. The authors also performed X-ray diffraction (XRD) measurements on the deposited aluminum alkoxide films, and showed that all the as-deposited films were amorphous [36, 37].

Zinc Alkoxide In addition to aluminum alkoxide films, a number of zinc alkoxide, or "zincone," thin films have been explored by MLD [38–42]. For instance, diethyl zinc (DEZ) and EG were used to deposit zinc alkoxide MLD films on planar silicon substrates [39], as well as on ZrO_2 [38] and TiO_2 [40] nanoparticles, with

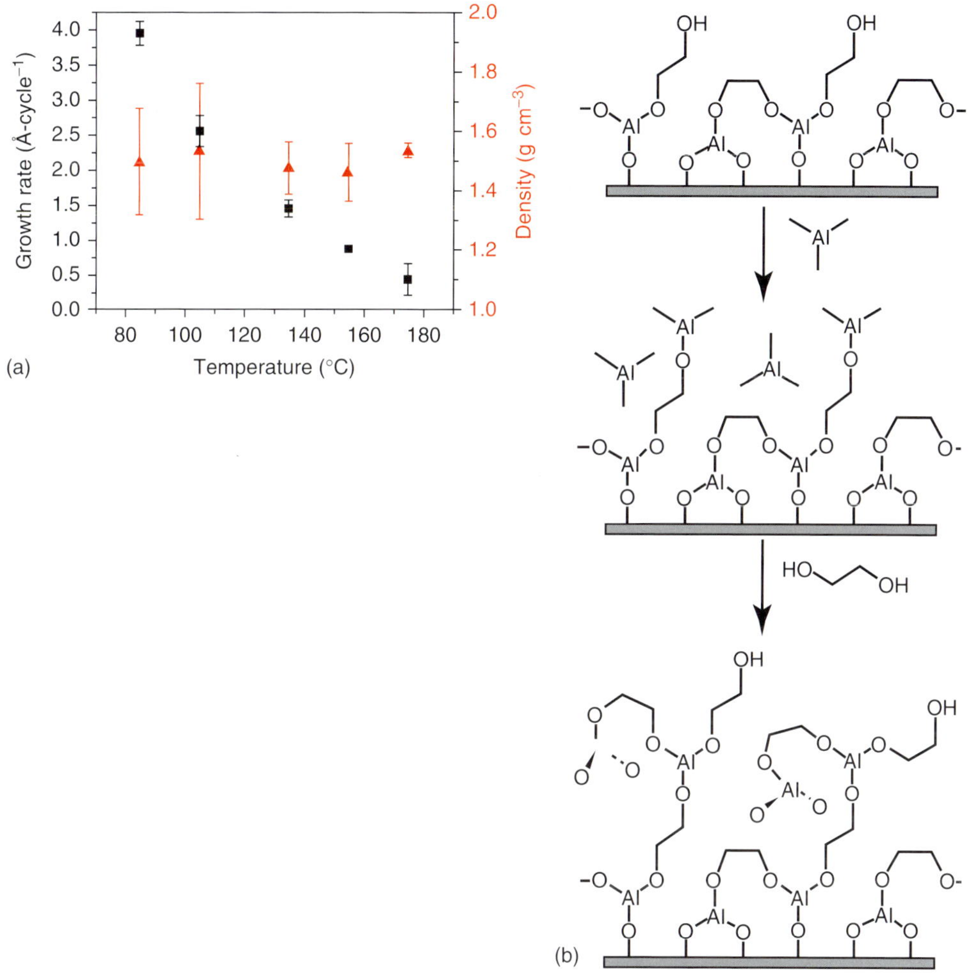

Figure 7.9 (a) Growth rates and film densities of aluminum alkoxide MLD films deposited at various temperatures. (b) Possible aluminum alkoxide MLD growth mechanism, where TMA can react with OH groups or diffuse into the film to facilitate further film growth. (Reprinted with permission from Dameron et al., Chem. Mater. 20, 3315 (2008). Copyright (2008) American Chemical Society.)

the overall MLD reaction shown in Figure 7.3i. Characteristic self-limiting growth behavior of MLD was observed during film deposition, and the growth rate was temperature dependent. The organic–inorganic hybrid film coating was highly conformal, as shown in Figure 7.10a. The detection of C–O vibrational absorption around 1100 cm^{-1} in the FTIR spectra of as-deposited films, as displayed in Figure 7.10b, confirmed incorporation of an $-O-CH_2-CH_2-O-$ backbone. Figure 7.10b also indicates that the zinc alkoxide films can uptake water from ambient air, similar to its aluminum analog mentioned in the previous section

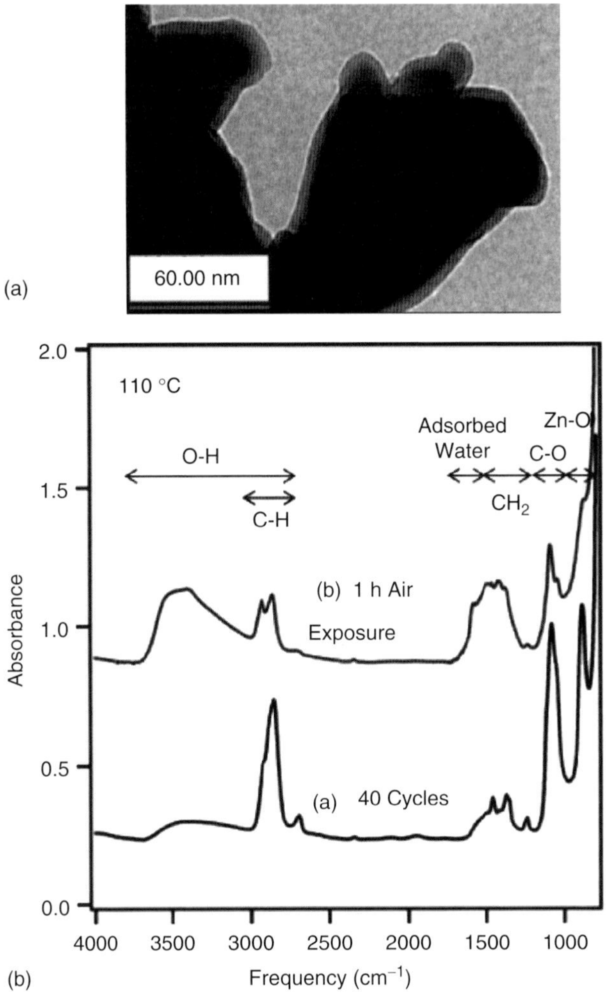

Figure 7.10 (a) TEM image of zinc alkoxide MLD film deposited on ZrO_2 nanoparticles. (b) FTIR spectra of zinc alkoxide MLD films as-deposited and after air exposure. (Reprinted with permission from Yoon et al., Chem. Vap. Deposit. 15, 112 (2009). Copyright (2009) Wiley-VCH.)

[35]. In addition to the EG precursor, hydroquinone was also used to deposit zinc alkoxide MLD films with DEZ through a very similar mechanism [42].

An advantage of MLD is that it allows for high flexibility in incorporating desired functionalities into the polymer chains by selection of suitable precursors, as previously described for the sulfide-containing polyurea MLD films [10]. For another demonstration, zinc alkoxide was deposited using DEZ and hexadiyne diol (HDD), leading to formation of a cross-linked network within the MLD film by UV (ultraviolet)-induced cross-linking of the imbedded alkyne functionalities after each HDD dose [41]. The MLD scheme is illustrated in

Figure 7.11a. TEM measurements (Figure 7.11b) showed that a linear growth rate of 5.2 Å/cycle was obtained with the cross-linked zinc alkoxide film, which was close to the expected monolayer thickness of 6 Å. Moreover, surface-enhanced Raman scattering (SERS) spectroscopy revealed characteristic vibrations of polydiacetylene in the deposited films, confirming the presence of the cross-linked network. The cross-linked zinc alkoxide films had good thermal stability up to 400 °C in air, benefiting from a more stable network formed via cross-linking of adjacent MLD polymer chains.

Titanium Alkoxide As an analog to the formation of organic-inorganic hybrid films through the reaction between TMA or DEZ and EG, titanium alkoxide films were also deposited using titanium tetrachloride ($TiCl_4$) and EG [43]. The overall reaction is shown in Figure 7.3j. Typical MLD growth characteristics such as self-saturating growth and linear growth rate were observed. In addition to EG, a triol precursor, glycerol (GL) was also used to deposit another titanium alkoxide film with more bridging between polymer chains. The MLD using $TiCl_4$ and GL also showed characteristic growth behaviors [43]. Despite a faster growth rate obtained with $TiCl_4$/EG (4.5 Å/cycle from 90 to 115 °C) than with$TiCl_4$/GL (2.8 Å/cycle at 130 °C), film densities for both MLD films were very similar, ~1.8 g cm^{-3} as measured by X-ray reflectivity (XRR). The film density of titanium alkoxide films is significantly lower than that of inorganic TiO_2 film, ~3.7 g cm^{-3}, due to the high content of organic components in the titanium alkoxide films. XPS measurements showed a Ti/O ratio of 1 : 4 for the $TiCl_4$/EG film and 1 : 5 for the $TiCl_4$/GL film, with the higher oxygen content indicating additional bridging between polymer chains in the $TiCl_4$/GL film [43]. The additional bridging between polymer chains can increase film strength and thermal stability, which was verified by the authors. As shown in Figure 7.12a, the $TiCl_4$/GL film had higher elastic modulus and hardness than did the $TiCl_4$/EG films. Figure 7.12b also shows that the $TiCl_4$/GL film could stand up to higher temperature when annealed in air.

Besides EG and GL, HDD was also used to deposit titanium alkoxide films by Yoon *et al.* [44] Combined with UV-induced cross-linking between adjacent polymer chains, a cross-linked titanium alkoxide film was formed, similar to its cross-linked zinc alkoxide film analog [41]. By alternating between TiO_2 ALD using $TiCl_4$ and water and titanium alkoxide MLD using $TiCl_4$ and HDD, nanolaminates consisting of pure inorganic TiO_2 layers and organic–inorganic hybrid titanium alkoxide layers were formed, as exhibited in Figure 7.12c.

Titanium Organo Nitride In addition to diols that were used to react with $TiCl_4$ and form titanium alkoxide MLD films [41, 43, 44], ODA, a diamine precursor, was explored as an MLD counter reactant for $TiCl_4$, and the resulting $(-Ti-N-C_6H_4-O-C_6H_4\text{-}N-)_n$ titanium organo nitride films were deposited on silicon substrate [45]. Free NH_2 vibrational modes were absent in the FTIR spectra of the deposited films, while absorptions from other functional groups of ODA-like C=C stretching and C-N stretching modes were retained, supporting the MLD film growth mechanism via reaction between the NH_2 groups of ODA

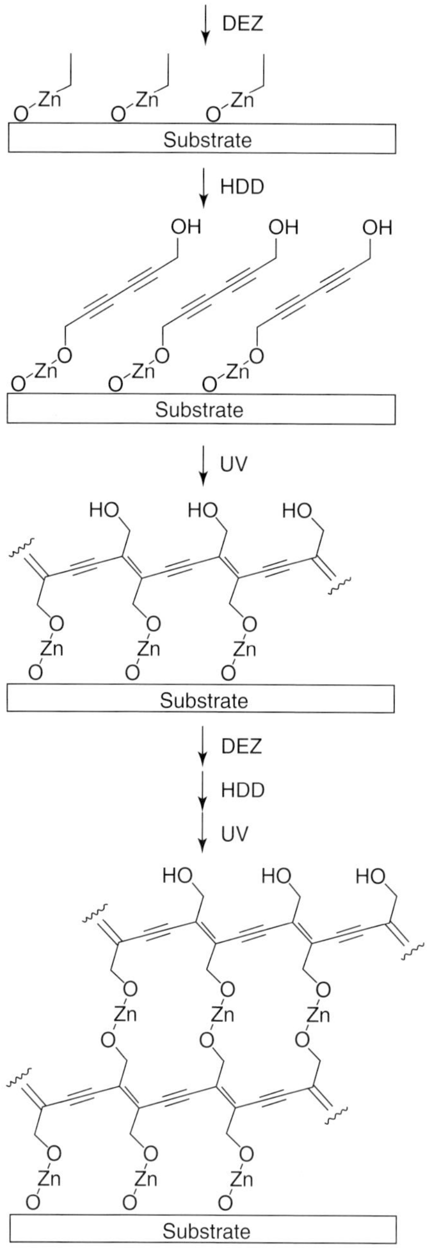

(a)

Figure 7.11 (a) MLD reaction scheme of cross-linked zinc alkoxide film. (b) TEM images of cross-linked zinc alkoxide film after different numbers of MLD cycles. (Reprinted with permission from Cho et al., Angew. Chem., Int. Ed. 50, 2742 (2011). Copyright (2011) Wiley-VCH.)

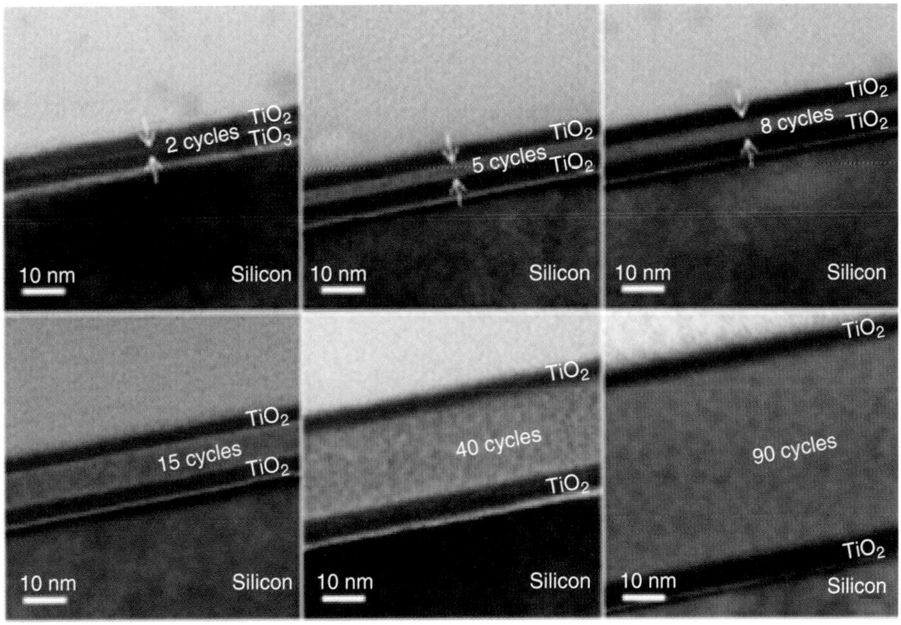

(b)

Figure 7.11 *(Continued)*

and $TiCl_4$. The deposited films were very smooth with an amorphous structure, consistent with other organic–inorganic hybrid MLD films [36, 37]. Interestingly, the growth rate of $TiCl_4$/ODA MLD increased from 0.3 to 1.1 Å/cycle as the deposition temperature increased from 160 to 190 °C, which is counter to the trend found in other MLD systems that MLD growth rate decreases as deposition temperature increases [35, 39, 43]. This abnormality was explained by Sood *et al.* by the complexity caused by the HCl byproduct in $TiCl_4$/ODA MLD, in which HCl may have an etching effect on the deposited titanium alkyl nitride film. A higher deposition temperature facilitates the removal of HCl from the MLD film. As a result, although the precursor molecules have a shorter residence time on the surface at high temperature, increasing the deposition temperature may still improve the growth rate.

7.2.2.2 Organic–Inorganic Hybrid MLD with Hetero-Bifunctional Precursors

As discussed above, various organic-inorganic hybrid MLD thin films have been deposited using homo-bifunctional precursors like diols [35, 39, 41] or dicarboxylic acids [36, 37]. However, the MLD processes with homo-bifunctional precursors have an inherent limitation, which is the “dual reaction” between functional groups on both ends with two surface reactive sites simultaneously. This dual reaction may also happen in the MLD process for depositing pure organic films [20]. The dual reaction imposes a “capping” molecule on the surface, reduces the amount of surface reactive sites and thus impedes film growth rate

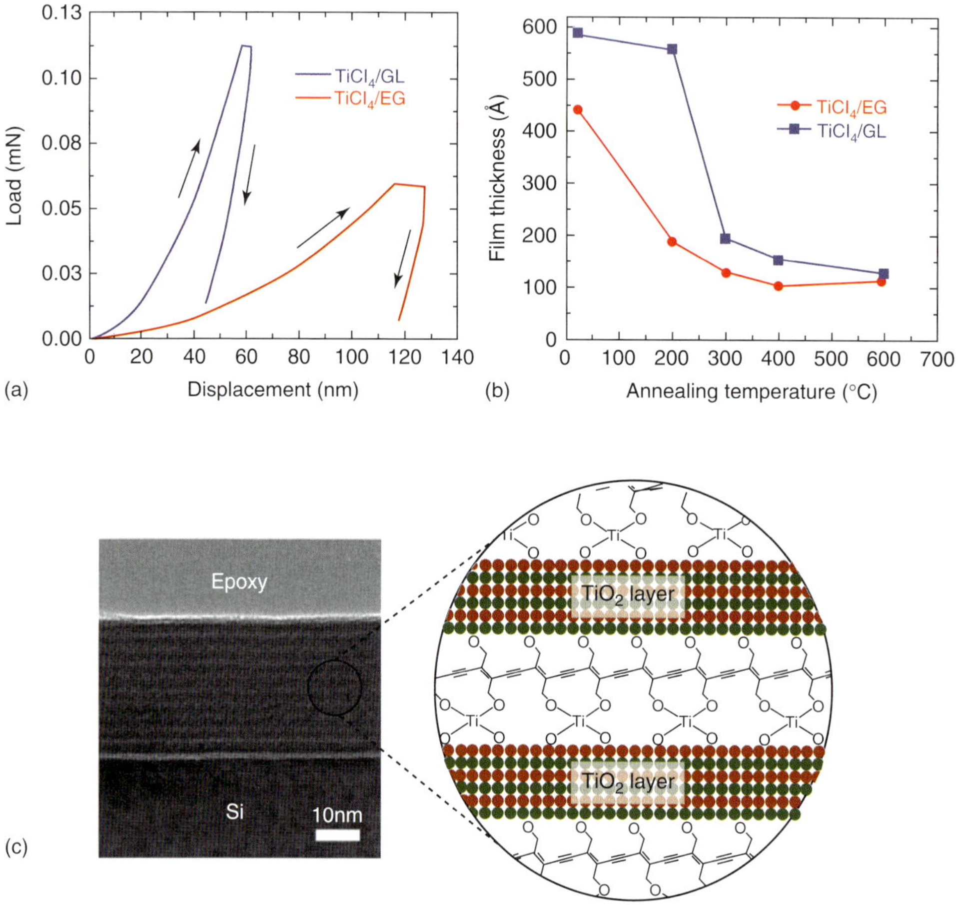

Figure 7.12 (a) Nanoindention measurements for $TiCl_4$/EG and $TiCl_4$/GL MLD films. (b) Film thicknesses of $TiCl_4$/EG and $TiCl_4$/GL MLD films versus annealing temperatures. (Reprinted with permission from Abdulagatov et al., Chem. Mater. 24, 2854 (2012). Copyright (2012) American Chemical Society.) (c) TEM image of nanolaminates formed by TiO_2 ALD and titanium alkoxide MLD using $TiCl_4$ and HDD. (Figure adapted with permission from Yoon et al., Nanoscale Res. Lett. 7, 71 (2012). Copyright (2012) Springer.)

and quality [14, 30]. To avoid the dual reaction associated with homo-bifunctional precursors, MLD reactions using hetero-bifunctional precursors and TMA were exploited to deposit aluminum alkoxides [46–48].

When reacting with surface Al-CH_3 sites, glycidol, with an OH group on one end and epoxy group on the other, can ideally be bound to the surface via only the OH groups, if Al-CH_3 groups have a higher reaction selectively to OH than to epoxy groups. Subsequently the epoxy groups of glycidol undergo a ring-opening process, yielding new surface OH groups to react with further TMA dose [46]. However, *in-situ* growth characterizations using FTIR and QCM showed that the aluminum alkoxide film growth using TMA and glycidol still proceeded with dual reaction of both of glycidol's functional groups, due to the insufficient selectivity of Al-CH_3 with OH and epoxy groups [46].

An alternative approach of a three-step MLD cycle using three precursors was adopted by George *et al.* [47, 48] The film deposition proceeded with a TMA dose, leaving the surface covered with Al-CH_3 groups, followed by a dose of ethanolamine (EA), which was bound to the surface via reaction between OH groups and surface Al-CH_3 groups and left the surface terminated with NH_2 groups. The third step consisted of a maleic anhydride (MA) dose, during which MA reacted with surface NH_2 groups, underwent a ring-opening reaction and left the surface with a COOH terminus to react further with TMA. The surface reactions of this three-step MLD process are shown in Figure 7.13. Dual reactions on both ends of the precursor molecules can be avoided, enabled by the use of hetero-bifunctional precursors like EA. The proposed MLD growth reaction was supported by *in situ* FTIR spectroscopic study. For example, the disappearance of CH_3 modes together with the presence of NH_2 vibrational modes after EA

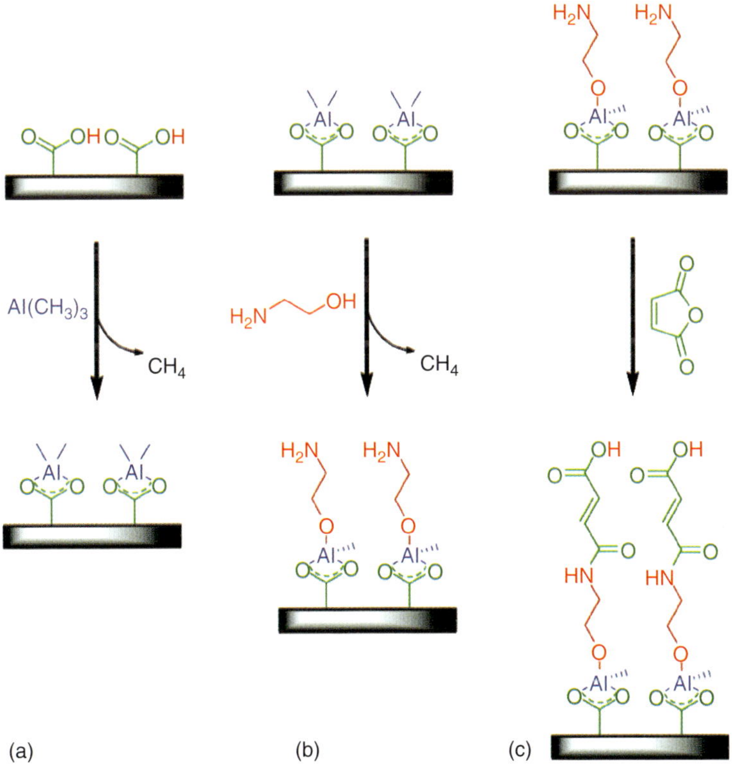

Figure 7.13 Surface reactions of the three-step MLD process for aluminum alkoxide thin films during (a) TMA, (b) EA, and (c) MA doses. (Reprinted with permission from Yoon et al., Chem. Mater. 21, 5365 (2009). Copyright (2009) American Chemical Society.)

exposure confirmed that the reaction between EA and Al-CH_3 proceeded via the OH group and that the surface was terminated with NH_2 groups [47]. A growth rate of 24 Å/cycle at 90 °C and 4 Å/cycle at 170 °C was measured for this three-step MLD process, consistent with the general trend observed in other MLD systems that a lower deposition temperature facilitates a higher growth rate. Like aluminum alkoxide films formed by TMA and EG [35], this new aluminum alkoxide film also showed moisture uptake when exposed to ambient air [47].

While the use of hetero-bifunctional precursors may mitigate the problem of dual reactions and reduced surface reactive sites, another complexity of MLD remains unsolved, which is the diffusion of precursor molecules into the porous deposited polymer films. The absorbed reactants can further react with other precursors, leading to additional film growth and affecting the controllability of the MLD process. A detailed *in situ* QCM study during the three-step MLD growth of aluminum alkoxide thin film revealed that TMA can diffuse into the film and introduce CVD-like growth behavior [48]. Further studies are needed to investigate the effect of precursor absorption on the MLD growth mechanism.

7.3 Applications of MLD Films

Because of the ability to add desired functionalities into the deposited films, the inherent tunability over both film thickness and composition, and high coating conformality, MLD shows promise for fabricating thin films with an exquisite level of control. This advantage of MLD can be highly desirable in many applications, especially those in which carefully designed ultrathin coatings or interfacial layers are required. As a result, there has been a lot of exploration for applying MLD in many areas, such as semiconductor processing [49, 50], catalysis [34, 51], and material modification [52, 53].

7.3.1 Applications of Pure Organic MLD Films

7.3.1.1 Organic Quantum Dots with Tunable Sizes and Bandgaps

MLD provides a very high level of control over film thickness and composition, since only one molecular layer is added at each reaction step. Taking advantage of this property, Yoshimura *et al.* used MLD to make organic quantum dots (QDs) with highly tunable sizes and absorption bands by choosing appropriate precursors. Such QDs may be used as sensitizers for photovoltaic materials [54]. The polyazomethine film deposited with TPA and PPDA has a conjugated electronic structure, which allows for electron delocalization over the polymer chain, as shown in Figure 7.14a. By replacing PPDA with oxalic dihydrazide (ODH) at certain steps during the polyazomethine MLD process, non-conjugated domains can be formed via the reaction between TPA and ODH, as Figure 7.14b illustrates. Thus, by alternating cycles of TPA/PPDA and TPA/ODH, laminated

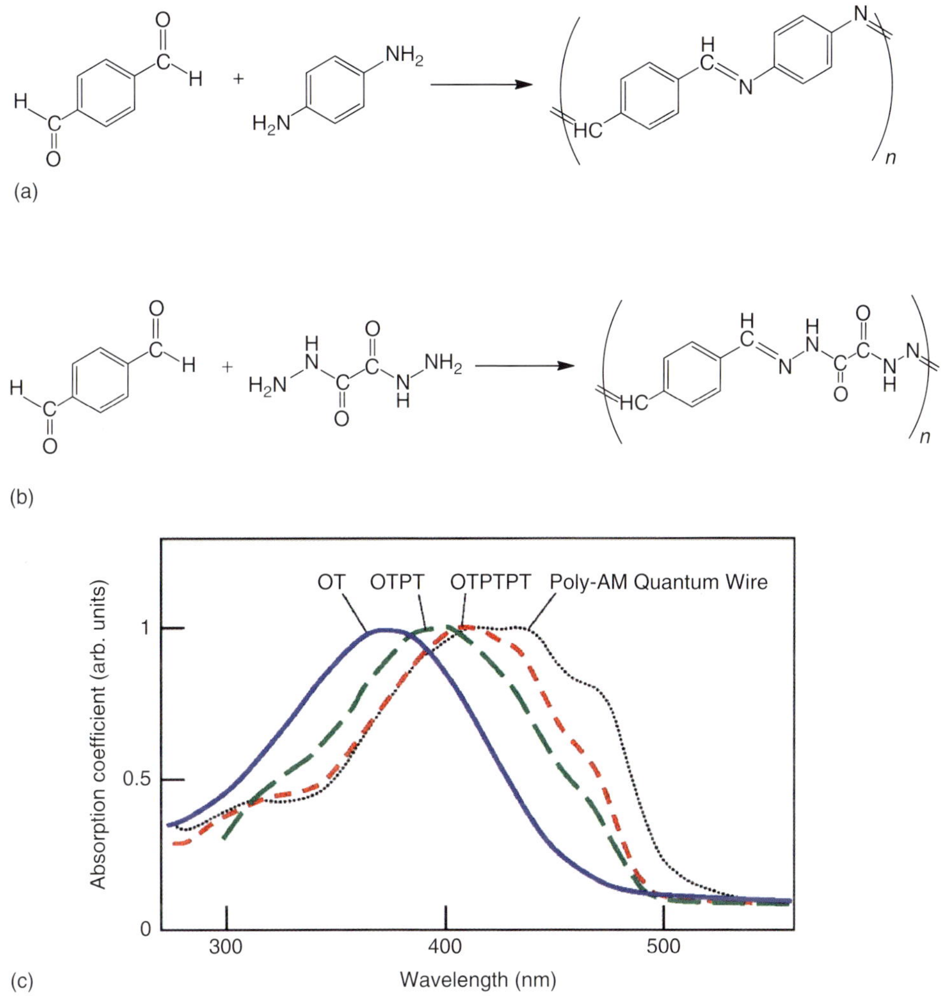

Figure 7.14 (a) MLD reaction between TPA and PPDA to form π-conjugated component, (b) MLD reaction between TPA and ODH to form non-conjugated component, and (c) absorption spectra of MLD films with different sizes of quantum dots. Poly-AM is a pure TPA/PPDA film, while OT, OTPT, and OTPTOT have quantum dots with lengths of 0.8, 2, and 3 nm, respectively. (Reprinted with permission from Yoshimura et al., J. Vac. Sci. Technol. A 29, 051510 (2011). Copyright (2011) American Vacuum Society.)

films were deposited with TPA/PPDA blocks serving as QDs and TPA/ODH blocks acting as spacers. Moreover, by controlling the number of TPA/PPDA cycles, the size of the QDs can be also tuned. Quantum dots of different sizes have different bandgaps, a property which is beneficial for solar cell sensitization [55]. Taking this approach, Yoshimura *et al.* deposited MLD thin films containing QDs with lengths of 0.8, 2, and 3 nm, respectively, as well as pure TPA/PPDA "quantum wire" films. The absorption spectra of the deposited films revealed absorption peak energy shifts due to the different sizes of QDs contained in the film, as shown in Figure 7.14c. In addition, incorporation of QDs with the three sizes in one MLD polymer film yielded a broadened absorption, attributed to the superposition of absorptions from different size QDs [54].

7.3.1.2 **Organic Films as Copper Diffusion Barrier**

The steady shrinking of semiconductor device sizes introduces new challenges throughout the integrated circuit (IC) fabrication process. One new requirement in back-end processing, for example, is for ultrathin (less than 1 nm) conformal, and pin-hole free copper diffusion barrier films by 2021 [56]. Because MLD provides precise control over film thickness together with high conformality, it can be viewed as a candidate for the deposition of organic films for copper diffusion barriers. A few different MLD organic thin films have been evaluated for their copper barrier properties by Loscutoff *et al.* [49], including a polyurea film deposited by PDIC and ethylenediamine (ED), a sulfide-containing polyurea film deposited by PDIC and TBEA, and a polythiourea film deposited by PDITC and ED. The incorporation of sulfur atoms in the latter two MLD films may be particularly beneficial for copper diffusion barriers due to the thiophilicity of copper species and consequently the sulfur atoms may serve as copper traps and prevent copper infiltration through the diffusion barriers. TEM studies were carried out to examine the effectiveness of MLD copper barriers against physical vapor deposition (PVD)-grown copper. As can be seen in Figure 7.15,

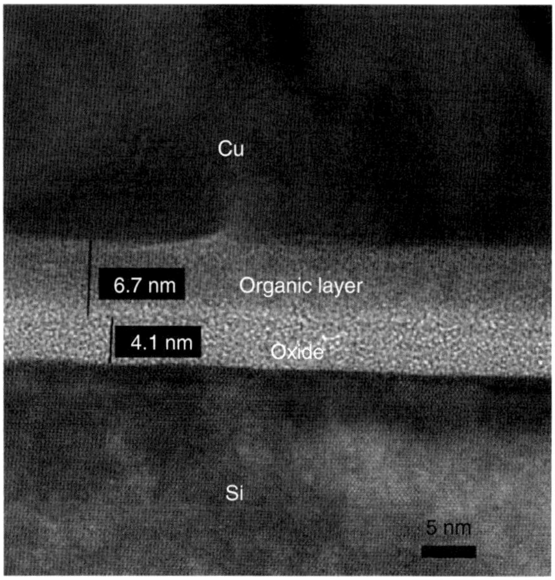

Figure 7.15 TEM images of PVD copper on sulfide-containing polyurea film deposited on SiO_2. (Figure adapted with permission from Loscutoff et al., Mater. Res. Soc. Symp. Proc. 1249, 1249-F02-03 (2010). Copyright (2010) Materials Research Society.)

for a sulfide-containing polyurea film with a total thickness of 6.7 nm, a darker, 6 nm thick, band was present after copper deposition, indicating significant copper infiltration [49]. Similar results were also observed for the polyurea and polythiourea films, suggesting that improvements like MLD film densification are needed before the MLD films can serve as effective copper diffusion barriers [49].

7.3.1.3 Organic Polyurea MLD Film as Photoresist

A high level of control over polymer film thickness and composition is desired in many applications, and photoresists are one of them [57]. To achieve good sensitivity, high resolution, and low patterning roughness simultaneously, chemically amplified photoresist thin films with uniform distributions of acid-labile groups and photoacid generators are desirable for next generation photoresists [58]. MLD polymeric thin films have a high potential to meet these requirements, since the composition can be precisely tuned at the molecular level. To explore the use of MLD to make chemically amplified photoresists, Zhou *et al.* deposited polyurea thin films with acid-labile ketal groups imbedded in the backbone of the polymer chains using PDIC and 2,2′-(propane-2,2-diylbis(oxy))diethanamine (PDDE) precursors, as shown in Figure 7.16a [50]. The deposited films had stoichiometric composition and high smoothness. After soaking photoacid generators into the polyurea films, directly patternable materials were obtained. Upon exposure to UV light, strong acids generated from photoacid generators reacted with the ketal groups in the polymer backbone and consequently decomposed the polyurea thin film in the exposed regions. Patterning demonstration was performed using UV photolithography. Figure 7.16b shows an image of the pattern with a feature size

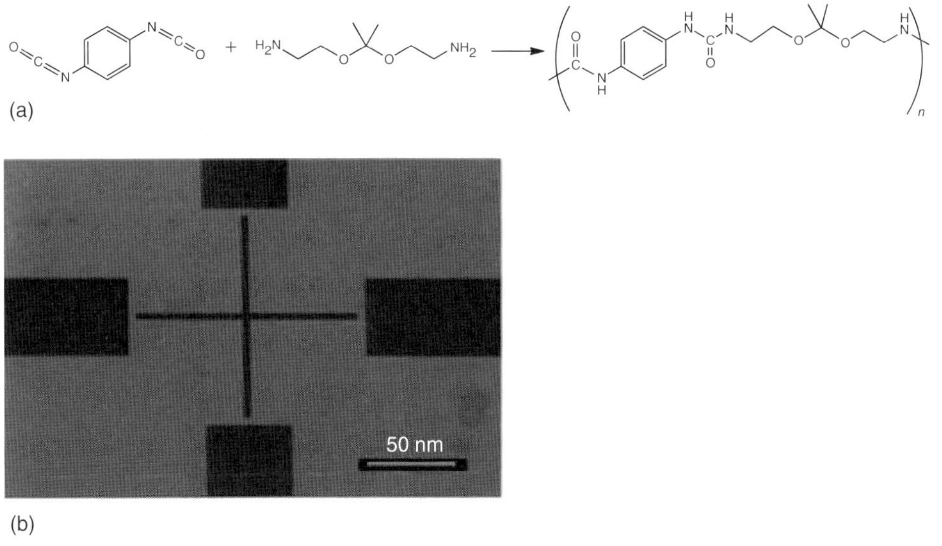

Figure 7.16 (a) MLD reaction for acid-labile polyurea film using PDIC and PDDE and (b) SEM image of the substrate after UV patterning. (Reprinted with permission from Zhou and Bent, ACS Appl. Mater. Interfaces 3, 505 (2011). Copyright (2011) American Chemical Society.)

of 4 μm obtained by using the polyurea thin films, and XPS study revealed silicon signal from the underlying substrate, proving the decomposition of the polyurea film in the exposed regions [50].

7.3.1.4 Organic Polyimide MLD Film with Switchable Conductivity

Polyimide MLD films deposited with PMDA and ODA exhibited a conductivity change when a voltage was applied to them using scanning probe microscopy (SPM) [59, 60]. The electrical modification by applying a voltage can greatly increase the conductivity of the polyimide film in the scanned area. Moreover, the dimension of the conductive domain of the film, on the order of tens of nanometers, can be tuned by the applied voltage and scanning speed. Yoshida *et al.* attributed the conductivity-switching behavior of the polyimide film to the charge transfer between its molecular orbitals or possibly the carrier trapping effect of the film [60]. With this unique property, the polyimide films may have the potential to be used as rewritable recording media [60].

7.3.2 Applications of Organic–Inorganic Hybrid MLD Films

7.3.2.1 Zeolite Modification Using Aluminum Alkoxide MLD Films

Among the materials used for gas separation, zeolite membranes are of particular interest because of their mesoporous structure and large surface area [61, 62]. To effectively separate gas mixtures, it is important to reduce or eliminate defects in zeolites, which may produce pathways larger than the pores and consequently affect the zeolite's selectivity [63]. Because MLD can form highly conformal, thickness-controlled coatings over various substrates, it may provide a novel method for zeolite surface passivation without significant seal-off of the pores. Yu *et al.* demonstrated that by applying aluminum alkoxide MLD films on zeolite membranes, the H_2/N_2 separation selectivity reached 1040 at 398 K, a great enhancement compared to the bare zeolite selectivity of 8 [34]. The aluminum alkoxide film was deposited with TMA and EG on SAPO-34 zeolite membranes, and a calcination step was performed at 573 K to remove the organic components from the aluminum alkoxide, leaving a microporous aluminum oxide coating over the zeolites. This process is illustrated in Figure 7.17a. A cross-sectional focused ion beam (FIB) SEM image (Figure 7.17b) of the coated zeolite confirmed a highly conformal MLD film coating. After MLD, the pore size of the zeolite was reduced slightly, from 0.38 to 0.36 nm, while the H_2/N_2 separation selectivity was greatly enhanced [34].

7.3.2.2 Polymer Modification Using Aluminum Alkoxide MLD Films

Polydimethylsiloxane (PDMS) is an important polymer in microfluidic devices [64] and micro-contact printing [65]. However, the hydrophobic nature of PDMS imposes limits on its compatibility with aqueous solutions. As a result, surface modification of PDMS to increase its hydrophilicity is necessary. Gong *et al.* performed studies of depositing an aluminum alkoxide MLD film on PDMS as a

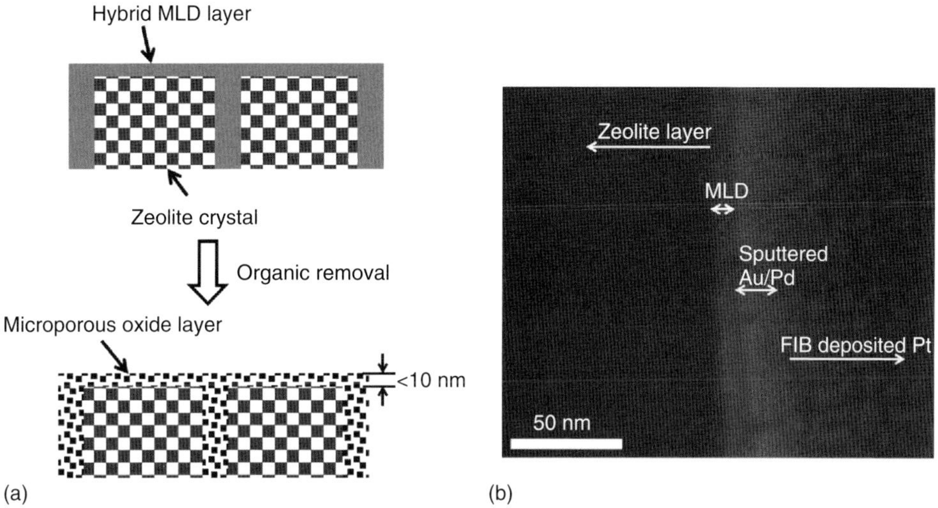

Figure 7.17 (a) Schematic illustration of preparation process for the MLD-modified zeolite membrane and (b) cross-sectional SEM image of the MLD-modified zeolite membrane. (Reprinted with permission from Yu et al., J. Am. Chem. Soc. 133, 1748 (2011). Copyright (2011) American Chemical Society.)

surface modification material [52]. Prior to aluminum alkoxide MLD using TMA and glycidol, a cycle of Al_2O_3 ALD using TMA and water was carried out with very long exposure times for both precursors (5 h for TMA and 30 min for water, respectively), followed by 100 cycles of a normal Al_2O_3 ALD process. This pre-treatment was found to be crucial for the modification of PDMS, because the long exposure time allowed precursors to infiltrate deeply into the polymer, forming an Al_2O_3 layer which can serve not only as a diffusion barrier preventing organic species of PDMS from outdiffusing and recovering surface hydrophobicity, but also as a mechanical buffer layer reducing cracks on the subsequently deposited MLD film. After the pre-treatment and 100 cycles of aluminum alkoxide MLD, the hydrophilicity of the PDMS substrate was greatly enhanced, with the water contact angle reduced to 40°. Moreover, this surface modification combining Al_2O_3 ALD and aluminum alkoxide MLD showed very good stability in air up to 14 days [52].

Another example of applying aluminum alkoxides as a polymer surface modification material is a study to increase the flexibility of Al_2O_3-coated biopolymer films [33]. Inorganic ALD-Al_2O_3 coatings are found to be beneficial for biopolymers, as they can reduce gas and vapor diffusion into the polymer substrates [66, 67]. However, inorganic Al_2O_3 coatings are often not strong enough mechanically and cracks can form upon straining. By blending aluminum alkoxide MLD thin film layers into inorganic Al_2O_3 coatings, laminated aluminum alkoxide/Al_2O_3 coating composites were deposited on biopolymer substrates and the flexibility of the coating was improved, without degrading the barrier properties [33].

7.3.2.3 Encapsulation of Cu Nanoparticles Using Aluminum Alkoxide MLD Films

While bare metal nanoparticles are often vulnerable to oxidation and form metal oxide hollow spheres, encapsulation using a thin layer of Al_2O_3 can greatly inhibit the oxidation, since inorganic Al_2O_3 coatings are good diffusion barriers against gases like oxygen [68]. The behavior is distinctly different with a hybrid MLD coating. Interestingly, after encapsulating Cu nanoparticles with an aluminum alkoxide MLD coating deposited with TMA and EG, it was found that upon annealing, Cu atoms can out-diffuse through the MLD encapsulation and form copper oxide particles supported on the MLD layer [53]. Figure 7.18a–c shows TEM images of Cu nanoparticles taken after annealing in the three above-mentioned conditions. Qin *et al.* [53] suggested that the unique oxidation behavior of Cu nanoparticles coated with aluminum alkoxide layer is likely to result from the porous nature of the organic–inorganic hybrid MLD films. The abundant pores in the MLD layer may form a number of micro channels. These channels may have size selectivity over Cu atoms and oxygen molecules, thus facilitate copper outward diffusion and inhibit oxygen inward diffusion since Cu atoms are smaller than oxygen molecules [69]. Moreover, by adjusting the thickness of the aluminum alkoxide encapsulation layer, different sizes of copper oxide nanoparticles can be formed. Since a thicker aluminum alkoxide layer contains more pores, the resultant copper oxide nanoparticles are smaller in size with a more even distribution over the surface [53]. The uniqueness of this controllable metal nanoparticle oxidation process may provide a novel method to make metal oxide nanoparticles.

7.3.2.4 Photocatalytic Layers Using Titanium Alkoxide MLD Films

Titanium oxide (TiO_2) has been studied extensively for its various applications, such as photocatalysis [70]. As an analog to TiO_2, titanium alkoxide deposited by MLD was also tested for its photocatalytic performance [51]. Ishchuk *et al.* deposited titanium alkoxide thin films using $TiCl_4$ and EG and showed that the photocatalytic activity of titanium alkoxide films was dependent on the post-deposition annealing conditions. After annealing at different temperatures, different crystallinity was found for the titanium alkoxide films, as the organic components can be decomposed to different levels. Combining TEM and XRD measurement, the authors found three stages of titanium alkoxide film evolution: 250 °C annealing resulted in an amorphous film, 850 °C annealing yielded a crystalline film with the anatase phase, and 650 °C produced an intermediate state of crystalline regions embedded in an amorphous film [51]. The photocatalytic performance of the titanium alkoxide films was evaluated using two metrics: (i) porphyrin dye loading and (ii) the illumination time needed to deplete half of the porphyrin dye absorption, as shown in Figure 7.19a,b, respectively. It can be seen that the intermediate state had the best performance as both high dye loading and high photocatalytic activity were achieved. Moreover, compared with TiO_2 ALD films annealed under the same conditions, a photocatalytic activity increase of fivefold was obtained with titanium alkoxide MLD film, attributed to the co-existence of a higher volume amorphous film with photocatalytically active crystalline regions [51].

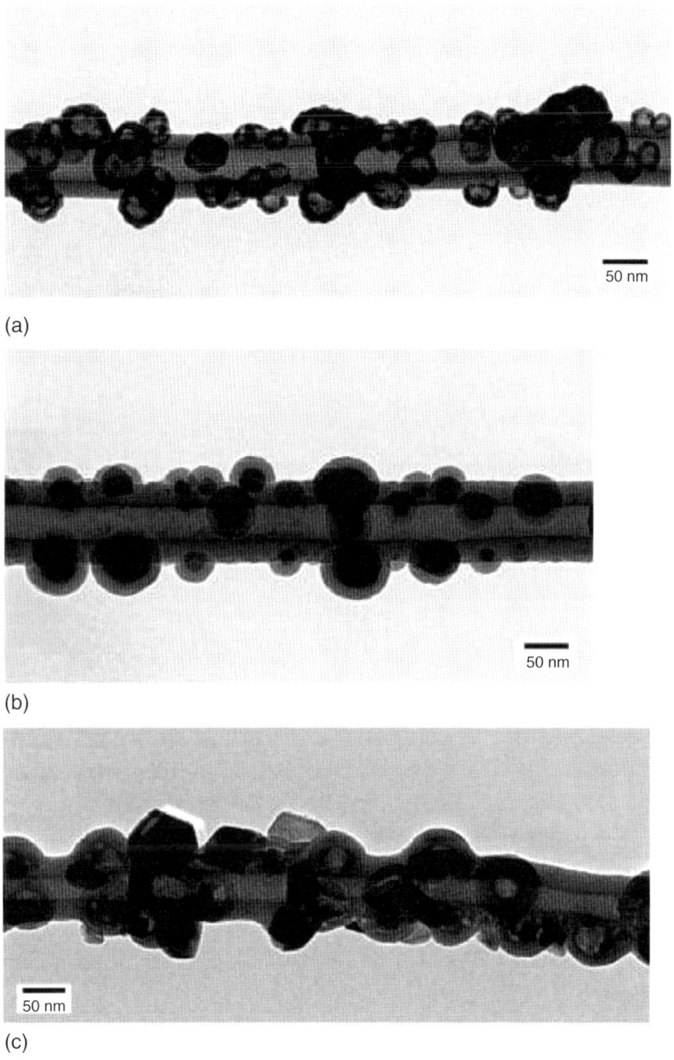

Figure 7.18 TEM images of Cu nanoparticles after annealing: (a) bare Cu nanoparticles supported on aluminum alkoxide MLD film with no encapsulation. Notice that copper oxide hollow spheres were formed. (b) Cu nanoparticles encapsulated with Al_2O_3 ALD film. Notice that no oxidation was observed. (c) Cu nanoparticles encapsulated with aluminum alkoxide MLD film. Notice that copper oxide formed outside the encapsulating layer. The Cu nanoparticles were prepared by reducing CuO nanowires precoated with aluminum alkoxide MLD film in H_2 (5% in air) at 650 °C. (Reprinted with permission from Qin et al., Nano Lett. 11, 2503 (2011). Copyright (2011) American Chemical Society.)

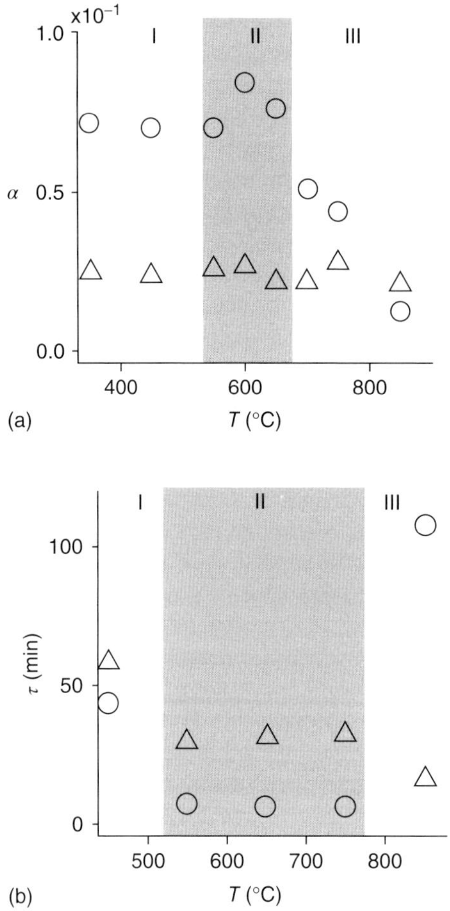

Figure 7.19 (a) Comparison of porphyrin dye loading, where α is the absorption of porphyrin dye and (b) comparison of illumination time needed to deplete half of porphyrin absorption of titanium alkoxide MLD (open circle) and TiO_2 ALD (open triangle) films as a function of annealing temperatures. Region I, II, and III refer to an amorphous film state, an intermediate film state with both amorphous and crystalline regions, and a crystalline film state, respectively. (Figures adapted with permission from Ishchuk et al., ACS Nano 6, 7263 (2012). Copyright (2012) American Chemical Society.)

7.3.2.5 **Organic Magnetic Materials Using Vanadium-Tetracyanoethylene Films**

Vanadium-tetracyanoethylene ($V[TCNE]_x$) thin films were previously deposited by CVD with $V(CO)_6$ and TCNE precursors and demonstrated magnetic properties [71]. However, CVD films do not meet the requirement of high thickness uniformity, which can be problematic for multilayer devices [72]. Leveraging the advantage of precise control over film thickness and conformality, Kao *et al.* used MLD to deposit $V[TCNE]_x$ thin films at room temperature also using $V(CO)_6$ and TCNE precursors [73]. The MLD reaction proceeded with ligand exchange

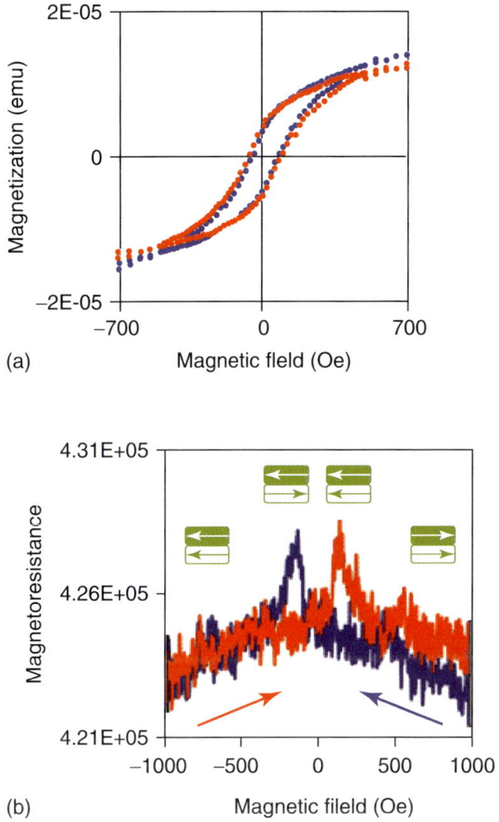

Figure 7.20 (a) Magnetization vs. magnetic field for a V[TCNE]$_x$ MLD film on a Si substrate, covered with 20 nm aluminum as oxidation barrier, measured at 5 K (blue dots, trace starting at the lower point at −700 Oe) and 300 K (red dots, trace starting at the higher point at −700 Oe) and (b) Magneto resistance curves of a V[TCNE]$_x$ (50 nm)/LAO (1.2 nm)/LSMO (50 nm) magnetic junction measured at 20 K with a bias field of 1.5 V. The red and blue dots, the traces with peaks at positive and negative Oe values are data collected with increasing and decreasing magnetic field, respectively. (Reprinted with permission from Kao et al., ACS Appl. Mater. Interfaces 4, 137 (2012). Copyright (2012) American Chemical Society.)

between the two precursors. Film characterization showed that V[TCNE]$_x$ MLD films had a growth rate of 0.98 nm/cycle with stoichiometric composition. Magnetometric measurements showed that the V[TCNE]$_x$ MLD film is magnetic at both 5 and 300 K, as illustrated by the hysteresis loops in Figure 7.20a. Moreover, the temperature dependence of magnetization showed that the V[TCNE]$_x$ MLD film remained magnetic at 400 K, suggesting that the V[TCNE]$_x$ MLD film had a higher thermal stability than the V[TCNE]$_x$ CVD film. After demonstrating the magnetic properties of V[TCNE]$_x$ MLD films, a standard magnetic tunnel junction structure was fabricated, which contained a 50 nm thick V[TCNE]$_x$

MLD film as the spin polarizing layer, a 1.2 nm thick $LaAlO_3$ (LAO) layer as the insulating barrier and an 80 nm thick $La_{0.67}Sr_{0.33}MnO_3$ (LSMO) film as the counter magnetic layer, respectively. Magnetoresistance was measured for the magnetic conjunction, as shown in Figure 7.20b, confirming the spin polarizing property of the $V[TCNE]_x$ MLD film and indicating its potential application in spintronics [73].

7.3.2.6 Electronic Device Applications

The organic–inorganic hybrid films also have been explored for applications in electronic devices, such as charge trap memory [74, 75], gate insulator in thin film transistor [76, 77], and transparent conductive films [42].

Charge trapping layers are important components in charge trap non-volatile flash memory devices [78]. A 3 nm-thick titanium alkoxide MLD film sandwiched with aluminum alkoxide MLD films showed hole-trapping behaviors [74]. In addition, Zhou *et al.* demonstrated charge trapping effects of laminated films composed of Al_2O_3 ALD layers and hybrid MLD layers deposited with TMA and PPDA [75]. However, problems of these MLD charge trapping layers, for instance, short retention time of the trapped charges which leads to a short data storage time, still remain to be solved.

A zirconium alkoxide MLD film was deposited using zirconium *tert*-butoxide $[Zr(OC)(CH_3)_4]$ and 7-octenyltrichlorosilane $[CH_2{=}CH(CH_2)_6SiCl_3]$, with the alkene group in the 7-octenyltrichlorosilane precursor converted to carboxylic acid group by O_3 exposure to sustain the film growth. After alloying the zirconium alkoxide MLD layers with ZrO_2 ALD layers, the laminated organic–inorganic hybrid film exhibited high dielectric constant ($\kappa \approx 16$) and good mechanical stability, making it a potential candidate for gate insulator in thin film transistor [76]. Taking a similar approach, another hybrid film containing Al_2O_3 and TiO_2 ALD layers with O_3-activated 7-octenyltrichlorosilane SAMs separating the Al_2O_3 and TiO_2 blocks was fabricated and exhibited gate insulator properties as well [77].

Zinc oxide-based organic-inorganic hybrid films composed of zinc alkoxide MLD layers, which was deposited with DEZ and hydroquinone, and ZnO ALD layers were deposited with various ratio of the MLD and ALD cycles [42]. With a 1 : 1 and 2 : 2 ratios of the MLD/ALD cycles, the hybrid films showed enhanced conductivity, 116 and 170 $S\,cm^{-1}$, respectively, compared to pure ZnO ALD film, which had a conductivity of 14 $S\,cm^{-1}$, and the nonconductive hybrid zinc alkoxide film. The approximately 10-fold increase in conductivity was attributed to a combination of high carrier density in the ZnO ALD block and high electron mobility due to the π-electron coupling effect of the conjugated phenylene rings in the backbone of the zinc alkoxide film. Besides the enhanced conductivity, the hybrid films also exhibited high transparency in the visible light range, leading to their potential application as transparent conducting films [42].

7.4 Study of MLD Film Structure

Compared to the abundance of research on the development and applications for various MLD films, much less work has been performed to study the microstructure of the MLD films, which may prove insightful for many applications such as diffusion barrier layers as well as for understanding film growth mechanisms. For example, the growth rate of MLD films ideally should be close to the molecular length of one repeating unit of the polymer, since a self-saturating molecular unit is added on the deposited film at each reaction step. However, for most MLD films developed so far, the growth rates are well below the unit length [10, 23]. This observation has been attributed to various reasons, including steric hindrance of the already-deposited polymer chains, insufficient surface reactive sites, and tilted orientation of the polymer chains. Recently, Peng *et al.* performed a near-edge X-ray absorption fine structure (NEXAFS) study to probe the orientation of polyamide polymer chains and showed that instead of forming linear vertical polymer brushes sticking out of the substrate surface, the oligomer units were indeed lying down on the surface [27]. In this study, 1,4-butylenediamine and terephthaloyl dichloride were used to deposited polyamide MLD films containing both alkyl and aryl moieties. Typical MLD growth characteristics such as self-saturating behavior, linear growth rate which decreased with increasing deposition temperature and amide IR absorptions were obtained, confirming the success of the MLD process. NEXAFS spectra of the deposited polyamide films at various incident angles are shown in Figure 7.21a. The change in absorption peak intensity as a function of incident X-ray angle revealed that the MLD polymer chains adopted a configuration that is nearly parallel to the substrate surface. Additionally, the effect of an amine-terminated promoting layer on the polymer orientation was

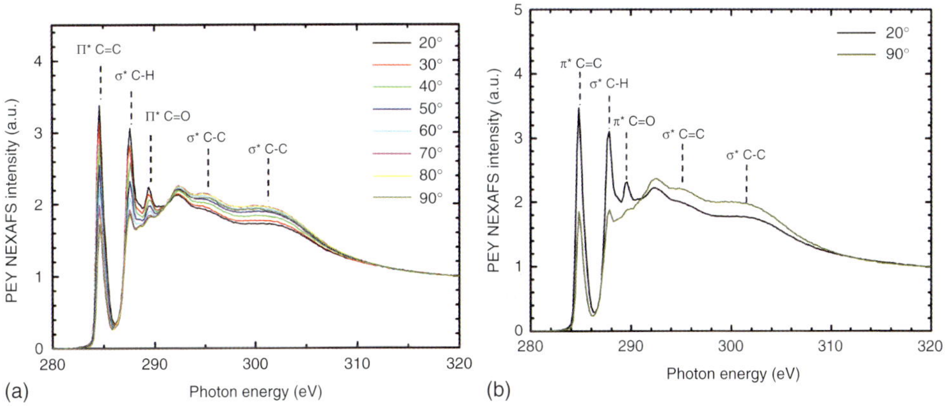

Figure 7.21 NEXAFS spectra of polyamide films at various incident angles, (a) film deposited on a Si substrate and (b) film deposited on an APTES-modified Si substrate. (Reprinted with permission from Peng et al., Langmuir 28, 10464 (2012). Copyright (2012) American Chemical Society.)

also investigated. Aminopropyltriethoxysilane (APTES) was used to form a SAM on the Si substrate prior to polyamide MLD. The results show that compared to polyamide films directly deposited on Si substrate, not only did the growth rate not change, the polymer chain orientation also remained parallel to the surface, as shown in Figure 7.21b [27].

7.5 Challenges and Opportunities for MLD

MLD has been shown to be a powerful method for depositing organic and organic–inorganic hybrid thin films. Its capabilities in control over film properties, the variety of functionality that can be incorporated, and its compatibility with existing vapor phase processes make it an attractive technique for future, nanoscale materials synthesis. However, MLD still faces challenges which should be addressed to further improve its robustness and effectiveness in various applications. Two important challenges in MLD are developing a better understanding of the growth behavior and developing new coupling chemistries.

Although characteristic growth behaviors, for example, self-saturating growth and linear deposition behavior, have been extensively studied alongside the development of various MLD systems, little understanding of the detailed growth mechanisms has been gained to date. First, the nucleation and growth processes of MLD films remain unclear. Although it has been proposed that, ideally, the reaction can take place at reactive surface sites, there is evidence showing that the reaction can also happen in the bulk of the deposited film [48]. It is still unclear to what extend the nucleation at absorbed excess precursor sites contributes to film growth. Second, in cases where "dual reaction" of a bifunctional precursor occurs, a chain termination takes place and it is not understood how re-initiation of reactive sites occurs to sustain constant film growth. Third, in contrast to the vertical chain growth model that has been used to illustrate MLD growth, a severely tilted orientation of the polymer chains was observed for a polyamide MLD film [27], while the configurations of other MLD polymeric films remain unclear. Gaining answers to these questions is important, since they may provide valuable insights to better control MLD growth and achieve higher film quality.

Another challenge of MLD is the limited number of linking reactions. Unlike the synthesis of polymers in the solution phase where solvents and catalysts can greatly facilitate the synthesis, these helpful components are largely absent in vapor-phase MLD processes. The lack of solvents and catalysts may impose limitations on the types of reactions that can be applied in MLD. So far, the pure organic MLD systems rely on the formation of polar linkages like urea bonds formed by diisocyanates and diamines [10, 19], or amide bonds formed by highly reactive precursors like diacyl chloride and diamines [20]. In contrast, deposition of some important polymers like polyethylene, polystyrene, and polyvinyl chloride using MLD may be difficult, as there are no polar linkages in the backbones of these polymers. New strategies need to be explored to expand the available linking chemistries. In addition to the limited number of linkages that can be used for MLD, the number of adequate precursors is another challenge. Highly volatile precursors are desired for a convenient MLD process, leading to difficulties in the use of precursors with low vapor pressures since excess heating may decompose the organic molecules before greatly enhancing their volatility. Opportunities therefore exist in the creative development of new precursors and new linking chemistries for MLD.

Looking forward, there are many opportunities for MLD to impact new applications. Increasingly, applications in areas such as molecular and nanoelectronics, optoelectronics, energy conversion, sensing, and tribology require controllable nanoscale organic thin films, and MLD is poised to make a contribution in these areas. For example, organic/inorganic hybrid solar cells may benefit from deposition of controllable, functional interfacial organic layers that can be grown using MLD. MLD films may also find wide use in processes for fabricating ICs, such as for pore sealing in porous low-κ dielectric materials, thereby taking advantage of the excellent coating conformality of MLD at the nanoscale. The exquisite control of thin film composition may also open the door for MLD to contribute to cutting-edge biomedical devices. For instance, stacks composed of different MLD layers can serve as drug-loading scaffolds, which can potentially achieve selective drug delivery and release at desired local targets. These examples are just a few possibilities that may be opened up for the MLD technique, and we anticipate more intriguing new opportunities of MLD to emerge in the future.

7.6 Conclusions

MLD has been shown to be a powerful technique for depositing a variety of pure organic and organic-inorganic hybrid thin films. Since MLD films are grown with only one molecular unit at each reaction step, an extremely high level of control over film thickness, composition and conformality can be achieved. MLD also has a high flexibility for incorporating desired functionalities into the deposited films. In addition to the development of deposition chemistries for MLD films, a number of novel applications have also been demonstrated, including barrier layers, photoresists, photocatalysts, and magnetic films. The number and types of applications for the MLD process will undoubtedly continue to grow. Although MLD is a relatively young technique that still faces some challenges, MLD polymeric films have very encouraging prospects and we anticipate their wide usage in the future.

Acknowledgments

We gratefully acknowledge support from the National Science Foundation (CHE 1213879). H.Z. was supported by the Intel Corporation.

References

1. Zhou, H. and Bent, S.F. (2013) *J. Vac. Sci. Technol., A*, **31**, 040801.
2. Ozaydin-Ince, G., Coclite, A.M., and Gleason, K.K. (2012) *Rep. Prog. Phys.*, **75**, 016501.
3. Hammond, P.T. (2012) *Mater. Today*, **15**, 196.
4. Tse, C.W., Man, K.Y.K., Cheng, K.W., Mak, C.S.K., Chan, W.K., Yip, C.T., Liu, Z.T., and Djurisic, A.B. (2007) *Chem. Eur. J.*, **13**, 328.
5. Hoivik, N.D., Elam, J.W., Linderman, R.J., Bright, V.M., George, S.M., and Lee, Y.C. (2003) *Sens. Actuators, A: Phys.*, **103**, 100.
6. Semaltianos, N.G. (2007) *Microelectron. J.*, **38**, 754.
7. Liang, J.X., Kohsaka, F., Matsuo, T., Li, X.F., and Ueda, T. (2008) *Microelectron. Eng.*, **85**, 1000.
8. Ouattara, L., Knutzen, M., Keller, S., Hansen, M.F., and Boisen, A. (2010) *Microelectron. Eng.*, **87**, 1226.
9. Alf, M.E., Asatekin, A., Barr, M.C., Baxamusa, S.H., Chelawat, H., Ozaydin-Ince, G., Petruczok, C.D., Sreenivasan, R., Tenhaeff, W.E., Trujillo, N.J., Vaddiraju, S., Xu, J.J., and Gleason, K.K. (2010) *Adv. Mater.*, **22**, 1993.
10. Loscutoff, P.W., Zhou, H., Clendenning, S.B., and Bent, S.F. (2010) *ACS Nano*, **4**, 331.
11. Asatekin, A., Barr, M.C., Baxamusa, S.H., Lau, K.K.S., Tenhaeff, W., Xu, J.J., and Gleason, K.K. (2010) *Mater. Today*, **13**, 26.
12. Salem, J.R., Sequeda, F.O., Duran, J., Lee, W.Y., and Yang, R.M. (1986) *J. Vac. Sci. Technol., A*, **4**, 369.
13. Kubono, A. and Okui, N. (1994) *Prog. Polym. Sci.*, **19**, 389.
14. George, S.M. (2010) *Chem. Rev.*, **110**, 111.
15. Bakke, J.R., Pickrahn, K.L., Brennan, T.P., and Bent, S.F. (2011) *Nanoscale*, **3**, 3482.
16. Yoshimura, T., Tatsuura, S., and Sotoyama, W. (1991) *Appl. Phys. Lett.*, **59**, 482.
17. Yoshimura, T., Tatsuura, S., Sotoyama, W., Matsuura, A., and Hayano, T. (1992) *Appl. Phys. Lett.*, **60**, 268.
18. Shao, H.I., Umemoto, S., Kikutani, T., and Okui, N. (1997) *Polymer*, **38**, 459.
19. Kim, A., Filler, M.A., Kim, S., and Bent, S.F. (2005) *J. Am. Chem. Soc.*, **127**, 6123.
20. Adamczyk, N.M., Dameron, A.A., and George, S.M. (2008) *Langmuir*, **24**, 2081.
21. Du, Y. and George, S.M. (2007) *J. Phys. Chem. C*, **111**, 8509.
22. Putkonen, M., Harjuoja, J., Sajavaara, T., and Niinisto, L. (2007) *J. Mater. Chem.*, **17**, 664.
23. Loscutoff, P.W., Lee, H.B.R., and Bent, S.F. (2010) *Chem. Mater.*, **22**, 5563.
24. Lee, J.S., Lee, Y.J., Tae, E.L., Park, Y.S., and Yoon, K.B. (2003) *Science*, **301**, 818.
25. Yoshimura, T., Ito, S., Nakayama, T., and Matsumoto, K. (2007) *Appl. Phys. Lett.*, **91**, 033103.
26. Ivanova, T.V., Maydannik, P.S., and Cameron, D.C. (2012) *J. Vac. Sci. Technol., A*, **30**, 01A121.
27. Peng, Q., Efimenko, K., Genzer, J., and Parsons, G.N. (2012) *Langmuir*, **28**, 10464.
28. Dillon, A.C., Ott, A.W., Way, J.D., and George, S.M. (1995) *Surf. Sci.*, **322**, 230.
29. Liu, X.Y., Ramanathan, S., Longdergan, A., Srivastava, A., Lee, E., Seidel, T.E., Barton, J.T., Pang, D., and Gordon, R.G. (2005) *J. Electrochem. Soc.*, **152**, L9.
30. George, S.M., Yoon, B., and Dameron, A.A. (2009) *Acc. Chem. Res.*, **42**, 498.
31. George, S.M., Lee, B.H., Yoon, B., Abdulagatov, A.I., and Hall, R.A. (2011) *J. Nanosci. Nanotechnol.*, **11**, 7948.
32. Leskela, M., Ritala, M., and Nilsen, O. (2011) *MRS Bull.*, **36**, 877.
33. Vaha-Nissi, M., Sundberg, P., Kauppi, E., Hirvikorpi, T., Sievanen, J., Sood, A., Karppinen, M., and Harlin, A. (2012) *Thin Solid Films*, **520**, 6780.
34. Yu, M., Funke, H.H., Noble, R.D., and Falconer, J.L. (2011) *J. Am. Chem. Soc.*, **133**, 1748.
35. Dameron, A.A., Seghete, D., Burton, B.B., Davidson, S.D., Cavanagh, A.S., Bertrand, J.A., and George, S.M. (2008) *Chem. Mater.*, **20**, 3315.
36. Klepper, K.B., Nilsen, O., Levy, T., and Fjellvag, H. (2011) *Eur. J. Inorg. Chem.*, **2011**, 5305.

37. Klepper, K.B., Nilsen, O., Hansen, P.A., and Fjellvag, H. (2011) *Dalton Trans.*, **40**, 4636.
38. Yoon, B., O'Patchen, J.L., Seghete, D., Cavanagh, A.S., and George, S.M. (2009) *Chem. Vap. Deposition*, **15**, 112.
39. Peng, Q., Gong, B., VanGundy, R.M., and Parsons, G.N. (2009) *Chem. Mater.*, **21**, 820.
40. Liang, X.H., Jiang, Y.B., and Weimer, A.W. (2012) *J. Vac. Sci. Technol., A*, **30**, 01A108.
41. Cho, S., Han, G., Kim, K., and Sung, M.M. (2011) *Angew. Chem. Int. Ed.*, **50**, 2742.
42. Yoon, B., Lee, B.H., and George, S.M. (2012) *J. Phys. Chem. C*, **116**, 24784.
43. Abdulagatov, A.I., Hall, R.A., Sutherland, J.L., Lee, B.H., Cavanagh, A.S., and George, S.M. (2012) *Chem. Mater.*, **24**, 2854.
44. Yoon, K.H., Han, K.S., and Sung, M.M. (2012) *Nanoscale Res. Lett.*, **7**, 71.
45. Sood, A., Sundberg, P., Malm, J., and Karppinen, M. (2011) *Appl. Surf. Sci.*, **257**, 6435.
46. Gong, B., Peng, Q., and Parsons, G.N. (2011) *J. Phys. Chem. B*, **115**, 5930.
47. Yoon, B., Seghete, D., Cavanagh, A.S., and George, S.M. (2009) *Chem. Mater.*, **21**, 5365.
48. Seghete, D., Hall, R.A., Yoon, B., and George, S.M. (2010) *Langmuir*, **26**, 19045.
49. Loscutoff, P.W., Clendenning, S.B., and Bent, S.F. (2010) *Mater. Res. Soc. Symp. Proc.*, **1249**, 1249-F02-03.
50. Zhou, H. and Bent, S.F. (2011) *ACS Appl. Mater. Interfaces*, **3**, 505.
51. Ishchuk, S., Taffa, D.H., Hazut, O., Kaynan, N., and Yerushalmi, R. (2012) *ACS Nano*, **6**, 7263.
52. Gong, B., Spagnola, J.C., and Parsons, G.N. (2012) *J. Vac. Sci. Technol., A*, **30**, 01A156.
53. Qin, Y., Yang, Y., Scholz, R., Pippel, E., Lu, X.L., and Knez, M. (2011) *Nano Lett.*, **11**, 2503.
54. Yoshimura, T., Ebihara, R., and Oshima, A. (2011) *J. Vac. Sci. Technol., A*, **29**, 051510.
55. Nozik, A.J. (2002) *Physica E*, **14**, 115.
56. *2009 International Technology Roadmap for Semiconductors: Emerging Materials*, SIA (Semiconductor Industry Association), Washington, DC.
57. Covert, K.L. and Russell, D.J. (1993) *J. Appl. Polym. Sci.*, **49**, 657.
58. Anderson, C.N., Naulleau, P.P., Niakoula, D., Hassanein, E., Brainard, R., Gallatin, G., and Dean, K. (2008) *J. Vac. Sci. Technol. B*, **26**, 2295.
59. Yoshida, S., Ono, T., and Esashi, M. (2010) *Micro Nano Lett.*, **5**, 321.
60. Yoshida, S., Ono, T., and Esashi, M. (2011) *Nanotechnology*, **22**, 335302.
61. Lai, Z.P. and Tsapatsis, M. (2004) *Ind. Eng. Chem. Res.*, **43**, 3000.
62. Gump, C.J., Tuan, V.A., Noble, R.D., and Falconer, J.L. (2001) *Ind. Eng. Chem. Res.*, **40**, 565.
63. Zhang, B.Q., Wang, C., Lang, L., Cui, R.L., and Liu, X.F. (2008) *Adv. Funct. Mater.*, **18**, 3434.
64. Whitesides, G.M. (2006) *Nature*, **442**, 368.
65. Wilbur, J.L., Kumar, A., Kim, E., and Whitesides, G.M. (1994) *Adv. Mater.*, **6**, 600.
66. Groner, M.D., George, S.M., McLean, R.S., and Carcia, P.F. (2006) *Appl. Phys. Lett.*, **88**, 051907.
67. Carcia, P.F., McLean, R.S., Reilly, M.H., Groner, M.D., and George, S.M. (2006) *Appl. Phys. Lett.*, **89**, 031915.
68. Hakim, L.F., Vaughn, C.L., Dunsheath, H.J., Carney, C.S., Liang, X., Li, P., and Weimer, A.W. (2007) *Nanotechnology*, **18**, 345603.
69. Jiang, Y.B., Xomeritakis, G., Chen, Z., Dunphy, D., Kissel, D.J., Cecchi, J.L., and Brinkertt, C.J. (2007) *J. Am. Chem. Soc.*, **129**, 15446.
70. Linsebigler, A.L., Lu, G.Q., and Yates, J.T. (1995) *Chem. Rev.*, **95**, 735.
71. Pokhodnya, K.I., Epstein, A.J., and Miller, J.S. (2000) *Adv. Mater.*, **12**, 410.
72. Pokhodnya, K.I., Bonner, M., and Miller, J.S. (2004) *Chem. Mater.*, **16**, 5114.
73. Kao, C.Y., Yoo, J.W., Min, Y., and Epstein, A.J. (2012) *ACS Appl. Mater. Interfaces*, **4**, 137.
74. Cha, S.H., Park, A., Lee, K.H., Im, S., Lee, B.H., and Sung, M.M. (2010) *Org. Electron.*, **11**, 159.

75. Zhou, W., Leem, J., Park, I., Li, Y., Jin, Z., and Min, Y.S. (2012) *J. Mater. Chem.*, **22**, 23935.
76. Lee, B.H., Im, K.K., Lee, K.H., Im, S., and Sung, M.M. (2009) *Thin Solid Films*, **517**, 4056.
77. Lee, B.H., Lee, K.H., Im, S., and Sung, M.M. (2008) *Org. Electron.*, **9**, 1146.
78. Lee, J.S., Cho, J., Lee, C., Kim, I., Park, J., Kim, Y.M., Shin, H., Lee, J., and Caruso, F. (2007) *Nat. Nanotechnol.*, **2**, 790.

Part II
Materials Chemistry

8
Reactive and Stimuli-Responsive Polymer Thin Films

Wyatt E. Tenhaeff

8.1
Introduction

Polymer thin films and coatings find demand in enumerable diverse applications. Examples include barrier layers in packaging, separation layers in membranes, photoresist and dielectric layers in semiconductor processing, insulation for flexible electronics, corrosion protection, and superhydrophobic surfaces. For many applications, polymer films are designed for passive, inactive roles. The physical and chemical properties are static. In existing and emerging applications, the exact opposite behavior is desired. Polymer thin films should respond dynamically to specific cues through changes in their physical or chemical properties. Reactive and responsive polymers are two classes of materials displaying this dynamic behavior. For this chapter, reactive polymer thin films are those polymers bearing chemical functionalities that can undergo post-polymerization reactions covalently linking target molecules to the film. In a sense, all polymers can be considered reactive because all polymers will react under extreme environmental conditions. Within the context of this chapter, though, reactive polymers are precisely engineered materials used in facile, intentional functionalization schemes. Responsive polymer thin films are distinct from reactive films as they undergo distinct physical property changes when exposed to a physical or chemical stimulus. The chemical composition of the film may change due to the stimulus, but the practical application of the material emphasizes the physical changes.

As will be discussed in the chapter, control of the polymers' chemical functionalities is central to developing useful reactive and responsive properties. The fabrication of polymer thin films with controlled chemical reactivities or stimuli-responsiveness via solution techniques is well established; synthetic strategies and applications are discussed thoroughly in many reviews [1–4]. As discussed in previous chapters, however, chemical vapor deposition of polymer thin films represents a significant processing alternative to solution-based techniques, creating unique opportunities for the implementation of these dynamic polymer films. This chapter will describe the synthesis of reactive and responsive polymer thin films

CVD Polymers: Fabrication of Organic Surfaces and Devices, First Edition. Edited by Karen K. Gleason.

by chemical vapor deposition techniques and highlight how their novel compositions and structures are used in a limited number of applications. Given the scope of this field, it is impossible to discuss every chemical system and application, but cases representative of the entire field will be emphasized.

8.2
Reactive Polymer Thin Films

8.2.1
Motivation

Surface engineering is critical to developing devices and materials for new applications as surface chemistry governs the interaction of a material with its environment. Often, the preferred materials of construction, having optimal chemical and physical properties, do not present the desired surface characteristics; surface energy is an important example. Surface functionalization technologies allow engineers to decouple surface and bulk properties. Surfaces bearing reactive functionalities are convenient as they can be further exploited to covalently attach new molecules to the surface, providing even greater control over surface properties. Certainly, an array of commercially implemented surface treatment technologies can introduce reactive functionalities using energetic photons, free radicals, ions, or electrons. Examples include wet chemical etching, UV/ozone treatments, corona discharges, and cold plasmas. These energetic processes lead to complex surface chemistries, presenting multiple reactive surface functionalities on the surface. For advanced applications, selective processes creating homogeneous surfaces bearing specific reactive functionalities are desirable.

Reactive polymer thin films can be considered a more general approach to surface engineering as these films can be applied to all kinds of surfaces including metals, ceramics, polymers, and even liquids. One solution-based approach to modify surfaces with polymers entails *in situ* polymerization of polymer brushes grafted to the surface. This popular "bottom-up" approach is the subject of many research articles and has been reviewed in several extensive review articles and monographs [5, 6]. Vapor phase deposition of reactive polymer thin films is distinct from these polymer grafting approaches. Grafting density is an important parameter for the polymer brushes, while vapor deposition excels where complete, conformal, pinhole-free coatings are required. Thus, it is important to understand the capabilities and limitations of reactive polymer thin films synthesized by these approaches.

The goal of depositing reactive polymer thin films by chemical vapor deposition is to polymerize monomers containing chemical functionalities that are inert to the polymerization conditions but can be chemically modified subsequent to the polymerization using facile reaction schemes. Ideally, the reactive functionality can be quantitatively converted under mild reaction conditions to the functionality demanded by the applications. Synthesizing these reactive thin films is important for several reasons:

1) It circumvents limitations of polymerizing functional monomers. Despite advances in controlled polymer synthesis, such as controlled radical living polymerizations, many classes of functionalities remain incompatible with polymerization conditions [3]. Moreover, the desired functionality to be attached to the polymer is often biological in nature, such as a peptide or protein. Conventional polymerization reaction conditions are often incompatible with these molecules.
2) Chemical vapor deposition techniques demand sufficient volatility of the precursors for efficient transport of their vapors into the deposition chamber. As the molecular weight of the required pendant functionalities increases or if the functional monomers develop strong intermolecular attractions, the volatilities of the monomers drop. Often, the volatility can be so low that heating the precursor becomes unfeasible as monomer will decompose or react (thermally polymerize) before sufficient vapor pressure is generated. Utilizing volatile monomers bearing reactive functionalities is preferred as they provide efficient, scalable deposition conditions. The desired molecule can then be introduced through reaction with the complementary functionality within the polymerized film.
3) They provide a stable reactive platform that minimizes the number of chemical structures and compositions that must be characterized. Instead of optimizing synthesis conditions and characterizing physicochemical properties of new polymer compositions for each and every application, established reactive polymer compositions that have been thoroughly studied and characterized can serve as reactive templates, providing access to hundreds/thousands of new compositions and structures through simple post-polymerization modification schemes [4].

8.2.2 Examples of Functionalization Reactions

With the breadth of organic synthetic possibilities, it is impossible to make strict generalizations about all functionalizations of reactive polymer thin films deposited by CVD techniques, but general trends within the research literature can be identified. Reactive polymer films deposited by CVD typically serve as frameworks for post-polymerization modifications. An example where a CVD polymer film is used directly for a sensing application without further chemical modification is described in Section 8.2.4.3. The purpose of the post-polymerization modifications being to introduce new functionalities into the films, most modification reactions can be classified as either nucleophilic addition or substitution reactions. Nucleophilic additions are preferred for biomedical applications or other applications where residual contamination in the film may

Table 8.1 Select examples of reactive polymer thin films prepared by CVD techniques.

Polymer functional group	Complementary reactivity	References	
		iCVD	Parylene
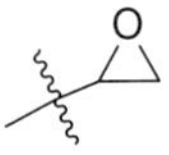	H_2N–R	[9–11]	—
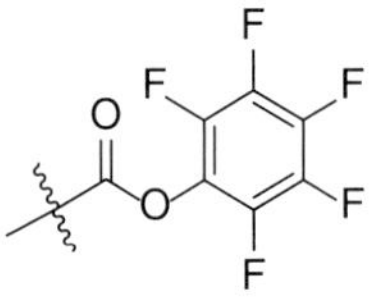	H_2N–R	[12–14]	[15, 16]
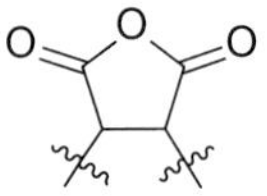	H_2N–R	[17–20]	[21]
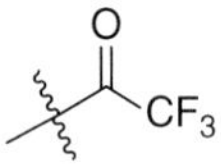	H_2N–NH–R	—	[22]
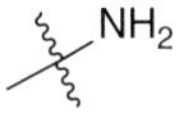	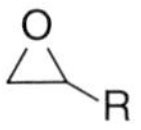	[9]	—
	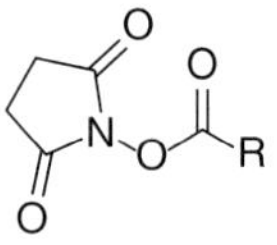	—	[23–25]
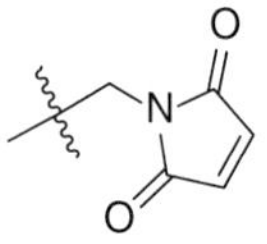	HS–R	—	[26]
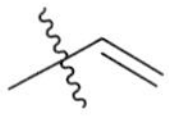	HS–R	—	[27]
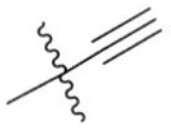	HS–R N_3–R	— [28, 29]	[27] [15, 30]
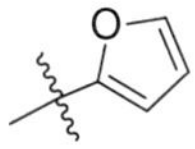	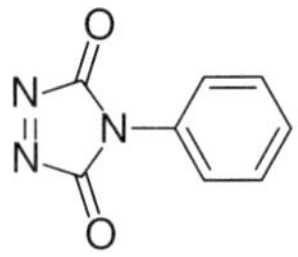	[31]	—

pose problems, as they do not generate by-products that must be thoroughly eliminated from the film. In order to minimize processing steps, modification reactions are desired that proceed with chemoselectivity in high yield via a single step. Furthermore, mild, near-ambient reaction conditions are often necessary. Surface modification with biological species (enzymes, antibodies, etc.,) place constraints on temperature, pH, and solvent. The physicochemical properties of the substrate will also constrain synthesis conditions. An example is the coating of water-soluble substrates [7, 8].

A large set of post-polymerization modification reaction schemes has been demonstrated. Table 8.1 provides examples of the reactive groups in polymer CVD films, along with the complementary reactive groups. These functionalities have been used in facile post-polymerization modifications. Typically, chemoselectivity of the modifications is achieved through kinetic effects. The first three polymer compositions listed in the table react with amines. All of these functional groups can also react with hydroxyls, yet due to the greater nucleophilicity of the amines, modification occurs primarily with the amine. Deng *et al.* successfully immobilized the epidermal growth factor (EGF) onto parylene thin films bearing pentafluorophenyl ester functionalities. The reaction medium consisted of $10\,\mu g\,ml^{-1}$ of EGF in phosphate buffered saline, yet the reaction between the amine in EGF and the pentafluorophenyl ester was dominant [15]. Successful functionalization of iCVD-synthesized poly(pentafluorophenyl methacrylate) (PFMA) with amine compounds in ethanol solutions has also been demonstrated [12]. Depositing poly(4-aminostyrene) (PAS) by iCVD, Xu *et al.* showed that it is possible to invert the modification approach, immobilizing the nucleophile rather than the electrophile in the polymer film [9]. Amine-bearing parylene-based films have also been developed [23, 24].

Important differences between nucleophilic substitution and addition reactions for post-polymerization modification becomes apparent by comparing the mechanisms of the amine reactions with two polymers from Table 8.1. Scheme 8.1 shows the nucleophilic addition of amines to poly(glycidyl methacrylate) (PGMA) and the nucleophilic substitution of amines in PFMA. Substitution reactions lead to a by-product that for many applications (e.g., biomedical implants) must be thoroughly eliminated. Addition reactions do not generate these byproducts. An interesting approach to addition functionalization reactions uses vaporized reactants, such that vacuum or a sweep gas can be applied after the reaction to eliminate any adsorbed, unreacted molecules. This concept has been demonstrated for cysteamine reactions with poly(styrene-*alt*-maleic anhydride) (PSMA). The primary amine reacts with anhydride functionality to form an amide bond, as confirmed by FTIR [32].

Thiol species (organic molecules bearing sulfhydryl functionalities) are also important nucleophiles for post-polymerization modifications, especially for bioconjugation applications. After amines, thiols are the second most common species used in bioconjugations [33]. Sulfhydryl functionalities, for example, can be found in cysteine residues in polypeptides. Tsai *et al.* developed maleimide-bearing parylene coatings, and demonstrated its post-polymerization

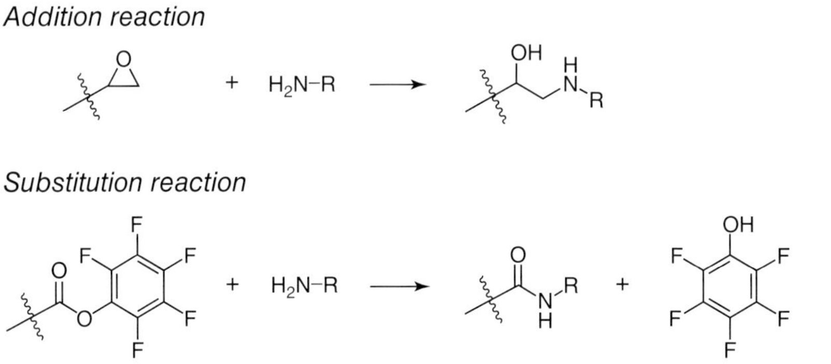

Scheme 8.1 Post-polymerization reaction of primary amines with reactive polymer films.

modification through alkylation modification of the coating using thiol addition, forming thioether bonds [26]. This is an example of a Michael addition reaction, showing high selectivity toward sulfhydryl conjugations. Thiols react with maleimides 1000 times faster than amines at pH 7 [33]. Tsai *et al.* reacted their maleimide-functional coating with fluorescein-labeled cysteine solution (aq) at pH 7.4 and room temperature without a catalyst [26]. This same group further extended their work on thiol-functional coatings by synthesizing poly(4-vinyl-*p*-xylylene-*co*-*p*-xylylene) and poly(4-ethynyl-*p*-xylylene-*co*-*p*-xylylene) using a modified parylene process. Using UV light and a photoinitiator to activate the modification reaction, the group was able to spatially pattern the reaction of thiol-functional poly(ethylene glycol) (PEG) with the pendant alkenyl and alkynyl functional groups within the parylene coatings [27].

These thiol-ene and thiol-yne reactions are examples of click chemistry. The term, click chemistry, is used to describe chemical reactions that (i) react efficiently and in high yields, (ii) are regio- and stereo-specific, (iii) are insensitive to oxygen and moisture, (iv) require mild reaction conditions, and (v) react orthogonally to other chemical functionalities [34]. The Huisgen 1,3-dipolar cycloaddition of an azide to a terminal alkyne to form a 1,2,3-triazole is the classic example of click chemistry [35]. The stability of the azide and alkenyl functionalities in typical organic synthesis conditions and toward biological molecules, makes this chemical synthesis particularly useful in biological applications [33]. In several iCVD studies, deposited films of Poly(propargyl methacrylate) (PPGMA) have been utilized in click chemistry reactions with azide-bearing target molecules [28, 29, 36]. In one example, PPGMA was deposited directly on the top of amine-bearing polymer using iCVD and then patterned into sub-micron features [28]. The orthogonal reactivity of the alkenyl and amine functional groups was demonstrated by reacting the alkenyls with an azide-bearing species and the amines with a *n*-hydroxysuccinimide ester-bearing species simultaneously in a one-pot aqueous solution.

Another important cycloaddition reaction is the Diels–Alder reaction. Diels–Alder is a 2,4-cycloaddition of an electron-rich diene and electron-poor alkene. It is an incredibly useful approach in synthetic organic chemistry to create six-membered rings with stereo- and regiochemical control, and it has been argued that some Diels-Alder systems can be classified as click chemical reactions [37]. Chen *et al.* used iCVD to deposit poly(furfuryl methacrylate) and then demonstrated the Diels–Alder reaction of the deposited film with 4-phenyl-1,2,3-triazolin-3,5-dione [31].

8.2.3
Important CVD Capabilities

The most important requirement in the synthesis of reactive polymer thin films is the retention of the chemical functional groups. Chemical functionalities define the reactivities of the films, and must be preserved to enable targeted, precise modification reactions. High degrees of functionality retention has been reported for at least 13 distinct polymer compositions derived from [2.2]paracyclophane in modified parylene processes [38]. iCVD also excels in applications where controlled film compositions are required: well over 20 distinct functionalities have been incorporated into polymer films. This number does not count the unique compositions synthesized through controlled copolymerization and crosslinking. Vapor deposition polymerization and molecular layer deposition (MLD) are additional techniques that provide high-fidelity polymer films. MLD can synthesize the following classes of polymers: polyamides, polyimides, polyureas, polyurethanes, polythioureas, polyesters, and polyimines [39].

The importance of the deposition process in defining the chemical functionality retention can be appreciated by comparing films deposited by iCVD and plasma enhanced chemical vapor deposition (PECVD). In PECVD, it is difficult to selectively control monomer activation in the plasma, resulting in functionality loss and the introduction of new moieties not originally present in the precursors. The anhydride functionality is highly reactive toward nucleophiles and is a versatile functionality for post-polymerization modifications. Figure 8.1 presents the polymer compositions of two maleic anhydride-bearing polymers: plasma-polymerized(maleic anhydride) [40] and iCVD-deposited

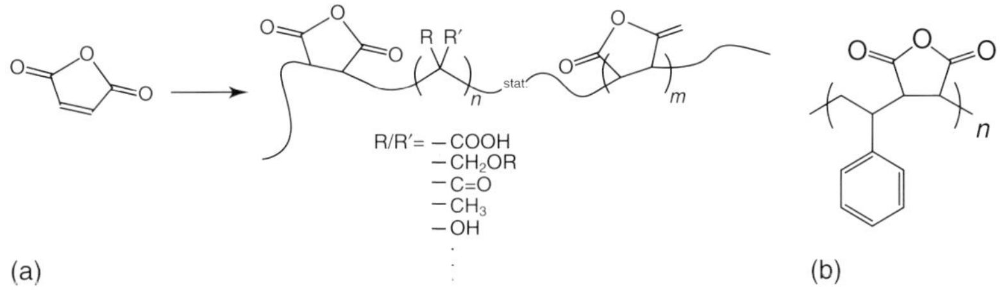

Figure 8.1 Polymer thin films bearing reactive anhydride functionalities deposited by (a) plasma-enhanced chemical vapor desposition and (b) initiated chemical vapor deposition. The plasma process (a) resulted in reduced anhydride concentrations and new moieties not present in the monomer. The alternating nature of poly(styrene-*alt*-maleic anhydride) deposited by iCVD was experimentally proven. ((a) Reprinted (adapted) with permission from Ref. [40]. Copyright 2001 American Chemical Society.)

poly(styrene-alt-maleic anhydride) (PSMA) [19]. FTIR characterization of the plasma-polymerized(maleic anhydride) revealed several new organic functionalities not present in the monomer structure. For applications where one needs to react the anhydride functionality with a nucleophile, these new functionalities reduce the potential functionalization density and can interfere with the functionalization reaction. Extensive FTIR, XPS, and ^{13}C NMR characterization of iCVD-deposited PSMA confirmed the composition and alternating nature of the PSMA film that was consistent with conventional solution-synthesized materials. Synthesis of a copolymer was necessary because maleic anhydride is known to undergo limited homopolymerization; the alternating nature of the copolymer is a consequence of the reactivity ratios of the comonomers [41]. Importantly, the high density of anhydride functionalities was maintained, which were used for post-polymerization modifications with amines [32].

Another important capability of chemical vapor deposition techniques is the ability to precisely control the flow rates of the precursors, which in turn governs film compositions. Mao and Gleason showed that it is possible to tune copolymer compositions and using the Fineman–Ross method, developed for conventional liquid phase copolymerizations, determined copolymerization reactivity ratios [42]. Control over copolymer chemistry for the parylene process has also been achieved. Elkasabi *et al.* demonstrated multi-potent parylene compositions containing two, orthogonally reactive chemical functionalities [23]. Deposition strategies were developed to create surface gradients of the two reactive functionalities [24]. Moreover, CVD offers the opportunity to copolymerize monomer pairs that would be difficult to achieve in solution. Because CVD does not use a solvent, comonomer miscibility and the existence of a common solvent is not a requirement. The fluorinated monomer perfluorododecylacrylate (PFA) is hydrophobic, while 2-hydroxyethyl methacrylate (HEMA) with a pendant hydroxyl group is hydrophilic, making solution-based copolymerization difficult. Yet, Baxamusa and Gleason were able to successfully copolymerize this system using piCVD [43]. piCVD is a variant of the iCVD process that uses UV light for initiator activation [44]. The advantage of solvent-free synthesis was also exploited in copolymerizing pentafluorophenyl methacrylate, which is also hydrophobic, with HEMA. Montero *et al.* described the synthesis of a graded film composition, where the bulk of the film was HEMA [13]. The HEMA component was important because it imparted hydrogel properties to the film. The authors wanted to create a reactive film composition that could be readily used to immobilize biomolecules, while maintaining the bulk hydrogel properties of the film. In the initial stages of the deposition, the copolymer composition was HEMA-rich; the precursor feed concentrations were then altered during the deposition to enrich the outer layers of the film in pentafluorophenyl methacrylate. The bulk of the film and the physicochemical properties were defined by HEMA because the reactive pentafluorophenyl functional group was confined to the region near the polymer/solution interface.

Adhesion of reactive polymer thin films to their substrates is an important property as the films may experience intense mechanical, thermal, and solvent forces. At the very least, robust adhesion is required for post-polymerization modifications of the films in liquid media. For some polymer compositions and microstructures, especially polymers with aromatic backbones, physical dispersion forces between the film and substrates can provide sufficient adhesion. In many applications, parylene coatings rely on these physical forces without further adhesion promoters. Enhanced adhesion, however, can be achieved by generating covalent bonds between the substrate and polymer moieties. MLD is predicated on the presence of reactive surface functionalities with complementary reactivity to one of the monomers for the formation of the first organic monolayer [39]. Organosilane compounds are effective and versatile adhesion promoters. The central Si atom can form four covalent bonds; at least one bond is to a functional organic component while the remaining bonds are hydrolyzable (e.g., alkoxy). After hydrolysis of the hydrolyzable groups, the resulting silanol functionalities condense with surface hydroxyls. Organic components are selected that will react with monomers during the vapor deposition process. Trichlorovinylsilane (TCVS) is a common choice for iCVD coatings, while Silane A174 (3-(trimethoxysilyl)propyl methacrylate) is popular for commercial parylene applications. The functional organic component in both TCVS and Silane 174 is an alkenyl (vinyl) group. Radicals generated during the iCVD and parylene process react with the surface-tethered vinyls incorporating them into the polymer chain – resulting in a covalent linkage between the substrate and the polymer film.

8.2.4 Applications of Reactive Films

8.2.4.1 Bioconjugation

Development of successful biomaterials and biomedical devices relies on controlling the physical and chemical properties of surfaces. Interactions between biological entities, such as proteins and cells, and synthetic substrates strongly influence the biological activities of said entities [14]. Understanding and controlling these interactions through synthetic techniques fall within the purview of biointerface engineering. Given the diversity of biomaterials (soft, water-laden polymer networks; hard and inert plastics; metals; etc.,), a substrate-independent interfacial engineering technique, such as polymer CVD, offers value because the same reactive platform can be applied across a spectrum of substrate materials.

Parylene coatings where the compositions have been engineered with reactive pendant functionalities have been thoroughly investigated for biomedical applications [15, 23, 16, 22, 27, 45, 46]. Lahann *et al.* developed reactive parylene coatings to decrease the thrombogenicity of Nitinol coronary stents [45]. The reactive parylene coatings with thicknesses of about 100 nm were deposited on the stents, then r-hirudin, a protein with anticoagulant activity, was covalently immobilized to the reactive coatings. The authors showed that this immobilization scheme was effective in improving blood compatibility, given the lower thrombogenicity and reduced platelet adhesion.

A popular biointerface engineering strategy exploits the strong, specific, and stable noncovalent interaction between the biological species biotin and streptavidin. Biotin is a small molecule target and streptavidin is a bacteria-derived protein that has four biotin-binding sites. Biotin–streptavidin binding is one of the strongest noncovalent interactions known, and the binding serves as the basis for highly specific bioconjugation reactions and assay designs. Chen *et al.* deposited a reactive parylene film inside the microchannels of a pre-assembled microfluidic system, demonstrating the capability to coat complex geometries having aspect ratios of at least 37 [22]. The parylene composition contained trifluoromethyl ketone moieties, which were reacted with biotin-hydrazide, as shown in Figure 8.2. Fluorescently labeled streptavidin was then introduced to bind with the biotin. The fluorescent dye confirmed that the biotin was present throughout the microfluidic device. Parylene copolymers with aldehyde and primary amine pendant functionalities were also deposited to generate a spatial compositional gradient. Using biotinylated reagents that were reactive toward either the amine or the aldehyde, Elkasabi *et al.* were able to create a gradient in the surface-tethered avidin [25]. This gradient in avidin then resulted in a gradient in biotinylated adenovirus that was incubated over the substrate. The ultimate objective of creating a gradient in human gingival fibroblasts on the surface in order to mimic natural cell signaling gradients was accomplished by culturing the cells over the adenovirus gradients. Beyond biotin–streptavidin conjugation approaches, several alternative strategies to immobilize proteins/enzymes on parylene thin films have been developed [15, 46].

8.2.4.2 Nonfouling Surface

Another important area of biointerface engineering is developing nonfouling surfaces that resist nonspecific, irreversible adsorption of proteins and other foulants. Hydrophilicity is a critical parameter in defining nonfouling surfaces, and several examples of thin film hydrogel layers have been demonstrated. Hydrophilic polymer compositions, including poly(2-hydroxyethyl methacrylate) (PHEMA) [44, 47, 48], poly(vinyl pyrrolidone) (PVP) [49–51], and poly(ethylene oxide) [52] have been synthesized by CVD. Surface functionalization with polymer brushes, especially those composed of PEG, is a common strategy for nonfouling applications. Polymer CVD techniques synthesize collapsed, dense polymer films – not polymer brushes. To realize the nonfouling properties of polymer brushes, both reactive parylene [26, 27] and iCVD [13] films have been functionalized with short-chain PEG molecules. Using two distinct reactive parylene-based films, patterns of nonfouling regions were prepared via microcontact printing [26, 27]. The masks were inked with sulfhydryl-bearing PEG coatings, which reacted with complementary functionalities within the parylene films, anchoring the PEG through the sulfhydryl group and allowing the ethylene oxide repeat units to extend away from the surface. The authors also demonstrated the brushes' effectiveness in preventing protein and cell adhesion. Another promising strategy for nonfouling surfaces uses zwitterionic brushes, which strongly bind water molecules. The results of zwitterionic chemistries for nonfouling layers on membranes will be discussed

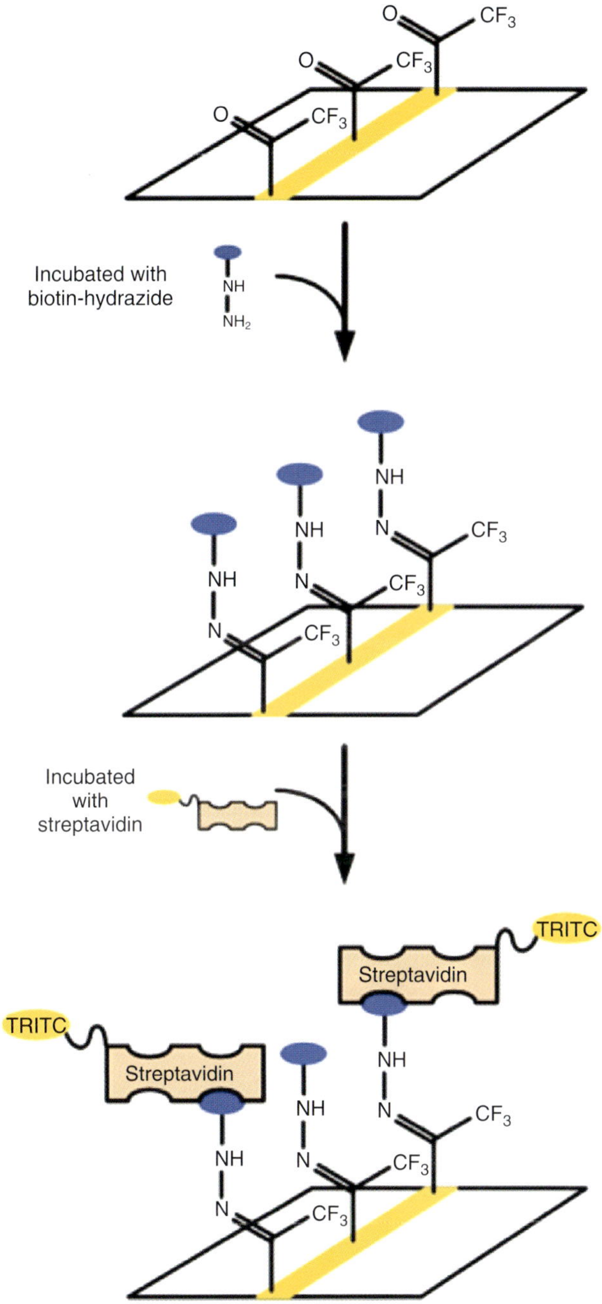

Figure 8.2 Functionalization of reactive parylene thin film with biotin, followed by binding of fluorescently labeled streptavidin. (Reprinted (adapted) with permission from Ref. [22]. Copyright 2006 American Chemical Society.)

in Chapter 13. These zwitterionic compositions are synthesized by reacting secondary amines of iCVD coatings with 1,3-propanesulfone [53].

8.2.4.3 Sensors

Highly selective chemical sensors for the detection of low concentrations of specific analytes requires the use of chemically selective transduction mechanisms. One strategy is to base the sensing transduction mechanism on a chemical reaction. The chemical or physical effect of the chemical reaction is transduced to a readily measured signal, such as an electrical current. The reaction ensures a high level of specificity as only analytes participating in the chemical reaction will be sensed. Ideally, the reaction would be reversible in order to regenerate the sensor, but there are applications where one-use sensors are desired.

Arora *et al.* developed selective chemical sensors based on an electromechanical switch to detect amine molecules [17]. The detection of amines is important for industrial/environmental monitoring, food quality control, and disease diagnosis [54]. Microcantilevers were fabricated in ultrathin Si_3N_4 membranes. One side of the cantilever was coated with gold, while the other side was coated with an iCVD copolymer bearing reactive anhydride functionalities. The microcantilevers were positioned above interdigitated electrodes (see Figure 8.3); the resistance across these interdigitated electrodes was initially greater than 500 MΩ (the limit of the resistance meter). Upon reaction of the anhydride functionalities with amines, the stress that developed in the iCVD coating led to the microcantilever's deflection, pushing the gold-coated side of the cantilever into contact with the electrodes. The shorted electrode array, shown in Figure 8.3, resulted in a drop in resistance greater than 6 orders of magnitude. The benefit of this sensing mechanism is that negligible power is required initially as the interdigitated electrodes form an open circuit. Upon a sensing event, the cantilever deflects and closes the circuit, at which point the electrical current can flow. The ability to precisely control the composition of the iCVD coating with the reactive anhydride functionality was instrumental to the success of this application as it was discovered that intentionally crosslinking the film results in enhanced stress and greater deflection due to the chemical reaction.

In another chemical sensing approach based on chemical reactions, the principles of the iCVD synthesis mechanism were adapted into a signal transduction mechanism for the detection of peroxide species, which are an important class of explosives [55]. Monomers were immobilized on the surface of a microcantilever. The peroxide analytes to be detected were passed over a heated filament array in order to homolytically cleave the peroxide chemical bond, yielding radical species (this is a key step in the synthesis of iCVD polymers). The radicals then reacted with the vinyl monomers on the surface of the microcantilever, resulting in the deflection of the microcantilever, which was then measured optically.

8.2.4.4 Adhesive Bonding of Microfluidics

Microfluidic devices, which are important research tools across many technological disciplines, require sealing the microfluidic channels, typically fabricated

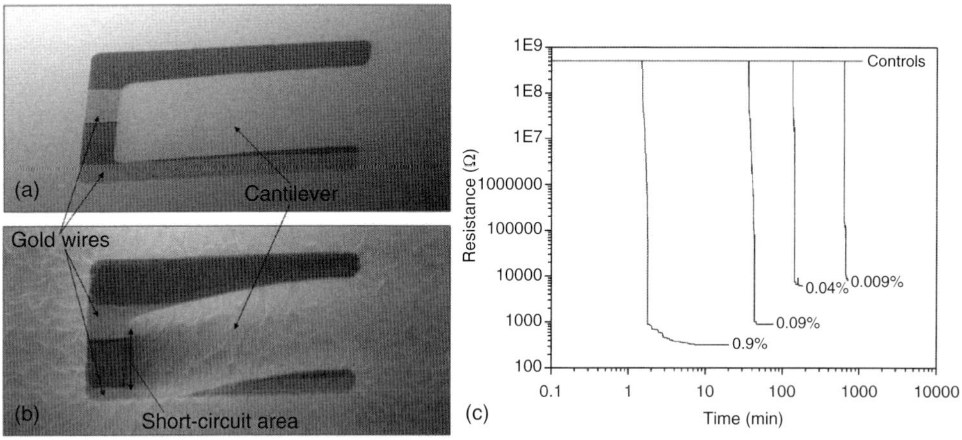

Figure 8.3 (a) Scanning electron micrographs of microcantilever coated with a polymer thin film bearing maleic anhydride functionalities positioned above an electrode array (a) before and (b) after exposure to hexylamine vapors. The hexylamine reacts with the maleic anhydride functionalities, creating film stress and consequently, cantilever deflection. (c) Electrical resistance measured across the electrodes, which are initially electrically isolated, as hexylamine is introduced. Copyright 2009 IEEE. Reprinted, with permission, from [17].

from poly(dimethylsiloxane) (PDMS), to a substrate to enclose the channels. The most popular materials for microfluidics systems are PDMS, glass, and Si. While several options exist for bonding the channels to the substrate, the most popular approach uses oxidative activation of the surface functionalities in the PDMS, glass, or Si using either an oxygen plasma or UV/ozone treatment. This approach is convenient because it is simple, solvent-free, and low temperature. The exclusion of solvent and high temperatures is important as they can lead to losses in fidelity of the exquisitely designed microchannels. However, the approach is not general; the plasma and UV/ozone-bonding process is not suitable for several alternative construction materials, such as PMMA, polyolefins, polystyrene, and so on.

The chemical vapor deposition of reactive polymer compositions as thin coatings on top of microfluidic components, has been investigated as an alternative bonding approach that is applicable to a broad swath of construction materials [9, 11, 56, 57]. The channel block is coated with one reactive polymer composition, while the substrate is coated by a polymer bearing complementary reactivity. Similar to plasma and UV/ozone bonding, this technique is completely dry (does not require the use of solvents) and requires low temperatures. Chen *et al.* deposited two parylene-based coatings with complimentary reactivities (one with primary amine functionality and the other with formyl functionality) onto PDMS microchannels and the following substrates: PDMS, PTFE, glass, Si, and stainless steel [57]. The tensile bonding strength of PDMS with the substrates was characterized, showing that the bond strengths between the parylene-treated PDMS and substrates, was higher than 2.44 MPa in each case, except PTFE. The bond strength of PDMS–PDMS conventionally treated with oxygen plasma was lower at 2.34 MPa.

iCVD has been investigated for this application, as well. The chemistries of these iCVD systems have been exclusively modeled after common epoxy chemistry: the reaction between primary amine and epoxide functionalities. Im *et al.* deposited PGMA onto the PDMS microchannels via iCVD and plasma polymerized allyl amine onto the substrates, which were Si, PTFE, polystyrene, polycarbonate, and poly(ethylene terephthalate) [11]. The strength of the bond was measured by determining the pressure at which the systems began to leak. For all substrates, leakage was not observed below 50 psi, which was considered sufficient for most microfluidic applications. Xu and Gleason developed an iCVD synthesis method for poly(4-aminostyrene), having a reactivity complementary to PGMA [9]. The new iCVD method for PAS was developed because it was hypothesized that the iCVD film should provide a higher surface concentration of amines relative to plasma-polymerized(allyl amine), thus leading to stronger bonding. They demonstrated burst pressures in excess of 150 psi. You *et al.* also used PGMA coatings prepared by iCVD for adhesive bonding, but instead of developing two coatings with complementary reactivities, PGMA was coated on both pieces of the microfluidic system [56]. Difunctional ethylenediamine, with its two primary amines, was used to create a covalent bridge between the two PGMA coatings. The minimum burst pressure for all the substrates tested was 363 psi, and the best burst pressure of 1700 psi was obtained for poly(ethylene terephthalate) bonded to glass. An additional benefit obtained from each approach described above is that the inner surfaces of the microfluidic channels are coated with the reactive polymer film. These functionalities remain chemically intact because destructive surface oxidation processes are not required for the bonding and can be later utilized for surface immobilization of molecules inside the channels after assembly [9, 57].

8.3
Responsive Polymer Thin Films

Stimuli-responsive polymers are polymers that undergo conformational or physical changes upon application of an external signal (stimulus). These changes lead to changes in the physical properties of the polymer. When designing these thin films, often called "reconstructable" or "smart" surfaces, the objective is to provide for dramatic transformations in wettability, film thickness, and/or mechanical, optical, and electrical properties. The key difference between reactive and stimuli-responsive polymer films is the emphasized property. When creating reactive polymer thin films, the emphasis is on designing specific surface compositions for functionalization schemes, whereas in stimuli-responsive polymers, even though compositional changes may occur, the physicochemical changes are most important. Polymers have been engineered to respond to stimuli such as light, electric and magnetic fields, pH, ionic strength, temperature, and chemical and biochemical species [1, 2].

The response of macroscopic polymers to stimuli typically occurs over relatively long time frames; response times can be on the order of several hours to days. By shrinking polymers to nanoscale dimensions, response times can be reduced to just seconds, opening the practicality of these materials for many applications. When the responsive film is used in a liquid environment, it is necessary to ensure that films will not dissolve or delaminate into the liquid. Polymer brushes have received quite a bit of attention for their responsive behavior [1]. Polymer brushes are synthesized either by "grafting-to" or "grafting-from" approaches, such that at least one end of the brush is immobilized physically or covalently to the surface, which provides stability of the surface layer in liquid environments. Polymer films deposited by CVD, on the other hand, are collapsed layers of polymer chains. For the polymer molecules not in immediate contact with the substrate provisions must be made to ensure stability in liquid environments. Introducing chemical crosslinking into the film effectively transforms the film into a single macromolecular network, thus preventing dissolution. For responsive polymer films, controlling the extent of crosslinking is integral to defining the stimuli-responsive behavior. Extensive crosslinking limits the amount by which the physical properties can change. Thus, CVD techniques that can provide precise control over crosslinking densities offer distinct advantages when engineering stimuli-responsive films. By carefully quantifying and controlling crosslink density in iCVD-deposited PHEMA hydrogel films, Chan and Gleason showed that these films can be engineered to increase in thickness by 10–55% due to moisture uptake [47]. Because composition controls function, the theme of chemical functionality retention is central in the discussion of stimuli-responsive polymer films by CVD as was the case for reactive polymer films.

8.3.1 Chemical-Responsive Polymers

The thermodynamics of mixing of linear polymers with low molecular weight species (solvents) can favor dissolution of the polymer, resulting in a polymer solution. The Flory–Huggins parameter for a solvent–polymer system governs whether mixing of the polymer/solvent system will occur. The Flory–Huggins parameters effectively describe specificity of this mechanism, as good solvents will mix with the polymer, while poor solvents will not or only to a limited extent. In polymer networks that are chemically crosslinked, the network cannot fully dissolve even though large amounts of solvent can be absorbed. This swelling behavior can be described by the Flory–Rehner equation [58].

When polymers are assembled as thin films and undergo swelling, the substrate constrains dimensional expansion in the plane parallel to the substrate, and expansion is primarily limited to the normal direction. This expansion is recognized as a film thickness increase, quantified through the swelling ratio, which is the ratio of the swollen thickness to the original thickness. The swelling induces stresses within the film, which have been successfully exploited for chemically sensitive transduction mechanisms in sensors. Polymer thin films are coated on microcantilevers, where analyte absorption and subsequent swelling of polymer thin film coatings leads to the deflection of the microcantilever. Optical [59], piezoelectric [60], piezoresistive [61], and capacitive [62] schemes have been used for measuring these deflections.

Exercising the conformal coating capability of iCVD, Tenhaeff *et al.* developed a new sensor design for the detection of nitroaromatics, where a high aspect ratio nanoscale trench (width was 600 nm) in a SiO_2-coated Si wafer was coated conformally with poly(4-vinylpyridine) (P4VP) [63]. A thin Au/Pd conductive layer was deposited on top of the P4VP. Initially, the Au/Pd conductive layers on the P4VP on the two sides of the trench were separated by approximately 75 nm (see Figure 8.4). Upon exposure to nitroaromatic vapors, which interacted strongly with the vinylpyridine pendant functionalities and absorbed into the films, the swelling led to the contact between the Au/Pd layers, which manifested as a dramatic decrease in resistance from the original open circuit condition. This was an off-on design, where negligible power was required to operate the sensor until a sensing event occurs, and the detection scheme is very simple and inexpensive to implement relative to more complex schemes, where optoelectronic components or circuits are required to detect even minor changes in electrical properties. It was shown that the sensor system offers a degree of specificity, as exposure to control vapors, including water, ethanol, and toluene, did not measurably alter the resistance because they do not swell P4VP by a large enough degree for contact between the sides.

Using the favorable mixing interactions between P4VP and nitroaromatics, Petruczok *et al.* developed another novel microscale sensing system for the detection of nitroaromatic explosives [64]. Standard microelectronic processing was employed to pattern polymer lines onto substrates with thicknesses varying 100–300 nm and widths from 500 nm to 100 μm. Patterned metal lines were then fabricated perpendicular to and on top of the polymers lines. Nitroaromatic vapors were absorbed by the polymer lines, and the resulting swelling caused the metal to plastically deform, resulting in increased resistance measured across the lines. The measured resistance increased by over 8%, and it was shown that the nanoscale cracks developed in the metal lines due to the swelling of the polymer. Consequently, the mechanism was irreversible and the resistance remained elevated as the nitrotoluene was purged out of the system. This sensing approach is envisioned being integrated into microelectronic chips in low cost, one-use devices. The entire active sensing area was less than 10 μm^2.

Karaman *et al.* showed that Bragg mirror devices fabricated from alternating layers of titania and iCVD-deposited PHEMA can function as sensitive and fast sensors for moisture vapor [48]. Alternating layers of titania and PHEMA were fabricated using the same hot filament CVD system by switching the precursor vapors. This led to sharp and uniform interfaces, with a large contrast in refractive index between the two layers. The measured refractive indexes

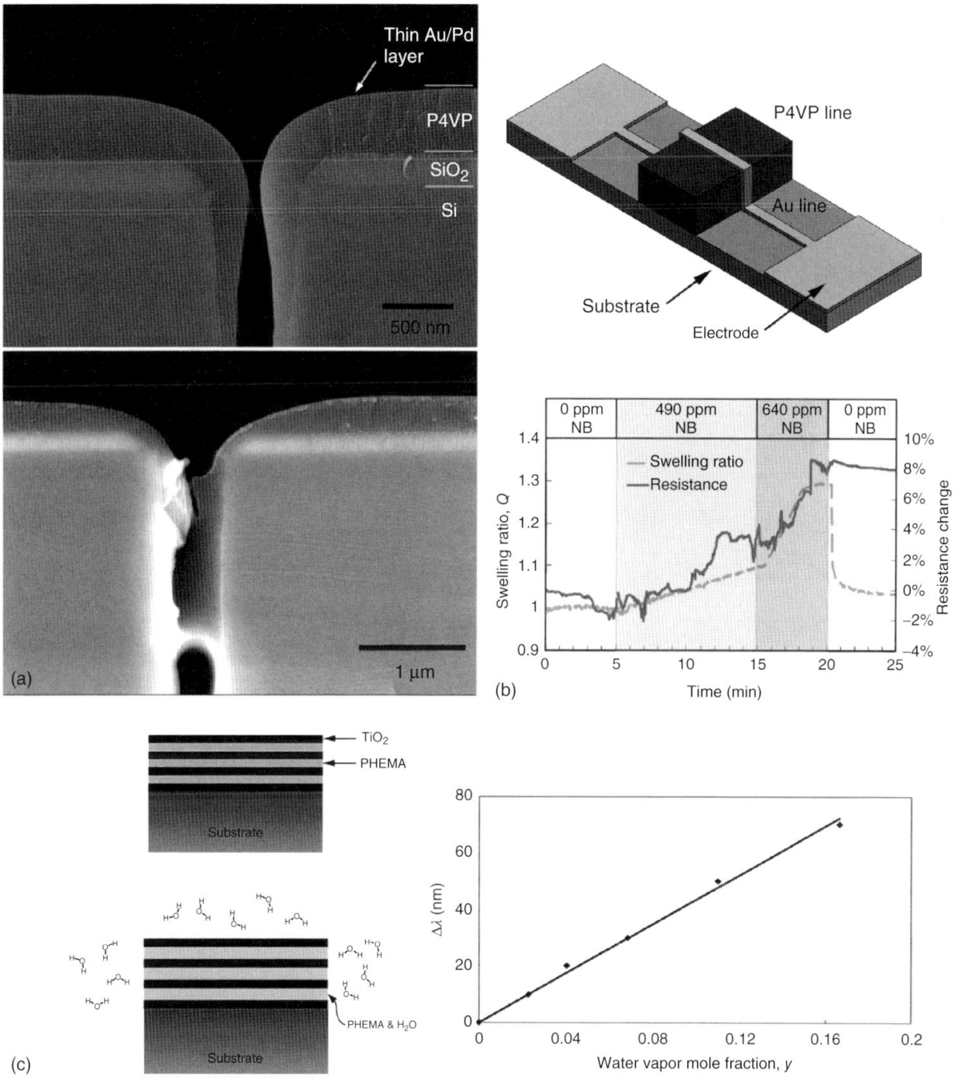

Figure 8.4 (a) 600 nm-wide trench fabricated in Si that has been coated with nitroaromatic-responsive poly(4-vinylpyridine) (P4VP) before (top) and (after) exposure to nitroaromatic vapors. A thin Au/Pd layer on top of the P4VP makes contact due to the swelling. (b) An alternative nitroaromatic detection system, where a nitroaromatic vapors are absorbed by a patterned line of P4VP. The change in resistance of the Au line as it plastically deforms due to P4VP swelling is measured. (c) Schematic of a Bragg mirror fabricated from alternating layers of vacuum deposited titania and PHEMA. Moisture absorption into the PHEMA layers, results in swelling, which alters the wavelength of maximum reflection as shown on the right. ((a) Reprinted (adapted) with permission from Ref. [63]. Copyright 2010 Wiley-VCH. (b) Copyright 2013 IEEE. Reprinted, with permission, from [64]. (c) Reprinted (adapted) with permission from Ref. [48]. Copyright 2008 American Chemical Society.)

before exposure to water vapor were 1.8 for titania and 1.5 for PHEMA. With seven layers of titania-PHEMA, the maximum reflected intensity occurred at 460 nm. Upon exposure to 6.7 mol% water vapor in nitrogen, the increased film thickness and reduced refractive index of the PHEMA layers resulted in a 30 nm shift in the wavelength of maximum intensity (see Figure 8.4). It was shown that the swelling is completely reversible; the maximum reflection intensity returned to its original wavelength and intensity after purging with dry nitrogen.

The swelling of PHEMA thin films by water has also been used for applications where the surface is submerged in aqueous solutions. PHEMA is an important hydrogel, and it has been thoroughly studied as a biomaterial. The water content in hydrated PHEMA and natural tissue is similar [65], making PHEMA especially suited for biomedical applications. Indeed, PHEMA has been used in many biomedical applications, including contact lenses, artificial cornea, drug delivery systems, tissue engineering, and composite materials for bone [66]. The extent of water absorption and swelling in PHEMA is of paramount importance as it governs the mechanical compliance of the gel, as well as the mesh size, which is the distance between crosslinks. The mesh size controls the diffusional transport of material through hydrogels. The work of Chan and Gleason showing the fine control of crosslinking in PHEMA thin films by iCVD was seminal work. Several papers have used the crosslink tunability to coat sensors with thin hydrogel coatings to prevent proteins from fouling the sensing mechanism, while allowing diffusion of the analyte [44, 65, 67]. The relationships between composition and gel properties have also been investigated in order to control protein adsorption [68] and cell behavior [14]. Further discussion of CVD polymers developed for bio-related applications can be found in Chapter 14.

8.3.2
pH Responsive Polymers

The maximum swelling ratio of 1.55 for iCVD-deposited PHEMA in water is achieved for uncrosslinked films [47]. For applications requiring greater extents of swelling, polyelectrolyte thin films have been prepared. In polyelectrolyte gels, the osmotic pressure effect of the counterions and, to a lesser extent, the electrostatic repulsion of the charged polymer chains, are dominant contributions to the enhanced water uptake and swelling [69, 70]. Owing to these ionization effects, polyelectrolyte gels display pH-dependent swelling behavior. An anionic polyelectrolyte gel will convert from a protonated to an ionized, deprotonated state as the pH of its medium is increased above its pK_a, resulting in enhanced degrees of swelling. The reverse behavior is observed for cationic polyelectrolytes, where protonation of the base at low pH levels results in positive charges on the backbone as well as increased swelling. Lau and Gleason were the first to demonstrate pH-responsive polymer thin films synthesized by iCVD [8]. Copolymers of methacrylic acid, ethyl acrylate, and ethyleneglycol dimethacrylate were prepared. At pH values of 1.2 and 5, a swelling ratio of 1.05 was reported. At neutral pHs, the methacrylic acids deprotonated, and swelling ratios in excess of 1.3 were observed.

Tenhaeff and Gleason developed high swellable hydrogel thin films by copolymerizing *N*,*N*-dimethylacrylamide with maleic anhydride; the films were lightly crosslinked with di(ethylene glycol) divinyl ether [18, 71]. *N*,*N*-dimethylacrylamide was chosen because its homopolymer is considered a "super-absorbent polymer," and hydrolysis of maleic anhydride yields two adjacent carboxylic acid groups. Swelling ratios in excess of 11.5 were achieved in basic solutions [18]. The adjacent carboxylic acids have two distinct dissociation constants ($pK_{a1} = 2.8$, $pK_{a2} = 6.1$), giving rise to three regions with characteristic swelling ratios. Below pH 3, the swelling ratio was approximately 5. Between pH 3 and 7, the swelling ratio was 9. The maximum swelling ratio exceeded 11 above a pH of 8. The swelling of a 150 nm thick film was shown to be both reversible, by cycling between pHs of 0.6 and 8, and extremely fast, reaching 95% of its equilibrium swelling ratio in less than 2.5 min. Karaman and Cabuk also characterized the pH-dependent behavior of iCVD-deposited poly(2-(diisopropylamino)ethyl methacrylate) [72]. Contact angle measurements were used to quantify the wettability of the deposited films. Alternating exposure of the films to pH 2.75 and 8.5 solutions, the contact angle reversibly cycled between 28° and 87°, respectively.

pH-responsive hydrogel films and coatings have been used in controlled release applications. Lau and Gleason coated 25 μm ibuprofen microcrystals with an iCVD-deposited, crosslinked copolymer coating, where the major component was methacrylic acid [8]. The molecular structure can be found in Figure 8.5a. When ionized, the copolymer coating was a negligible barrier to diffusion, and there was an insignificant difference in the amount of time required for ibuprofen to completely dissolve into pH 7 solutions. At pH 1.2, however, the copolymer is protonated and absorbs less water, thus providing greater diffusional resistance. Approximately 70 min was required for the complete dissolution of the uncoated ibuprofen in pH 1.2 solution, while dissolution of the coated ibuprofen required over 160 min. Using a different device design, McInnes *et al.* were able to control and modify the diffusional transport of another drug, camptothecin, using iCVD-deposited poly(methacrylic acid-*co*-ethylene glycol dimethacrylate) [poly(MAA-*co*-EGDA)] [73]. Camptothecin was loaded into a porous Si substrate, and then coated with the iCVD polymer. The controlled release system is depicted in Figure 8.5. Drug release was measured at pH 1.8, where the methacrylic acids were protonated, and at pH 7, where the methacrylic acids were deprotonated, leading to extensive swelling and a larger hydrogel mesh size. The effect of swelling and mesh size on diffusion out of the porous Si was experimentally confirmed: 13% of the loaded drugs was released from the iCVD-coated porous Si in 10 h at pH 7, while only 3% was released over the same time frame at pH 1.8. In control experiments on uncoated porous Si, the kinetics of drug release was identical at pH 1.8 and 7.

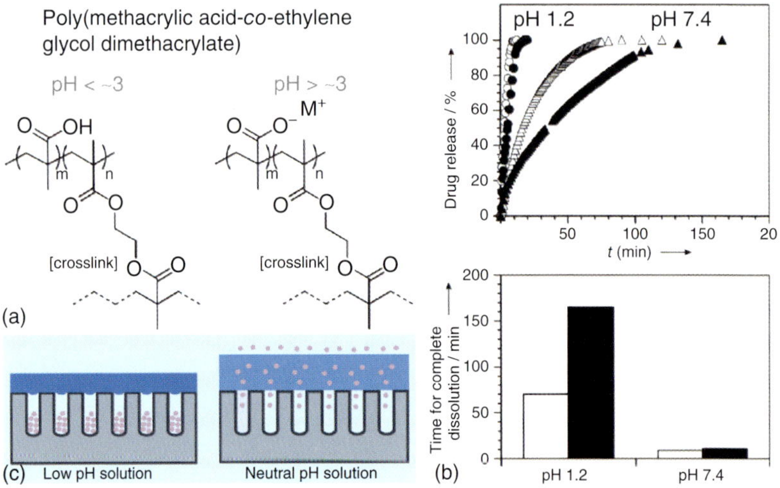

Figure 8.5 (a) Composition of poly(methacrylic acid-*co*-ethylene glycol dimethacrylate) [poly(MAA-*co*-EGDA)] in the protonated and deprotonated states. (b) Evaluation of poly(MAA-*co*-EGDA) deposited by iCVD onto 25 μm ibuprofen particles. The kinetics of dissolution of uncoated (white) and iCVD-coated (black) particles were compared. (c) Control of camptothecin release from a porous substrate using the pH-responsive behavior of poly(MAA-*co*-EGDA). (b) Reprinted (adapted) with permission from Ref. [8]. Copyright 2007 Wiley-VCH. (c) Reprinted (adapted) with permission from Ref. [73]. Copyright 2012 American Chemical Society.)

8.3.3
Temperature-Responsive Polymers

In solution, several polymers show interesting phase behavior defined by a lower critical solution temperature (LCST) or upper critical solution temperature (UCST). Of the many polymer/solvent systems showing this behavior, poly(*N*-isopropylacrylamide) (PNIPAM) hydrogels are among the most extensively studied systems because the LCST of 32 °C is conveniently positioned between room temperature (circa 20 °C) and physiological temperatures (37 °C). Hydrogen bonding is largely responsible for the phase behavior of PNIPAM in water [74]. Below the LCST, hydrogen bonding between PNIPAM's pendant group and water lead to water infusion into the hydrogel. As temperature is raised above the LCST, the reduced entropy of the system dominates over enthalpic effects and the system phase separates, expelling water from the hydrogel. Slight deviations in the energetics of hydrogen bonding can dramatically influence phase behavior. Copolymerization of NIPAM provides a handle for fine tuning the LCST [75]. However, where the composition cannot be precisely controlled, the melee of organic functionalities leads to broadening of the temperature range over which phase transition occurs. This has been observed in plasma polymerized NIPAM, where it is difficult to independently control the crosslinking, which ensures stability of the surface immobilized gel, and retention of the pendent chemical moieties, which govern hydrogen bonding interactions [76–78].

Surface-tethered PNIPAM hydrogels have been deposited by iCVD, providing functional retention and controllable crosslinking [79, 80]. Compared to linear PNIPAM, iCVD-deposited PNIPAM has a slightly broadened phase transition and reduced LCST (approximately 28.5 °C) due to the interaction of the polymer with the substrate and necessary crosslinking with a divinyl comonomer. Alf and Gleason, however, demonstrated that iCVD provided sharper phase transitions relative to other film formation techniques [80]. On planar Si wafers coated with PNIPAM, the phase transition above the LCST resulted in the contact angle changing by more than 30 °C and a 3 times decrease in film thickness. They also synthesized the coating on electrospun nanofiber mats and multi-walled carbon nanotube forests, providing even more dramatic changes in contact angles, as shown in Figure 8.6.

The capability of iCVD to deposit PNIPAM on microstructured surfaces has been used for tissue engineering applications [81]. Microstructured PDMS was coated with iCVD PNIPAM and seeded with cells, which were cultured for tissue formation at 37 °C. The PNIPAM layer at 37 °C was compatible with the tissue. Later, to release the tissue from the PDMS mold for harvesting, the temperature was reduced below the LCST, which increased the hydrophilicity of the PNIPAM, thereby releasing the tissue.

8.3.4
Piezoelectric Polymers

Piezoelectric polymers are dielectric materials that develop an electric field when subjected to mechanical stress. This is termed the direct dielectric effect. Alternatively, the indirect dielectric effect describes the application of an electric field to create mechanical strain. Piezoelectric polymers have a permanent dipole, the magnitude of which changes with applied stress. Piezoelectric polymers are often semi-crystalline, where the crystalline phase lacks inversion symmetry, but amorphous polymers can also be piezoelectric. Poling processes, consisting of heating the piezoelectric material *in vacuo* and applying a large electric field across its boundaries (10–100 MV m^{-1}), can be used to align dipoles and enhance the piezoelectricity of polymers. The piezoelectric properties of thin films and bulk polymers are different due to distinct crystallite growth during film deposition and the role of substrate constraints on film microstructure [82].

Poly(vinylidene difluoride) (PVDF) and its copolymers are important semicrystalline piezoelectric polymers because, they have the largest piezoelectric coefficients compared to other piezoelectric polymers. PVDF thin films consisting primarily of oligomeric chains can be synthesized through vacuum evaporation techniques; the substrate temperature during the evaporation controls the crystalline phases in the deposited film [83]. Poling the substrate during the evaporation is effective in further aligning the crystallites and enhancing the piezoelectric coefficient. Rastogi and Desu developed a chemical vapor deposition method for

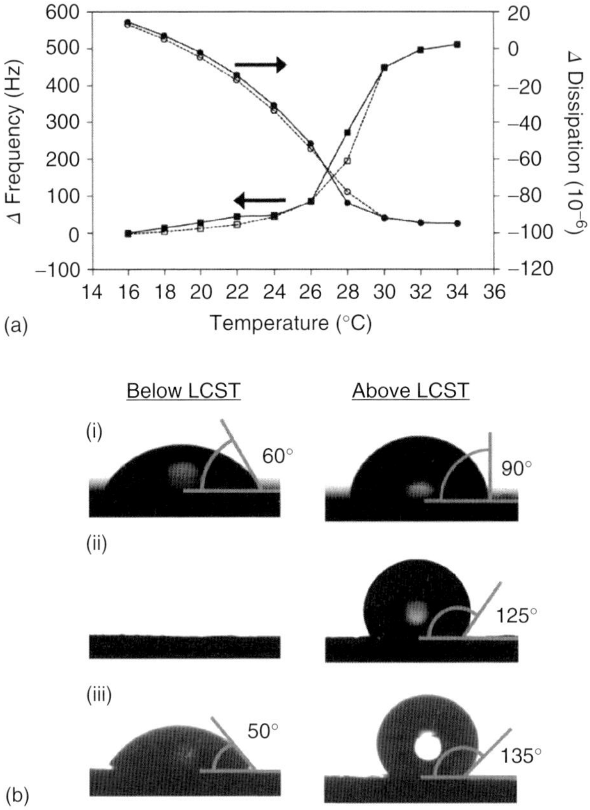

Figure 8.6 (a) Temperature-responsive behavior of iCVD-deposited PNIPAM characterized by quartz crystal microbalance monitoring. (b) The change in hydrophilicity above and below the LCST evaluated on a (i) Si wafer, (ii) mat of electrospun polymer fibers, and (iii) multi-walled carbon nanotube forest. (Reprinted (adapted) with permission from Ref. [80]. Copyright 2010 Wiley-VCH.)

PVDF where vinylidene fluoride is adsorbed on the surface and impinging carbene molecules generated through the thermal decomposition of hexafluoropropropylene oxide at a resistively heated filament resulted in polymerization [84]. The semicrystalline PVDF films were primarily in the α crystal form, which displays negligible piezoelectricity. Post-polymerization thermal treatments were shown to generate more β phase.

Polyureas [85–90] and polyamides (nylons) [91, 92] grown by vapor deposition polymerizations have also been shown to be piezoelectric. In the vapor deposition polymerization of polyureas, diisocyanate monomers (MDIs), and diamine monomers (MDAs) evaporate simultaneously from their respective precursor containers and then adsorb on the substrate surface where they react through polyaddition reactions. Poling at temperatures of 200 °C and electric fields up to 250 MV m^{-1} are required to align the dipoles. Piezoelectricity was first demonstrated in aromatic polyureas [85], and films with piezoelectric coefficients up to 10 pC N^{-1} have been developed [88, 92]. This is about half of the typical values for PVDF, but the benefit of polyureas is that they are stable up to 200 °C, showing negligible loss in piezoelectricity. The piezoelectric effect has also been demonstrated with aliphatic polyureas synthesized from monomers with an odd number of methylene groups between the reactive chain ends [87, 93]. The piezoelectric effect is highest in polyurea-5, where 5 is the number of methylene groups in the aliphatic chain, but the piezoelectricity of aliphatic polyureas is lower than the aromatic compositions.

Polyamides thin films can be synthesized by vapor deposition polymerization of diamine and diacid chloride monomers [91]. Similarly to polyureas, polyamides show enhanced temperature stability relative to PVDF but a reduced piezoelectric coefficient. Vapor-deposited parylene-C, in which a single chlorine atom is substituted onto the aromatic repeat unit, has been shown to be piezoelectric [82]. However, its piezoelectric coefficient (d_{33}) is quite low at 2.0 pC N^{-1}.

8.4 Conclusions

Developing reactive and responsive polymer thin films for targeted applications requires synthetic methods that provide acute control over film composition and structure. Several distinct polymer CVD techniques satisfying this requirement have been highlighted. Moreover, applications exploiting the additional advantages of polymer CVD, including solvent-free synthesis conditions, simultaneous and tunable crosslinking, and conformality over nano- and microscopic features, have been emphasized. Reactive polymer film and coating platforms that can be functionalized through facile, one-step post-polymerization modification reactions are highly desirable as they simplify product development work. The polymer CVD literature is replete with examples of reactive polymer thin films, which can be modified through simple nucleophilic addition and substitution reactions. In applications of responsive polymer thin films chemical functionality retention is also critical as film composition and structure govern physicochemical behavior. A host of stimuli-responsive behaviors have been introduced into polymer films by CVD and employed in varied applications. In summary, the field of reactive and responsive polymer films synthesized by CVD is quite broad, finding implementations as varied as sensors for nitroaromatic explosives to reactive platforms for the immobilization of biological species. As the field continues to mature and researchers make further progress in understanding the fundamentals of CVD polymer synthesis and processing, novel reactive and responsive chemistries will emerge to meet the demands of an always-evolving technological landscape.

References

1. Stuart, M.A.C., Huck, W.T.S., Genzer, J., Muller, M., Ober, C., Stamm, M., Sukhorukov, G.B., Szleifer, I., Tsukruk, V.V., Urban, M., Winnik, F., Zauscher, S., Luzinov, I., and Minko, S. (2010) *Nat. Mater.*, **9** (2), 101–113.
2. Tokarev, I. and Minko, S. (2009) *Soft Matter*, **5** (3), 511–524.
3. Gauthier, M.A., Gibson, M.I., and Klok, H.A. (2009) *Angew. Chem. Int. Ed.*, **48** (1), 48–58.
4. Buck, M.E. and Lynn, D.M. (2012) *Polym. Chem.*, **3** (1), 66–80.
5. Edmondson, S., Osborne, V.L., and Huck, W.T.S. (2004) *Chem. Soc. Rev.*, **33** (1), 14–22.
6. Zhao, B. and Brittain, W.J. (2000) *Prog. Polym. Sci.*, **25** (5), 677–710.
7. Lau, K.K.S. and Gleason, K.K. (2006) *Adv. Mater.*, **18** (15), 1972–1977.
8. Lau, K.K.S. and Gleason, K.K. (2007) *Macromol. Biosci.*, **7** (4), 429–434.
9. Xu, J.J. and Gleason, K.K. (2010) *Chem. Mater.*, **22** (5), 1732–1738.
10. Lau, K.K.S. and Gleason, K.K. (2007) *Surf. Coat. Technol.*, **201** (22–23), 9189–9194.
11. Im, S.G., Bong, K.W., Lee, C.H., Doyle, P.S., and Gleason, K.K. (2009) *Lab Chip*, **9** (3), 411–416.
12. O'Shaughnessy, W.S., Mari-Buye, N., Borros, S., and Gleason, K.K. (2007) *Macromol. Rapid Commun.*, **28** (18-19), 1877–1882.
13. Montero, L., Baxamusa, S.H., Borros, S., and Gleason, K.K. (2009) *Chem. Mater.*, **21** (2), 399–403.
14. Mari-Buye, N., O'Shaughnessy, S., Colominas, C., Semino, C.E., Gleason, K.K., and Borros, S. (2009) *Adv. Funct. Mater.*, **19** (8), 1276–1286.
15. Deng, X.P., Eyster, T.W., Elkasabi, Y., and Lahann, J. (2012) *Macromol. Rapid Commun.*, **33** (8), 640–645.
16. Lahann, J., Balcells, M., Lu, H., Rodon, T., Jensen, K.F., and Langer, R. (2003) *Anal. Chem.*, **75** (9), 2117–2122.
17. Arora, W.J., Tenhaeff, W.E., Gleason, K.K., and Barbastathis, G. (2009) *J. Microelectromech. Syst.*, **18** (1), 97–102.
18. Tenhaeff, W.E. and Gleason, K.K. (2009) *Chem. Mater.*, **21** (18), 4323–4331.
19. Tenhaeff, W.E. and Gleason, K.K. (2007) *Langmuir*, **23** (12), 6624–6630.
20. Xu, J.J., Asatekin, A., and Gleason, K.K. (2012) *Adv. Mater.*, **24** (27), 3692–3696.
21. Lahann, J., Balcells, M., Rodon, T., Lee, J., Choi, I.S., Jensen, K.F., and Langer, R. (2002) *Langmuir*, **18** (9), 3632–3638.
22. Chen, H.Y., Elkasabi, Y., and Lahann, J. (2006) *J. Am. Chem. Soc.*, **128** (1), 374–380.
23. Elkasabi, Y., Chen, H.Y., and Lahann, J. (2006) *Adv. Mater.*, **18** (12), 1521–1526.
24. Elkasabi, Y. and Lahann, J. (2009) *Macromol. Rapid Commun.*, **30** (1), 57–63.
25. Elkasabi, Y.M., Lahann, J., and Krebsbach, P.H. (2011) *Biomaterials*, **32** (7), 1809–1815.
26. Tsai, M.Y., Lin, C.Y., Huang, C.H., Gu, J.A., Huang, S.T., Yu, J.S., and Chen, H.Y. (2012) *Chem. Commun.*, **48** (89), 10969–10971.
27. Wu, J.T., Huang, C.H., Liang, W.C., Wu, Y.L., Yu, J.S., and Chen, H.Y. (2012) *Macromol. Rapid Commun.*, **33** (10), 922–927.
28. Im, S.G., Bong, K.W., Kim, B.S., Baxamusa, S.H., Hammond, P.T., Doyle, P.S., and Gleason, K.K. (2008) *J. Am. Chem. Soc.*, **130** (44), 14424–14425.
29. Im, S.G., Kim, B.S., Lee, L.H., Tenhaeff, W.E., Hammond, P.T., and Gleason, K.K. (2008) *Macromol. Rapid Commun.*, **29** (20), 1648–1654.
30. Nandivada, H., Chen, H.Y., Bondarenko, L., and Lahann, J. (2006) *Angew. Chem. Int. Ed.*, **45** (20), 3360–3363.
31. Chen, G.H., Gupta, M., Chan, K., and Gleason, K.K. (2007) *Macromol. Rapid Commun.*, **28** (23), 2205–2209.
32. Tenhaeff, W.E. and Gleason, K.K. (2007) *Surf. Coat. Technol.*, **201** (22-23), 9417–9421.
33. Hermanson, G.T. (2013) *Bioconjugate Techniques*, 3rd edn, Acadmic Press, St. Louis, Mo.
34. Hoyle, C.E. and Bowman, C.N. (2010) *Angew. Chem. Int. Ed.*, **49** (9), 1540–1573.
35. Kolb, H.C., Finn, M.G., and Sharpless, K.B. (2001) *Angew. Chem. Int. Ed.*, **40** (11), 2004–2021.

36. Im, S.G., Kim, B.S., Tenhaeff, W.E., Hammond, P.T., and Gleason, K.K. (2009) *Thin Solid Films*, **517** (12), 3606–3611.
37. Tasdelen, M.A. (2011) *Polym. Chem.*, **2** (10), 2133–2145.
38. Lahann, J. and Langer, R. (2002) *Macromolecules*, **35** (11), 4380–4386.
39. Sundberg, P. and Karppinen, M. (2014) *Beilstein J. Nanotechnol.*, **5**, 1104–1136.
40. Schiller, S., Hu, J., Jenkins, A.T.A., Timmons, R.B., Sanchez-Estrada, F.S., Knoll, W., and Förch, R. (2001) *Chem. Mater.*, **14** (1), 235–242.
41. Odian, G. (2004) *Principles of Polymerization*, 4th edn, John Wiley and Sons, Inc, New York.
42. Mao, Y. and Gleason, K.K. (2006) *Langmuir*, **22** (4), 1795–1799.
43. Baxamusa, S.H. and Gleason, K.K. (2009) *Adv. Funct. Mater.*, **19** (21), 3489–3496.
44. Baxamusa, S.H., Montero, L., Dubach, J.M., Clark, H.A., Borros, S., and Gleason, K.K. (2008) *Biomacromolecules*, **9** (10), 2857–2862.
45. Lahann, J., Klee, D., Pluester, W., and Hoecker, H. (2001) *Biomaterials*, **22** (8), 817–826.
46. Elkasabi, Y., Yoshida, M., Nandivada, H., Chen, H.Y., and Lahann, J. (2008) *Macromol. Rapid Commun.*, **29** (11), 855–870.
47. Chan, K. and Gleason, K.K. (2005) *Langmuir*, **21** (19), 8930–8939.
48. Karaman, M., Kooi, S.E., and Gleason, K.K. (2008) *Chem. Mater.*, **20** (6), 2262–2267.
49. Chan, K., Kostun, L.E., Tenhaeff, W.E., and Gleason, K.K. (2006) *Polymer*, **47** (20), 6941–6947.
50. Chen, G.H., Lau, K.K.S., and Gleason, K.K. (2009) *Thin Solid Films*, **517** (12), 3539–3542.
51. Sedransk, K.L., Tenhaeff, W.E., and Gleason, K.K. (2010) *Chem. Vap. Deposition*, **16** (1-3), 23–28.
52. Bose, R.K., Nejati, S., Stufflet, D.R., and Lau, K.K.S. (2012) *Macromolecules*, **45** (17), 6915–6922.
53. Yang, R., Xu, J.J., Ozaydin-Ince, G., Wong, S.Y., and Gleason, K.K. (2011) *Chem. Mater.*, **23** (5), 1263–1272.
54. Gao, T., Tillman, E.S., and Lewis, N.S. (2005) *Chem. Mater.*, **17** (11), 2904–2911.
55. Lock, J.P., Geraghty, E., Kagumba, L.C., and Mahmud, K.K. (2009) *Thin Solid Films*, **517** (12), 3584–3587.
56. You, J.B., Min, K.I., Lee, B., Kim, D.P., and Im, S.G. (2013) *Lab Chip*, **13** (7), 1266–1272.
57. Chen, H.Y., McClelland, A.A., Chen, Z., and Lahann, J. (2008) *Anal. Chem.*, **80** (11), 4119–4124.
58. Sperlling, L.H. (2006) *Introduction to Physical Polymer Science*, 4th edn, Wiley-Interscience, Hoboken, NJ.
59. Battiston, F.M., Ramseyer, J.P., Lang, H.P., Baller, M.K., Gerber, C., Gimzewski, J.K., Meyer, E., and Guntherodt, H.J. (2001) *Sens. Actuators, B*, **77** (1-2), 122–131.
60. Adams, J.D., Parrott, G., Bauer, C., Sant, T., Manning, L., Jones, M., Rogers, B., McCorkle, D., and Ferrell, T.L. (2003) *Appl. Phys. Lett.*, **83** (16), 3428–3430.
61. Abedinov, N., Grabiec, P., Gotszalk, T., Ivanov, T., Voigt, J., and Rangelow, I.W. (2001) *J. Vac. Sci. Technol., A*, **19** (6), 2884–2888.
62. Baselt, D.R., Fruhberger, B., Klaassen, E., Cemalovic, S., Britton, C.L., Patel, S.V., Mlsna, T.E., McCorkle, D., and Warmack, B. (2003) *Sens. Actuator, B*, **88** (2), 120–131.
63. Tenhaeff, W.E., McIntosh, L.D., and Gleason, K.K. (2010) *Adv. Funct. Mater.*, **20** (7), 1144–1151.
64. Petruczok, C.D., Choi, H.J., Yang, S.Y., Asatekin, A., Gleason, K.K., and Barbastathis, G. (2013) *J. Microelectromech. Syst.*, **22** (1), 54–61.
65. Yague, J.L. and Gleason, K.K. (2012) *Soft Matter*, **8** (10), 2890–2894.
66. Casadio, Y.S., Brown, D.H., Chirila, T.V., Kraatz, H.-B., and Baker, M.V. (2010) *Biomacromolecules*, **11** (11), 2949–2959.
67. Ozaydin-Ince, G., Dubach, J.M., Gleason, K.K., and Clark, H.A. (2011) *Proc. Natl. Acad. Sci. U.S.A.*, **108** (7), 2656–2661.
68. Ince, G.O., Demirel, G., Gleason, K.K., and Demirel, M.C. (2010) *Soft Matter*, **6** (8), 1635–1639.
69. Rubinstein, M., Colby, R.H., Dobrynin, A.V., and Joanny, J.F. (1996) *Macromolecules*, **29** (1), 398–406.
70. Katchalsky, A., Kunzle, O., and Kuhn, W. (1950) *J. Polym. Sci.*, **5** (3), 283–300.

71. Tenhaeff, W.E. and Gleason, K.K. (2009) *Thin Solid Films*, **517** (12), 3543–3546.
72. Karaman, M. and Cabuk, N. (2012) *Thin Solid Films*, **520** (21), 6484–6488.
73. McInnes, S.J.P., Szili, E.J., Al-Bataineh, S.A., Xu, J.J., Alf, M.E., Gleason, K.K., Short, R.D., and Voelcker, N.H. (2012) *ACS Appl. Mater. Interfaces*, **4** (7), 3566–3574.
74. Urban, A.M. and Urban, M.W. (2005) Stimuli-responsive macromolecules and polymeric coatings, in *Stimuli-Responsive Polymeric Films and Coatings* (ed M.W. Urban), American Chemical Society, Washington, DC.
75. Shimada, N. and Maruyama, A. (2013) Thermoresponsive polymers with functional groups selected for pharmaceutical and biomedical applications, in *Tailored Polymer Architectures for Pharmaceutical and Biomedical Applications* (eds C. Scholz and J. Kessler), American Chemical Society, Washington, DC.
76. Tamirisa, P.A., Koskinen, J., and Hess, D.W. (2006) *Thin Solid Films*, **515** (4), 2618–2624.
77. Teare, D.O.H., Barwick, D.C., Schofield, W.C.E., Garrod, R.P., Beeby, A., and Badyal, J.P.S. (2005) *J. Phys. Chem. B*, **109** (47), 22407–22412.
78. Cheng, X.H., Canavan, H.E., Stein, M.J., Hull, J.R., Kweskin, S.J., Wagner, M.S., Somorjai, G.A., Castner, D.G., and Ratner, B.D. (2005) *Langmuir*, **21** (17), 7833–7841.
79. Alf, M.E., Hatton, T.A., and Gleason, K.K. (2011) *Langmuir*, **27** (17), 10691–10698.
80. Alf, M.E., Godfrin, P.D., Hatton, T.A., and Gleason, K.K. (2010) *Macromol. Rapid Commun.*, **31** (24), 2166–2172.
81. Tekin, H., Ozaydin-Ince, G., Tsinman, T., Gleason, K.K., Langer, R., Khademhosseini, A., and Demirel, M.C. (2011) *Langmuir*, **27** (9), 5671–5679.
82. Ramadan, K.S., Sameoto, D., and Evoy, S. (2014) *Smart Mater. Struct.*, **23** (3), 26.
83. Kubono, A. and Okui, N. (1994) *Prog. Polym. Sci.*, **19** (3), 389–438.
84. Rastogi, A.C. and Desu, S.B. (2006) *Chem. Vap. Deposition*, **12** (12), 742–750.
85. Takahashi, Y., Ukishima, S., Iijima, M., and Fukada, E. (1991) *J. Appl. Phys.*, **70** (11), 6983–6987.
86. Fukada, E. (1994) *Ferroelectr. Polym. Ceram.-Polym. Compos.*, **92–99**, 143–159.
87. Hattori, T., Iijima, M., Takahashi, Y., Fukada, E., Suzuki, Y., Kakimoto, M.A., and Imai, Y. (1995) *Jpn. J. Appl. Phys., Part 1 - Regul. Pap. Short Notes Rev. Pap.*, **34** (1), 395–395.
88. Wang, X.S., Takahashi, Y., Iijima, M., and Fukada, E. (1995) *Jpn. J. Appl. Phys., Part 1 - Regul. Pap. Short Notes Rev. Pap.*, **34** (3), 1585–1590.
89. Hattori, T., Takahashi, Y., Iijima, M., and Fukada, E. (1996) *J. Appl. Phys.*, **79** (3), 1713–1721.
90. Tajitsu, Y., Date, M., and Fukada, E. (1999) *Jpn. J. Appl. Phys., Part 1 - Regul. Pap. Short Notes Rev. Pap.*, **38** (9B), 5653–5656.
91. Hattori, T., Takahashi, Y., Iijima, M., and Fukada, E. (1996) *Jpn. J. Appl. Phys., Part 1 - Regul. Pap. Short Notes Rev. Pap.*, **35** (11), 5763–5764.
92. Fukada, E. (2006) *IEEE Trans. Dielectr. Electr. Insul.*, **13** (5), 1110–1119.
93. Fukada, E. (1998) *Jpn. J. Appl. Phys. Part 1 - Regul. Pap. Short Notes Rev. Pap.*, **37** (5B), 2775–2780.

9
Multifunctional Reactive Polymer Coatings

Xiaopei Deng, Kenneth C. K. Cheng, and Joerg Lahann

9.1
Introduction

Recent developments in the area of biotechnology have sparked multidisciplinary research efforts in biology, chemistry, and materials science. Developing bioactive surfaces with accurate spatial arrangements at the submicrometer scale is required for numerous biomedical applications ranging from immunoassays and biosensors to drug delivery and development of novel medical implants. To date, many methods and surface chemistries have been used for controlled surface modification [1–4]. Most immobilization methods fall into two categories: physical adsorption or covalent immobilization [1, 4]. Physical adsorption is typically straightforward to carry out and can be applied to most biomedical material surfaces and even complex devices. However, physically adsorbed surface layers have limited stability and lack the same control over molecular composition when compared to covalent immobilization [1]. Covalent binding of biomolecules to a surface is generally considered superior for long-term biomedical applications. However, covalent binding requires adequate reactive sites on the substrate surface, which are not available on most standard materials surfaces. Therefore, the choices of available conjugation chemistries for covalent immobilization are typically very limited and are highly substrate-dependent. Promising examples include silane chemistry for glass/silica/silicone substrates or organosulfur compounds for gold and coinage metal surfaces [5]. Surface functionalization techniques that are compatible with polymers include plasma or corona treatments, but similarly limit the choice of available functional groups.

Like other chemical vapor deposition polymerization techniques, CVD polymer coatings based on reactive poly(*p*-xylylene) (PPX) are solvent-free, substrate-independent, based on low-temperature deposition, and allow for easy integration into micro-/nanofabrication schemes [6–8]. The basic precursor used for CVD polymerization is [2.2]paracyclophane (PCP), which forms two diradicals at high temperature under vacuum driven by the spontaneous release of ring strain in the paracyclophane molecule [9]. The radicals polymerize on solid substrates at or below room temperature and form conformal and pinhole-free

CVD Polymers: Fabrication of Organic Surfaces and Devices, First Edition. Edited by Karen K. Gleason.

polymer films [10]. The Lahann group has established an entire library of reactive coatings with a range of different functional groups (Scheme 9.1). Reactive PPX coatings constitute a versatile and substrate-independent surface functionalization platform [11]. Functional groups compatible with the above-described paracyclophane polymerization include alkyne, vinyl, amine, hydroxyl, aldehyde, ketone, benzophenone, maleimide, anhydride, pentafluorophenyl (PFP) ester, and 2-bromoisobutyryl groups. Paracyclophane molecules carrying one of these functional groups have been successfully synthesized and polymerized via the CVD process. We note that some unlisted functional groups, such as carboxylic acid, underwent partial elimination under the conditions of pyrolysis, that is, temperatures above 500 °C. The polymer coatings carrying functional groups listed in the library have been used in a range of different biomedical studies such as immobilizing proteins, peptides, sugar, DNA, and polyethylene glycol (PEG) [12–17].

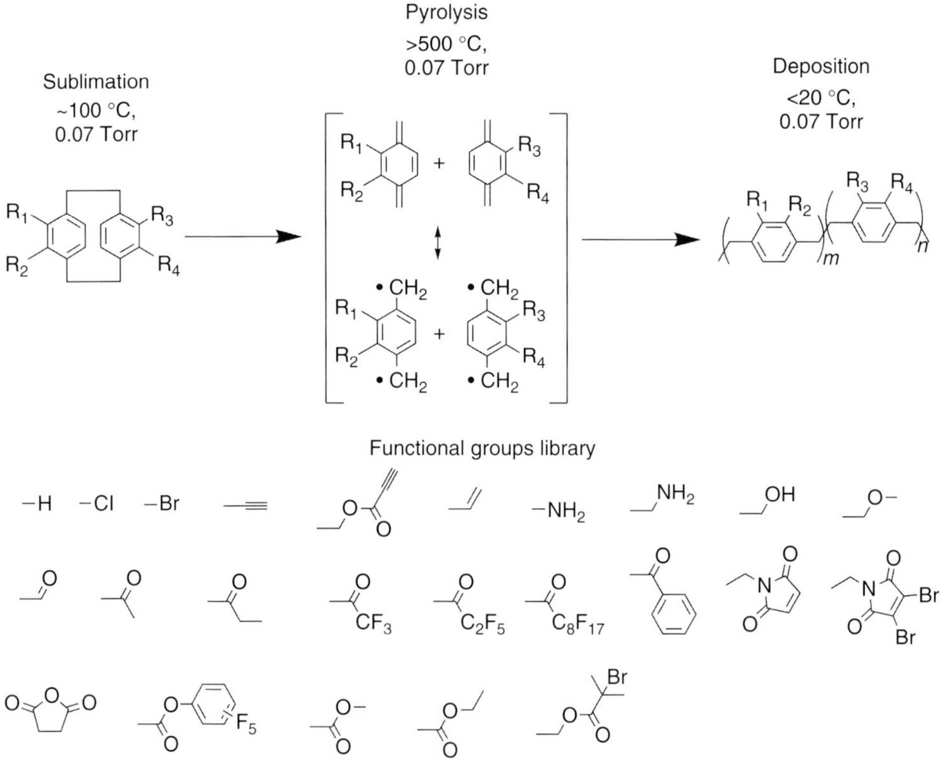

Scheme 9.1 Scheme of polymerizing different substituted [2.2]paracyclophanes via the CVD polymerization process and the functional groups library for functional PPX coatings. The functionalized [2.2]paracyclophanes sublimate around 100 °C under vacuum, flow with the argon carrier gas to the pyrolysis zone (>500 °C) and break into radicals. The radicals adsorb on the substrates placed on the cooled deposition stage (typically <20 °C) and polymerize to form functionalized polymer coatings. (Reproduced from Ref. [11], with permission from Wiley (Copyright 2014 Wiley).

To accommodate the requirements of high-precision biomedical surface engineering, the reactive polymer coatings provide many options for efficient conjugation of biomolecules and orthogonal immobilization strategies [11]. Moreover, the spatial distributions of the functional groups could be accurately controlled in many cases. Three different possible spatial arrangements for multiple functional groups are shown in Scheme 9.2: gradients, random distribution, and a micropattern. The three different spatial arrangements can enable different biomedical applications. The random distribution of functional groups (Scheme 9.2a) is suitable for applications in need of multiple immobilized molecules evenly distributed on the substrate. Moreover, more than two reactive groups can be presented with adjustable and predefined surface ratios. This feature is ideal for preparing complex biomimetic surfaces. Gradient surfaces (Scheme 9.2b) can provide continuous and gradual changes to the surface density of two orthogonal functional groups, which can in turn create a biological gradient of immobilized molecules. The gradient surfaces present interesting possibilities for cell migration or surface-directed molecular transport. A patterned surface (Scheme 9.2c) can provide features for miniaturized high-throughput biomedical devices. This chapter reviews the recent progress made in using multifunctional PPX CVD coatings for high-precision biofunctionalization, with accurate control over chemical functionality as well as spatial arrangements.

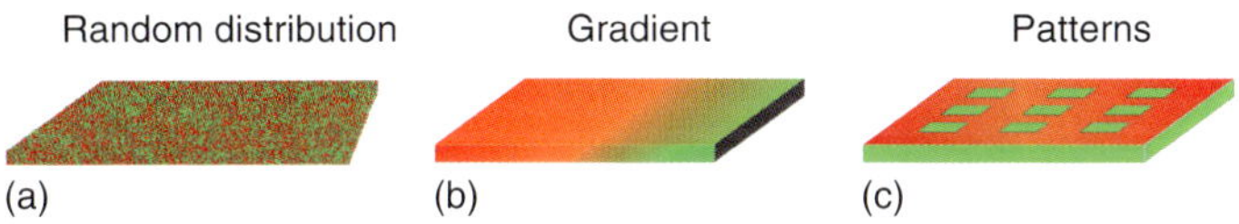

Scheme 9.2 (a–c) Different functional group spatial arrangements for multifunctional polymer coatings with orthogonal functional groups. Red and green pixels each represents a different type of functional group. (Reproduced from Ref. [11], with permission from Wiley (Copyright 2014 Wiley).

9.2 CVD Copolymer Coatings with Randomly Distributed Functional Groups

As summarized above, CVD copolymer coatings with randomly distributed functional groups are ideal for preparing biomimetic surface, which typically requires simultaneous presentation and manipulation of more than one biomolecule on the surface. Two or more functional groups for orthogonal surface reactions are required for avoiding chemical cross-reactions. Ideally, the surface density and ratio of each of the different functional groups can be changed. The adjustment of the surface density of the immobilized molecules can lead to different applications. Most functionalized [2.2]paracyclophane molecules carry only one single functional group. Hence, preparing multifunctional PPX CVD coatings usually involves feeding multiple types of functional [2.2]paracyclophane molecules through the CVD copolymerization process.

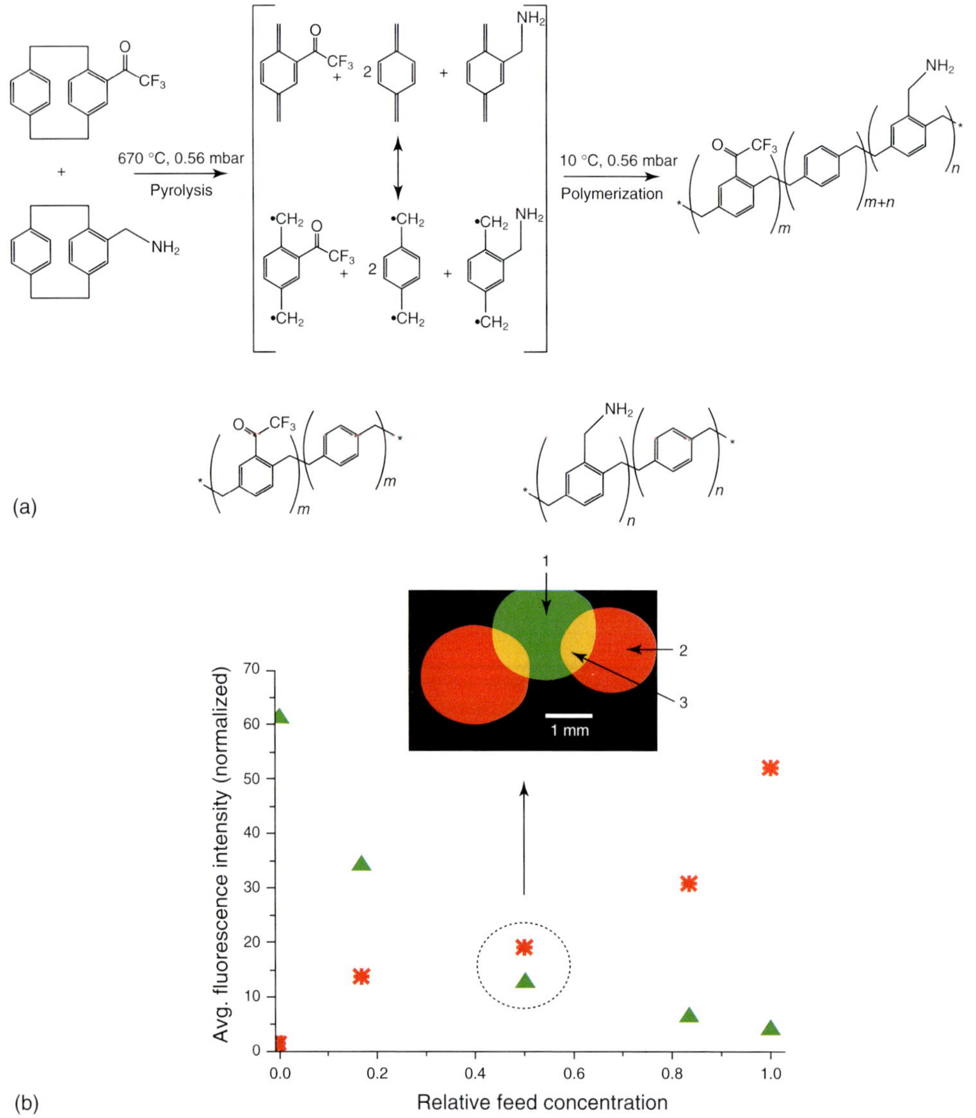

Figure 9.1 (a) Scheme of synthesizing a multifunctional copolymer via the CVD copolymerization of different functionalized [2.2]paracyclophanes. (b) Characterization of the chemical composition change of copolymers with different precursor ratios by fluorescent ligand binding. Biotin hydrazide followed by TRITC-streptavidin (green area, No. 1) and Atto 655 NHS ester (red area, No. 2) were applied to specifically target the ketone groups and amine groups, respectively. No cross reaction was observed in the overlapped areas (yellow area, No. 3). (Reproduced from Ref. [18], with permission from Wiley (Copyright 2006 Wiley).

Figure 9.1 presents a typical example of the process underlying the deposition of multifunctional coatings via CVD copolymerization [18]. In this study, two different substituted [2.2]paracyclophanes (4-trifluoroacetyl [2.2]paracyclophane and 4-aminomethyl-[2.2]paracyclophane) were introduced into the CVD system for copolymerization. Copolymer coatings with different functional group ratios were prepared by changing the ratios of the two different precursors. Fourier transform infrared spectroscopy (FTIR), X-ray photoelectron spectroscopy (XPS), and X-ray Diffraction (XRD) were used to characterize the chemical composition of the copolymers for different functional group ratios [18]. As shown in Figure 9.1b, the presence of amine groups was confirmed by reacting with Atto 655 NHS ester. Biotin hydrazide was used to react with the ketone group (trifluoroacetyl) followed by specific binding of tetramethylrhodamine-labeled streptavidin (TRITC-streptavidin) to the immobilized biotin. We found no cross-reaction for the two surface reactions for detecting different functional groups. The graph showing the change of normalized fluorescence intensity versus the precursor feed ratio demonstrates that the relative ratio of immobilized ligands can be changed by varying the feed ratio of different substituted [2.2]paracyclophanes.

More recently, we reported the preparation a copolymer carrying PFP ester and alkyne groups to orthogonally immobilize the growth factor and adhesion peptide on the surface (Scheme 9.3) [16]. The PFP ester is an active ester commonly used for peptide synthesis [19]. It can readily react with amine under physiological conditions and is yet more hydrolytically stable than the more widely used *N*-hydroxysuccinimide (NHS) ester [11]. In the specific study shown in Scheme 9.3, epidermal growth factor was immobilized by the highly efficient reaction of PFP-ester and amine groups. An orthogonal and highly efficient reaction, azide-alkyne click reaction, was used to immobilize a cyclic arginyl-glycyl-aspartic acid (RGD) peptide with an azide end group. The orthogonally co-immobilized growth factor and adhesion peptide remained bioactive as shown in the study [16].

Similarly, a trifunctional PPX multifunctional CVD coating has been developed and reported by Chen *et al.* [20]. The trifunctional coating has an alkyne, a maleimide, and ketone groups, which were reacted with biomolecules functionalized with azide, thiol, and hydrazide, respectively. Simultaneous presentation of multiple ligands was demonstrated on the trifunctional coating deposited on a flat substrate as well as a stent surface with complex geometry.

9.3 Multifunctional Gradient Coatings

Surface gradients with a gradual change in chemical composition play an important role in many applications as well as in the study of living cell migration and of combinatorial materials discovery [21–23]. In light of this, many gradient fabrication methods have been developed including those using microfluidics, photochemical methods, electron/X-ray irradiation, gel diffusion, and so

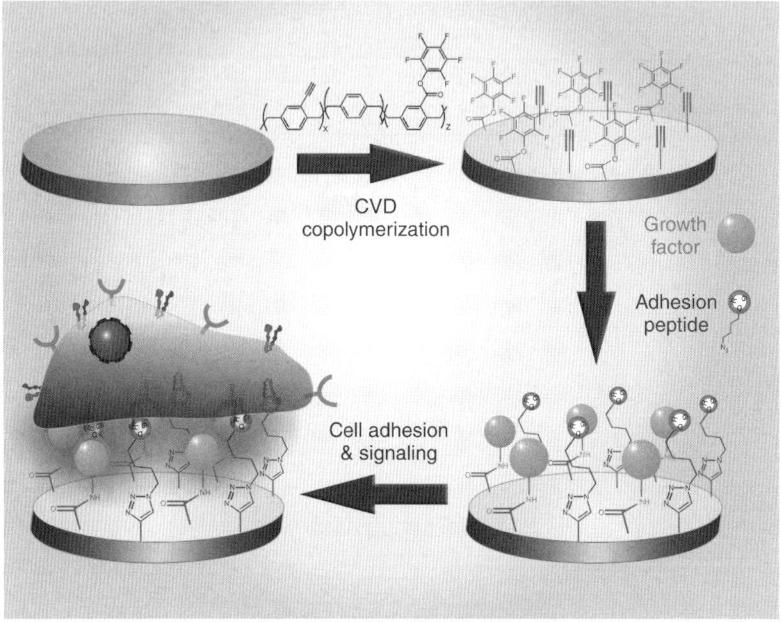

Scheme 9.3 Scheme of orthogonal co-immobilization of growth factor and adhesion peptide on a CVD copolymer by PFP ester-amine reaction and alkyne-azide click reaction. (Reproduced from Ref. [16], with permission from Wiley (Copyright 2012 Wiley).

on.[23–26] Nevertheless, many of these methods require relatively complicated procedures and are only applicable to certain substrates. In addition, they are not all biocompatible. In this following section, we will discuss the use of chemical vapor deposition polymerization of substituted [2.2]paracyclophanes for generating biocompatible, continuous, and substrate-independent chemical composition gradients.

9.3.1 Composition Gradient Preparation and Biomedical Applications

Elkasabi *et al.* [27] have demonstrated the use of a modified CVD system to generate polymer films bearing continuous chemical composition gradients. This CVD system for gradient fabrication consists of two separate sources that are connected, in a counter-flow setup, to the deposition chamber, as shown in Figure 9.2. With this counter-flow setup, radicals generated from two differently substituted precursors deposited anisotropically onto the substrates. The radicals copolymerized in the form of two composition gradients. In this proof-of-concept study, trifluoroacetyl-[2.2]paracyclophane and aminomethyl-[2.2]paracyclophane were used as the two precursors for the counter-flow CVD process. Both Fourier-transformed infrared spectroscopy (FTIR) and XPS confirmed a gradual change in chemical composition along the longitude direction

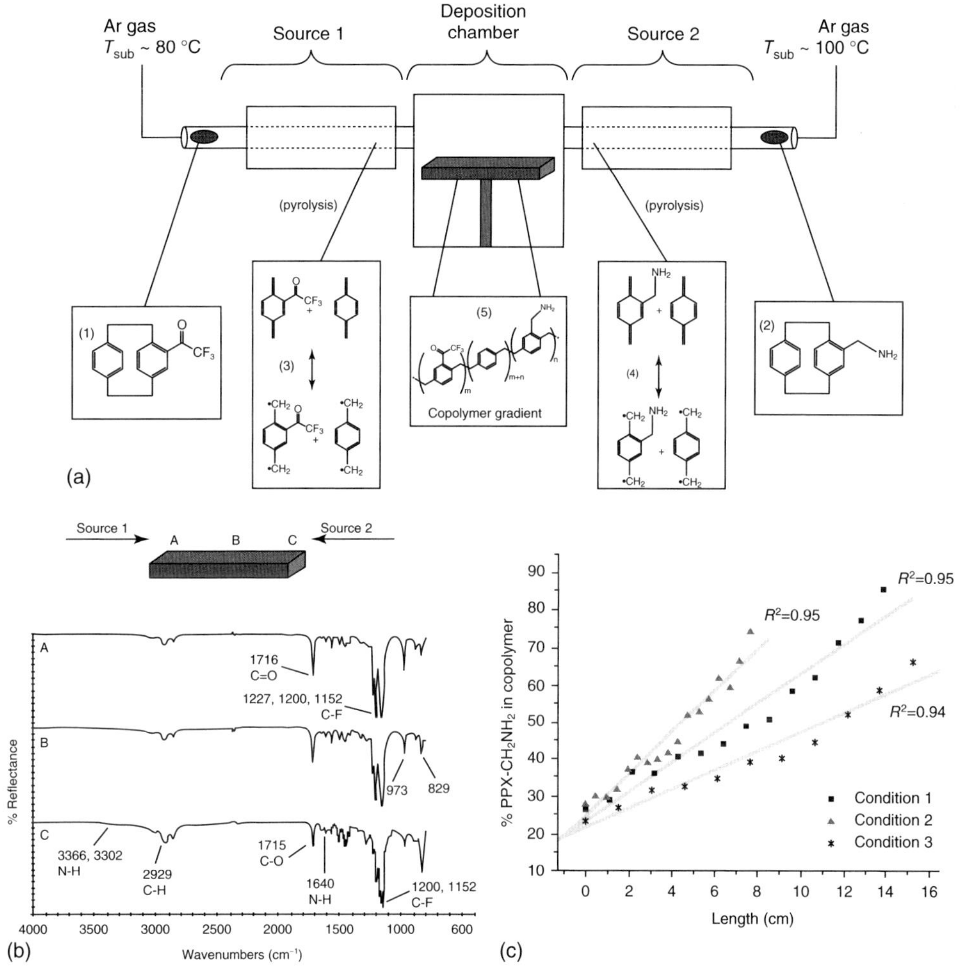

Figure 9.2 (a) Scheme of the custom-built two-source CVD system. Each source has a quartz tube passing through a three-zone furnace and connecting with the deposition chamber. The whole system is under vacuum during the CVD process. For gradient generation, the sample holder does not rotate. In this specific study, the trifluoroacetyl and the aminomethyl groups form a chemical composition gradient [27]. (b) FTIR spectra of the CVD polymer gradient. (c) Copolymer compositions along the surface gradient based upon XPS survey spectra. Changing sublimation rates and argon flow rates can be used to control the slope of the compositional change within a certain distance. (Reproduced from Ref. [27], with permission from Wiley (Copyright 2009 Wiley).

of the gradient (Figure 9.2). The composition gradients were found to be in the centimeter scale. Fluorescence-labeled ligands were also immobilized on the composition gradient to visualize and quantify the orthogonal reactivity of the two composition gradients [27]. These characterizations all confirmed that continuous composition gradients are successfully prepared using this counter-flow CVD system. Because CVD polymerization is a substrate-independent process, the composition gradients can be applied onto a wide range of materials. In addition, the materials to be coated do not require any pretreatments, making this counter-flow CVD process a simple yet universal method for generating continuous composition gradients.

In a more recent study, Elkasabi *et al.* [28] prepared a wider range of composition gradients and demonstrated the preparation of biomimetic gradients

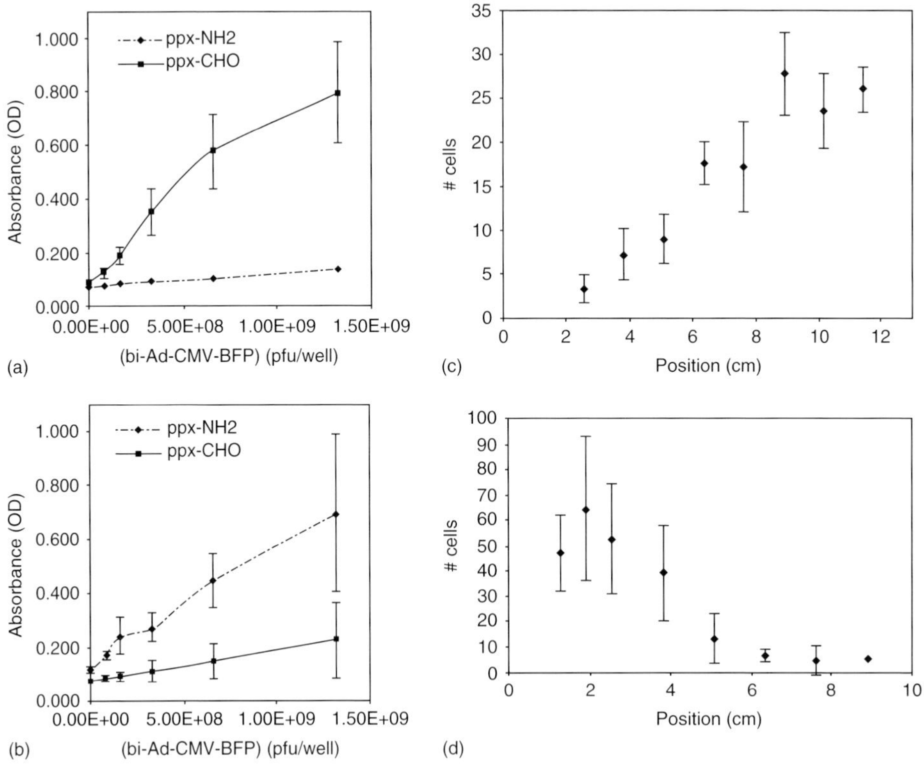

Figure 9.3 (a,b) Sandwich enzyme-linked immunosorbent assays (ELISAs) were used to evaluate the binding efficiency and selectivity of the VBABM immobilization method. Using the functionalized biotins, biotinylated adenovirus encoding blue florescence protein adhered selectively on either (a) aldehyde or (b) amine-functionalized poly(*p*-xylylene). *Y*-values are plotted with respect to the reaction medium titer. A gradient of transduced cells were generated within the confluent cell population, with the gradient direction determined by (c) biotin hydrazide attachment to CHO groups, or (d) sulfo-NHS-LC-biotin attachment to NH2 groups. (Reproduced from Ref. [28], with permission from Elsevier (Copyright 2011 Elsevier BV).

(Figure 9.3). Specifically, composition gradients bearing amino and aldehyde groups were prepared. The virus-biotin-avidin-biotin-method (VBABM) was used to chemically immobilize cell-signaling adenovirus onto gradients surfaces, resulting in a gradient of adenovirus. When cultured on these adenovirus gradients, human gingival fibroblasts exhibited asymmetric transduction with full confluency, showing that a controlled gradient of cellular activity can be achieved with these biomimetic gradients. These bioactive gradients have potential use for mimicking biological gradients to study cell migration and differentiations, as well as in other advanced delivery studies that involve gradients of therapeutic genes.

9.3.2
Formation of Steep Surface Gradient

The length scale of the composition gradients, prepared using the counter-flow CVD system described above, are typically in the centimeter range. To expand the use of these composition gradients for other applications, a steeper change over smaller length scales may be desired. Elkasabi *et al.* [29] recently developed an improved counter-flow CVD process to reduce the scale of the composition gradients by an order of magnitude. This improved process involves the use of a simple plate placed on top of the substrate during the CVD process. The plate is fixed on a polydimethylsiloxane (PDMS) mold and features millimeter-scale channels, as shown in Figure 9.4. With the substrates sitting under the confined channels, the diffusion of vapor toward the substrates can be controlled, resulting in a much steeper/shorter composition gradient deposited onto the substrates, as confirmed via XPS analysis (Figure 9.4). A mathematical model was also established to predict the preparation of the gradient under confined channels. The model was found to correlate well with the experimental data. With this mathematical model, one can precisely design composition gradients with the desired gradient profile using this improved counter-flow CVD system to meet the requirements of specific applications.

Since CVD polymerization is a substrate-independent process, it has been used to generate composition gradients on three-dimensional (3D) objects [29]. As a proof-of-concept, a circular tube, sectioned from a 10-μl-pipette tip, was chosen as the 3D object. This circular tube has a diameter of roughly 9 mm. After the same deposition process as described above, the tube was cut in half along the length, and the inside of the tube was analyzed using XPS. The CVD polymer film was detected inside the tube and the amount of amino groups was found to gradually increase along the longitude direction of the tube, thus confirming the presence of the composition gradient inside the tube. The ability to generate continuous chemical composition gradients on 3D objects could be useful in a variety of biomedical applications, such as in the development of 3D biomimetic gradients and in the preparation of composition gradients on microfluidic devices.

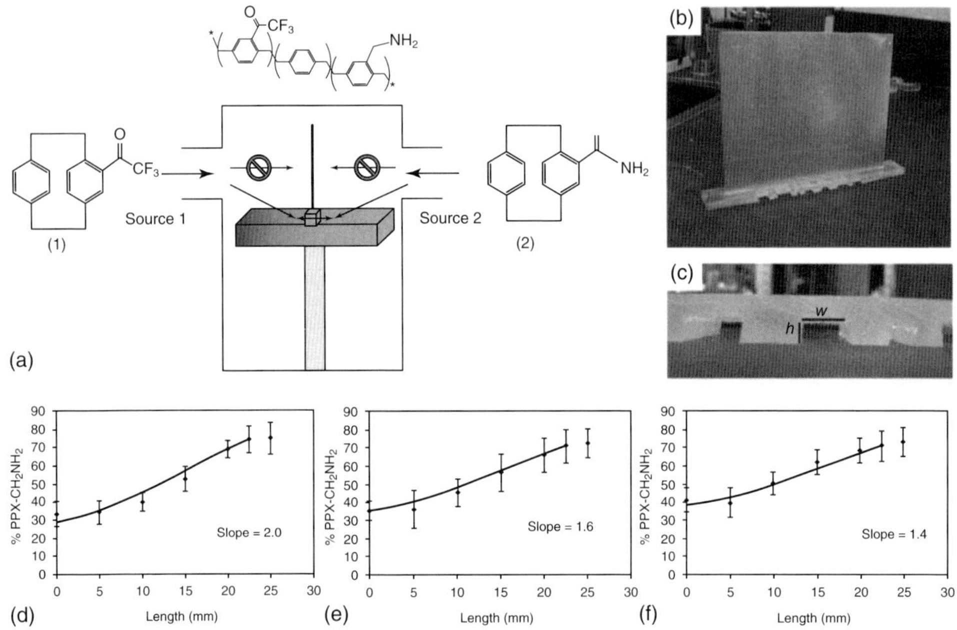

Figure 9.4 (a) Scheme showing a dual-source counter-flow CVD system with the apparatus for generating steeper gradients. (b) Photo of the apparatus for generating steeper gradient. (c) Magnified view of the PDMS channel at the bottom of the apparatus. (d–f) Copolymer ratios calculated from the results of XPS. Substrates are placed within rectangular channels $h = 3.3$ mm and (d) $w = 3.3$ mm, (e) $w = 6.8$ mm, and (f) $w = 12$ mm. There is strong agreement between computational modeling (solid) and experimental (dots) results [29]. (Reproduced from Ref. [29], with permission from Wiley (Copyright 2014 Wiley).

9.4
Functional Coatings with Micro- and Nanopatterns

Micro- and Nanopatterning techniques are widely used in electronics and a lot of the patterning techniques are also applicable for biomedical devices [30–32]. They are of vital importance to the development of high-throughput bioassays, miniaturized biomedical devices, and precise surface design and engineering for cell culture scaffolds/substrates. These techniques include soft lithography, photolithography, stencil-assisted patterning, jet patterning, laser-guided writing, and so on [31].

We have applied several different micro-/nanopatterning techniques to the functional PPX CVD coatings. As shown in Figure 9.5, the micro-patterns can be generated post CVD (e.g., microcontact printing, photo patterning) or during CVD (e.g., applying microstencils during CVD) [33]. Nanopatterning technique such as dip-pen nanolithography (DPN) were also applied in our recent work [34]. In this section, we will discuss each technique in more detail with specific examples.

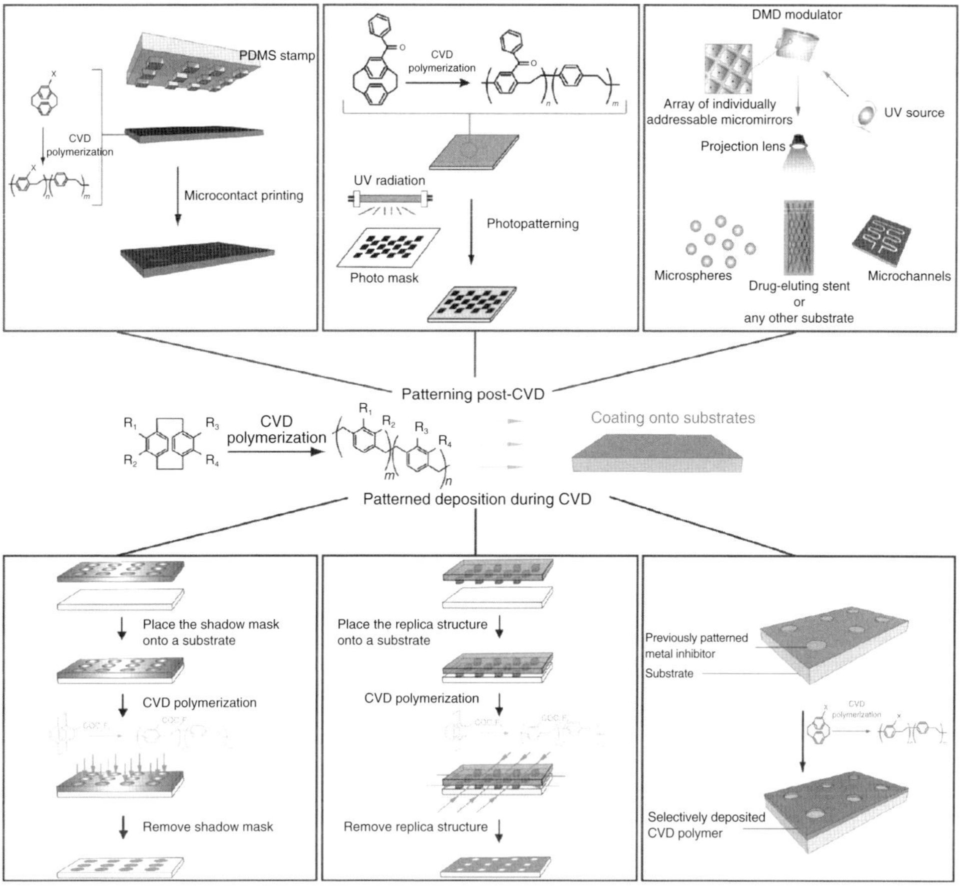

Figure 9.5 Scheme of surface patterning techniques using reactive CVD coatings. Methods can be categorized into two groups: patterning post-CVD (e.g., μCP and photopatterning processes), and patterning deposition during CVD with the assistance of masking molds/microstencils or metal inhibitors applied before the CVD process [33]. (Reproduced from Ref. [33], with permission from American Chemical Society (Copyright 2011 American Chemical Society).

9.4.1
Microcontact Printing (μCP)

Microcontact printing (μCP), since first introduced by Whitesides and coworkers, has become the most widely used soft lithographic technique [31, 35]. It typically involves preparing an elastomeric stamp by casting a mixture of liquid prepolymer and curing agent over a master/mold with microstructures followed by curing/hardening. After complete curing, the stamp can be removed from the master and inked with solutions of molecules to print. Subsequently, the printing/stamping process can transfer the "ink" to the substrate [31]. μCP has been used to pattern different types of substrates in a wide variety of different applications such as printing proteins, DNAs, self-assembled monolayers, catalytic colloids, nanoparticles, and so on [36–40]. Multifunctional orthogonal ligation by μCP printing have also been reported [41]. By μCP, pattern feature size in the range of micrometer or sub-micrometer can be fabricated. The μCP technique is popular because it is easy to use, flexible, low cost, and applicable to many different types of inks (molecules to print) and substrates.

We applied the μCP technique to print biomolecules on reactive PPX coatings in many studies. To give just one example, amine functionalized biotin molecules were printed by μCP on PPX CVD coatings with PFP-ester groups [42]. The printed surfaces were then washed and backfilled with 2-(2-aminoethoxy)ethanol to passivate the unpatterned area. In order to generate cell patterns, a cell-binding antibody was immobilized by the biotin-streptavidin conjugation. As shown in Figure 9.6, cells recognized the strip patterns by μCP and formed corresponding cell patterns along the strips.

In another work, we applied supramolecular nanostamping (SuNS) to a poly(4-formyl-*p*-xylylene-co-*p*-xylylene) film for controlled immobilization of DNA molecules with submicron pattern feature sizes [43]. We also demonstrated that the SuNS nanopatterns could be fabricated on different substrates (PDMS, polystyrene, acrylic) without changing the DNA immobilization protocol. The combination of the μCP technique and the substrate-independent functional

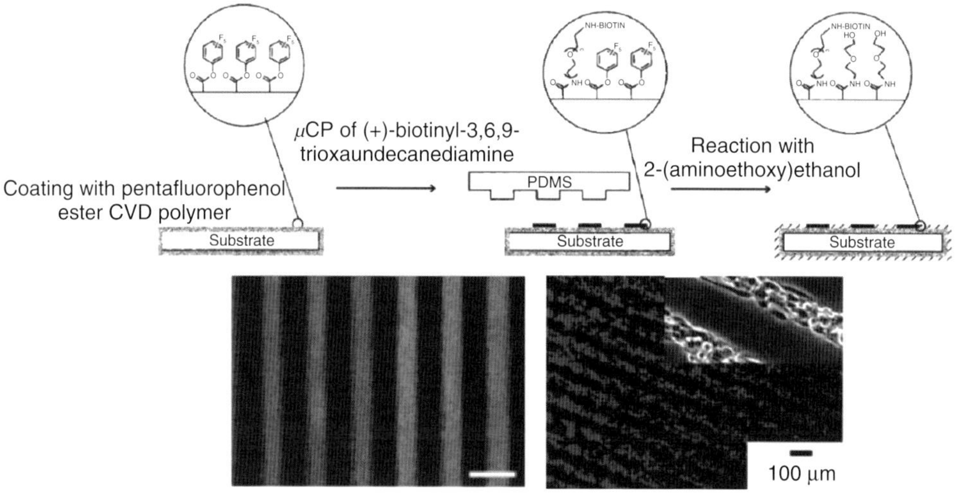

Figure 9.6 Example of μCP biomolecules onto reactive CVD coatings. The amino-functionalized biotin was printed onto PFP ester-containing CVD polymer coating. The bottom left fluorescence image shows the fluorescein conjugated streptavidin self-assembled on the biotin patterns. The bottom right fluorescence image shows the bovine aortic endothelial cells (BAECs) seeded for 24 h on a surface modified with biotin-streptavidin-biotin conjugated cell binding antibody [42]. (Reproduced from Ref. [42], with permission from American Chemical Society (Copyright 2002 American Chemical Society).

CVD coatings enables the functionalizing and patterning of many miniaturized biomedical devices.

9.4.2
Photopatterning

Poly(4-benzoyl-pxylylene-co-*p*-xylylene) is a CVD polymer coating functionalized with a photo-reactive functional group. The advantage of photopatterning is that it can be applied not only on flat substrates but also on 3D substrates. Ultraviolet light can be patterned onto the photo-reactive coatings with the help of a photomask or by projection lithography.

A representative work of using the benzophenone functionalized CVD coating for photopatterning and immobilizing biomolecules on 3D substrates is shown in Figure 9.7 [44]. Conformal poly(4-benzoyl-pxylylene-co-*p*-xylylene) coatings were first deposited onto 3D substrates such as microspheres, fibers, microchannels, and stents. Subsequently, a projection lithography process (UV light 365 nm) using a programmable digital micromirror was used to generate micropatterns on the 3D substrates. Though the whole 3D substrate was deposited with the photoreactive CVD coating, only the preprogrammed illuminated areas had photochemical reactions. For the proof-of-concept experiment, biotin-hydrazide was photoimmobilized on the CVD coated substrates followed by incubation in TRITC-streptavidin to visualize the photo-generated patterns. The projection lithography technique combining with the substrate-independent photo-reactive CVD coatings offer a convenient surface modification and patterning method for devices with complex geometry. A recent publication by Chen and coworkers shows that the poly(4-benzoyl-pxylylene-co-*p*-xylylene) CVD coating can also be used as a photoresist [45].

9.4.3
Vapor-Assisted Patterning During CVD

Reactive CVD coatings can be deposited within predefined pattern areas with the help of partially closed micro-structured stencils made by the PDMS elastomer. This soft lithographic technique provides a straightforward method to fabricate patterns with topological and/or chemical differences, simply by masking some predefined areas of the substrate during the vapor deposition and polymerization. We defined this patterning technique as "vapor-assisted micropatterning in replica structures" (VAMPIRs) [15, 46].

To give a representative example, poly(*p*-xylylene-4-methyl-2-bromoisobutyrate-co-*p*-xylylene) was deposited on the substrate with a PDMS microstencil placing on top of the substrate to mask predefined areas (Figure 9.8) [15]. This coating has the functional group for initiating atom-transfer radical polymerization (ATRP). Subsequently, poly(oligo(ethylene glycol) methyl ether methacrylate) (poly(OEGMEMA)) polymer brushes were grafting from the initiator CVD coating. The poly(OEGMA) brush, if successfully grown, can prevent

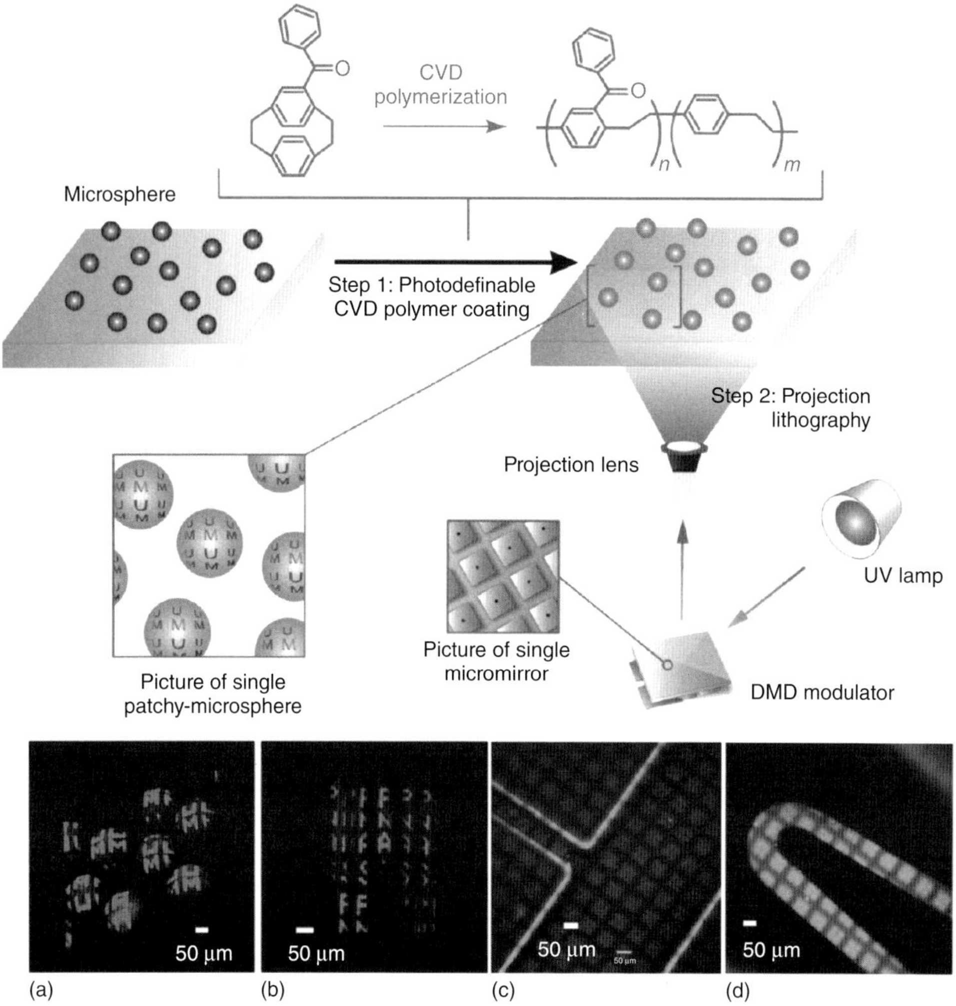

Figure 9.7 Scheme of photopatterning photo-reactive CVD polymer coated 3D surfaces. The technique can be applied to 3D complex substrates such as (a) microspheres, (b) fibers, (c) microchannels, and (d) stent. (Reproduced from Refs. [33, 44], with permission from American Chemical Society and National Academy of Sciences of the United States of America. (Copyright 2011 American Chemical Society and Copyright 2007 National Academy of Sciences of the United States of America).

protein adsorption and cell adhesion. Thus, fluorescence-labeled fibrinogen proteins and fibroblast cells (NIH3T3) were used to evaluate if the polymer brushes were successfully grown from the CVD coating. The results (Figure 9.8) show that fibrinogen or the fibroblast cells only present on the areas without the poly(OEGMA) brushes (these areas were initially masked by PDMS and thus had no CVD coating to initiate the ATRP brush growth).

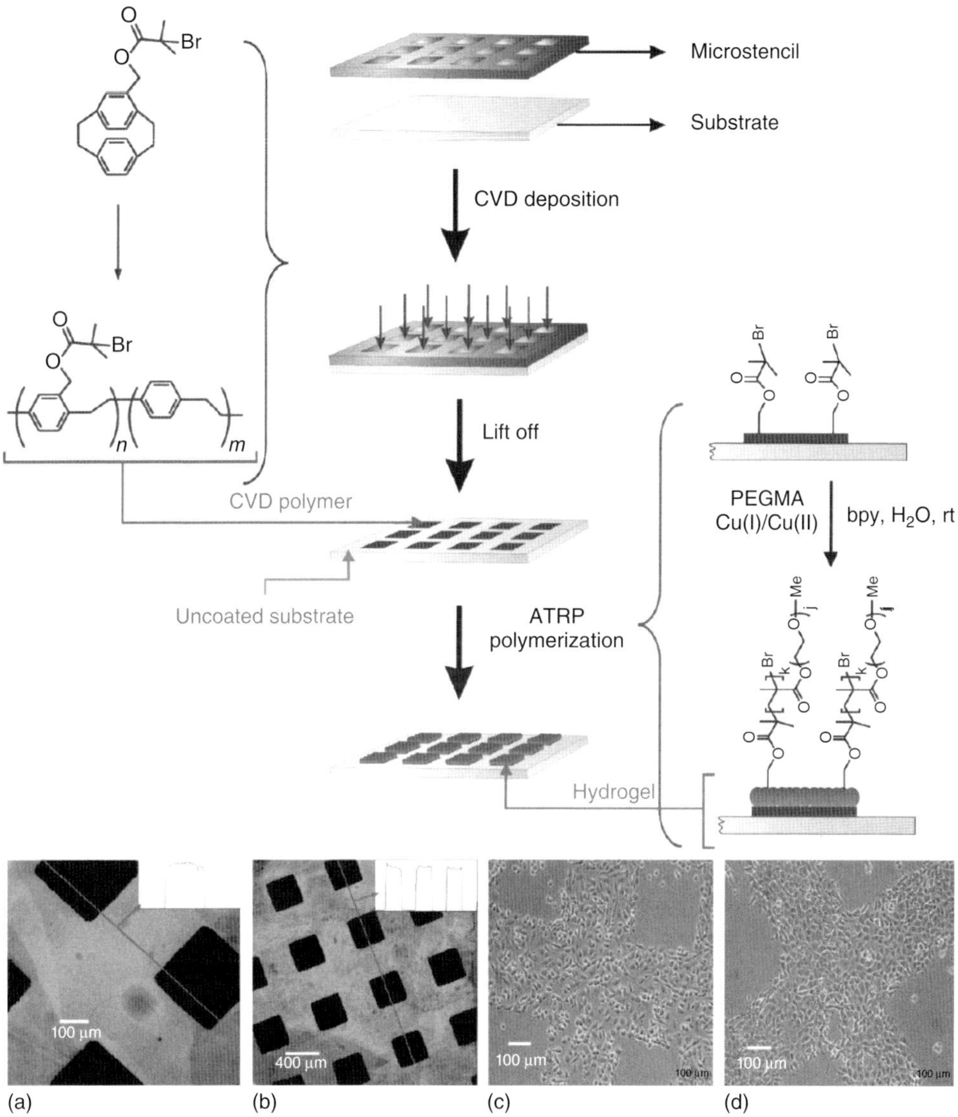

Figure 9.8 Atom transfer radical polymerization initiator functionalized PPX coating was patterned by vapor-assisted patterning during CVD followed by growing poly[oligo(ethylene glycol) methyl ether methacrylate] brushes. The patterned polymer brushes can prevent protein adsorption (a,b) and cell adhesion (c,d) on selected areas [15]. (Reproduced from Ref. [15], with permission from Wiley (Copyright 2008 Wiley).

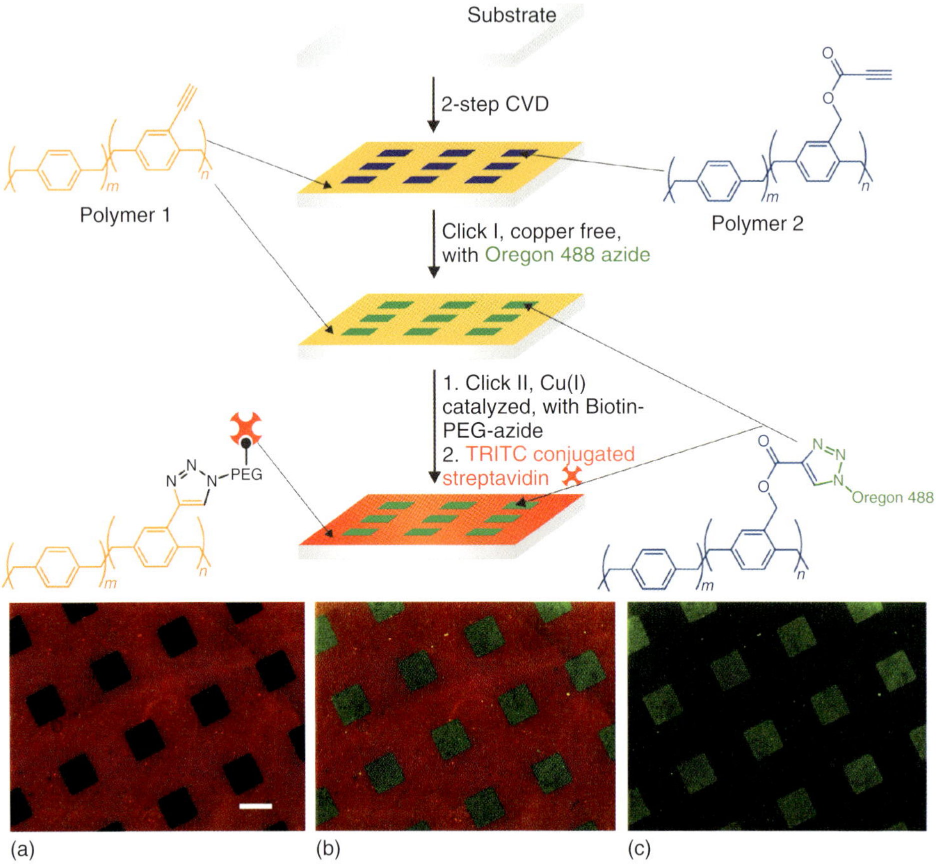

Figure 9.9 Scheme and fluorescent micrographs of the "Double-click" reactions using different reactivities of activated and nonactivated alkyne groups toward azide groups. The surface is coated with differently functionalized CVD polymers in different defined areas. (a–c) Fluorescence micrographs of samples prepared exactly as shown in the schematic illustration. (b) An overlay image of green and red channels shown in (a) and (c). Scale bar represents 200 μm. (Reproduced from Ref. [47], with permission from Wiley (Copyright 2011 Wiley).

In a more recent work (Figure 9.9), we applied the VAMPIR technique to fabricate a surface with the activated alkyne (methyl propiolate group) inside the pattern squares and nonactivated alkyne outside the squares [47]. The activated alkyne with an electron-withdrawing ester group can react with azide molecules under copper-free condition and the nonactivated alkyne needs copper(I) as catalyst for the reaction to proceed at room temperature. This kind of micro-engineered surface allowed us to carry out two-step sequential "double-click" reactions to immobilize two different azide-functionalized molecules, first under copper free condition and then with the copper(I) catalyst.

9.4.4
Nanopatterning by Dip-Pen Lithography (DPN)

DPN is a nanolithography technique based on scanning probe microscopy [48, 49]. It can deliver "chemical inks" (solutions of target molecules for printing) to substrates and generate nanopatterns by direct writing without using stamps. The nanopatterning can be carried out on a large surface by parallel writing with arrays of numerous tips [50]. DPN does not involve radiations or treatment with harsh chemicals typically used for conventional photolithography [51]. With optimized process conditions, DPN can generate pattern features with lateral resolution less than 50 nm [52].

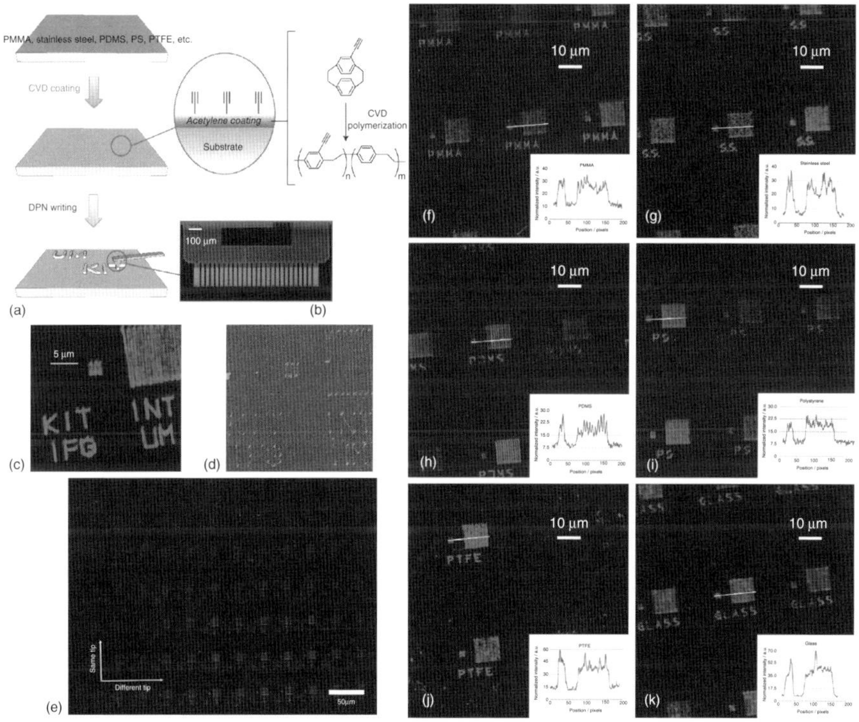

Figure 9.10 (a) Scheme of the DPN writing on alkyne functionalized CVD polymer coated substrates. "Ink" molecules were covalently immobilized by alkyne-azide click chemistry. (b) Micrograph of the 1D cantilever array (26 pens) used for this work. (c) Fluorescence image of an alkyne functionalized PPX coated gold surface after writing lines (300 nm) of biotin azide to form rectangular boxes and texts and applying fluorescence dye conjugated streptavidin to visualize the patterns. (d) Atomic force microscopy (AFM) image of the patterns shown in (c). (e) Fluorescence micrograph of a large scale DPN pattern. (f–k) Fluorescence images showing DPN patterns on the alkyne functionalized CVD polymer coating on different substrates as marked with the pattern text. Fluorescence intensity profiles along the white lines are shown in the images. (Reproduced from Ref. [34], with permission from American Chemical Society (Copyright 2010 American Chemical Society).

DPN can be applied with many types of "inks" but the selections of substrates are limited (mostly model substrates such as silicon and gold) [34]. The substrate-independent multifunctional CVD coatings can significantly expand the selection of substrates for DPN. We recently reported using DPN for modifying six different types of substrates (silicon, glass, stainless steel, Teflon, polystyrene, PDMS elastomer, etc.) with the same printing chemistry and mechanism (Figure 9.10) [34]. This is achieved by first depositing a thin layer of reactive CVD coating (with alkyne groups) on all the substrates followed by printing/writing azide molecules with a copper catalyst on the coated substrates by DPN. Pattern qualities are all good on the different substrates with only a slight loss of quality on the surface with higher roughness (stainless steel, polytetrafluoroethylene (PTFE)). The study demonstrates that combining the substrate-independent reactive CVD coating with the DPN technique results in a widely applicable surface modification and nanopatterning method.

9.5 Summary and Future Outlook

In summary, multifunctional reactive polymer coatings provide a versatile substrate-independent surface functionalization platform. The functional groups can serve as anchor sites for a broad range of biomolecules and present designable biointerfaces with precise control of the biomolecule immobilization chemistries as well as spatial arrangements.

The development of the multifunctional reactive CVD polymer coatings can result in copolymers, gradients, and even micro- and nanopatterns and thus has great potential in a wide spectrum of biomedical areas, such as in the preparation of biomimetic surfaces, the study of cell or biomolecule interactions with surfaces, the development of biosensors, or the surface modification of miniaturized biomedical devices. Improving commercial availability of different substituted [2.2]paracyclophanes can greatly accelerate the applications of the powerful surface functionalization platform. Future improvements in this field may include the development of more highly efficient and biocompatible surface chemistries, better control of molecule distribution on the surface, and integration of the coatings into novel biomaterials and highly sophisticated devices.

Acknowledgments

The authors gratefully acknowledge support from Defense Threat Reduction Agency (DTRA), under project HDTRA1-12-1-0039 and the Army Research Office (ARO) under Grant W911NF-11-1-0251.

References

1. Castner, D.G. and Ratner, B.D. (2002) Biomedical surface science: foundations to frontiers. *Surf. Sci.*, **500** (1–3), 28–60.
2. Cosnier, S. (1999) Biomolecule immobilization on electrode surfaces by entrapment or attachment to electrochemically

polymerized films. A review. *Biosens. Bioelectron.*, **14** (5), 443–456.
3. Siow, K.S. *et al.* (2006) Plasma methods for the generation of chemically reactive surfaces for biomolecule immobilization and cell colonization – a review. *Plasma Processes Polym.*, **3** (6–7), 392–418.
4. Shin, H., Jo, S., and Mikos, A.G. (2003) Biomimetic materials for tissue engineering. *Biomaterials*, **24** (24), 4353–4364.
5. Ulman, A. (1996) Formation and structure of self-assembled monolayers. *Chem. Rev.*, **96** (4), 1533–1554.
6. Lahann, J. (2006) Vapor-based polymer coatings for potential biomedical applications. *Polym. Int.*, **55** (12), 1361–1370.
7. Tawfick, S. *et al.* (2010) Nanocomposite microstructures with tunable mechanical and chemical properties. *Phys. Chem. Chem. Phys.*, **12** (17), 4446–4451.
8. Yoshida Kozai, T.D. *et al.* (2012) Ultrasmall implantable composite microelectrodes with bioactive surfaces for chronic neural interfaces. *Nat. Mater.*, **11** (12), 1065–1073.
9. Greiner, A. *et al.* (1997) Poly(p-xylylene)s: synthesis, polymer analogous reactions, and perspectives on structure–property relationships. *Acta Polym.*, **48** (1–2), 1–15.
10. Hou, H.Q. *et al.* (2002) Poly(p-xylylene) nanotubes by coating and removal of ultrathin polymer template fibers. *Macromolecules*, **35** (7), 2429–2431.
11. Deng, X.P. and Lahann, J. (2014) Orthogonal surface functionalization through bioactive vapor-based polymer coatings. *J. Appl. Polym. Sci.*, **131** (14), 40315.
12. Lahann, J. *et al.* (2001) A new method toward microengineered surfaces based on reactive coating. *Angew. Chem. Int. Ed.*, **40** (17), 3166–3169.
13. Lahann, J. *et al.* (2001) Bioactive immobilization of r-hirudin on CVD-coated metallic implant devices. *Biomaterials*, **22** (8), 817–826.
14. Bally, F. *et al.* (2013) Co-immobilization of biomolecules on ultrathin reactive chemical vapor deposition coatings using multiple click chemistry strategies. *ACS Appl. Mater. Interfaces*, **5** (19), 9262–9268.
15. Jiang, X. *et al.* (2008) Vapor-based initiator coatings for atom transfer radical polymerization. *Adv. Funct. Mater.*, **18** (1), 27–35.
16. Deng, X. *et al.* (2012) Bio-orthogonal polymer coatings for co-presentation of biomolecules. *Macromol. Rapid Commun.*, **33** (8), 640–645.
17. Yu, A.A. *et al.* (2005) Supramolecular nanostamping: using DNA as movable type. *Nano Lett.*, **5** (6), 1061–1064.
18. Elkasabi, Y., Chen, H.Y., and Lahann, J. (2006) Multipotent polymer coatings based on chemical vapor deposition copolymerization. *Adv. Mater.*, **18** (12), 1521–1526.
19. Han, S.Y. and Kim, Y.A. (2004) Recent development of peptide coupling reagents in organic synthesis. *Tetrahedron*, **60** (11), 2447–2467.
20. Chen, H.Y. *et al.* (2013) Vapor-based trifunctional coatings. *Chem. Commun.*, **49** (40), 4531–4533.
21. Ridley, A.J. *et al.* (2003) Cell migration: integrating signals from front to back. *Science*, **302** (5651), 1704–1709.
22. Liu, L.Y. *et al.* (2007) Endothelial cell migration on surface-density gradients of fibronectin, VEGF, or both proteins. *Langmuir*, **23** (22), 11168–11173.
23. Jayaraman, S. and Hillier, A.C. (2004) Construction and reactivity screening of a surface composition gradient for combinatorial discovery of electro-oxidation catalysts. *J. Comb. Chem.*, **6** (1), 27–31.
24. Jeon, N.L. *et al.* (2000) Generation of solution and surface gradients using microfluidic systems. *Langmuir*, **16** (22), 8311–8316.
25. Burdick, J.A., Khademhosseini, A., and Langer, R. (2004) Fabrication of gradient hydrogels using a microfluidics/photopolymerization process. *Langmuir*, **20** (13), 5153–5156.
26. Ballav, N. *et al.* (2007) A flexible approach to the fabrication of chemical gradients. *Adv. Mater.*, **19** (7), 998–1000.
27. Elkasabi, Y. and Lahann, J. (2009) Vapor-based polymer gradients. *Macromol. Rapid Commun.*, **30** (1), 57–63.
28. Elkasabi, Y.M., Lahann, J., and Krebsbach, P.H. (2011) Cellular transduction gradients via vapor-deposited

polymer coatings. *Biomaterials*, **32** (7), 1809–1815.

29. Elkasabi, Y. *et al.* (2014) Design strategies for reduced-scale surface composition gradients via CVD copolymerization. *Chem. Vap. Deposition*, **20** (1–3), 23–31.
30. Menard, E. *et al.* (2007) Micro- and nanopatterning techniques for organic electronic and optoelectronic systems. *Chem. Rev.*, **107** (4), 1117–1160.
31. Falconnet, D. *et al.* (2006) Surface engineering approaches to micropattern surfaces for cell-based assays. *Biomaterials*, **27** (16), 3044–3063.
32. Nie, Z. and Kumacheva, E. (2008) Patterning surfaces with functional polymers. *Nat. Mater.*, **7** (4), 277–290.
33. Chen, H.Y. and Lahann, J. (2011) Designable biointerfaces using vapor-based reactive polymers. *Langmuir*, **27** (1), 34–48.
34. Chen, H.Y. *et al.* (2010) Substrate-independent dip-pen nanolithography based on reactive coatings. *J. Am. Chem. Soc.*, **132** (51), 18023–18025.
35. Kumar, A., Biebuyck, H.A., and Whitesides, G.M. (1994) Patterning self-assembled monolayers – applications in materials science. *Langmuir*, **10** (5), 1498–1511.
36. Bernard, A. *et al.* (2000) Microcontact printing of proteins. *Adv. Mater.*, **12** (14), 1067–1070.
37. Lange, S.A. *et al.* (2004) Microcontact printing of DNA molecules. *Anal. Chem.*, **76** (6), 1641–1647.
38. Wilbur, J.L. *et al.* (1994) Microfabrication by microcontact printing of self-assembled monolayers. *Adv. Mater.*, **6** (7–8), 600–604.
39. Hidber, P.C. *et al.* (1996) Microcontact printing of palladium colloids: micron-scale patterning by electroless deposition of copper. *Langmuir*, **12** (5), 1375–1380.
40. Santhanam, V. and Andres, R.P. (2004) Microcontact printing of uniform nanoparticle arrays. *Nano Lett.*, **4** (1), 41–44.
41. Wendeln, C. *et al.* (2012) Rapid preparation of multifunctional surfaces for orthogonal ligation by microcontact chemistry. *Chem. Eur. J.*, **18** (19), 5880–5888.
42. Lahann, J. *et al.* (2002) Reactive polymer coatings: a platform for patterning proteins and mammalian cells onto a broad range of materials. *Langmuir*, **18** (9), 3632–3638.
43. Thévenet, S. *et al.* (2007) A generic approach towards nanostructured surfaces based on supramolecular nanostamping on reactive polymer coatings. *Adv. Mater.*, **19** (24), 4333–4337.
44. Chen, H.Y. *et al.* (2007) Colloids with high-definition surface structures. *Proc. Natl. Acad. Sci. U.S.A.*, **104** (27), 11173–11178.
45. Wu, M.G. *et al.* (2012) Vapor-deposited parylene photoresist: a multipotent approach toward chemically and topographically defined biointerfaces. *Langmuir*, **28** (40), 14313–14322.
46. Chen, H.Y. and Lahann, J. (2007) Vapor-assisted micropatterning in replica structures: a solventless approach towards topologically and chemically designable surfaces. *Adv. Mater.*, **19** (22), 3801–3808.
47. Deng, X., Friedmann, C., and Lahann, J. (2011) Bio-orthogonal "double-click" chemistry based on multifunctional coatings. *Angew. Chem. Int. Ed.*, **50** (29), 6522–6526.
48. Piner, R.D. *et al.* (1999) "Dip-pen" nanolithography. *Science*, **283** (5402), 661–663.
49. Hong, S.H., Zhu, J., and Mirkin, C.A. (1999) Multiple ink nanolithography: toward a multiple-pen nano-plotter. *Science*, **286** (5439), 523–525.
50. Salaita, K. *et al.* (2006) Massively parallel dip-pen nanolithography with 55000-pen two-dimensional arrays. *Angew. Chem. Int. Ed.*, **45** (43), 7220–7223.
51. Ginger, D.S., Zhang, H., and Mirkin, C.A. (2004) The evolution of dip-pen nanolithography. *Angew. Chem. Int. Ed.*, **43** (1), 30–45.
52. Salaita, K., Wang, Y.H., and Mirkin, C.A. (2007) Applications of dip-pen nanolithography. *Nat. Nanotechnol.*, **2** (3), 145–155.

10
CVD Fluoropolymers[1)]

Jose L. Yagüe

10.1
Introduction

Fluorinated polymers have attracted great interest due to the unique properties conferred by the carbon–fluorine bond. The outstanding chemical and physical properties of fluoropolymers have found applications in a wide variety of fields, such as optics [1], anti-fouling [2], electronics [3], and especially in the design of hydrophobic surfaces [4–6]. Typical synthesis of these polymers involves solvent-based processes. However, the use of CVD methods enables certain difficulties that entail the use of solvents to be overcome, and at the same time introduces attractive features, including the following:

- synthesis of true linear homopolymers, including polytetrafluoroethylene [(PTFE), Teflon®];
- no solvents required, favorable for depositing polymer coatings which do not dissolve in common solvents;
- no surfactants required, thus avoiding the need to use compounds which are persistent in the environment, such as perfluorooctanoic acid (PFOA), a common surfactant for PTFE;
- one-step formation of robust, insoluble, cross-linked networked layers from the copolymerization of fluorinated species, such as 1H,1H,2H,2H-perfluorodecyl acrylate (PFDA), with a cross-linking multivinyl monomer, such as ethylene glycol diacrylate (EGDA);
- coating objects which swell, dissolve, or otherwise degrade during solution and melt processing;
- ability to form ultrathin contiguous films that are <20 nm thick;
- ability to form pinhole free films that are >10 μm thick at high deposition rates;
- graded composition throughout the thickness of the film produced by varying the feed ratio of the reactants over the course of the deposition period;

1) This text was adapted from the chapter "Vapor Deposition of Fluoropolymer Surfaces" published in "Handbook of Fluoropolymer Science and Technology" with permission from Wiley.

CVD Polymers: Fabrication of Organic Surfaces and Devices, First Edition. Edited by Karen K. Gleason.

- grafting to the substrate, by building films from the substrate up, resulting in robust adhesion;
- conformal coating over the entire surface of 3D objects and penetration to micro- and nano-scale features, resulting from the high diffusivity of vapor phase reactants at low pressure;
- automated and reproducible processing in reactors scaled up to dimensions of >1 m;
- real-time thickness monitoring and end-point control;
- absence of residual solvent in CVD, which obviates the need for curing. Avoiding residual solvent is important for achieving biocompatibility and desirable dielectric properties.

Synthesis of fluoropolymers by means of CVD is based on the generation of reactive species inside of the vacuum chamber, which induces a free-radical polymerization. These species consist of the monomer itself or a free radical initiator molecule that undergoes thermal decomposition over the heated filaments. Table 10.1 contains typical fluoropolymers synthesized using CVD methods and the experimental conditions employed. It should be noted that the inclusion of a free radical initiator allows for decreasing the temperature of the filaments while preserving the chemical functionality of the monomers.

Selecting the proper monomer, or comonomers, and adjusting the experimental conditions is possible to control the physical and chemical properties of the fluoropolymers. The hydrophobicity and texture of these polymers are key features when developing a new coating for a specific application and dictate the viability of the process. Thus, this chapter is focused on the hydrophobic character of the fluoropolymers, and the tuning of their morphology and crystallinity that can be achieved by means of CVD.

10.2 Polytetrafluoroethylene (PTFE)

PTFE is the most common engineering material to fabricate hydrophobic materials. Plasma enhanced chemical vapor deposition (PECVD) has been a conventional method to synthesize PTFE films via vapor phase [7–9]. Because of the plasma exposure, these polymers show high variation in film composition and structure [10]. Alternatively, hot wire chemical vapor deposition (HWCVD) produces spectroscopically indistinguishable coatings from bulk PTFE. Extensive research has been conducted on HWCVD PTFE using hexafluoropropylene oxide (HFPO) as the only gas feed in the reactor [11, 12]. Thermal decomposition of the perfluorooctane sulfonyl fluoride (PFOSF) over the heated filament produces the initiating radicals that yield the polymer via free radical polymerization (see Chapter 2 for more details of the complete mechanism of the reaction).

Exhaustive characterization conducted on HWCVD PTFE confirms that this method, as compared to bulk methods, yields chemically identical polymers

Table 10.1 Examples of iCVD fluoropolymer reactants and example process conditions.

Fluoropolymer	Monomer	Initiator	T_f (°C)	T_s (°C)	P (mTorr)
Poly(tetrafluoroethylene) (PTFE)	Difluorocarbene (CF_2:) generated *in situ* from the pyrolysis of hexafluoropropylene oxide (HFPO)	Perfluorooctane sulfonyl fluoride (PFOSF) *or* perfluorobutane sulfonyl fluoride (PFBSF)	400	25	500
Poly(1H,1H,2H,2H-perfluorodecyl acrylate (PPFDA) and its copolymers	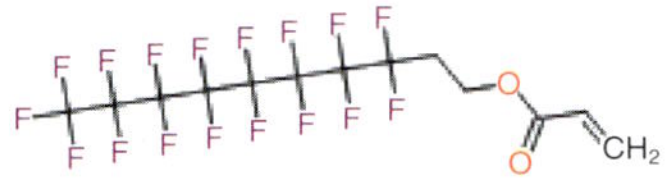	Tert butyl peroxide (TBPO)	200	35	125
Copolymers w/pentafluorophenyl methacrylate (PFM)		Tert butyl peroxide (TBPO)	280	30	200
Polyvinylidene fluoride	CH_2=CF_2	Difluorocarbene (CF_2:) generated from the pyrolysis of hexafluoropropylene oxide (HFPO)	500	−125	1000

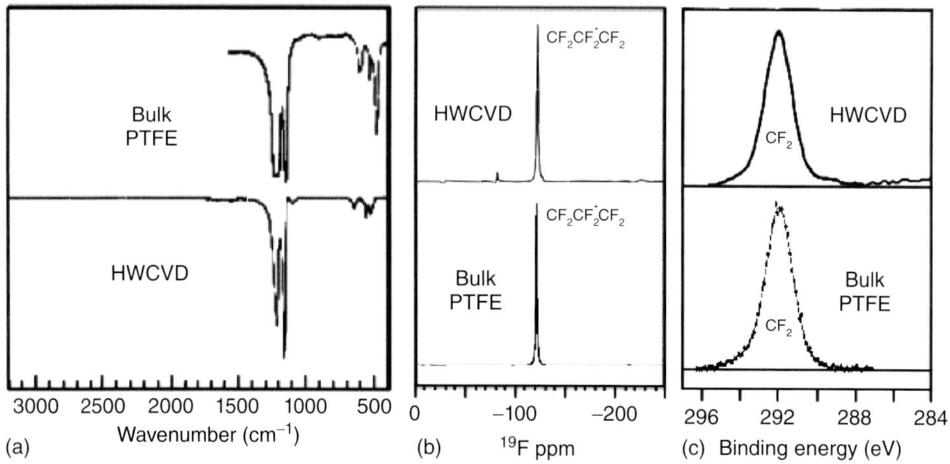

Figure 10.1 (a) FTIR spectra of bulk PTFE (top) and HWCVD PTFE (bottom) showing the characteristic bands for CF_2 stretching at 1155 and 1215 cm^{-1}. (b) [16] F NMR spectra of HWCVD fluorocarbon film (top) and bulk PTFE (bottom). The HWCVD film shows the $CF_2CF_2CF_2$ as the dominant species at −123 ppm, chemically identically to the bulk PTFE. (c) XPS spectra of HWCVD PTFE (top) and bulk PTFE (bottom) presenting virtually the same chemical environment for both polymers.

[11, 13–15]. The Fourier Transform Infrared (FTIR) spectra of bulk PTFE and the HWCVD film are compared in Figure 10.1a. Both of them present the symmetric and asymmetric bands at 1155 and 1215 cm^{-1}, which is characteristic of the CF_2 stretching of the PTFE polymer. The chemical environment of the polymer was also studied by F Nuclear Magnetic Resonance (NMR) [16]. The spectrum of the HWCVD PTFE (Figure 10.1b) shows the $CF_2CF_2{}^*CF_2$ peak at −123 ppm representative of the PTFE bulk. The predominance of the $-CF_2-$ bonding environment in the film was confirmed by X-ray Photoelectron Spectroscopy (XPS) (Figure 10.1c). The most intense peak at 292.1 eV corresponds to the CF_2 group and represents the 90% content of the polymer background. The other two peaks at 294.0 and 289.5 eV belong to the CF_3 and CF groups respectively with a content of 5% each. Tribological studies on HWCVD PTFE thin films to measure its coefficient of friction (COF) showed a similar behavior to that of bulk PTFE. Yet, using a vapor deposition polymerization method, the chemistry and morphology of amorphous and crystalline domains in the thin film resembles to that of bulk PTFE.

CVD PTFE coatings enable the ability to modify the wettability of any kind of substrate. Conformal coating, 16 μm thick, was achieved around the circumference of a 25-μm diameter stainless-steel wire (Figure 10.2a) [9]. Film morphology and crystallinity of HWCVD PTFE films can be controlled by fine-tuning of the experimental conditions during the polymerization process. The presence of PTFE crystallites was observed by X-ray diffraction (XRD) [14]. Crystallization can be enhanced by annealing, producing spherulites up to 1 mm in diameter

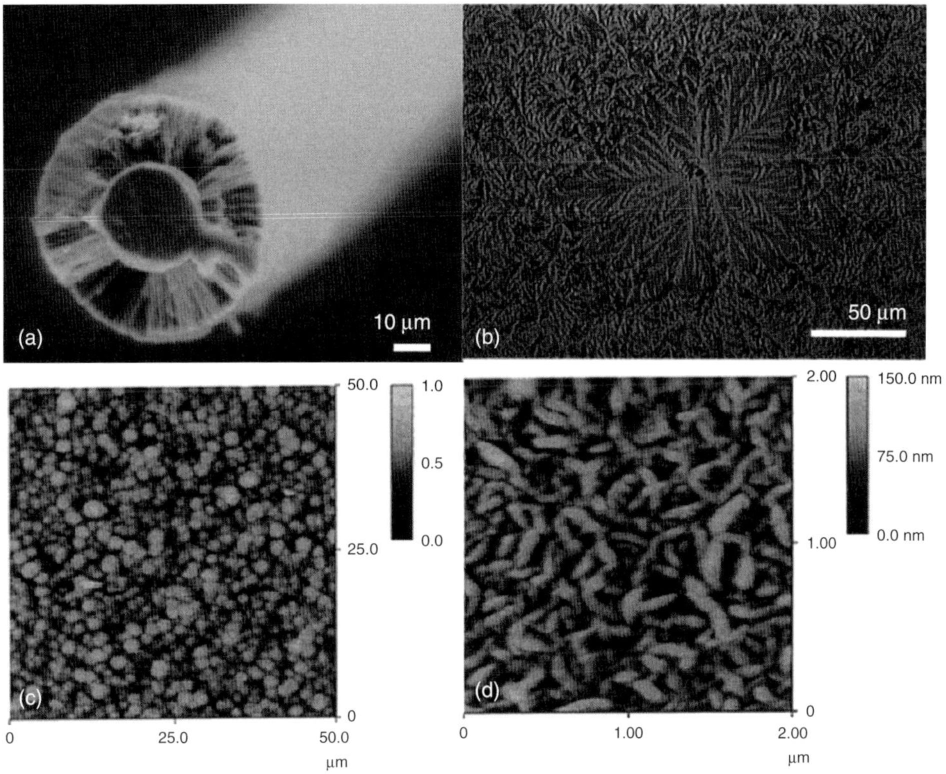

Figure 10.2 (a) Cross-section of a stainless-steel wire coated with a uniform 16 µm thick coating of PTFE. (b) Optical micrograph of crystalline spherulites of up to 1 mm in diameter produced by annealing of a PTFE film. AFM images of a PTFE film with spherical nodule morphology produced without pre-heating of the filaments (scan area 50 µm × 50 µm) (c) and with rod-like grain structure as a consequence of pre-heating the filaments (scan area 2 µm × 2 µm) (d).

(Figure 10.2b) [17]. Corresponding changes in the intensity of the CF_2 wagging ($641/629\,cm^{-1}$), deformation ($555\,cm^{-1}$), and rocking ($513/530\,cm^{-1}$) modes in the FTIR spectra provided additional evidence supporting changes in film crystallinity. Inefficient packing of anisotropic crystallites was observed to lead to micro- and nano-porous films [18]. Porosity up to 30 vol% was quantified by variable angle spectroscopic ellipsometry (VASE). More rapidly deposited films had spherical nodules and lower porosity [19]. The difference in morphology is readily imaged by atomic force microscopy (AFM). When the filaments are not pre-heated, the deposition leads to spherical nodules (Figure 10.2c), while when the filaments are pre-heated the resulting films showed rod-like grain structures (Figure 10.2d).

Research continues on the thermal decomposition of HFPO to yield transparent, superhydrophobic PTFE films by CVD. Different filament materials, including nickel, have been explored [16, 20, 21]. An rf-sputtered PTFE film as a buffer layer was used to synthesize HWCVD PTFE having a contact angle of 150° [22].

Selective deposition of fluorinated films has emerged after the development of novel patterning processes. Patterning CVD PTFE films have been achieved through hard masks with 50 μm spaces [21]. E-beam lithography has resulted in line resolution as low as 1 μm. Furthermore, the integration of supercritical CO_2 as the developer has lead to a new and friendly environmental process [23].

Kim *et al.* [24] used HFPO as monomer and tungsten (W) as catalyst to deposit PTFE films via catalyzer chemical vapor deposition (Cat-CVD) method. These layers showed a 90% transmittance in the visible range independent of the film thickness, which could be used as a coating for pollution-free surfaces in opto-electronics. The deposition rate of PTFE films in Cat-CVD can be adjusted by appropriate selection of the metal wires that are used as catalyzers [25].

Further research demonstrated the viability of initiated chemical vapor deposition (iCVD) to synthesize PTFE films. The first ever iCVD process used the initiator PFOSF for the deposition of PTFE thin films [26]. The addition of PFOSF accelerated the deposition rates, which in some cases exceeded 1 μm min^{-1}. Also, the addition of PFOSF increases the concentration of CF_3 groups as a result of end-capping. The CF_3 groups have a lower surface energy resulting in polymers with higher hydrophobicity.

Ultrathin PTFE (<50 nm) films were deposited conformally by iCVD deposition to surface modify vertically aligned multiwalled carbon nanotube forests (mean diameter ~50 nm, mean height ~2 μm) (Figure 10.3b) [27, 28]. Conversely, conventional spray and bake PTFE coatings are generally non-conformal and yield >10 μm-thick films. The HFPO polymerization was initiated by PFOSF over a stainless steel filament. The low surface energy of the iCVD PTFE film (~20 mN m^{-1}) and the surface geometry of the nanotube forest led to a Wenzel state displaying advancing and receding contact angles with water of 170 and 160°, respectively. Essential spherical water drops were suspended on the surface (Figure 10.3a). The coated carbon nanotube forest remained unalterable after several cycles of condensation and evaporation preventing the bundling caused by the surface tension effect when drying in uncoated samples.

More recently, to avoid the problem of bioaccumulation related to C8 species, the C4 initiator, perfluorobutane sulfonyl fluoride ($CF_3(CF_2)_3SO_2F$, PFBSF) has been used[29]. With PFBSF as the initiator, ultrathin iCVD PTFE conformal coatings of single-wall carbon nanotubes (SWCNTs) show lamellar structures spaced on average 13 nm apart parallel to the SWCNT axis (Figure 10.3c,d).

10.3 Poly(vinylidene fluoride) (PVDF)

Rastogi and Desu [30] described the synthesis of poly(vinylidene fluoride) (PVDF) by a low-pressure chemical vapor polymerization (CVP) process. Owing to the high vapor pressure of the VDF monomer (26 200 Torr at 25 °C), the stage of the

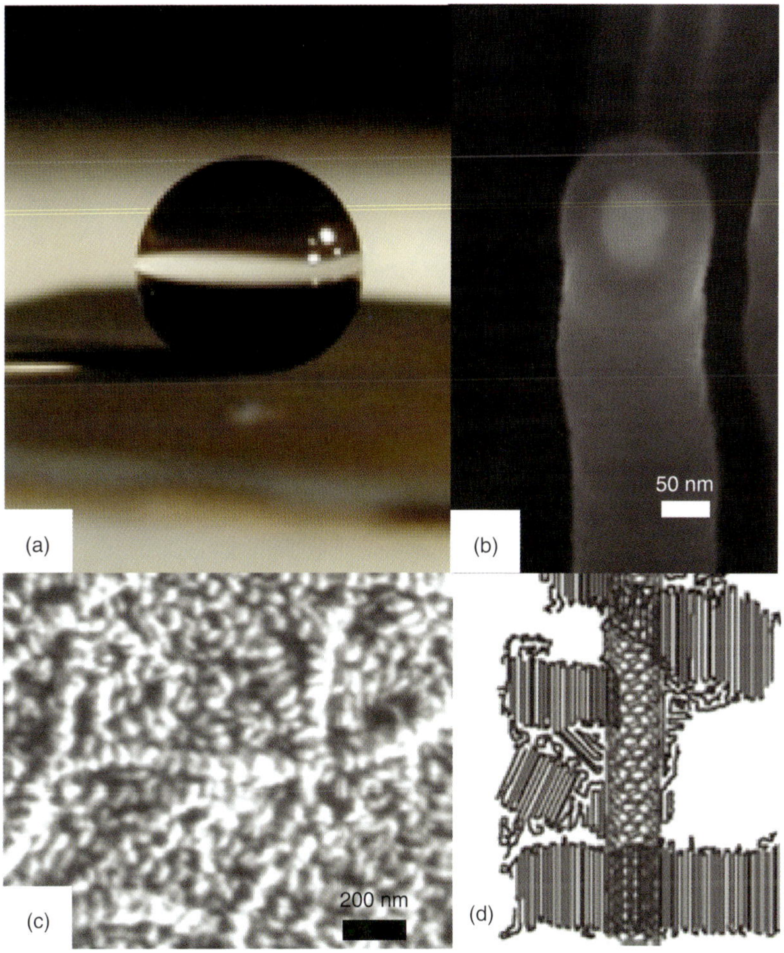

Figure 10.3 (a) Droplet of water with a near-spherical shape deposited on top of a carbon nanotubes forest PTFE coated by iCVD. (b) SEM image of a single carbon nanotube coated with a 40 nm thick PTFE film. (c) SEM image of SWCNT conformally coated with PTFE showing a "shish kebab" morphology. The SWCNT acts as a nucleation site for the growth of crystalline PTFE. (d) Scheme of the growth of the PTFE parallel to the carbon nanotube axis with lamellar structure.

chamber was back cooled down to 150 K by passing nitrogen liquid to enhance the adhesion on the substrate. Once a layer of VDF has been physisorbed, a flow of HFPO is introduced into the reactor and the temperature of the filaments is raised to 500–600 °C. Decomposition of the oxide generates CF_2 radicals, which in turn, act as the initiator for the polymerization of the VDF. The polymerization can be carried out in continuous, as explained, or pulsed, alternating cycles of VDF condensation and formation of the radical initiator, to yield ferroelectric PVDF thin films.

10.4 Poly(1H,1H,2H,2H-perfluorodecyl acrylate) [*p*(PFDA)]

As a result of the their $-CF_3$ terminated side groups and comb-like structure, the typical range of surface energies for fluorinated acrylic polymers (5.6–7.8 mN m^{-1}) are even lower than that of PTFE (~20 mN m^{-1}) [31]. The *p*(PFDA) side chains with eight perfluorinated carbons, so-called "C8," provide highly desirable iCVD surface properties but there is concern over the bioaccumulation factor of the monomer [32], suggesting that future research would benefit from using monomers with shorter perfluorocarbon side chains.

Chemical characterization of iCVD *p*(PFDA) synthesized using tert-butyl peroxide (TBPO) as the initiator, confirmed that the monomer's pendant fluorocarbon group is completely retained while the vinyl bond is complete consumed [33]. Deposition rate increased the weight up to 375 nm min, with increasing P/P_{sat}, where P is the monomer's partial pressure and P_{sat} is its vapor pressure. The number average molecular weight, Mn, also increased up to 177 300 Da with increasing P/P_{sat}. The measured absorption isotherm of the monomer and model-derived rate constants of film deposition verified that *p*(PFDA) film growth kinetics are mechanistically similar to the free radical polymerization of other non-fluorinated iCVD polymers [34, 35].

Conformal coverage of iCVD *p*(PFDA) was also demonstrated on electrospun mats to create superhydrophobic fabrics [31]. Decreasing the fiber diameter from 2200 nm down to 600 nm, produced a systematic increase in the static water contact angle, which agreed with the predictions of a theoretical model. Hierarchical morphology added in the form of beaded fibers, further increased the static water contact angle, up to a value of 175° and decreased the sliding angle to <2.5° for a 20 mg droplet. In one case, grade 8 oleophobicity was also exhibited by the iCVD *p*(PFDA)-coated electrospun mat. Adding nanoscale pores to the fibers is an alternative means of adding hierarchical structure. A 30 nm-thin conformal iCVD coating preserved the nanopores and fiber structure, while successfully lowering the surface energy [36].

Deposition conditions affect both the degree of crystallinity of iCVD *p*(PFDA) and the orientation of the crystallites (perpendicular or parallel to the surface) [37]. The XRD characterization shows a smectic B structure with a periodicity of 3.24 nm in the iCVD *p*(PFDA) films, displaying crystallinity. This is the same crystal structure observed in conventionally synthesized *p*(PFDA). The perfluorinated side chains form alternating chains face toward each other forming a bilayer. The orientation of the side chains is orthogonal to the plane of the lamellar, which exhibits hexagonal packing with a lattice parameter of 0.64 nm.

At the highest filament temperature, 300 °C, diffraction peaks are clearly seen when the X-ray beam oriented perpendicular to the substrate (Figure 10.4b) but are absent when the X-ray beam is parallel to the substrate (Figure 10.4a), indicating a vertical alignment of the perfluoro side-chains. The converse is observed at the lowest filament temperature, 270 °C, indicating a change to a horizontal alignment. Figure 10.4 schematically shows the two different crystalline

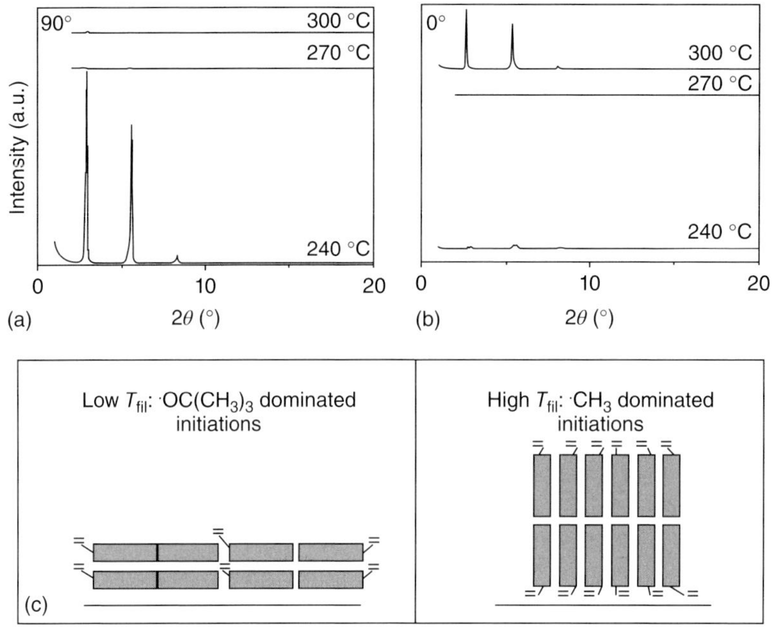

Figure 10.4 XRD profiles of iCVD *p*(PFDA) films synthesized using different filament temperature (240, 270, and 300 °C). At low temperature the perfluoro side chains arrange in a Smectic B structure parallel to the surface showing a diffraction pattern only for an irradiation angle of 90° with respect to the surface normal (a). While at high temperatures the perfluoro side chains are aligned perpendicular to the surface and interacting only with the X-ray beam at 0° (b). (c) Scheme of the orientation of the perfluoro side chains depending on the filament temperature.

textures. The difference in orientation is hypothesized to arise from a change in the dominant initiation mechanism. At low temperature, the bulky tert butoxy radical predominates, while at high temperatures, β-scission of the tert butoxy radical leads to smaller methyl radicals becoming the dominant initiation species [38]. For films with parallel orientation, the highest advancing water contact angle observed is 130°. Films with perpendicular orientation have advancing water contact angles as high as 150°. The lowest degree of hysteresis in the water contact angle is observed in films with the highest degree of crystallinity. Grafting of the iCVD *p*(PFDA) using a vinyl silanol coupling agent pretreatment, resulted in an even higher advancing water contact angle of 160°, with a hysteresis of only 5°. The grafted film was also oleophobic, having an advancing contact angle with mineral oil of 120° [39].

Bifunctional surfaces displaying areas of hydrophobic iCVD *p*(PFDA) next to hydrophilic regions composed of iCVD hydrogel, polyhydroxyethylmethacrylate, p(HEMA), were patterned down to ~7 μm resolution using transmission electron microscope grids as lift-off masks, thereby allowing control over microcondensation. Water droplets preferentially condense on the hydrophilic region allowing potential applications in microfluidics or water-harvesting [40].

The substrate-independent nature of iCVD enabled to encapsulate substrates as delicate as a paper-based photovoltaic (PV) array by deposition of a *p*(PFDA) layer to protect the device and extend its life time (see Chapter 16 for more information about the device). Since no solvents are involved in the deposition method, the polymer electrodes were gently coated with the hydrophobic polymer. The coated array was folded and submerged into a water bath during operation without affecting the performance of the device [41].

10.5 Copolymerization of Fluorinated Monomers

Copolymerization of fluorinated monomers has resulted in the synthesis of new materials with very unique properties. Owing to its solventless nature, CVD methods allow the use of monomers with a lack of a common solvent, such as the hydrophobic vinyl monomer PFDA, together with the hydrophilic monomer, 2-hydroxyethyl methacrylate (HEMA). Reacting two different monomers allows benefiting from the different properties of each one of them. As a consequence, the copolymer shows an enhancement in its physical and/or chemical properties and allows engineering the fluoropolymer for new applications. A list of fluorinated monomers includes PFDA, 2,2,3,3,4,4,5,5,6,6,7,7-dodecafluoroheptyl acrylate (DFHA), pentafluorophenyl methacrylate (PFM), or (perfluoroalkyl)ethyl methacrylate (PFEMA), among others. In addition, some examples of PFDA and HFPO copolymers are described, where the fluoromonomer benefits from the introduction of a comonomer (see Chapter 4 for the complete description and application of the rest of copolymers).

10.5.1 Copolymers with 1H,1H,2H,2H-perfluorodecyl acrylate (PFDA)

Copolymerization with divinyl benzene (DVB) resulted in a loss of the crystalline structure of *p*(PFDA). The characteristic lamellar structure of *p*(PFDA) disappears as a consequence of the cross-linking function of DVB. The intercross linking among the chains prevents the bilayer organization of the perfluorinated side chains yielding an amorphous polymer. Annealing of the copolymer films *p*(PFDA-*co*-DVB) increased the cross-linking density by reaction of the pendant vinyl bonds of DVB. The thermal treated films displayed a contact angle hysteresis of less than 10° and 1° for water and mineral oil respectively. In comparison, the untreated films showed a contact angle hysteresis of 20° for both water and mineral oil. This effect is due to the fact that the cross-linking enhancement acts as a steric barrier to prevent the surface reconstruction on the surface. The fluorinated groups are fixed on the outermost surface polymer layer independently of the media exposed to the polymer, thus reducing the hysteresis contact angle [42]. Dropwise condensation of steam was achieved depositing the annealed copolymers on metal heat transfer surfaces. The hydrophobicity and low hysteresis of the coating caused the formation of small droplets of water that shed rapidly from the metallic surface (Figure 10.5a). The resulting heat transference coefficient was found to be $35\,kW\cdot m^{-2}\cdot K^{-1}$, seven times greater than when film condensation occurs on typical heat transfer substrates (Figure 10.5c). Furthermore, the iCVD films showed an outstanding durability compared to a fluorosilane coating, which was degraded after several minutes (Figure 10.5b) [43].

Hydrophilic monomers were combined with PFDA to obtain a dual functionality in the polymer chain. Enhanced resistance to protein adsorption was achieved reacting HEMA and PFDA to yield a randomly distributed copolymer film *p*(HEMA-*co*-PFDA) [44]. This heterogeneity at molecular scale turns the surface energetically unfavorable inhibiting either hydrophilic or hydrophobic interactions (see Chapter 13 for more details).

10.5.2 Copolymers with Organosilicons

Organosilicon polymers have found widespread application including biocompatible coatings for medical implants, barrier films for food packaging, and as dielectric films in integrated circuits [45–48]. Copolymerization of fluoropolymers with silicon-based polymers allows profiting from the richness of the C–F and Si–O bonds to explore the synthesis of new materials. The properties of these copolymers can be tuned for controlling the chemical structure of the deposited layer.

The possibility to copolymerize fluoromonomers with organosilicon monomers was explored first by means of HWCVD [11, 49]. After heating the filaments to 620 °C, the monomers HFPO and hexamethylcyclotrisiloxane (D3) were decomposed into the active species difluorocarbene, dimethylsiloxane, and linear and cyclic units coming from the ring structure to form the copolymer. In addition, the cross-linking of the polymer was found to be due only to the siloxane units. Further research demonstrated that the final chemical structure of the copolymer depends intimately on the filament temperature. Copolymers thin films of PFOSF and 1,3,5-trivinyl-1,3,5-trimethylcyclotrisiloxane (V_3D_3) were synthesized at low filament temperature (350 °C) and high temperature (540 °C). At low temperature, the backbone of the polymer is composed of a C–C chain with siloxane rings as pendant groups. This structure presents high thermal stability and flexibility. Conversely, high filament temperatures yield linear polymers with a siloxane-backbone cross-linked among them. This chemical structure presents lower thermal stability and the films deposited are more brittle and are prone to cracking.

Also, using an iCVD approach, PFDA was combined with (1,3,5,7-tetravinyl-1,3,5,7-tetramethylcyclotetrasiloxane) (V4D4) to form a stacked polymer deposited on fabrics (see Chapter 6 for more details) [50]. The iCVD process

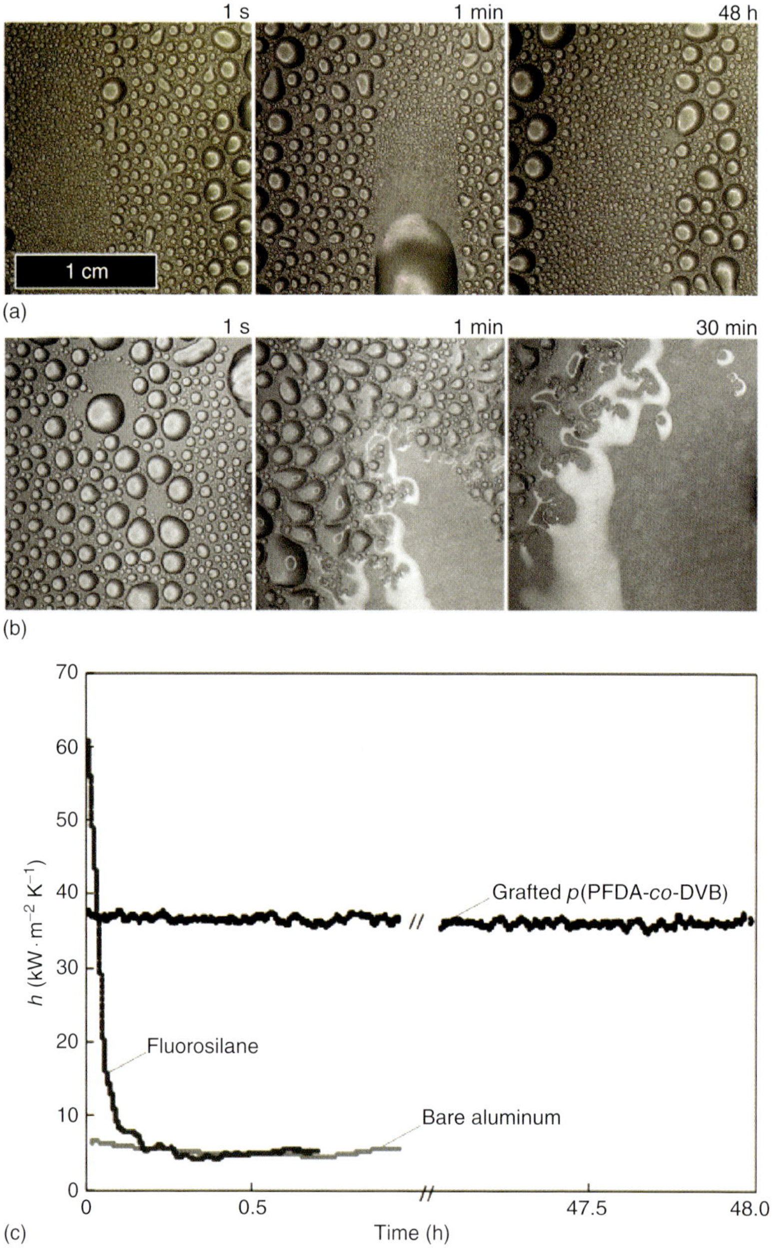

Figure 10.5 Photographs during condensation of saturated steam at 100 °C and 101 kPa of (a) prolonged dropwise condensation on grafted coating over a period of 48 h and (b) degradation of fluorosilane coating over a period of 30 min. (c) Heat transfer coefficient of aluminum substrates with no coating, with a fluorosilane coating, and with a grafted *p*(PFDA-*co*-DVB) coating, plotted versus time.

induced a hierarchical structure on the surface without altering the texture morphology of the fabric showing superhydrophobicity. The superhydrophobic properties of the fabrics remained unalterable when exposed to harsh environments, including chemical and mechanical attacks and typical cycles of laundry test.

10.6 Summary

CVD methods are highly desirable for modifying surfaces with fluoropolymers. In this chapter we have reviewed the most commonly synthesized polymers and the ability to tailor their properties and morphology by adjusting the working conditions. CVD methods produce fluoropolymer thin films chemically identical to those prepared by bulk. These thin films are characterized by a low surface energy. Furthermore, inducing the proper texture during the deposition results in superhydrophobic materials with extremely low contact angle hysteresis. The conformal and gentle nature of iCVD enables to tune the wettability of any kind of surface or 3D substrates. The crystallinity and grain shape of these polymers can be adjusted during the synthesis process leading to controlled modification of their physical properties, and creating diverse morphologies and structures on demand. Current investigations address toward the development of coatings with long-lasting and enhanced hydrophobic properties, new designs and implementation of the polymers, and the use of environmentally friendly monomers to benefit from all the advantages of fluorinated films.

References

1. Dalton, L. (2002) *Book, Nonlinear Optical Polymeric Materials: From Chromophore Design to Commercial Applications*, Springer.
2. Imbesi, P.M., Gohad, N.V., Eller, M.J., Orihuela, B., Rittschof, D., Schweikert, E.A., Mount, A.S., and Wooley, K.L. (2012) *ACS Nano*, **6**, 1503.
3. Seyrat, E. and Hayes, R.A. (2001) *J. Appl. Phys.*, **90**, 1383.
4. Han, D. and Steckl, A.J. (2009) *Langmuir*, **25**, 9454.
5. Im, M., Im, H., Lee, J.-H., Yoon, J.-B., and Choi, Y.-K. (2010) *Soft Matter*, **6**, 1401.
6. Liu, Y., Chen, X., and Xin, J. (2009) *J. Mater. Chem.*, **19**, 5602.
7. Yasuda, H. (1985) *Plasma Polymerization*, Academic Press.
8. Flamm, D.L., Auciello, O., and d'Agostino, R. (2012) *Plasma Deposition, Treatment, and Etching of Polymers: the Treatment and Etching of Polymers*, Elsevier.
9. Limb, S.J., Gleason, K.K., Edell, D.J., and Gleason, E.F. (1997) *J. Vac. Sci. Technol., A: Vac. Surf. Films*, **15**, 1814.
10. Stelmashuk, V., Biederman, H., Slavínská, D., Zemek, J., and Trchová, M. (2005) *Vacuum*, **77**, 131.
11. Lau, K.K.S., Lewis, H.G.P., Limb, S.J., Kwan, M.C., and Gleason, K.K. (2001) *Thin Solid Films*, **395**, 288.
12. Lau, K.K.S., Murthy, S.K., Lewis, H.G.P., Caulfield, J.A., and Gleason, K.K. (2003) *J. Fluorine Chem.*, **122**, 93.

13. Karnath, M.A., Sheng, Q., White, A.J., and Mueftue, S. (2010) *Tribol. Trans.*, **54**, 36.
14. Limb, S., Lau, K.S., Edell, D., Gleason, E., and Gleason, K. (1999) *Plasmas Polym.*, **4**, 21.
15. Limb, S.J., Labelle, C.B., Gleason, K.K., Edell, D.J., and Gleason, E.F. (1996) *Appl. Phys. Lett.*, **68**, 2810.
16. Yasuoka, H., Yoshida, M., Sugita, K., Ohdaira, K., Murata, H., and Matsumura, H. (2008) *Thin Solid Films*, **516**, 687.
17. Lau, K.K.S. and Gleason, K.K. (2001) *J. Phys. Chem. B*, **105**, 2303.
18. Lau, K.K.S., Caulfield, J.A., and Gleason, K.K. (2000) *J. Vac. Sci. Technol., A: Vac. Surf. Films*, **18**, 2404.
19. Lau, K.K.S., Caulfield, J.A., and Gleason, K.K. (2000) *Chem. Mater.*, **12**, 3032.
20. Cha, J.O., Yeo, S.J., Pode, R., and Ahn, J.S. (2011) *J. Nanosci. Nanotechnol.*, **11**, 5829.
21. Takachi, M., Yasuoka, H., Ohdaira, K., Shimoda, T., and Matsumura, H. (2009) *Thin Solid Films*, **517**, 3622.
22. Kim, H.-M., Sohn, S., and Ahn, J.S. (2012) *Surf. Coat. Technol.*, **228**, S389.
23. Lewis, H.G.P., Weibel, G.L., Ober, C.K., and Gleason, K.K. (2001) *Chem. Vap. Deposition*, **7**, 195.
24. Kim, D., Bae, K., Seo, S.-B., Sohn, S., Kim, H.-M., and Hong, J.-S. (2011) Meeting Abstracts, The Electrochemical Society, 2011, p. 1508.
25. Kieu, N.T.T., Ohdaira, K., Shimoda, T., and Matsumura, H. (2010) *J. Vac. Sci. Technol., B*, **28**, 775.
26. Lewis, H.G.P., Caulfield, J.A., and Gleason, K.K. (2001) *Langmuir*, **17**, 7652.
27. Lau, K.K.S. and Gleason, K.K. (2007) *Surf. Coat. Technol.*, **201**, 9189.
28. Lau, K.K.S., Bico, J., Teo, K.B.K., Chhowalla, M., Amaratunga, G.A.J., Milne, W.I., McKinley, G.H., and Gleason, K.K. (2003) *Nano Lett.*, **3**, 1701.
29. Laird, E.D., Bose, R.K., Wang, W., Lau, K.K.S., and Li, C.Y. (2013) *Macromol. Rapid Commun.*, **34**, 251.
30. Rastogi, A.C. and Desu, S.B. (2006) *Chem. Vap. Deposition*, **12**, 742.
31. Ma, M.L., Mao, Y., Gupta, M., Gleason, K.K., and Rutledge, G.C. (2005) *Macromolecules*, **38**, 9742.
32. Kumar, V., Pulpytel, J., and Arefi-Khonsari, F. (2010) *Plasma Processes Polym.*, **7**, 939.
33. Gupta, M. and Gleason, K.K. (2006) *Langmuir*, **22**, 10047.
34. Lau, K.K.S. and Gleason, K.K. (2006) *Macromolecules*, **39**, 3688.
35. Lau, K.K.S. and Gleason, K.K. (2006) *Macromolecules*, **39**, 3695.
36. Ma, M.L., Gupta, M., Li, Z., Zhai, L., Gleason, K.K., Cohen, R.E., Rubner, M.F., and Rutledge, G.C. (2007) *Adv. Mater.*, **19**, 255.
37. Coclite, A.M., Shi, Y.J., and Gleason, K.K. (2012) *Adv. Funct. Mater.*, **22**, 2167.
38. Ozaydin-Ince, G. and Gleason, K.K. (2009) *J. Vac. Sci. Technol., A*, **27**, 1135.
39. Coclite, A.M., Shi, Y.J., and Gleason, K.K. (2012) *Adv. Mater.*, **24**, 4534.
40. Baxamusa, S.H., Montero, L., Borros, S., and Gleason, K.K. (2010) *Macromol. Rapid Commun.*, **31**, 735.
41. Barr, M.C., Rowehl, J.A., Lunt, R.R., Xu, J.J., Wang, A.N., Boyce, C.M., Im, S.G., Bulovic, V., and Gleason, K.K. (2011) *Adv. Mater.*, **23**, 3500.
42. Yagüe, J.L. and Gleason, K.K. (2013) *Macromolecules*, **46**, 6548.
43. Paxson, A.T., Yagüe, J.L., Gleason, K.K., and Varanasi, K.K. (2014) *Adv. Mater.*, **26**, 418.
44. Baxamusa, S.H. and Gleason, K.K. (2009) *Adv. Funct. Mater.*, **19**, 3489.
45. Colas, A. and Curtis, J. (2004) Biomaterials Science: An Introduction to Materials in Medicine, p. 80.
46. Lamendola, R. and d'Agostino, R. (1998) *Pure Appl. Chem.*, **70**, 1203.
47. Tien, P., Smolinsky, G., and Martin, R. (1972) *Appl. Opt.*, **11**, 637.
48. Loboda, M. (2000) *Microelectron. Eng.*, **50**, 15.
49. Murthy, S.K. and Gleason, K.K. (2002) *Macromolecules*, **35**, 1967.
50. Yoo, Y., You, J.B., Choi, W., and Im, S.G. (2013) *Polym. Chem.*, **4**, 1664.

11
Conjugated CVD Polymers: Conductors and Semiconductors

Rachel M. Howden

11.1
Overview

Chemical vapor deposition (CVD), a widely used technique, presents a viable alternative to solution-based methods for fabricating thin organic layers. A specific type of CVD processing called oxidative chemical vapor deposition (oCVD) allows for the deposition of conjugated polymers onto a wide range of substrates. A variety of electrically conducting and semiconducting polymers can be fabricated using this technique, including polymers that are difficult to process using solution-based methods due to their insolubility. The oCVD process, which takes place in a controlled environment within a vacuum deposition chamber, brings together one or more monomers and an oxidizing agent, both in the vapor phase, to achieve in one step simultaneous polymer synthesis and deposition of a thin film.

There are many techniques available for depositing conjugated polymers, but there are many benefits by depositing layers via oCVD, especially when working with textiles or fragile substrates that may be damaged by solvents, and, therefore, the focus of this chapter. Readers may refer to a review by Alf *et al.* [1] for more detail on other CVD techniques. Another feature of oCVD is the ability to conformally coat onto a variety of surfaces, even those with highly complex or otherwise difficult to coat features such as fibers or trenches.

The optical and electronic properties of the films may be tuned by varying deposition process conditions as well as performing post-treatment on the films. The ability to enhance or tune properties such as conductivity, stability, work function, and bandgap, is important for integrating and optimizing these films within various applications.

CVD Polymers: Fabrication of Organic Surfaces and Devices, First Edition. Edited by Karen K. Gleason.

11.2
Reactors and Process

The design of an oCVD reactor, in which chemical synthesis proceeds via step growth polymerization, differs from designs for processes with different synthetic mechanisms, such as initiated chemical vapor deposition (iCVD), which occurs by free radical polymerization. The oCVD mechanism is a spontaneous reaction and does not require filaments to drive the gas-phase chemistry, as with iCVD, or plasma or photoexcitation of the reactants to proceed with polymerization.

The materials react in a controlled environment within a vacuum deposition chamber to achieve in one step simultaneous polymer synthesis and thin-film deposition. This process is shown in Figure 11.1a. The monomer, frequently in liquid form, is heated in a vacuum jar outside the chamber and fed into the system through a heated line at a controlled flow rate. Typical monomer flow rates are in the range 1–10 sccm. An oxidant that is solid at room temperature, such as iron(III) chloride or copper(II) chloride, is typically used. The oxidant is placed in a crucible at the bottom of the chamber oriented upward at the inverted stage. The crucible is heated to sublime the solid oxidant (e.g., $FeCl_3$ is heated to ~340 °C). Oxidants that are liquid at room temperature, such as bromine and vanadium oxytrichloride, have also been used [3, 4]. Liquid oxidants can be introduced into the chamber using a method similar to that used for introducing monomer species, which can allow greater flow rate control. Substrates are placed on the temperature-controlled stage, which is typically inverted and facing down toward the oxidant source, to limit contamination by particulates of the solid oxidant. Modest stage temperatures (25–150 °C) are used, with the value depending on the polymer being deposited. During a deposition, the pressure is maintained at a value (typically ~100 mTorr) using a butterfly valve.

Figure 11.1b,c shows a recently designed oCVD reactor that has improved capabilities for depositing films. Stage rotation helps achieve even thickness over a larger substrate surface area. The reactor is fitted with four heated crucibles that can allow materials to be coevaporated, enabling the use of solid monomers with solid oxidants or the ability to incorporate other materials into a growing film, for example, growing a bulk heterojunction photovoltaic by evaporating a small molecule acceptor during the polymerization of a semiconducting polymer.

11.3
Chemistry and Mechanism

Due to their rigid backbone structure, conjugated polymers are often insoluble, motivating the use of vapor deposition techniques to synthesize these materials in the form of electrically conducting, flexible, thin films. oCVD replicates the mechanism of oxidative polymerization in solution, in which a desired monomer reacts with an oxidizing species, such as $Fe(III)Cl_3$ or $CuCl_2$, resulting in step growth polymerization.

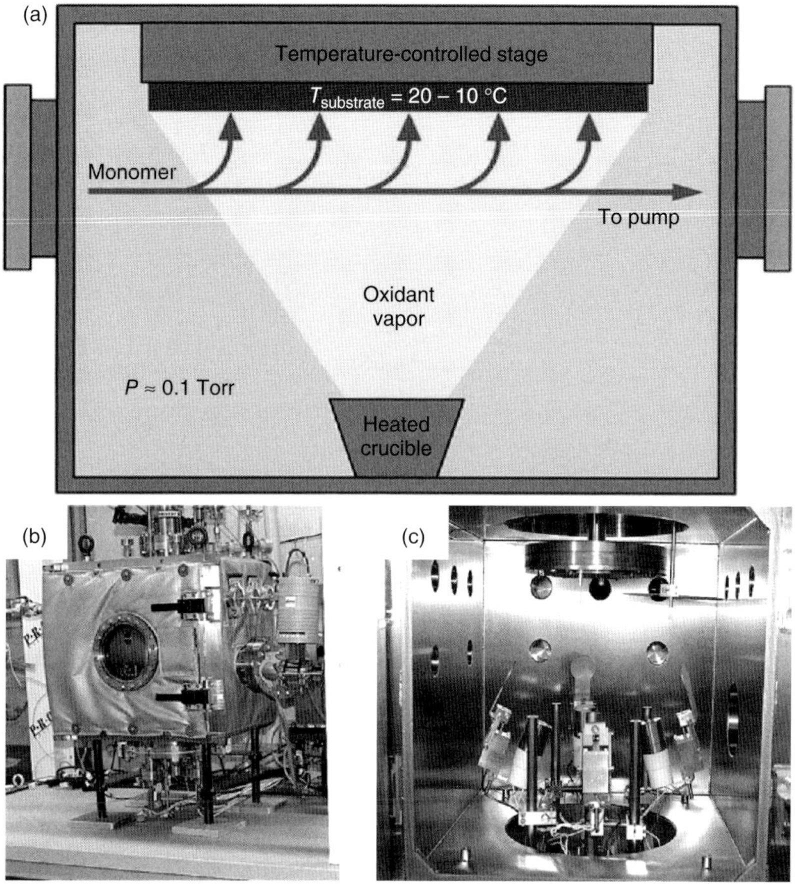

Figure 11.1 (a) Schematic of oCVD process chamber. (b) Picture of an oCVD chamber. (c) Interior of oCVD chamber showing inverted stage and multiple crucible sources for solid materials. (a) Reproduced with permission [2]. Copyright 2013, Wiley.

Figure 11.2 shows the reaction pathway for a typical oCVD synthesis of poly(3,4-ethylenedioxythiophene) (PEDOT) using Fe(III)Cl_3 as the oxidizing agent. The oxidizing agent reacts with monomers to form cation radicals. Pairs of cation radicals then dimerize and deprotonate. Subsequent stepwise reactions with the oxidizing agent and the cation radicals grow the polymer chain. It is important to note that in this case, the $FeCl_3$ is not only oxidizing the PEDOT but chlorine is also getting incorporated into the growing polymer chain as a dopant molecule. The maximum theoretical dopant concentration has long been assumed to be approximately one in every three monomer units [5–9]. However, a dopant to monomer unit ratio of 1 : 1 has recently been measured [10]. Each additional dopant adds a charge carrier that also acts as a scattering center. These two competing effects result in a limited variation in the resulting conductivity at doping levels higher than the 1 : 3 ratio. Note that the addition of each monomer requires reaction with

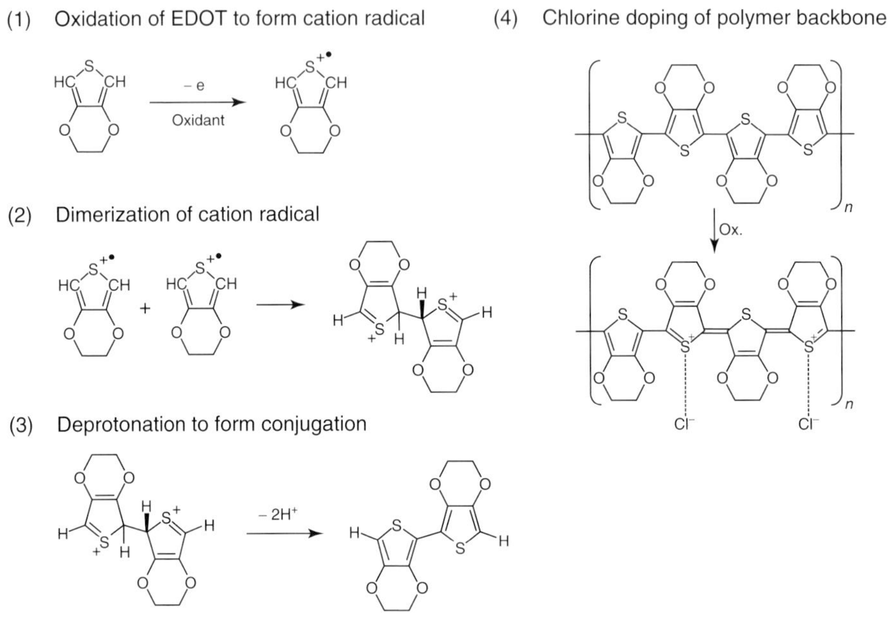

Figure 11.2 Proposed polymerization mechanism for oxidative polymerization of PEDOT with $FeCl_3$ oxidant: (1) oxidation of EDOT to form cation radical; (2) dimerization of cation radicals; (3) deprotonation to form conjugation; and (4) doping of the PEDOT chain.

several oxidants, quite a different situation from free radical polymerization where one initiation radical can induce the polymerization of tens, hundreds, or even thousands of units.

11.3.1 Monomers

In order to be compatible with the oCVD process, liquid oxidants and monomers must have a high enough vapor pressure that they can be delivered to the substrate surface. Similarly, solid oxidants and monomers must be able to be heated to the vapor phase without first degrading. The required monomers, such as thiophene, aniline, pyrrole, and their derivatives, are often reasonably volatile and can be readily introduced into a vacuum deposition chamber. For example, at 25 °C, the monomer 3,4-ethylenedioxythiophene (EDOT, Table 11.1) has a vapor pressure of 0.278 Torr. A dimer of EDOT is formed in the first step in the polymerization process and is estimated to have an extremely low vapor pressure of ~10^{-7} Torr at 25 °C, and even lower volatilities are expected for longer oligomers and polymeric chains [11]. Hence, the polymerization reaction is hypothesized to occur on the surface rather than in the gas phase.

The first demonstrations of oCVD were used for forming conducting films of PEDOT [12, 13]. Subsequently, the oCVD technique has expanded to the deposition of other conjugated polymers, including semiconducting layers and

Table 11.1 Monomers that have been used in oCVD [46].

Monomer	Molecular structure	Molecular weight ($g\,mol^{-1}$)	Vapor pressure (Torr) at 25 °C
Thiophene		84.14	81.7
Pyrrole		67.09	12.2
Selenophene		131.03	27.8
Aniline	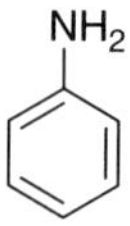	93.13	0.733
3,4 Ethylenedioxythiophene (EDOT)	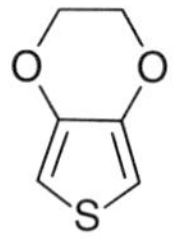	142.18	0.278
1,3-Dihydroisothianapthene (DHITN)		136.22	—
2-Thiophene acetic acid (TAA)	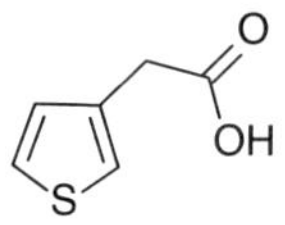	142.18	2.65×10^{-3}
3-Thiopheneethanol (3TE)	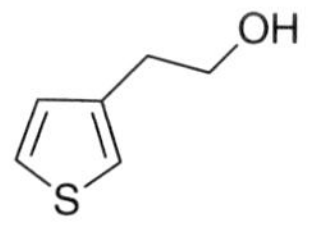	128.19	0.0545

A high vapor pressure facilitates introduction into the vacuum chamber. High vapor pressures also lower the equilibrium surface coverage of adsorbed monomer at the growth surface.
DHITN:1,3-dihydroisothianapthene.
Source: SciFinder (calculated using Advanced Chemistry Development (ACD/Labs) Software V11.02 (©1994-2012 ACD/Labs)).

functionalizable conducting copolymers. Table 11.1 lists monomers that have been used in oCVD polymerization and copolymerization. To date, these oCVD polymers have been applied to fulfill various roles in organic photovoltaics (OPVs), electrochromic devices, and sensors. A comprehensive review comparing oCVD and other methods for the vapor deposition of conjugated polymers, such as vapor-phase polymerization (VPP) and plasma-enhanced chemical vapor deposition (PECVD), and their applications has recently been published [1].

11.3.2 Oxidants and Dopants

Because many oxidants have low volatility, introduction of oxidants through the vapor phase presents a much larger challenge than delivering the monomer. Some previous researchers have avoided this difficultly by first solution casting the oxidant and then drying the substrate. The substrate with the dried oxidant is placed into a vacuum chamber and simply exposed to monomer delivered via the vapor phase. This sequential two-step solution/vacuum method has produced high-quality PEDOT films, with conductivities exceeding 1000 $S\,cm^{-1}$ by using Fe(III) tosylate as an oxidant [14–16].

The term oCVD denotes the process in which both the oxidant and the monomer are introduced through the vapor phase. One means to obtain sufficient oxidant flux to the substrate for oCVD is to mount a heated crucible inside the vacuum chamber and supply an inert gas flow over the heated powder to improve mass transport. This method has proved successful with the oxidant, $Fe(III)Cl_3$, which is a solid of limited volatility. Using a heated crucible with the oxidant Fe(III)tosylate has proved unsuccessful, most likely because decomposition occurred prior to sublimation at the conditions explored [17]. In addition to volatility and thermal stability, the ability of the chosen oxidant to drive the oCVD reaction is also critical. These three criteria are essential to consider when selecting an appropriate oxidant.

In the first report of oCVD PEDOT deposition, the crucible was located above the wafer. In this configuration, microscopy revealed that some oxidant particulates were transferred to the surface of the film [17]. Inverting the configuration, as shown in Figure 11.1a, resulted in the more standard arrangement for evaporation of a solid material onto an inverted substrate and produced films with few observable particulates [13]. PEDOT films with conductivities over 1600 $S\,cm^{-1}$ have been synthesized using $FeCl_3$ as the oxidant in this configuration [17]. Other solid oxidants that may be sublimated in the same way as $FeCl_3$, such as $CuCl_2$, are compatible with the standard oCVD process. More recently, liquid oxidants have been used in oCVD to synthesize conjugated polymer films [3, 18]. Liquid oxidants can be heated and controllably introduced into the vacuum chamber in a similar process as liquid monomers are introduced. Table 11.2 shows a list of oxidants that have been used in oCVD.

11.4 Grafting and Patterning

The vapor-phase delivery of oxidants in oCVD builds the film from the surface up and permits covalent bonds to form at the interface between the substrate and the film of the oCVD conjugated polymer. When the bulk substrate itself contains aromatic groups, no linker molecule or surface treatment is required for grafting the oCVD polymer films (Figure 11.3c) [19]. Many flexible plastic

Table 11.2 List of oxidants used in oCVD and the corresponding monomers that have been used with.

Oxidant	Phase of oxidant at 25 °C	Monomer
Iron(III) chloride	Solid	EDOT [12, 13, 19, 20, 21, 27, 40, 43, 44, 47, 48, 49, 50] Pyrrole [25] Thiophene [18] 3-Thiophene acetic acid (TAA) [25, 44] 3-Thiopheneethanol (3TE) [43] Selenophene [37] PITN [36]
Iron(III) *p*-toluenesulfonate or iron(III) tosylate	Solid	EDOT [13]
Copper(II) chloride	Solid	EDOT [20, 49]
Bromine	Liquid	EDOT [3, 45] 3-Thiophene acetic acid (TAA) [45]
Vanadium oxytrichloride	Liquid	Thiophene [4]

substrates contain these groups, such as polystyrene, polyethylene terephthalate, polycarbonate, polyethylene naphthalate, polyurethane, and poly(acrylonitrile-butadiene-styrene). During the oCVD process, through a Friedel–Craft reaction at the surface, a radical cation site is created on an aromatic ring that subsequently participates in reactions that form brushes of conducting polymer chains (Figure 11.3a). The resulting adhesion from this single-step grafting process is strong, as confirmed by the onset of periodic wrinkling in bilayer structures that comprise oCVD PEDOT layers grafted to swellable substrates containing aromatic groups [19].

Grafting can also be applied to inorganic substrates using silane-coupling agents. For instance, a silicon substrate with a native silicon oxide layer can be treated with oxygen plasma to form hydroxyl groups. The sample is then exposed to trichlorovinylsilane (TCVS) to form a vinyl-terminated surface to which PEDOT can graft. Other silane agents may be used as well. Figure 11.3e clearly shows that pretreatment with a phenyltrichlorosilane linker is a robust methodology for patterning the conducting polymer layers on silicon wafer substrates at micron-scale resolution [19]. Grafting improves interfacial adhesion, enabling pattern formation. Without grafting, failure due to delamination is often observed. X-ray diffraction (XRD) reveals the crystallinity and preferential ordering of oCVD PEDOT with respect to a (100)-oriented silicon substrate. For non-grafted samples only (010) type reflections are observed, whereas for grafted samples (100) reflections also appear (Figure 11.3d) [2].

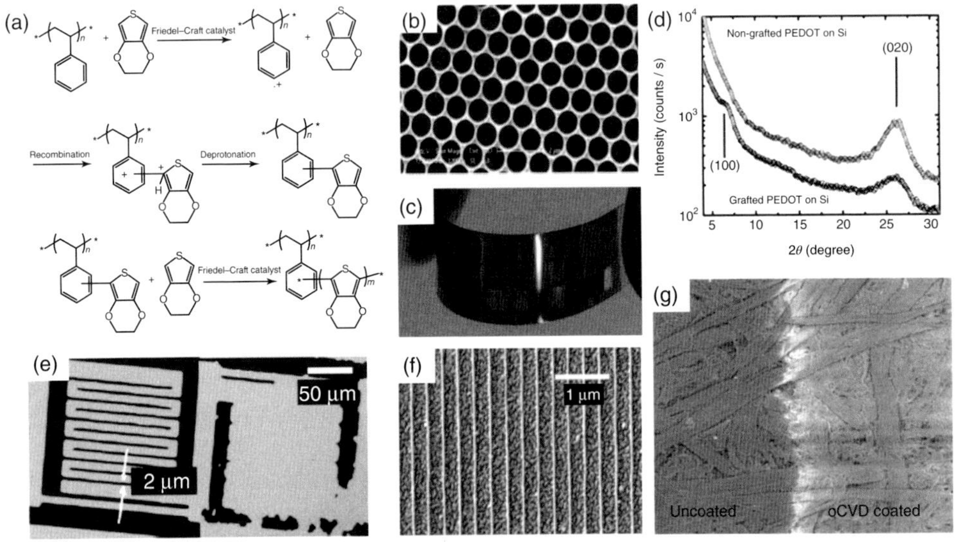

Figure 11.3 (a) Schematic of reaction mechanism involved in grafting PEDOT films onto aromatic polymeric substrates. (b) Grafted oCVD PEDOT patterns obtained by colloidal patterning. (c) Image of patterned PEDOT film on flexible PET substrate. (d) Grafting produces a (100) X-ray diffraction peak in oCVD PEDOT grown on Si(100), in addition to the (010) peak seen without grafting. (e) Conventional lift-off photolithography pattern on Si wafers with grafted PEDOT (left) and with ungrafted PEDOT (right). (f) 60-nm line pattern of PEDOT film grafted on flexible PET substrate by e-beam lithography. (g) Scanning electron microscope image of a 100-nm-thick PEDOT electrode on newsprint. The left-hand side is bare newsprint and the right-hand side is coated with the CVD PEDOT layer. (a,c,e,f) Reproduced with permission [19]. Copyright 2007, Wiley. (b) Reproduced with permission [20]. Copyright 2010, Royal Society of Chemistry. (d) Reproduced with permission [2]. Copyright 2013, Wiley. (g) Reproduced with permission [21]. Copyright 2011, Wiley.

The ability to control the interface of the polymer and the substrate offers the potential to open up new device technologies where the electrical and thermal properties could be tuned independently. Excellent interfacial properties are also essential for the fabrication of flexible devices, where adhesion and durability are crucial. Although many proposed flexible transparent electrodes require a complex transfer step from another substrate, the low deposition temperature of oCVD allows for direct deposition onto plastic substrates, a characteristic that is highly favorable for manufacturing. Novel nanostructures with organic–inorganic material interfaces could be designed in such a way where charge transport is enhanced and phonon transport is obstructed.

Patterns of oCVD PEDOT have been directly formed by vapor printing on paper substrates, including the patterns already printed with tradition inks. In vapor printing, the synthesis of the polymer chains from vapor-phase reactants, thin-film formation, and patterning occur in a single step [21]. The scanning electron micrograph in Figure 11.3g shows the edge of 100-nm-thick vapor-printed PEDOT electrode on newsprint obtained with shadow-masking. The image also shows how well PEDOT conforms to the geometry of the paper fibers. Well-ordered conducting polymer nanostructures (Figure 11.3b) were achieved by colloidal lithography combined with the highly conformal grafted oCVD PEDOT [20]. At the thinnest junction, the PEDOT wall between the nanobowls is only ~25 nm wide. Other traditional patterning techniques, such as e-beam lithography, have also been demonstrated as compatible with oCVD PEDOT with well-defined lines of 60 nm obtained (Figure 11.3f).

11.5 Conformality

Vapor-phase processing enables conformal coating of polymer layers on substrates with a wide range of surface characteristics. Vapor-phase reactants are able to diffuse into complex microstructured and nanostructured surfaces. Such 3D surface architectures are desired to enhance light absorption in organic photovoltaic devices, where the exciton diffusion length limits active layer thicknesses. Howden *et al.* [22] showed that oCVD PEDOT is compatible with wide range of nanoscale to macroscale textured surfaces, enhancing the light absorption (light trapping) in photovoltaic structures. Figure 11.4c shows the ability of the oCVD PEDOT to precisely follow the geometry of the substrate interface. The conformal nature oCVD PEDOT is also useful when coating larger feature sizes. Figure 11.4b shows oCVD PEDOT conformally coating a microtrench in a silicon wafer with even thickness on the top, bottom, and side walls.

The oCVD method is able to penetrate and coat the inner surfaces of porous media, such as foams and membranes. Subsequent dissolution of the membrane leaves behind the templated oCVD features. In the case of membranes with cylindrical pores, nanotubes of conjugated polymers are formed. When applied to fiber-base substrates, such as paper, electrospun nanofiber mats, or textiles, the oCVD conjugated polymers completely surround each fiber [26]. Poly(acrylonitrile) fiber mats conformally coated with oCVD PEDOT (Figure 11.4d) creates a high surface area platform for resistance-based sensing [27]. Conformal coating of carbon nanotubes (CNTs) by oCVD PEDOT of ~6 nm thick was demonstrated by Vaddiraju *et al.* [25]. By lowering the activation energy required for creation of mobile carriers, an enhancement in radial conductivity was obtained. Figure 11.4a shows another example of a CNT conformally coated with PEDOT. More recently, oCVD PEDOT has been conformally coated onto CNTs for energy storage applications, with the coated nanotubes investigated and applied as a supercapacitor electrode [23, 28, 29]. This approach can be used to create many functional materials and devices.

The ability to form conformal layers of conductive films on substrates is useful in a variety of applications. The oCVD process is compatible with substrates having highly complex geometries and nonplanar surface features and enables the

Figure 11.4 (a) TEM images of A-CNT conformally coated with PEDOT. (b) SEM cross section showing conformality of oCVD PEDOT on micron-sized trenches in silicon. (c) Cross-sectional SEM of conformal coating of oCVD PEDOT on imprinted PMMA features. (d) SEM micrograph showing the morphology of PEDOT-coated poly(acrylonitrile) fiber mat. (a) Reproduced with permission [23]. Copyright 2013, Elsevier. (b) Reproduced with permission [24]. Copyright 2008, Wiley. (c) Reproduced with permission [17]. Copyright 2013, Elsevier. (d) Reproduced with permission [25]. Copyright 2008, Wiley.

deposited polymer to form a high surface area interface with substrate surfaces or other underlying material layers.

11.6 Dopants, Rinsing, Stability

The choice of oxidant is another important factor for the film characteristics. The fact that thermally evaporated oxidants such as $FeCl_3$ leave residual materials behind, necessitating rinsing steps and perhaps hindering tightly packed film growth, may make moving to a more volatile liquid oxidant as an attractive option. Chelawat *et al.* [3] showed that films grown using a liquid oxidant, Br_2, had lower roughness than films grown using $FeCl_3$ as the oxidant. The size of the dopant molecules should also be considered as they may impact chain packing and stability.

The post-deposition rinsing of oCVD films has been used as a method to improve film properties, such as conductivity and stability. Films are often rinsed in methanol as a process of removing residual oxidant or unreacted monomer. Rinsing also results in a much smoother film surface. More recently, acid rinsing has been explored as a process to not only remove residual oxidant more effectively but also improve the final film conductivity [17, 10]. The reason for the observed conductivity enhancement was thought to be a combination of increased doping and improved chain packing.

Rinsing PEDOT films, deposited using Fe(III)Cl_3 as the oxidant, in even weak acid solutions (<0.1 M HCl, HBr, and H_2SO_4) was successful at removing residual oxidant from the deposited films. A dopant exchange occurred for the films rinsed in HBr and H_2SO_4, as evidenced by X-ray photoelectron spectroscopy (XPS) spectra. In addition, conductivity of the films increased by 117 and 135% for H_2SO_4 and HBr rinsed PEDOT films, respectively [17]. Rinsing in acid resulted in films with lower roughness than unrinsed or MeOH-rinsed films, which is especially important for conductive polymer electrode applications where excessive roughness can cause defects in devices.

Another important consideration for polymer electrode materials is the film stability. PEDOT films post-treated by acid rinse showed better conductivity stability over time (degradation accelerated by keeping the films at elevated temperatures). A variety of factors may account for the observed stability improvement. Researchers have shown that two primary mechanisms of PEDOT degradation are exposure to oxygen and exposure to water vapor [30, 31]. The acid-rinsed films may have tighter chain packing, which would provide a better barrier to the atmosphere [32]. The removal of the excess oxidant, which is hygroscopic, could also reduce water content within the films. The size and reactivity of the dopant molecules probably play a role in the film degradation and conductivity loss over time as well. Chelawat *et al.* [3] suggest that the size of the dopant molecules can have an impact on conductivity loss over time.

Control over film optoelectronic properties such as conductivity and transparency and optimization of film stability by varying doping counterions and performing different chemical post-treatments is beneficial for various applications and enables tunability of the interfaces between oCVD conjugated polymers and other materials.

11.7 Semiconductors

Semiconducting polymers are of great interest as active layer materials in organic electronic and optoelectronic devices, such as OPVs [33], organic light-emitting diodes [34], and organic field-effect transistors [35]. Researchers have recently used oCVD for the deposition of several semiconducting polymers. Unsubstituted polythiophene (PT), which is insoluble and typically difficult to process, was easily prepared by oCVD. Using iron(III) chloride as the oxidizing agent, the as-deposited oCVD PT films were blue in appearance [18]. Excess oxidizing agent present during the oCVD process doped the films, resulting in conductivities up to 20 S cm^{-1}. Positions of the mid-gap absorption peaks (0.8 and 1.6 eV) in the doped films demonstrate that the polymer films are heavily doped and bipolarons are present (Figure 11.5a). Reduction with methanol dedoped the polymer and resulted in neutral, red PT films. Changes in the absorption spectra confirmed the semiconducting nature of the rinsed PT films. oCVD PT was then successfully integrated as the electron-donor layer in bilayer heterojunction OPVs, reaching efficiencies up to 0.8% [18]. Further exploration of vapor-deposited PT films, for use in organic thin-film transistors (TFTs), demonstrated that changing the chamber pressure during deposition resulted in changes to the PT film structure [38]. Films deposited at the lowest pressure (1 mTorr) are blue-shifted in comparison to films deposited at higher pressures and have shorter conjugation of the polymer chains.

Properties of oCVD semiconducting polymers are easily tuned by controlling the deposition parameters. A separate study of oCVD PT demonstrated the use of vanadium oxytrichloride ($VOCl_3$) as the oxidizing agent [4]. Since $VOCl_3$ is a volatile liquid, controlling its flow rate and the stage temperature allowed for the control of the surface concentration of $VOCl_3$ on the stage in the reactor chamber. Thus, by controlling these parameters as well as the monomer flow rate, the resulting PT film properties, such as conductivity and absorption maximum, were tuned. High-resolution XPS scans (Figure 11.5b,c) of the C 1s and S 2p peaks of the PT films revealed the expected 1 : 1 ratio of α and β carbons as well as a single sulfur species in the thiophene ring, suggesting that the thiophene ring structure is preserved resulting in stoichiometric PT even after rinsing. A cyclic voltammetry sweep (Figure 11.5f) of the acid-washed PT film showed anodic and cathodic peaks at 1.1 and 0.6 V, respectively, which is in good agreement with the reported data for PT. The film color changes from orange in its dedoped state to blue representing the redoped state during the sweep. Figure 11.5g shows the relationship between the peak absorption wavelength versus the number of thiophene monomer units, where the peak shifts to higher wavelengths with increasing chain length. On the basis of this trend, the absorption peak for the rinsed PT film indicates a long conjugation length, with many of the shorter oligomer chains being washed away during rinsing.

Work on oCVD polyisothianaphthene (PITN) has shown the importance of the stage temperature during deposition [36]. PITN is of interest for its various appealing properties, including its low bandgap of 1 eV in the neutral state. In the doped state, PITN displays high optical transmission in the visible part of the spectrum. As shown in Figure 11.5d, the color and transparency of the as-deposited oCVD PITN films greatly depends on the deposition temperature. Higher deposition temperatures were shown to result in greater conjugation and longer polymer

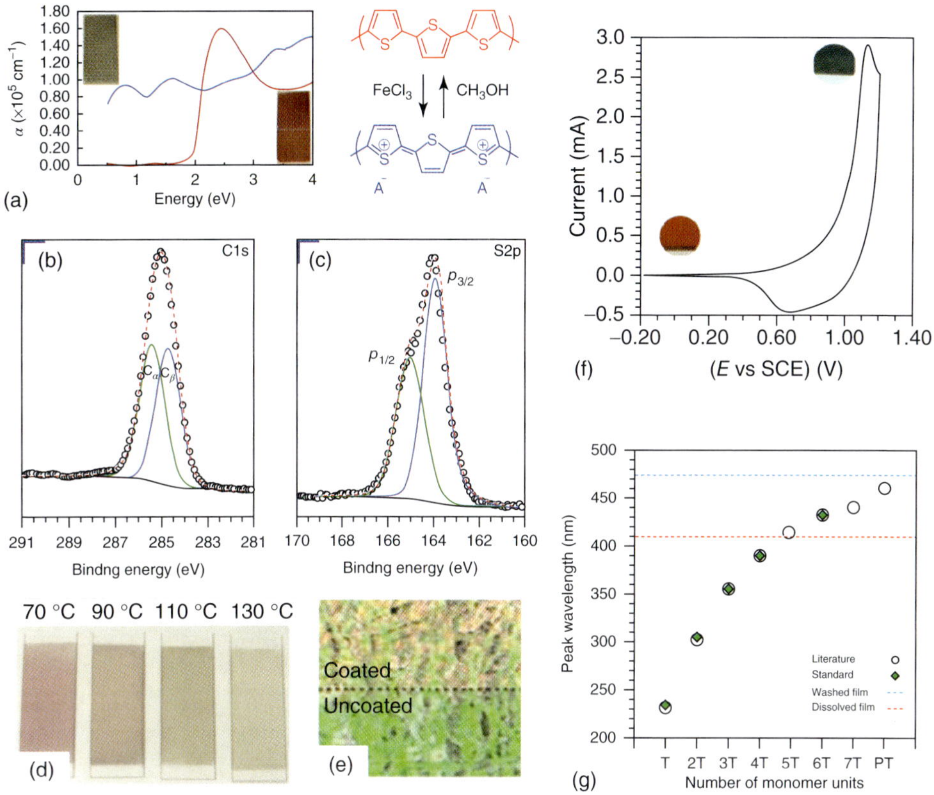

Figure 11.5 (a) Absorption spectra and photographs of oCVD unsubstituted polythiophene (PT) in doped (blue) and undoped states (red). (b,c) High-resolution (b) C 1s and (c) S 2p XPS spectra of rinsed PT. (d) Polyisothianaphthene (PITN) films prepared by oCVD at various temperatures. (e) Scanning electron microscope-energy dispersive spectroscopy (SEM-EDS) image of polyselenophene deposited on paper, showing the presence of selenium (indicated by red) only on the coated portion of the paper while only carbon (indicated by green) was observed on the uncoated masked portion. (f) Cyclic voltammogram of washed PT in a supporting electrolyte solution of acetonitrile and tetraethylammonium hexafluorophosphate. The corresponding film colors in the oxidized and reduced forms are also shown. (g) Correlation of peak wavelength with the number of thiophene chain units, comparing literature data, and experimental data of undoped standard oligothiophenes (T to 7 T) and polythiophene with washed PT and its dissolves soluble fraction after washing. (a) Reproduced with permission [2]. Copyright 2013, Wiley. (b,c,f,g) Reproduced with permission [4]. Copyright 2011, American Chemical Society. (d) Reproduced with permission [36]. Copyright 2013, American Chemical Society. (e) Reproduced with permission [37]. Copyright 2012, Royal Society of Chemistry.

chains. The increased conjugation length was manifested in various optical and electronic properties of the oCVD PITN films.

Polyselenophene, an analog of PT with a slightly lower bandgap, has also been prepared using oCVD [39]. Figure 11.5e exhibits the ability to conformally deposit the polymer on rough paper substrates.

The versatility of oCVD makes it a promising technique for processing semiconducting polymers with desirable properties for many different organic electronic applications.

11.8 Electrical Properties

PEDOT is one of the most promising conjugated polymers for flexible organic electronics because of its combination of high electrical conductivity, optical transparency, and mechanical properties [40]. Since oCVD has low substrate temperature requirements, there is no need for elaborate processing strategies, as with microcrystalline silicon or graphene, where the materials are first grown at high temperatures on a handling substrate and then transferred to the surface of less heat resistant materials. In addition, the conformal coverage oCVD PEDOT layers allows them to be grown directly on top of flexible substrates with rough surfaces, such as ordinary plastic, and even fiber-based papers and textiles. Thus, oCVD PEDOT represents a new fabrication process that is compatible with substrates other than glass, silicon, and smooth plastics. The process only requires low-energy input, utilizes earth-abundant elements (e.g., no indium), and is scalable to large area processing (e.g., roll-to-roll).

The substrate temperature of the oCVD process is a critical parameter for controlling the electrical conductivity of PEDOT thin films. Im *et al.* [40] reported that increasing substrate temperature from 15 to 110 °C while holding all other process conditions constant resulted in the electrical conductivity of oCVD PEDOT polymer films monotonically increasing from 9.1×10^{-1} to 348 S cm^{-1}, a range of more than five orders of magnitude (Figure 11.6a). The improvement in conductivity is attributed to longer conjugation length and higher doping levels [12, 40]. Since higher doping concentration moves the Fermi energy level in the band gap, the work function of PEDOT can also be controlled [40] from ~5.1 to ~5.4 eV during oCVD by simply adjusting substrate temperature as shown in Figure 11.6b. The tunability of the work function of oCVD PEDOT films is advantageous for achieving favorable band alignment between dissimilar layers in organic devices, thus lowering the energy barrier to charge transport at the contact interface. More recently, Howden *et al.* [17] used diluted acids as post-process rinsing solutions, resulting in improved electrical conductivity. Increase in average conductivity by 117 and 135% was observed for H_2SO_4 and HBr rinses, respectively, accompanied by dopant exchange. A maximum conductivity of 1620 S cm^{-1} was obtained by rinsing in HBr.

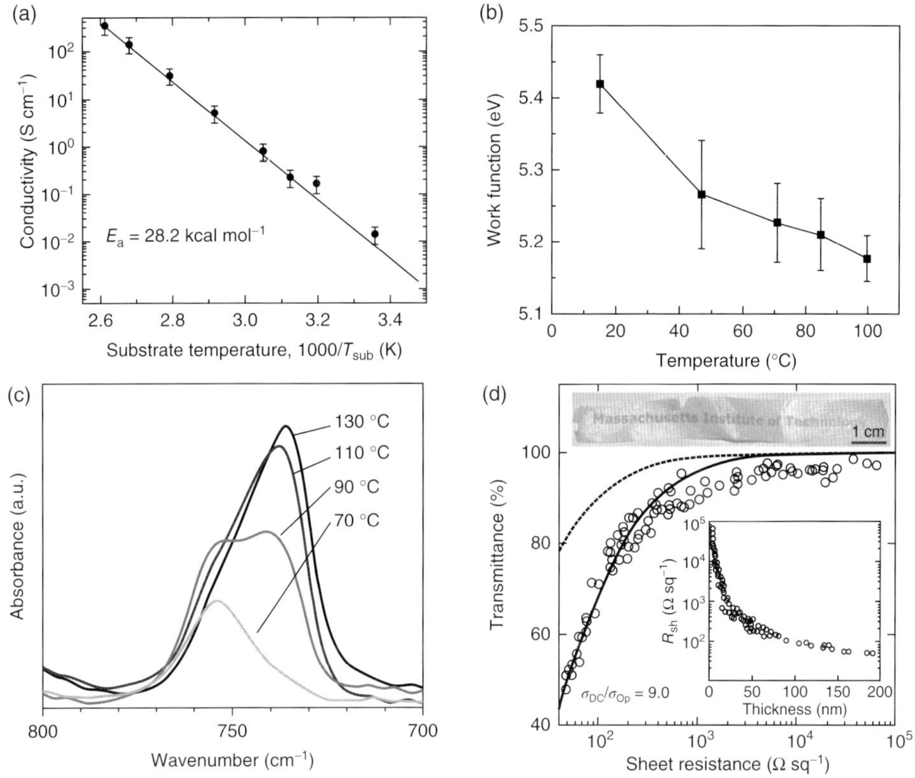

Figure 11.6 (a) Arrhenius plot of oCVD substrate temperature versus PEDOT conductivity. (b) Effect of the oCVD PEDOT deposition temperature on work function. (c) Absorption peaks corresponding to the C–H vibration of the four adjacent hydrogens in PITN films deposited at varying stage temperatures. (d) Relationship between percent transmittance (at 550 nm) and sheet resistance in ohms per square ($\Omega\,sq^{-1}$) for vapor-printed oCVD PEDOT. The solid gray line is a fit to Eq. (11.1) giving $\sigma_{dc}/\sigma_{op} = 9$ (the dashed black line is for reference and corresponds to $\sigma_{dc}/\sigma_{op} = 35$, representative for traditional metal oxide electrodes). Lower inset: sheet resistance plotted versus film thickness. Upper inset: 200-nm thick PEDOT film vapor printed on tissue paper. (a) Reproduced with permission [12]. Copyright 2007, Wiley. (b) Reproduced with permission [40]. Copyright 2007, American Institute of Physics. (c) Reproduced with permission [36]. Copyright 2013, Royal Society of Chemistry. (d) Reproduced with permission [21]. (Barr 2011) Copyright 2011, Wiley.

Temperature also plays an important role in the deposition of semiconducting polymer films using oCVD. Figure 11.6c shows absorption peaks for PITN films corresponding to the C–H vibration of the four adjacent hydrogens. As the deposition temperature of the oCVD PITN films increases, a peak emerges at lower wave numbers, suggesting a change in the bonding environment, possibly due to a transition to a more quinoid form. Tunability of the absorption characteristics (both the intensity and the band gap) is important for designing semiconducting materials for a variety of applications.

For implementing transparent conducting materials as an electrode or a buffer layer in optoelectronic devices, high optical transparency ($T > \sim 85\%$) in the visible regime is required while maintaining a low sheet resistance ($R_{sh} < \sim 100\,\Omega\,sq^{-1}$). Since there is a trade-off between transparency and electrical conductivity in this class of materials, the thickness has to be controlled to limit losses from optical absorption and, in parallel, to provide sufficiently low sheet resistance for the efficient charge transport or extraction. The sheet resistance and optical transparency can be related by Eq. (11.1):

$$T = \left(1 + \frac{Z_0}{2R_{sh}} \frac{\sigma_{op}}{\sigma_{dc}}\right)^{-2} \tag{11.1}$$

where $Z_0 = 377\,\Omega$ is the impedance of free space, and σ_{op} and σ_{dc} are the optical and dc conductivities, respectively [41]. Figure 11.6d shows this relation for oCVD PEDOT films deposited on glass substrates at temperatures of 80 °C and a working chamber pressure of 0.1 Torr [21]. The measured data give the σ_{dc}/σ_{op} ratio of 9, which is comparable to the commercially available solution-processed conducting polymer and is approximately threefold lower than the conventional metal oxide (i.e., ITO) of $\sigma_{dc}/\sigma_{op} \approx 35$ [21]. With HBr acid rinsing, the σ_{dc}/σ_{op} ratio of oCVD PEDOT increased to 12 [17]. Further work to increase σ_{dc}/σ_{op} is currently in progress.

11.9 Functional oCVD Copolymers

The oCVD process also enables introduction of functional groups in conducting polymers films using commercially available monomers, for instance through the copolymerization of pyrrole with thiophene-3-acetic acid [25]. Various conductive oCVD copolymers have been successfully applied in chemiresistive sensors for detecting chemical and biological species. Chemiresistive sensors are an economical sensing technology that operates by measuring changes in the resistance (or conductance) of the sensing element upon exposure [42]. Many conductive polymers are not responsive or selective enough alone to sense low chemical concentrations. However, by covalently bonding a molecule or nanoparticle that is analyte specific to the conductive polymer backbone, larger changes in the composite resistance will occur upon exposure. To accomplish this, oCVD copolymers have been developed that combine the benefits from both monomer species. For example, copolymers of pyrrole and 3-thiophene acetic acid (TAA) combine the high conductivity and good mechanical properties of pyrrole with the desirable –COOH functional group of TAA [25]. It was then possible to covalently bond silver nanoparticles to the oCVD poly(pyrrole-*co*-TAA) polymer (Figure 11.7a). This was done using a linker molecule containing a $-NH_2$ group (4-aminothiophenol) that formed an amide bond with the –COOH group of TAA. This approach was then extended to deposit oCVD copolymers of poly(EDOT-*co*-TAA) [44]. The attachment of nickel and palladium nanoparticles allowed for selective detection of toluene and acetone, respectively. Adsorption of vapors of volatile organic compounds onto the metal surfaces caused changes in the charge distribution that resulted in higher resistance (lower conductivity) of the composite films. The resistive response of the hybrid films was found to be proportional to the concentration of analyte present, and thinner films were more responsive due to lower shunt resistances. Detection limits for the composites were estimated as being 10–20 ppm of toluene for Ni/poly(EDOT-*co*-TAA) and 40–50 ppm of acetone for Pd/poly(EDOT-*co*-TAA). A significant advantage of this approach of attaching metal nanoparticles to a conductive polymer is its versatility. Instead of requiring to develop a completely new chemistry, changes in the chemical being detected can be accomplished by simply changing the metal nanoparticle used.

oCVD copolymers with –COOH and –OH groups have been used for sensing biological agents as well. Copolymers of EDOT and TAA were prepared on glass substrates with bromine as the oxidizing agent [45]. A biomolecule, bovine serum albumin (BSA), was shown to covalently attach to the poly(EDOT-*co*-TAA) films through carbodiimide chemistry with the –COOH functional group of TAA. This resulted in the sheet resistance of the copolymer more than doubling, thus providing an effective way to detect the presence of BSA. In another study, EDOT was copolymerized with 3-thiopheneethanol (3TE) using $FeCl_3$ as the oxidizing agent on 2 cm × 2 cm electrospun nylon fiber mats [43]. Avidin is a protein with high affinity and specificity toward biotin. By reacting a crosslinker molecule (*p*-maleimidophenylisocyanate) with the $-NH_2$ group on avidin and the –OH group of 3TE, the avidin biomolecule was covalently attached to the poly(EDOT-*co*-3TE). Upon exposure of the avidin/copolymer films to solutions of biotin, the biotin quickly attached to the avidin and caused an increase in the resistance of the composite film. Figure 11.7d,e shows polymer-coated nanofibers that are fluorescing due to the attachment of fluorescently labeled avidin (green) and biotinylated red quantum dots that reacted with the avidin. Exposure of the composite films to 5 nM to 5 μM solutions of biotin showed faster response times and higher steady-state responses with higher biotin concentrations. The use of high surface area electrospun fiber mats allowed for significantly improved signal and response time compared to the use of flat substrates. The response of the biosensors on the electrospun fiber mats demonstrated a sixfold increase in response and a 30% decrease in response time relative to the respective values for sensors on planar substrates. Solution-based methods cannot fully coat such high surface area structures due to wetting effects, whereas oCVD provides conformal coverage that enables the use of these substrates. This approach demonstrates the potential to obtain inexpensive, field-deployable, lightweight, and flexible biosensors that can provide fast detection of various analytes.

Bhattacharyya *et al.* [26] copolymerized EDOT with 3TE and used the functional –OH groups to attach silica nanoparticles to fabricate high surface area hybrid nanomaterials. Figure 11.7f shows the process for the covalent attachment of the nanoparticles to the copolymer nanostructures. Figure 11.7b,c shows a scanning electron microscope-energy dispersive spectroscopy (SEM-EDS) of the conducting polymer bundles after the silica nanoparticles were covalently

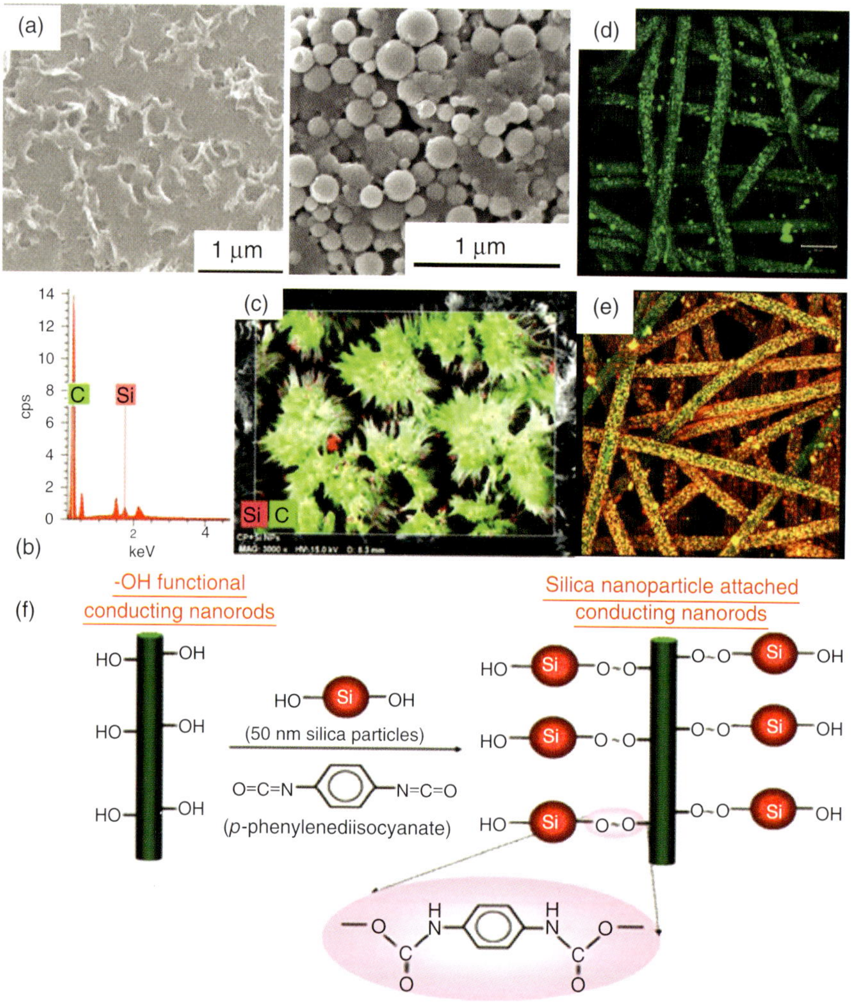

Figure 11.7 (a) Scanning electron micrographs of poly(PPY-*co*-TAA) film (left) as-obtained and (right) after silver nanoparticle attachment followed by 5-min ultrasonication treatment. (b) Scanning electron microscope-energy dispersive spectroscopy (SEM-EDS) and (c) image of conducting copolymer poly(EDOT-*co*-3TE) nanobundles after 50-nm silica particles were covalently attached. Red represents silicon while green represents carbon. (d) and (e) Laser scanning confocal microscope images of the electrospun fiber mats coated with copolymers of EDOT and 3-thiopheneethanol (3TE) after covalent attachment of FITC-avidin (green fluorescence) to the fibers and reaction of biotinylated red quantum dots to the FITC-avidin on the fibers, respectively (scale bar: 50 µm). (f) Covalent attachment of silica nanoparticles to conducting copolymer poly(EDOT-*co*-3TE) nanostructures. (a) Reproduced with permission [25]. Copyright 2008, Wiley. (b,c,f) Reproduced with permission [26]. Copyright 2012, Royal Society of Chemistry. (d,e) Reproduced with permission [43]. Copyright 2011, Wiley

attached. Fabricating these conductive polymer nanostructures by oCVD creates a high surface area for post-deposition functionalization and may be useful in binding inorganic, organic, or biomolecules for various applications.

11.10 Concluding Remarks

oCVD of conducting polymers is a technique that offers many advantages, enabling synthesis and deposition of conjugated polymers that are often insoluble and difficult to process by other methods. A variety of monomers have been used to create both homopolymers and copolymers for various applications. The oCVD process allows for conformal, non-destructive deposition of conjugated polymers onto a variety of substrate surfaces, even those that are fragile and may be damaged by solution processing or high temperatures. The reaction occurs at the substrate surface with both the monomer and oxidizing agent in the vapor phase, which enables the coating of substrates with complex geometries (such as nanorods or fibers) or those with intricate surface features (trenches or other micropatterns and nanopatterns). Simple shadow-masking or other common lithography patterning techniques can be applied to create patterns of conjugated polymers for various applications, and grafting of the films to the substrate surface, for both inorganic and organic materials, creates an interface with strong adhesion.

The properties of the conjugated films deposited via CVD may be tuned and optimized a number of ways. Modifications can be made to the process parameters within the vacuum deposition system during the reaction. Properties may be tuned by selecting different polymers or varying concentrations during copolymerization. The films themselves may also be chemically or physically treated after the deposition, as was seen with applying acid rinsing or annealing steps. The highly tunable nature of these materials is optimal for their application in various devices, as characteristics such as band gap, conductivity, work function, and optical transparency can be modified for individual application requirements.

The oCVD method enables the synthesis of not only highly conductive conjugated polymers but also undoped conjugated polymers than can act as semiconductors or charge transport layers in device applications. By combining different monomers together, the oCVD technique has been used to create copolymers with functional groups that are useful in different applications, such as sensors. More detail on the application of these conjugated CVD polymers in different devices and in combination with other novel materials is discussed later in Chapter 16.

References

1. Alf, M.E., Asatekin, A., Barr, M.C., Baxamusa, S.H., Chelawat, H., Ozaydin-Ince, G., Petruczok, C.D., Sreenivasan, R., Tenhaeff, W.E., Trujillo, N.J., Vaddiraju, S., Xu, J., and Gleason, K.K. (2010) *Adv. Mater.*, **22** (18), 1993–2027.
2. Coclite, A.M., Howden, R.M., Borrelli, D.C., Petruczok, C.D., Yang, R., Yagüe, J.L., Ugur, A., Chen, N., Lee, S., Jo, W.J.,

Liu, A., Wang, X., and Gleason, K.K. (2013) *Adv. Mater.*, **25** (38), 5392–5423.
3. Chelawat, H., Vaddiraju, S., and Gleason, K. (2010) *Chem. Mater.*, **22** (9), 2864–2868.
4. Nejati, S. and Lau, K.K.S. (2011) *Langmuir*, **27** (24), 15223–15229.
5. Elschner, A., Kirchmeyer, S., Lovenich, W., Merker, U., and Reuter, K. (2011) *PEDOT: Principles and Applications of an Intrinsically Conductive Polymer*, CRC Press, Boca Raton, FL.
6. Randriamahazaka, H., Noel, V., and Chevrot, C. (1999) *J. Electroanal. Chem.*, **472**, 103–111.
7. Hwang, J., Tanner, D.B., Schwendeman, I., and Reynolds, J.R. (2003) *Phys. Rev. B*, **67**, 115205.
8. Zotti, G., Zecchin, S., Schiavon, G., Louwet, F., Groenendaal, L., Crispin, X., Osikowicz, W., Salaneck, W., and Fahlman, M. (2003) *Macromolecules*, **36** (9), 3337–3344.
9. Aasmundtveit, K.E., Samuelsen, E.J., Pettersson, L.A.A., Inganäs, O., Johansson, T., and Feidenhans'l, R. (1999) *Synth. Met.*, **101** (1-3), 561–564.
10. Lee, S., Paine, D.C., and Gleason, K.K. (2014) *Adv. Funct. Mater.*, **24**, 7187–7196. doi: 10.1002/adfm.201401282.
11. Odian, G. (2004) *Principles of Polymerization*, 4th edn, Wiley-Interscience, Hoboken, NJ.
12. Im, S.G. and Gleason, K.K. (2007) *Macromolecules*, **40**, 6552–6556.
13. Lock, J.P., Im, S.G., and Gleason, K.K. (2006) *Macromolecules*, **39** (16), 5326–5329.
14. O'Shaughnessy, W.S., Murthy, S.K., Edell, D.J., and Gleason, K.K. (2007) *Biomacromolecules*, **8**, 2564–2570.
15. Yang, R., Xu, J., Ozaydin-Ince, G., Wong, S.Y., and Gleason, K.K. (2011) *Chem. Mater.*, **2011**, 1263–1272.
16. Coclite, A.M., Ozaydin-Ince, G., Palumbo, F., Milella, A., and Gleason, K.K. (2010) *Plasma Processes Polym.*, **7**, 561–570.
17. Howden, R.M., McVay, E.D., and Gleason, K.K. (2013) *J. Mater. Chem. A*, **1**, 1334–1340.
18. Borrelli, D.C., Barr, M.C., Bulovic, V., and Gleason, K.K. (2012) *Sol. Energy Mater. Sol. Cells*, **99**, 190–196.
19. Im, S.G., Yoo, P.J., Hammond, P.T., and Gleason, K.K. (2007) *Adv. Mater.*, **19** (19), 2863–2867.
20. Trujillo, N.J., Barr, M.C., Im, S.G., and Gleason, K.K. (2010) *J. Mater. Chem.*, **20** (19), 3968–3972.
21. Barr, M.C., Rowehl, J.A., Lunt, R.R., Xu, J., Wang, A., Boyce, C.M., Im, S.G., Bulović, V., and Gleason, K.K. (2011) *Adv. Mater.*, **23** (31), 3500–3505.
22. Howden, R.M., Flores, E.J., Bulović, V., and Gleason, K.K. (2013) *Org. Electron.*, **14** (9), 2257–2268.
23. Ghaffari, M., Kosolwattana, S., Zhou, Y., Lachman, N., Lin, M., Bhattacharya, D., Gleason, K.K., Wardle, B.L., and Zhang, Q.M. (2013) *Electrochim. Acta*, **112**, 522–528.
24. Tenhaeff, W.E. and Gleason, K.K. (2008) *Adv. Funct. Mater.*, **18**, 979–992.
25. Vaddiraju, S., Seneca, K., and Gleason, K.K. (2008) *Adv. Funct. Mater.*, **18** (13), 1929–1938.
26. Bhattacharyya, D., Yang, R., and Gleason, K.K. (2012) *J. Mater. Chem.*, **22** (33), 17147–17152.
27. Vaddiraju, S., Cebeci, H.I., Gleason, K.K., and Wardle, B.L. (2009) *ACS Appl. Mater. Interfaces*, **1** (11), 2565–2572.
28. Lachman, N., Xu, H., Zhou, Y., Ghaffari, M., Lin, M., Bhattacharyya, D., Ugur, A., Gleason, K.K., Zhang, Q.M., and Wardle, B.L. (2014) *Adv. Mater. Interfaces*, **1**, 1400076. doi: 10.1002/admi.201400076.
29. Zhou, Y., Lachman, N., Ghaffari, M., Xu, H., Bhattacharya, D., Fattahi, P., Abidian, M.R., Wu, S., Gleason, K.K., Wardle, B.L., and Zhang, Q.M. (2014) *J. Mater. Chem. A*, **2** (26), 9964–9969.
30. Kawano, K., Pacios, R., Poplavskyy, D., Nelxon, J., and Bradley, D.D.C. (2006) *Sol. Energy Mater. Sol. Cells*, **90** (20), 3520–3530.
31. Norrman, K., Madsen, M.V., Gevorgyan, S.A., and Krebs, F.C. (2010) *J. Am. Chem. Soc.*, **132**, 16883–16892.
32. Nardes, A.M., Kemerink, M., de Kok, M.M., Vinken, E., Maturova, K., and Janssen, R.A.J. (2008) *Org. Electron.*, **9**, 727–734.

33. Gunes, S., Neugebauer, H., and Sariciftci, N.S. (2007) *Chem. Rev.*, **107** (4), 1324–1338.
34. Kulkarni, A.P., Tonzola, C.J., Babel, A., and Jenekhe, S.A. (2004) *Chem. Mater.*, **16** (23), 4556–4573.
35. Zaumseil, J. and Sirringhaus, H. (2007) *Chem. Rev.*, **107** (4), 1296–1323.
36. Borrelli, D.C. and Gleason, K.K. (2013) *Macromolecules*, **46** (15), 6169–6176.
37. Bhattacharyya, D. and Gleason, K.K. (2012) *J. Mater. Chem.*, **22** (2), 405–410.
38. Borrelli, D.C., Lee, S., and Gleason, K.K. (2014) *J. Mater. Chem. C*, **2** (35), 7223–7231.
39. Bhattacharyya, D., Yang, R., and Gleason, K.K. (2012) *J. Mater. Chem.* doi: 10.1039/C2JM32473C
40. Im, S.G., Gleason, K.K., and Olivetti, E.A. (2007) *Appl. Phys. Lett.*, **90** (15), 152112–152113.
41. Scardaci, V., Coull, R., and Coleman, J.N. (2010) *Appl. Phys. Lett.*, **97** (2), 023114.
42. Bai, H. and Shi, G. (2007) *Sensors*, **7**, 267–307.
43. Bhattacharyya, D., Senecal, K., Marek, P., Senecal, A., and Gleason, K.K. (2011) *Adv. Funct. Mater.*, **21** (22), 4328–4337.
44. Vaddiraju, S. and Gleason, K.K. (2010) *Nanotechnology*, **21** (12), 125503.
45. Bhattacharyya, D. and Gleason, K.K. (2011) *Chem. Mater.*, **23** (10), 2600–2605.
46. Bhattacharyya, D., Howden, R.M., Borrelli, D., and Gleason, K.K. (2012) *J. Polym. Sci., Part B: Polym. Phys.*, **50**, 1329–1351.
47. Im, S.G., Olivetti, E.A., and Gleason, K.K. (2007) *Surf. Coat. Technol.*, **201** (22-23), 9406–9412.
48. Lock, J.P., Lutkenhaus, J.L., Zacharia, N.S., Im, S.G., Hammond, P.T., and Gleason, K.K. (2007) *Synth. Met.*, **157** (22-23), 894–898.
49. Im, S.G., Kusters, D., Choi, W., Baxamusa, S.H., van de Sanden, M.C.M., and Gleason, K.K. (2008) *ACS Nano*, **2** (9), 1959–1967.
50. Barr, M.C., Carbonera, C., Po, R., Bulovic, V., and Gleason, K.K. (2012) *Appl. Phys. Lett.*, **100** (18), 183301.

Part III
Applications

CVD Polymers: Fabrication of Organic Surfaces and Devices, First Edition. Edited by Karen K. Gleason.

12 Controlling Wetting with Oblique Angle Vapor-Deposited Parylene

Melik C. Demirel and Matthew J. Hancock

12.1 Introduction

In the plant and animal kingdoms, directional textured surfaces with microscale and nanoscale features provide key functions for survival [1]. Directional surfaces enable butterflies to cast water from their wings, water striders to walk on water, and plants to trap pollen and insects. Directional textured surfaces on gecko, spider, and lizard feet allow the animals to climb smooth and rough vertical walls, which require that they controllably adhere to and detach from smooth and rough surfaces. The intestine and the lung are lined with microvilli that direct flow. All these natural surfaces derive their directional properties from asymmetric micro- or nanostructures such as ratchets or hairy forests, the asymmetric placement of such structures, or an emergent property due to a surface being coated with many copies of such structures. At the macroscopic scale, directionality manifests itself as anisotropic or unidirectional materials or processes. At the molecular scale, thermal Brownian motion and molecular polarity fuel spatial or dynamical symmetry breaking. A system exhibiting such directionality is often called a molecular ratchet [2].

Inspired by natural directional surfaces, engineers have developed a myriad of synthetic surfaces with precisely tuned physicochemical properties to transport water droplets and soft materials, control liquid spreading, provide directional adhesion, and exhibit directional friction [3, 4]. Due to the wide range of applications, there has been an increasing interest in studying the anisotropy of biological and synthetic materials. Reviews already exist for the large body of research on natural [5, 6] and engineered [7] superhydrophobic surfaces including the water-repellent hairy exteriors of plant and animal species. Another review highlighted methods for synthesizing materials for dry adhesion that mimic the lizard or insect footpads [7]. Our own reviews have focused on anisotropic wet and dry adhesion properties of surfaces from experimental and theoretical perspectives [3, 4]. In this chapter, we begin by defining anisotropy in Section 12.2 and then outlining fabrication methods for CVD (chemical vapor deposition) films, specifically focusing on the OAP method, in Section 12.3. In Section 12.4, we outline an array

CVD Polymers: Fabrication of Organic Surfaces and Devices, First Edition. Edited by Karen K. Gleason.

of applications for CVD films. Finally, in Section 12.5, we highlight theoretical approaches for modeling wet and dry adhesion on CVD films.

12.2 Definition of Anisotropy in Materials Science

The physicochemical properties of materials often depend on direction. Most materials have anisotropic physical properties because of their asymmetric micro- and nanoscale features. To determine the anisotropy of a system, a host of factors must be considered: the asymmetry of the material's structure and geometry; the asymmetry of the external field or stress; the asymmetry of the surface or boundary conditions; the asymmetry of the physical property relating stress to strain; and the asymmetry of the resulting body motion or deformation. These are schematically shown in Figure 12.1. Here, the stress and strain could be external (e.g., mechanical, optical, electrical, magnetic) or internal (e.g., self-oscillatory chemical reactions [8]). Examples include the walking of an asymmetric gel strip fueled by an oscillating chemical reaction (i.e., Belouzov [9], Zhabotinsky [10]), and a polymer-magnetic composite swimmer actuated by a magnetic field [11]. Figure 12.1a illustrates types of material asymmetries, which could be due to structural or compositional variations. A recent review on nanomaterials assembly classified these variations in detail [12]. An example of such assembly is the anisotropic superstructures formed by uniformly amphiphile-coated spherical nanoparticles to balance the energy gain between particle coalescence and the entropy of the distorting grafted polymers [13]. A schematic of field asymmetry is shown in Figure 12.1b. A particular example of field asymmetry is the anisotropic lateral vibration on a hydrophobic surface that induces unidirectional drop motion [14]. The influence of boundary

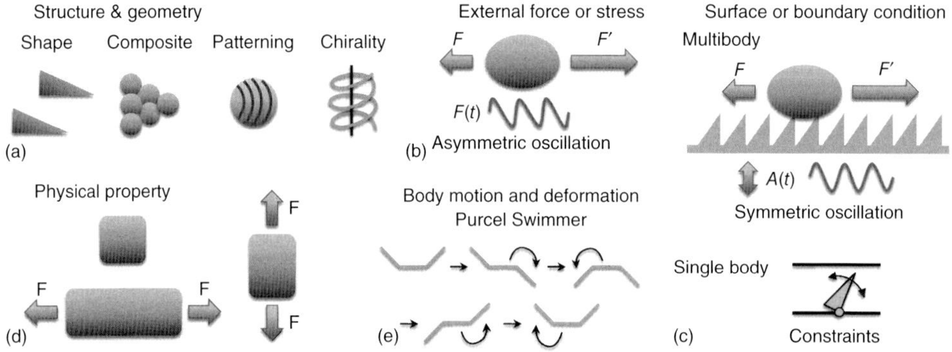

Figure 12.1 Sources of asymmetry in materials science. The asymmetry of (a) material's structure or geometry, (b) the external field or stress, (c) surface or boundary condition, (d) the physical property relating stress to strain, and (e) strain or deformation. Source: Hancock *et al.* [3] copyright (2012) John Wiley & Sons.

conditions and geometry is illustrated in Figure 12.1c. The material itself may be isotropic, but its behavior may be anisotropic due to, for example, surface texture. Also, anisotropy may arise from multibody and single-body interactions. In many cases, the asymmetric nature of a material is the result of a many-body emergent property. For example, a forest of carbon nanotubes [15] or a textured polymer nanofilm [16, 17] will have different mechanical responses than the individual nanotubes or fibers. The most commonly known anisotropy of material systems is sketched in Figure 12.1d: an anisotropic stretch or contraction can be created if the modulus has internal asymmetry. Finally, strain asymmetry is illustrated in Figure 12.1e. An example is Purcell's swimmer [18], made of three slender rods, that performs an asymmetric stroke to translate at low speeds in viscous fluids.

12.3
OAP Surfaces: Fabrication

The poly(*p*-xylylene) (i.e., parylene or PPX) films discussed herein consist of an array of poly(*p*-xylylene) (i.e., parylene or PPX) nanorods fabricated via a bottom-up vapor-phase technique called as oblique angle polymerization (OAP). OAP involves vaporization and pyrolysis of a *p*-cyclophane precursor to generate a diradical vapor flux (Figure 12.2a). The deposition starts with a monomer (i.e., xylylene) or a dimer (i.e., paracyclophane), and the monomers are converted to a polymer (poly(*p*-xylylene), PPX) during the growth process by radical polymerization (Figure 12.2b). The flux is directed at a controlled shallow angle onto a substrate, where surface diffusion, in combination with deposition geometry, leads to shadowing and selective growth of anisotropic structures during polymerization [19]. The equipment necessary to perform OAP is shown in Figure 12.2c, including a computer-controlled stepper motor and a nozzle assembly. Planar PPX films may be fabricated without nanostructured morphology, as shown via the cross-sectional scanning electron microscope image in Figure 12.2d. Figure 12.2e shows the cross section of a PPX nanofilm of ~150-nm diameter nanorods fabricated by OAP of dichloro-[2,2]-*p*-cyclophane (chemical process outlined in Figure 12.2b).

PPX nanofilms have been functionalized to form composites with metals [21, 22] and ceramics [23] exhibiting unique properties, such as mechanical anisotropy [19], catalytic activity [24], and enhanced activity for surface Raman spectroscopy [25]. Other PPX nanorod films synthesized with hydrophilic groups [16] (e.g., $R = -CH(OH)CF_3$) promote water penetration into the interstitial regions between nanorods, leading to droplet impalement that renders the surfaces nonideal for droplet transport applications.

The morphology of PPX nanofilms may be altered by rotating the substrate during oblique angle deposition. Computerized control of substrate rotation introduces a rotation axis normal to the substrate surface, which changes the relative

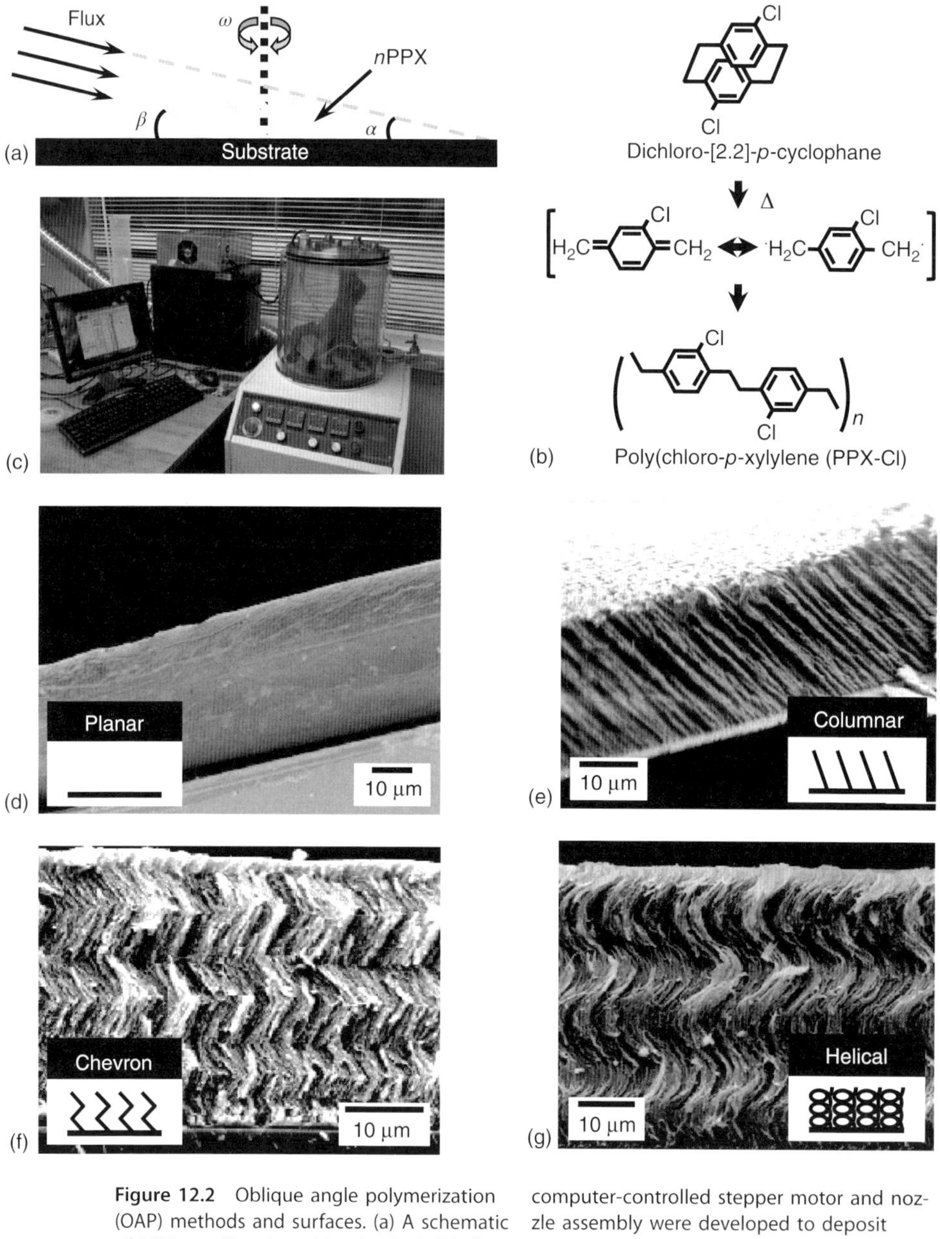

Figure 12.2 Oblique angle polymerization (OAP) methods and surfaces. (a) A schematic of PPX nanofilm deposition by OAP. (b) The deposition starts with a monomer (i.e., xylylene) or a dimer (i.e., paracyclophane), and the monomers are converted to a polymer (poly(*p*-xylylene), PPX) during the growth process by radical polymerization. (c) A computer-controlled stepper motor and nozzle assembly were developed to deposit structured PPX. (d–g) Cross-sectional SEM micrographs of (d) planar, (e) columnar, (f) helical, and (g) chevron PPX films. Planar PPX films do not possess nanostructured morphology. Source: Demirel [17], copyright (2008), with permission from Elsevier.

direction of the vapor flux and manipulates the growth direction of the nanotextured PPX film. For example, nanotextured chevron and helical PPX films are shown in Figure 12.2f,g, respectively, are fabricated by a rotation normal to the substrate surface.

12.4
Directional OAP Surfaces: Form and Function

Inspired by the structure and utility of biological surfaces, various directional surfaces have been engineered with micro- and nanoscale features. A number of nanostructured surfaces have recently been developed with directional wet adhesion, in other words, the force of adhesion on a drop is less in the drop's direction of motion than in other directions. The source of this directional adhesion is related to contact angles and is discussed at length in the theory section (Section 12.5). The advantage of using nanostructures is the 10-fold or more increase in droplet adhesion forces over microstructured surfaces.

Recently, we reported an engineered nanofilm composed of an array of hydrophobic PPX nanorods (poly-chloro-*p*-xylylene, aka, parylene [26]), which is both hydrophobic and exhibits anisotropic wet adhesion via a ratchet mechanism (Figure 12.3) [20]. Such a nanofilm also provides a microscale smooth surface for transporting microliter droplets.

Inspired by the gecko footpad, surfaces have been engineered for directional dry adhesion. One class of our nanoPPX films demonstrates directional dry adhesion [19]. By measuring the contact friction on these films, we found larger contact depths and higher friction coefficients for sliding in the direction of nanorod tilt compared to sliding against the tilt. In comparison, planar PPX films did not exhibit either frictional anisotropy or depth hysteresis [19].

Microfluidic devices require smart surface coatings for flow control. So far, such control has been achieved using chemical patterns [27, 28] and topologically textured surfaces. The unidirectional surfaces described here may be patterned in microfluidic channels. The geometry of roughness elements, for example, the tilt of nanorods, may be tuned to achieve a particular fluidic resistance in continuous systems or contact line motion in drop-based (digital) systems. A preferential direction of transport may be achieved in a channel by aligning the unidirectional surface with the channel axis. Fluid droplets may either slide or roll on textured surfaces depending on the viscosity, droplet size, wetting properties, and forcing characteristics [14, 29–31]. Droplet motion may be powered by gravitational forces, alternating electric fields, or mechanical vibration [14, 29–31]. A recent example is the unidirectional wetting behavior of OAP fabricated hydrophobic nanofilms. Such nanofilms exhibit a pin-release droplet ratchet mechanism (Figure 12.3) [20]. Moreover, applying low-amplitude, high-frequency ($\sim$30–80 Hz) vibrations on the boundary of such nanofilms induces droplet transport in the direction the asymmetric asperities (Figure 12.4) [20]. The maximum droplet translation

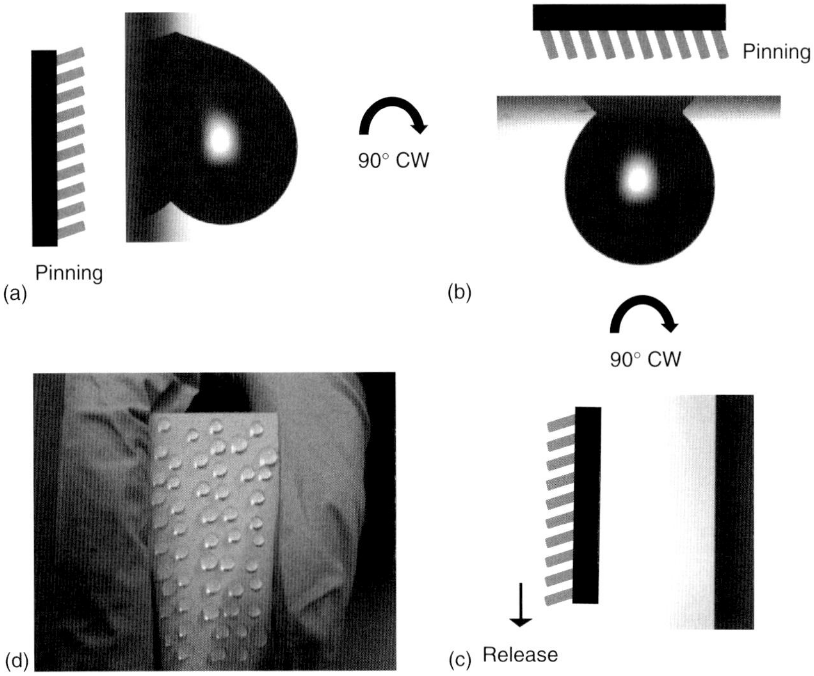

Figure 12.3 The pin-release wetting mechanism on a PPX nanofilm. (a–c) Water adhesion and release in three orientations of the nanofilm. Schematics illustrate the nanorod inclination at each orientation and correspond to photographs showing the pin-release wetting behavior of the nanofilm. (d) Picture of the water droplets pinned to a PPX nanofilm. Source: Malvadkar *et al.* [20] copyright (2010) Nature Publishing Group.

speed was approximately $3.5\,mm\,s^{-1}$ and depended on both the drop volume and vibration frequency, discussed in detail in Section 12.5. By adjusting the vibration frequency, drops of different sizes could be selectively moved and merged (Figure 12.5a) [32]. Cargo such as microgels can be carried by such droplets (Figure 12.5a), which, combined with selective drop motion and merging, offers a pathway toward high-throughput microgel assembly [32].

The directional release of gas bubbles may also be generated on anisotropic surfaces. We generated H_2 bubbles by the catalytic release from alkaline aqueous solutions of sodium borohydride on cobalt-functionalized directional PPX matrices [3]. Cobalt catalysts were prepared on PPX nanofilm substrates via electroless plating with adsorbed pyridine ligand as an adhesion promoter [24, 34]. The force of retention on gas bubbles was larger when the nanorods were pointing down than when pointing up (Figure 12.5b) [3].

A rich class of phenomena arises when a droplet impacts a textured surface. For particular impact speeds, drops may rebound from superhydrophobic surfaces without actually wetting the surface [35]. For a unidirectional surface,

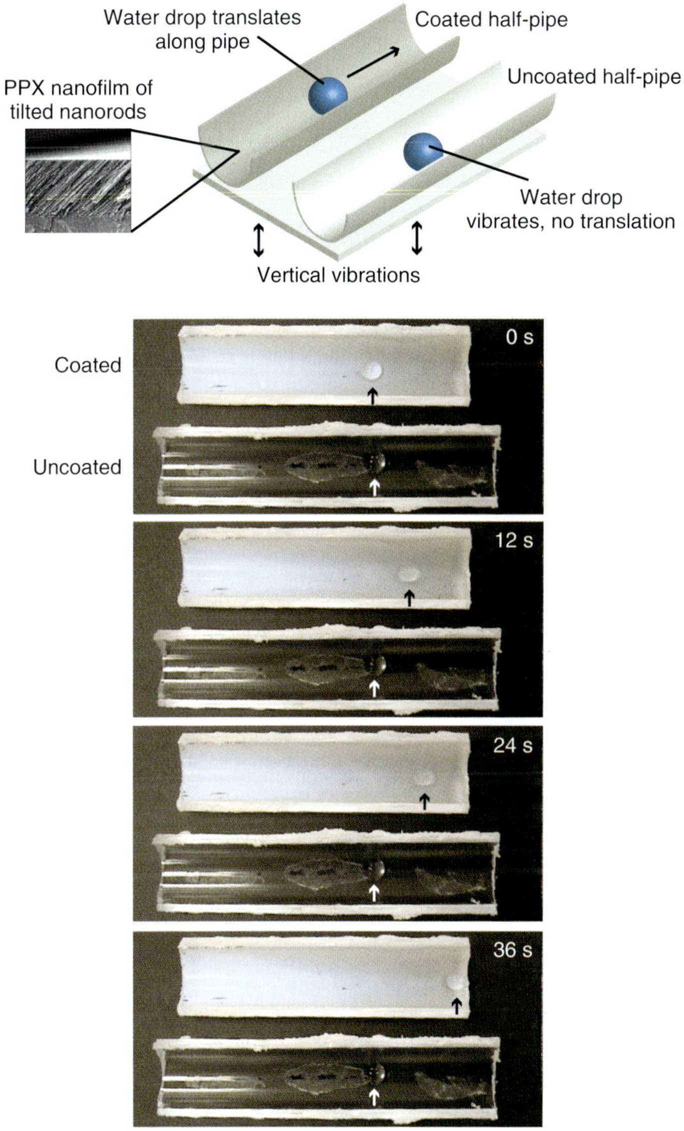

Figure 12.4 Unidirectional drop motion on a PPX nanofilm. (a) Schematic of experimental setup. Two half-pipes, one coated with a columnar PPX nanofilm and the other uncoated, were glued to a base. Low-amplitude ~85 Hz vibrations of the base caused translational droplet motion on the coated half-pipe, but not on the uncoated pipe. (b) Time lapse of droplet motion in the coated and uncoated half-pipes. Water drops translated axially on the half-pipe coated with a PPX nanofilm, whereas those on the uncoated half-pipe merely vibrated in-place. Glue is visible beneath uncoated half-pipe; the surface is smooth and clean. The arrows indicate droplet position. Source: Malvadkar *et al.* [20] copyright (2010) Nature Publishing Group.

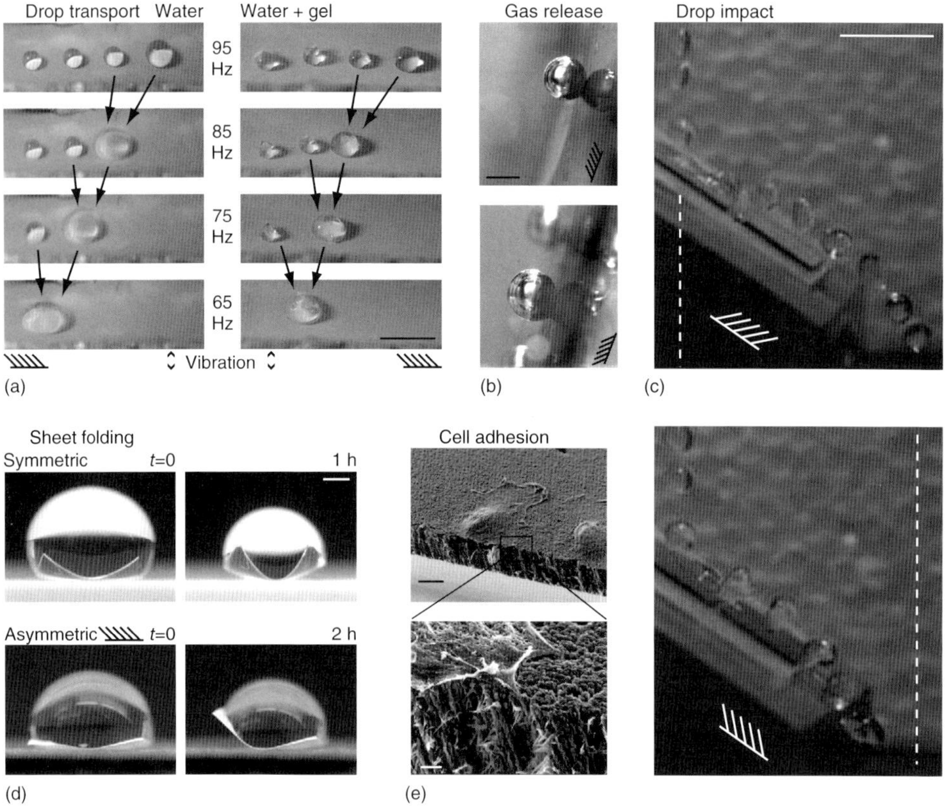

Figure 12.5 Applications of directional surfaces fabricated with OAP. (a) Drops and drops containing cargo being transported on directional surfaces forced with mechanical vibration. The forcing frequency was chosen close to the resonant frequency of the largest drop, and amplitude low enough so that smaller drops did not move. When two drops coalesced, their size increased and resonant frequency decreased. Thus, the forcing frequency was also reduced to keep the largest drop moving. Scale bar 5 mm. Reprinted with permission from Sekeroglu *et al*. [32] Copyright 2011, American Institute of Physics. (b) Directional control of gas release. Gravity and buoyancy force point in (above) and against (below) the direction of nanorod tilt, as indicated by the schematic. Scale bar 1 mm. Source: Hancock *et al.* [3] copyright (2012) John Wiley & Sons. (c) Drops impacting a directional surface with the grain (top) bounce further than those impacting against the grain (bottom). Dashed vertical lines demark the extent of the trajectory in the top image; schematics indicate the grain, that is, nanorod direction. Scale bar 1 cm. Source: Hancock *et al.* [3] copyright (2012) John Wiley & Sons. (d) Directional folding. Polymer sheets were folded by evaporating droplets. Folding was symmetric and asymmetric for sheets with isotropic and directional surfaces, respectively. Scale bar 4 mm. Source: Hancock *et al.* [3] copyright (2012) John Wiley & Sons. (e) Directional cell adhesion on a directional surface. Zoomed image indicates cell filopodia penetrating between the nanorods. Scale bars 10 μm (top) and 1 μm (bottom). Reprinted with permission from Christophis *et al*. [33] Copyright 2011, American Vacuum Society.

we expect that contact line motion and the rebound trajectory will depend on not only impact speed but also the angle of attack. Preliminary experiments performed at our laboratory showed that when drops impacted an inclined directional nanorod surface in the direction of nanorod tilt, the horizontal extent of the trajectory was larger than when drop impact was against the nanorod direction (Figure 12.5c) [3]. In addition to the rebound trajectory, we suspect that the critical impalement speed, rebound angle, and distorted drop shape will all depend on the orientation of droplet impact with respect to the asymmetric roughness features of the directional textured surface.

Capillary wrinkling and folding of thin sheets of soft material by droplets is another well-documented phenomenon [36]. Two-dimensional soft polymer sheets will fold into 3D structures as a function of the number and direction of folds [37]. Anisotropic textured surfaces could control the directionality of a folding sheet by creating an asymmetry in the contact angles along the contact line of the water drop folding the sheet. In preliminary experiments, we showed that the anisotropy of the nanoPPX can guide sheet folding (Figure 12.5d) [3]. Folding was driven by water evaporation following droplet deposition; the shrinking contact line of the evaporating droplet pulled the edges inward to drive the folding. The orientation of the nanorods determined the direction of folding: when the nanorod tilt was to the right, the receding contact angle along the right edge was smaller than that on the left, making the component of force larger on the right edge than on all other edges. Hence, the thin film bent to the left (Figure 12.5d) [3].

Bioadhesion properties have also been observed on synthetic directional nanofilms [33], making them potentially useful for tissue engineering applications. In a previous study, we used fluidic shear stress to remove fibroblast cells from a microfluidic channel whose substrate was coated with our nanoPPX film (Figure 12.5e) [33]. The fibroblast cells were removed with lower shear stresses when the flow was in the direction of nanorod tilt, compared to the flow against the tilt. We have also studied the growth cone adhesion and axon alignment for neurons cultured on asymmetric nanotextured surfaces [38, 39]. We believe that axons preferentially extend along the asymmetry of the surface texture (Figure 12.6a) and have demonstrated such unidirectional growth experimentally (Figure 12.6b,c) [38]. Furthermore, we varied the nanotexture asymmetry by tilting the nanorods under shear using a rubber strip [39]. The shear plastically deforms the surfaces and organizes the nanorods into ratchets, shown schematically in Figure 12.6d. The surfaces are then fully characterized and quantified by a single parameter; we call the ratchet angle ratio, C_α, defined in Figure 12.6d as the ratio of the angle of the ratchet face perpendicular to the nanorods to that parallel to the nanorods. For example, $C_\alpha = 1$ indicates a symmetric ratchet. We have shown experimentally that the strength of the axon surface coupling increases linearly with increasing ratchet asymmetry (Figure 12.6e), demonstrating that axonal outgrowth could be controlled by the asymmetry of nanoPPX surfaces [39].

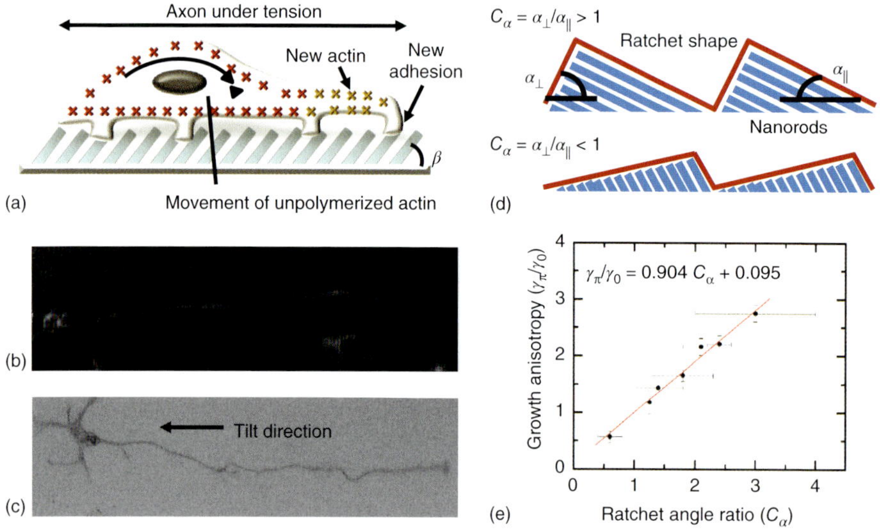

Figure 12.6 Directional neuron growth on a PPX nanofilm. (a) Schematic of growth cone adhesion to an asymmetric nanotextured surface composed of nanorods tilted at an angle β. (b,c) Axons extend preferentially in a direction opposite to the direction of the tilted nanorods, as shown by (b) fluorescence and (c) bright field images. The length of the axons shown in (b,c) is approximately 60 μm. Source for (a–c) is Beighley *et al.* [38] copyright (2012) American Institute of Physics. (d) Schematic defining the ratchet angle ratio $C_\alpha = \alpha_\perp/\alpha_{||}$, where $\alpha_\perp$ and $\alpha_{||}$ are the angles of ratchet faces perpendicular and parallel to nanorods, respectively. (e) Cell-surface coupling increases linearly with increasing ratchet asymmetry as quantified by the ratchet angle ratio C_α. Source for (d–e) is Spedden *et al.* [39] (open access).

12.5
Modeling Adhesion, Wetting, and Transport on Directional Surfaces

An integral part of understanding and designing directional surfaces involves modeling their behavior and interaction with fluids and solids. In this section, we review an array of modeling approaches with a hierarchy of sophistication to model adhesion, wetting, and transport on directional surfaces. The five sources of anisotropy outlined in the introduction and in Figure 12.1 arise in theoretical models from asymmetries in the geometry or governing equations, such as conservation laws (i.e., mass, momentum, energy), kinematic relations, and constitutive equations with boundary conditions [40]. The salient anisotropic mechanism must be included in theoretical models to achieve accurate predictions. For example, simulations of the microtextures within metallic thin films agree well with experimentally characterized microtextures when anisotropic interface properties are incorporated in the calculations, while agreement is poor when isotropic properties are used instead [41]. Similarly, at the nanoscale, fluctuations about the native conformation of proteins have been suitably reproduced using an anisotropic network model, which enables evaluation of the directional preferences of proteins [42].

12.5.1 Modeling Dry Adhesion

Models for dry adhesion are based on contact mechanics balancing elastic and surface energies and include crack-like mechanisms to model separation [43]. Contaminants or roughness reduce the strength of dry adhesion, while hierarchical structures can be used to increase it [43, 44]. Modeling approaches range from analytic formulas to finite element method (FEM) simulations to calculate the contact mechanics in the presence of normal and shear loads among solids, viscoelastic material, and biological matter [43, 45]. A recent example is the contact mechanics-based analytic model used to elucidate the robust and releasable adhesion of the gecko foot [43]. When a thin layer of lubricating liquid exists between two solids, additional fluidic forces must be coupled to the solid mechanical model to describe contact and adhesion. For example, an analytic model was developed to describe the oscillations and net motion in a gel cylinder transported on a polydimethylsiloxane (PDMS) layer with angled incisions [46].

12.5.2 Modeling Wetting, Adhesion, and Transport in Solid–Fluid Systems

The wetting of solid surfaces by fluid and the adhesion and transport of fluid on solid surfaces depends on both the topology of the solid surface and the chemical and material properties of the fluid and solid phases [47–50]. In what follows, we provide an overview based on existing works [47, 49] of two important concepts related to wetting, adhesion, and transport: contact angles and contact angle hysteresis. We then review theoretical tools for calculating and quantifying wetting, adhesion, and transport in solid–fluid systems. Although such tools have been applied mainly to isotropic surfaces, they may be readily generalized to directional surfaces. As reviewed below, thus far three groups have put forth models of wetting and adhesion on directional surfaces [20, 51, 52].

The contact angle θ is defined as the angle measured through the liquid between the meniscus (liquid–gas interface) and the substrate (liquid–solid interface) at the three-phase (solid–liquid–gas) contact line. The equilibrium contact angle between a drop and a flat smooth surface is given by Young's relation [53], $\cos\,\theta_e = (\gamma_{SG} - \gamma_{SL})/\gamma$, where γ, γ_{SL}, and γ_{SG} are the surface tensions of the gas–liquid, solid–liquid, and solid–gas interfaces, respectively. In reality, a range of static contact angles, referred to as contact angle hysteresis, may be observed for a given solid–fluid combination [54], $\theta_r \leq \theta \leq \theta_a$, bounded above and below by the receding and advancing contact angles. Contact angle hysteresis is directly related to adhesion, because differences in contact angles around the drop perimeter can yield a net surface tension force that resists drop transport. The force of wet adhesion, also called the droplet retention force, is estimated by [55] $F_r = k\gamma R(\cos\ \theta_r - \cos\ \theta_a)$, where R is the radius of the drop and k is a coefficient depending on the shape of the contact line. Such a formula may be used to estimate the maximum force on a raindrop sticking to a window, where the advancing and receding contact angles θ_a and θ_r are the contact angles on the drop's lower and upper edges, respectively. Contact angle hysteresis is due to contact lines pinning on microscopic irregularities on the surface [56]; consequently, the most effective way to reduce it is by removing such irregularities via polishing or conversely by minimizing the interaction between the interface and solid by roughening or texturing the surface.

On chemically heterogeneous or rough surfaces, the microscopic contact angle θ will generally differ from the apparent angle θ^* observed at a macroscopic scale. In cases where the fluid fully wets a rough surface, the apparent macroscopic contact angle θ^* is given by Wenzel's relation [57], $\cos\theta^* = r\cos\theta_e$ where the roughness parameter r equals the real surface area per planar area of the surface. When air pockets are trapped between roughness elements and the liquid, θ^* is given by the Cassie–Baxter [58] relation, $\cos\theta^* = 1 + \phi_s(1 + \cos\theta_e)$, where ϕ_s is the area solid fraction. The Cassie–Baxter formula reveals that as the area solid fraction vanishes, $\phi_s \rightarrow 0$, the contact angle θ^* approaches 180°. For nanotextured surfaces, air pockets are trapped at the interface between water and nanoporous hydrophobic surfaces. Such air pockets may be generated when a hydrophobic surface is immersed in a liquid [59] and may be quantified by X-ray scattering and contact angle measurements [60]. It has been shown that the layer of air at the solid–aqueous interfaces is less than 0.1 nm thick for several hydrophobic solids [61].

12.5.2.1 Analytic Models of Contact Angle Hysteresis

An array of analytic models exist to predict contact angle hysteresis including simple averages of the intrinsic contact angles on a material surface and in the voids, to scaling models incorporating the shape of the meniscus. Extrand's [62] geometrical scaling argument for the macroscopic advancing and receding contact angles, θ_a and θ_r, on a pillared substrate of a material with advancing angle over 90°, $\theta_{a0} > \pi/2$, gives $\theta_a = \pi$ and $\theta_r = \lambda\,\theta_{r0} + (1-\lambda)\pi$, where λ is the solid fraction, defined differently for different types of configurations. For example, for square arrays, $\lambda = d/\delta$, where d is the post-diameter and δ is the spacing. The contact angle hysteresis for a drop in a Cassie state on a pillared substrate has also been calculated [63] using a slice approach [64]. Other estimates have been made for the hysteresis on 2D substrates consisting of grooves with semicircular surface undulations and on 3D substrates consisting of pillars made up of many spheres [65].

The hysteresis on more complicated surfaces such as fiber layers with regular or random spacing may be estimated using the more general framework of de Gennes and colleagues [48, 49, 66–68]. In the limit of small contact angle, the hysteresis for a flat surface with single and distributed chemical defects has been derived [66].

The word "defect" refers to a perturbation in the wetting energy localized near a particular point. The pinning argument has been repeated for $\theta = \pi/2$, that is, a vertical meniscus on a flat surface passing a single chemical defect or a distribution of defects [49]. Similar results were found for the contact angle hysteresis due to weak random roughness [67]. The framework developed in these works may be used to determine the contact angle hysteresis of many textured surfaces including directional surfaces.

Recently, Reyssat and Quéré [68] generalized de Gennes' spring-like model to the case of a drop in a Fakir (Cassie) state on square array of cylinders. Assuming the resistance was entirely due to the receding contact line [69], they calculated the force to depin the receding surface of the drop and found that the hysteresis scaled as the solid fraction ϕ_s times its log. Their model included a fitting parameter a quantifying the effects of contact line distortion: $\cos\theta_r - \cos\theta_a = (a\phi_s/4)\log(\pi/\phi_s)$, where $a = 2$ (scaling) and $a = 3.8$ (data fit) [68].

We recently reported a simple ratchet model for the contact angle hysteresis to rationalize the directional wetting response observed in Figures 12.3–12.5 [20]. The model was based on the directional dependence of the advancing and receding contact angles, θ_a and θ_r, of a sessile droplet on a tilted pillared substrate [20]. We estimated θ_a and θ_r by generalizing Extrand's [51, 62] approach for 2D ratchets and 3D vertical pillars. Following Extrand, we introduced the linear fraction of the receding or advancing contact lines on the pillars, λ. The contact angles θ_r and θ_a were approximated as weighted averages, in terms of λ, of the respective contact angles between the meniscus and pillar and 180° in air. The calculations are now repeated as an example for the reader. The contact angle between the advancing or receding meniscus is related geometrically to the pillar surface in terms of the intrinsic contact angles θ_{a0} and θ_{r0} of the material, the nanorod tilt angle β and diameter d, and the perpendicular spacing δ (Figure 12.7). In the release direction, the advancing angle is $\theta_{aREL} = \lambda\min(\pi, \max(\theta_{a1}, \theta_{a2})) + (1-\lambda)\pi$, where $\theta_{a1} = 3\pi/2 - \beta - \arctan(\cot(\beta)/(1-s))$ (Figure 12.7, Case 1) and $\theta_{a2} = \theta_{a0} + \pi/2 - \beta$ (Figure 12.7, Case 2), and the receding angle is $\theta_{rREL} = \lambda(\theta_{r0} + \beta - \pi/2) + (1-\lambda)\pi$. The solid fraction in the longitudinal direction is defined as $s = d/(d+\delta)$. In the pinning direction, $\theta_{aPIN} = \lambda\min(\pi, \theta_{a0} + \beta) + (1-\lambda)\pi$ and $\theta_{rPIN} = \lambda(\theta_{r0} - \beta) + (1-\lambda)\pi$. For a single droplet, the ratio of the retention forces in the pin and release directions is

$$\frac{F_{PIN}}{F_{REL}} = \frac{\cos\theta_{rPIN} - \cos\theta_{aPIN}}{\cos\theta_{rREL} - \cos\theta_{aREL}} \tag{12.1}$$

For the case of a droplet moving parallel to longitudinal ratcheted bars ($\delta = 0$, $s = 1$),

$$\frac{F_{PIN}}{F_{REL}} = \frac{\sin\left(\lambda\left(\beta + {}^{1}\!/\!_{2}\Delta\theta_0\right)\right)}{\sin\left(\lambda\left(\pi/2 - \beta + {}^{1}\!/\!_{2}\Delta\theta_0\right)\right)} \tag{12.2}$$

where $\Delta\theta_0 = (\theta_{a0} - \theta_{r0})$ is the difference in the intrinsic advancing and receding contact angles of the underlying material. For a 2D ratcheted grating ($\delta = 0$, $s = 1$, $\lambda = 1$), Eq. (12.2) reduces to Extrand's ratchet result [51]. For a droplet on a

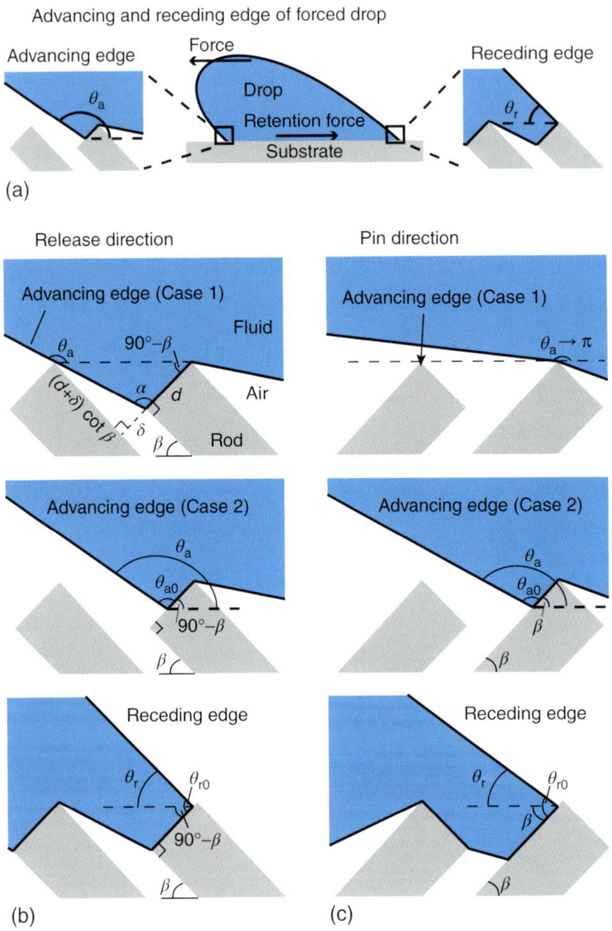

Figure 12.7 Analytic model of advancing and receding contact angles on a columnar PPX nanofilm. (a) Schematic of drop and its advancing and receding edges on a textured surface. Geometry of contact between the drop and roughness elements when the drop is forced in (b) the release and (c) the pinning directions. For the drop to advance in either direction, its advancing edge must contact the next roughness element (Case 1) and move along the roughness top (Case 2). The macroscopic contact angles θ_a and θ_r between the advancing and receding menisci and the substrate axis are functions of the intrinsic contact angles θ_{a0} and θ_{r0} of the material, the nanorod tilt angle β and diameter d, and the perpendicular spacing λ. Note that the substrate constrains θ_a and θ_r to be between 0° and 180°. Source: Malvadkar *et al.* [20] copyright (2010) Nature Publishing Group.

substrate of vertical pillars ($\beta = 90°$), the advancing and receding angles reduce to Extrand's vertical pillar result. The parameters β, λ, s, and $\Delta\theta_0 = (\theta_{a,0} - \theta_{r,0})$ are fixed and measurable for a given nanofilm. To recap, Eq. (12.1) predicts that the ratio of the retention forces for a drop on the nanofilm will be a constant, C_0, as observed in our experiments [20]. In addition, as the retention force equals the critical gravitational force ρVg sinα, we have

$$\frac{V_{\mathrm{PIN}}R_{\mathrm{REL}}\sin\alpha_{\mathrm{PIN}}}{V_{\mathrm{REL}}R_{\mathrm{PIN}}\sin\alpha_{\mathrm{REL}}}=\frac{\cos\theta_{\mathrm{rPIN}}-\cos\theta_{\mathrm{aPIN}}}{\cos\theta_{\mathrm{rREL}}-\cos\theta_{\mathrm{aREL}}}=C_0 \tag{12.3}$$

The factors R originate from the equation for the retention force; we have neglected the changes in the constant k due to droplet shape [55b, 70]. Assuming that R scales as $V^{1/3}$, then for a fixed tilt angle α, $(V_{\mathrm{PIN}}/V_{\mathrm{REL}})^{2/3}=C_0$. For example, our PPX nanofilm reported in Malvadkar *et al.* [20] consists of nanorods of diameter $d\sim 150$ nm, spacing $\delta\sim 40$ nm, solid fraction $s\sim 0.80$, tilt angle $\beta=55°$, and intrinsic advancing and receding angles (of flat PPX) of $\theta_{\mathrm{a0}}=95°$ and $\theta_{\mathrm{r0}}=75°$. Approximating our nanofilm as a square array of circular posts, $\lambda=\pi s/(2(1+(\pi/2-1)s)\cong 0.85$. With these parameter values, Eq. (12.3) gives $C_0=1.22$, close to the measured value $(V_{\mathrm{PIN}}/V_{\mathrm{REL}})^{2/3}=(1.22\pm 0.06)^{2/3}=1.14\pm 0.04$ [20]. In the pinning direction, our model predicts the advancing and receding angles 154° and 43°, respectively, compared to the measured angles 132° and 47° ± 5°; in the release direction, the predictions 147° and 60° compared with the measured values 128° and 60° ± 5° [20]. The receding angles are predicted much more accurately than the advancing angles, which still yield reasonable estimates of the hysteresis due to the mild slope of the cosine function near 180° [20]. Moreover, the differences in hysteresis approximately cancel when calculating the ratio C_0. The agreement between model and experiment is encouraging, given the modeling simplifications and variations associated with PPX nonuniformities, evaporative losses, especially important for smaller drops, and spurious mechanical vibrations that promote drop release. For more details and analysis of this contact angle hysteresis model, the reader is referred to Malvadkar *et al.* [20].

A large class of the directional textured surfaces considered in Figures 12.3–12.5 consist of asperities with long and thin features that bend when in contact with droplets [20]. The tips of these flexible asperities can pull back on the trailing surface of the drop as it passes, enhancing hysteresis. A future goal is to integrate the effects of such flexibility into analytic models of contact angle hysteresis to improve predictions of wetting, adhesion, and transport.

12.5.2.2 **Analytic Models of Drop Transport on Textured Surfaces**

The relation between drop speed and vibration frequency, amplitude, drop volume, and surface parameters may be rationalized theoretically [14, 29–31]. The power required to move a droplet at a speed U on a surface with retention force F_{r} scales as $F_{\mathrm{r}}U$ in addition to the rate of energy dissipation in the vibration and translation drops. The retention force on a moving droplet is somewhat less than that for a static droplet, but the static calculation, proportional to contact angle hysteresis, gives a reasonable upper bound. Steady droplet motion is achieved when the energy transferred to the droplet is balanced by dissipation (e.g., Figures 12.4 and 12.5a). For drops with length scales smaller than the capillary length $l_{\mathrm{c}}=(\gamma/\rho g)^{1/2}$, energy dissipation due to internal viscous motions is confined to a small dissipation zone near the region of contact with the substrate (elsewhere, the drop is in solid body motion) and scales as $\mu U^2 R^4/(l_c)^3$ [30]. The dissipative losses due to vibration may be estimated from Stokes third problem [71] as $\mu k(a\omega)^2\, l^2$, where $l \sim R^2/l_c$ is the radius of the contact region [30], a and ω are the amplitude and angular frequency of vibration, and $k = (\omega\rho/2\mu)^{1/2}$. Drop speeds have been measured on isotropic textured surfaces for motion powered by gravity [29, 30], anisotropic mechanical vibration [14], and on directional surfaces by mechanical vibration [20, 32]. Drop speed is highly dependent on the forcing frequency. For the same input energy, larger drop speeds are associated with the Rayleigh modes of the droplet [14, 32]. Estimates and scaling arguments of drop speed have been derived based on the simple estimates of contact angle hysteresis and dissipation listed earlier [14, 29–31].

12.5.2.3 Finite Element Models of Static Drops on Textured Surfaces

The simplest computational approach for modeling droplet behavior on textured surfaces involves FEM calculations of the static droplet meniscus and the corresponding contact angle hysteresis. A popular public domain FEM package for such simulations is Surface Evolver [72], which evolves a surface to achieve a state of minimal energy. The user creates an initial surface of vertices, edges, and faces, and has broad control over the mesh refinement, surface evolution, energy functionals, and convergence. The initial surface is iteratively evolved and refined until reaching a local energy minimum (Figure 12.8a). The package also includes basic programming constructs such as loops and conditional statements. Surface Evolver has been used to calculate free surface profiles on isotropic surfaces with chemical patterns [27, 80, 81, 76], parallel grooves [73], pillars [75, 82, 74] (Figure 12.8b,c), and inverted pillar surfaces [83]. Surface Evolver has also been used to calculate droplet profiles in electric fields [84].

To calculate the contact angle between a sessile drop and a rough surface, the size of the computational domain may be significantly reduced when the roughness scale is much smaller than the drop radius and the capillary length (Figure 12.8b,c). In this regime, a good approximation is that the curvature of the liquid–gas interface near the roughness is effectively zero compared to the scale of the roughness. A zero-curvature constraint may then be imposed within Surface Evolver to simulate the meniscus near the asperities. Note that zero curvature does not imply planar; a saddle shape in which the component curvatures in perpendicular directions cancel has zero curvature. Such saddle-like topology characterizes the meniscus geometry near the contact line of a drop on a rough surface. However, at the scale of the roughness, the far-field condition (away from the roughness) is that the receding surface is in fact planar. In this regime, using such an approach allows the shape of the three-phase contact line and the local contact angle to be calculated to a good approximation with a computational domain that includes only the nearby free surface, and not the entire drop surface. Symmetries in the roughness configuration may also be used to further reduce the size of the computational domain [75, 74].

The key to modeling adhesion and transport is to move beyond calculations of contact angles to the limiting cases of advancing and receding angles. In one of the

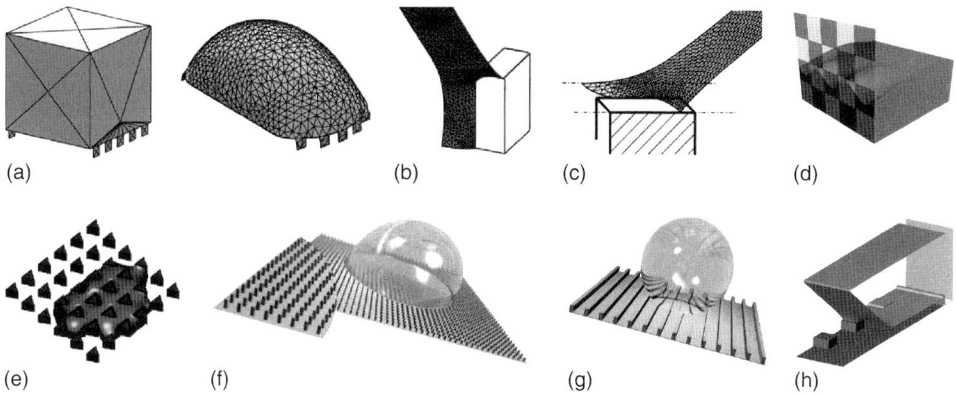

Figure 12.8 Modeling wet adhesion on textured surfaces. (a) Initial and final Surface Evolver simulation profiles for a static drop on a pillared surface. Reprinted from Chen *et al.* [73] copyright (2005), with permission from Elsevier. (b) Surface Evolver simulation of a portion of a droplet in a Wenzel state on a pillared surface. Reprinted with permission from Dorrer and Rühe [74]. Copyright (2008), American Chemical Society. (c) Surface Evolver simulation of a portion of a droplet in a Cassie state on a pillared surface. Reprinted with permission from Dorrer and Rühe [75]. Copyright (2007), American Chemical Society. (d) Surface Evolver simulation of a droplet advancing or receding on a checkerboard chemically patterned surface. Reprinted with permission from Kwon *et al.* [76] Copyright (2010), American Chemical Society. (e) Lattice Boltzmann simulation of directional spreading through an array of triangular pillars. Reprinted with permission of Blow *et al.* [77] Copyright 2009, Institute of Physics. (f,g) Lattice Boltzmann simulations of drops sliding on pillared and grooved surfaces. Source: Hyväluoma *et al.* [78] with kind permission of The European Physical Journal (EPJ). (h) Lattice Boltzmann simulation of a portion of a drop in a Cassie state receding on a pillared surface. Reprinted with permission from Mognetti and Yeomans [79]. Copyright (2010) American Chemical Society.

few studies to report such a calculation, Kwon *et al.* [76] calculated the advancing and receding angles by gradually increasing or decreasing the volume of a drop, respectively, on a chemically patterned surface and averaging the contact angle along the contact line just prior to depinning events (Figure 12.8d). When the contact line moved, local contact angles were enforced by a phase field model coupled to the Surface Evolver calculations. The concept of coupling the phase field model to the FEM simulations not only maintained accurate contact angles during contact line motion, but also allowed the intrinsic advancing and receding angles of the material to be included in the simulations. Such an approach could be extended to model contact angle hysteresis for surfaces with hierarchical scales of roughness, such as those of the Lotus leaf.

The previous contact angle hysteresis models, both analytic and finite elements, have all calculated the static meniscus. As noted earlier, the static calculations are useful to provide upper bounds on the wet adhesive property of a surface. However, accurately modeling dynamic situations such as a drop sliding along a surface or depinning events require full numerical treatment, discussed next.

12.5.2.4 Full Numerical Simulation of Fluid Flow and Free Surface on Directional Surfaces

An array of numerical approaches exists to generate full dynamical simulations of free surface flows in and over complex geometries. At the microscopic scale, molecular dynamics (MDs) simulations have modeled contact line dynamics and shed light on wetting hysteresis due to chemical defects and roughness [85]. At the mesoscale, phase field models employ a free energy field variable that acts as a marker for the different continuum phases [86]. Phase field models have been used to model a variety of dynamic multiphase flow problems, including droplet hysteresis and contact line pinning and motion on chemically and structured heterogeneous surfaces [86, 87]. Phase field models have been implemented with both sharp and diffuse treatments of the interface to model drop, free surface, and contact line phenomena [88]. A popular phase-field diffuse-interface approach is the lattice Boltzmann method (LBM) [89]. In LBM, a mean-field lattice-gas cellular automata model is implemented with a set of distribution functions to represent the average population of fluid particles [90]. Free surfaces are implemented using a free energy functional that is minimized in equilibrium [90]. The model can be extended to complex and multiphase fluids and complex geometries. Thus, LBM is ideal for simulating chemically patterned and rough surfaces. An array of dynamic wetting problems have been solved including the spreading of a fluid through a post-array (Figure 12.8e) [77], droplets sliding on inclined pillared and grooved substrates (Figure 12.8f,g) [78], periodic droplet formation within chemically patterned microchannels [91], anisotropic adhesion on ratcheted surfaces [52], the impalement of a droplet on a superhydrophobic textured surface [92], the advancing of a meniscus through microchannels patterned by posts [93], and receding contact lines on superhydrophobic surfaces (Figure 12.8h) [79]. For static drops, FEM and LBM simulations produce comparable drop profiles [81].

12.6 Conclusions

Inspired by directional surfaces in Nature, engineers have employed recent innovations in microfabrication and materials synthesis to produce an array of synthetic directional surfaces with tunable anisotropic chemical, mechanical, physical, morphological, and topological properties. Such directional surfaces may exhibit a host of emergent properties such as friction and wetting. These directional surfaces not only mimic biological surfaces but also provide new avenues to discover surface phenomena beyond those in Nature. In concert with innovative synthesis techniques, theoretical and computational tools link the structure to the behavior of directional surfaces. The next generation of directional surfaces will find use in syringes, hydropower turbines, micro- and nanoscale microfluidic devices for biomedical research, drag reducing and nonfouling ship hull coatings, self-cleaning solar-cell panels, high-efficiency thermal cooling systems for microchips, and friction-control coatings for tires.

A future challenge is to engineer dynamic directional surfaces that change wetting and friction properties in real time and enact such changes in response to external stimuli. In Nature, various stimuli response mechanisms exist to modify the surface shape or wetting properties. For example, water striders release surfactants to break through the free surface while laying eggs [5]. Certain insects generate emergency propulsion by releasing surface-active fluids to generate Marangoni stresses [5]. Many surface dwelling insects lift or pierce the free surface using retractable wetting claws at the ends of their hydrophobic legs [94]. Inspired by these natural systems, engineered surfaces exist that change their wetting and adhesion properties on command or in response to stimuli [95]. Such systems also employ surface clumping to lower the surface solid fraction and temperature gradients to induce surface tension gradients. Under the right conditions, cycling the stimulus may induce transport, offering a method to propel cargo or droplets in addition to mechanical vibration, acoustics, and electrowetting. Marangoni flows induced by surfactant release could also be used to create fluid motion on a unidirectional surface. New theoretical tools must also be developed to link the structure and stimuli response mechanisms to the overall behavior of the directional surface and to better inform design criteria for optimal behavior.

Acknowledgments

We gratefully acknowledge financial support for this work from National Institutes of Health and the Pennsylvania State University.

References

1. Vogel, S. (1988) *Life's Devices: The Physical World of Animals and Plants*, Princeton University Press, Princeton, NJ.
2. Reimann, P. (2002) *Phys. Rep.*, **361**, 57.
3. Hancock, M.J., Sekeroglu, K., and Demirel, M.C. (2012) *Adv. Funct. Mater.*, **22**, 2223.
4. Hancock, M.J. and Demirel, M.C. (2013) *MRS Bull.*, **38**, 391.
5. Bush, J.W.M. and Hu, D.L. (2006) *Annu. Rev. Fluid Mech.*, **38**, 339.
6. Hsu, S.-H., Woan, K., and Sigmund, W. (2011) *Mater. Sci. Eng., R*, **72**, 189.
7. Kwak, M.K., Jeong, H.-E., Kim, T.-I., Yoon, H., and Suh, K.Y. (2010) *Soft Matter*, **6**, 1849.
8. Epstein, I.R. and Pojman, J.A. (1998) *An Introduction to Nonlinear Chemical Dynamics: Oscillations, Waves, Patterns, and Chaos*, Oxford University Press, New York.
9. Belousov, B.P. (1959) *Collection of Short Papers on Radiation Medicine for 1958*, Medicine Publications, Moscow, p. 145.
10. Zaikin, A.N. and Zhabotinsky, A.M. (1970) *Nature*, **225**, 535.
11. Zimmermann, K., Naletova, V.A., Zeidis, I., Turkov, V.A., Kolev, E., Lukashevich, M.V., and Stepanov, G.V. (2007) *J. Magn. Magn. Mater.*, **311**, 450.
12. Glotzer, S.C. and Solomon, M.J. (2007) *Nat. Mater.*, **6**, 557.
13. Akcora, P., Liu, H., Kumar, S.K., Moll, J., Li, Y., Benicewicz, B.C., Schadler, L.S., Acehan, D., Panagiotopoulos, A.Z., Pryamitsyn, V., Ganesan, V., Ilavsky, J., Thiyagarajan, P., Colby, R.H., and Douglas, J.F. (2009) *Nat. Mater.*, **8**, 354.
14. Daniel, S., Chaudhury, M.K., and de Gennes, P.-G. (2005) *Langmuir*, **21**, 4240.

15. Ok, J.G., Tawfick, S.H., Juggernauth, K.A., Sun, K., Zhang, Y., and Hart, A.J. (2010) *Adv. Funct. Mater.*, **20**, 2470.
16. Boduroglu, S., Cetinkaya, M., Dressick, W.J., Singh, A., and Demirel, M.C. (2007) *Langmuir*, **23**, 11391.
17. Demirel, M.C. (2008) *Colloids Surf., A*, **321**, 121.
18. Purcell, E.M. (1977) *Am. J. Phys*, **45**, 3.
19. So, E., Demirel, M.C., and Wahl, K.J. (2010) *J. Phys. D: Appl. Phys.*, **43**, 045403.
20. Malvadkar, N.A., Hancock, M.J., Sekeroglu, K., Dressick, W.J., and Demirel, M.C. (2010) *Nat. Mater.*, **9**, 1023.
21. Demirel, M.C., Cetinkaya, M., Singh, A., and Dressick, W.J. (2007) *Adv. Mater.*, **19**, 4495.
22. Malvadkar, N.A., Sekeroglu, K., Dressick, W.J., and Demirel, M.C. (2010) *Langmuir*, **26**, 4382.
23. Malvadkar, N., Dressick, W.J., and Demirel, M.C. (2009) *J. Mater. Chem.*, **19**, 4796.
24. Malvadkar, N., Park, S., Urquidi-MacDonald, M., Wang, H., and Demirel, M.C. (2008) *J. Power Sources*, **182**, 323.
25. Demirel, M.C., Kao, P., Malvadkar, N., Wang, H., Gong, X., Poss, M., and Allara, D.L. (2009) *Biointerphases*, **4**, 35.
26. (a) Cetinkaya, M., Malvadkar, N., and Demirel, M.C. (2008) *J. Polym. Sci., Part B: Polym. Phys.*, **46**, 640; (b) Demirel, M.C., Boduroglu, S., Cetinkaya, M., and Lakhtakia, A. (2007) *Langmuir*, **23**, 5861.
27. Darhuber, A.A., Troian, S.M., Miller, S.M., and Wagner, S. (2000) *J. Appl. Phys.*, **87**, 7768.
28. Zhao, B., Moore, J.S., and Beebe, D.J. (2001) *Science*, **291**, 1023.
29. Lv, C., Yang, C., Hao, P., He, F., and Zheng, Q. (2010) *Langmuir*, **26**, 8704.
30. Mahadevan, L. and Pomeau, Y. (1999) *Phys. Fluids*, **11**, 2449.
31. Richard, D. and Quéré, D. (1999) *Europhys. Lett.*, **48**, 286.
32. Sekeroglu, K., Gurkan, U.A., Demirci, U., and Demirel, M.C. (2011) *Appl. Phys. Lett.*, **99**, 063703.
33. Christophis, C., Sekeroglu, K., Demirel, G., Thome, I., Grunze, M., Demirel, M.C., and Rosenhahn, A. (2011) *Biointerphases*, **6**, 158.
34. Malvadkar, N.A., Sekeroglu, K., Dressick, W.J., and Demirel, M.C. (2011) *J. Power Sources*, **196**, 8553.
35. (a) Bartolo, D., Bouamrirene, F., Verneuil, É., Buguin, A., Silberzan, P., and Moulinet, S. (2006) *Europhys. Lett.*, **74**, 299;(b) Reyssat, M., Pépin, A., Marty, F., Chen, Y., and Quéré, D. (2006) *Europhys. Lett.*, **74**, 306.
36. (a) Huang, J., Juszkiewicz, M., de Jeu, W.H., Cerda, E., Emrick, T., Menon, N., and Russel, T.P. (2007) *Science*, **317**, 650;(b) Py, C., Reverdy, P., Doppler, L., Bico, J., Roman, B., and Baroud, C.N. (2007) *Phys. Fluids*, **19**, 91104.
37. Py, C., Reverdy, P., Doppler, L., Bico, J., Roman, B., and Baroud, C.N. (2007) *Phys. Rev. Lett.*, **98**, 156103.
38. Beighley, R., Spedden, E., Sekeroglu, K., Atherton, T., Demirel, M.C., and Staii, C. (2012) *Appl. Phys. Lett.*, **101**, 143701.
39. Spedden, E., Wiens, M.R., Demirel, M.C., and Staii, C. (2014) *PLoS One*, **9**, e106709.
40. Newnham, R.E. (2005) *Properties of Materials: Anisotropy, Symmetry, Structure*, Oxford University Press, Oxford.
41. (a) Demirel, M.C., Kuprat, A.P., George, D.C., and Rollett, A.D. (2003) *Phys. Rev. Lett.*, **90**, 016106; (b) Kuprat, A., George, D., Straub, G., and Demirel, M.C. (2003) *Comput. Mater. Sci.*, **28**, 199.
42. Atilgan, A.R., Durell, S.R., Jernigan, R.L., Demirel, M.C., Keskin, O., and Bahar, I. (2001) *Biophys. J.*, **80**, 505.
43. Yao, H. and Gao, H. (2006) *J. Mech. Phys. Solids*, **54**, 1120.
44. Kamperman, M., Kroner, E., del Campo, A., McMeeking, R.M., and Arzt, E. (2010) *Adv. Eng. Mater.*, **12**, 335.
45. (a) Autumn, K., Sitti, M., Liang, Y.A., Peattie, A.M., Hansen, W.R., Sponberg, S., Kenny, T.W., Fearing, R., Israelachvili, J.N., and Full, R.J. (2002) *Proc. Natl. Acad. Sci. U.S.A.*, **99**, 12252;(b) Autumn, K., Liang, Y.A., Hsieh, S.T., Zesch, W., Chan, W.P., Kenny, T.W., Fearing, R., and Full, R.J. (2000) *Nature*, **405**, 681; (c) Gao, H. and Yao, H. (2004) *Proc. Natl. Acad. Sci. U.S.A.*, **101**, 7851; (d) Gao, H., Wang, X., Yao, H., Gorb, S., and Arzt, E. (2005) *Mech. Mater.*, **37**,

275;(e) Persson, B.N.J. (2003) *J. Chem. Phys.*, **118**, 7614;(f) Spolenak, R., Gorb, S., Gao, H., and Arzt, E. (2005) *P. Roy. Soc. A - Math. Phy.*, **461**, 305; (g) Huber, G., Gorb, S.N., Spolenak, R., and Arzt, E. (2005) *Biol. Lett.*, **1**, 2.

46. Mahadevan, L., Daniel, S., and Chaudhury, M.K. (2004) *Proc. Natl. Acad. Sci. U.S.A.*, **101**, 23.
47. Bush, J.W.M., Prakash, M., and Hu, D.L. (2007) *Adv. Insect Physiol.*, **34**, 117.
48. de Gennes, P.-G. (1985) *Rev. Mod. Phys.*, **57**, 827.
49. de Gennes, P.-G., Brochard-Wyart, F., and Quéré, D. (2003) *Capillarity and Wetting Phenomena: Drops, Bubbles, Pearls, Waves*, Springer, Berlin.
50. (a) Adamson, A.W. (1982) *Physical Chemistry of Surfaces*, John Wiley & Sons, Inc., New York; (b) Dussan, E.B.V. (1979) *Annu. Rev. Fluid Mech.*, **11**, 371.
51. Extrand, C.W. (2007) *Langmuir*, **23**, 1867.
52. Balazs, A.C. and Yeomans, J.M. (2010) *Soft Matter*, **6**, 703.
53. Young, T. (1805) *Philos. Trans. R. Soc. London, Ser. A*, **95**, 65.
54. (a) Johnson, R.E. and Dettre, R.H. (1964) *Contact Angle, Wettability, and Adhesion*, Advances in Chemistry Series, vol. **43**, American Chemical Society, Washington, DC, p. 112;(b) Dettre, R.H. and Johnson, R.E. (1964) in *Contact Angle, Wettability, and Adhesion*, Advances in Chemistry, vol. **43** (ed F.M. Fowkers), American Chemical Society, Washington, DC, p. 136;(c) Dettre, R.H. and Johnson, R.E. (1965) *J. Phys. Chem.*, **69**, 1507.
55. (a) Dussan, V.E.B. and Chow, R.T.P. (1983) *J. Fluid Mech.*, **137**, 1;(b) Extrand, C.W. and Kumagai, Y. (1995) *J. Colloid Interface Sci.*, **170**, 515.
56. Nadkarni, G.D. and Garoff, S. (1992) *Europhys. Lett.*, **20**, 523.
57. Wenzel, R.N. (1936) *Ind. Eng. Chem.*, **28**, 988.
58. (a) Cassie, A.B.D. and Baxter, S. (1944) *Trans. Faraday Soc.*, **40**, 546;(b) Cassie, A.B.D. and Baxter, S. (1945) *Nature*, **155**, 21.
59. Krasowska, M., Krastev, R., Rogalski, M., and Malysa, K. (2007) *Langmuir*, **23**, 549.
60. Checco, A., Hofmann, T., DiMasi, E., Black, C.T., and Ocko, B.M. (2010) *Nano Lett.*, **10**, 1354.
61. Mao, M., Zhang, J., Yoon, R.H., and Ducker, W.A. (2004) *Langmuir*, **20**, 1843.
62. Extrand, C.W. (2002) *Langmuir*, **18**, 7991.
63. Shastry, A., Case, M.J., and Böhringer, K.F. (2006) *Langmuir*, **22**, 6161.
64. Brochard, F. (1989) *Langmuir*, **5**, 432.
65. Nosonovsky, M. (2007) *Langmuir*, **23**, 3157.
66. Joanny, J.F. and de Gennes, P.G. (1984) *J. Chem. Phys.*, **81**, 552.
67. Pomeau, Y. and Vannimenus, J. (1985) *J. Colloid Interface Sci.*, **104**, 477.
68. Reyssat, M. and Quéré, D. (2009) *J. Phys. Chem. B*, **113**, 3906.
69. (a) Gao, L. and McCarthy, T.J. (2006) *Langmuir*, **22**, 2966;(b) Dorrer, C. and Rühe, J. (2006) *Langmuir*, **22**, 7652.
70. ElSherbini, A. and Jacobi, A. (2006) *J. Colloid Interface Sci.*, **299**, 841.
71. Acheson, D.J. (1990) *Elementary Fluid Dynamics*, Oxford University Press.
72. Brakke, K.A. (1992) *Exp. Math.*, **1**, 141.
73. Chen, Y., He, B., Lee, J., and Patankar, N.A. (2005) *J. Colloid Interface Sci.*, **281**, 458.
74. Dorrer, C. and Rühe, J. (2008) *Langmuir*, **24**, 1959.
75. Dorrer, C. and Rühe, J. (2007) *Langmuir*, **23**, 3179.
76. Kwon, Y., Choi, S., Anantharaju, N., Lee, J., Panchagnula, M.V., and Patankar, N.A. (2010) *Langmuir*, **26**, 17528.
77. Blow, M.L., Kusumaatmaja, H., and Yeomans, J.M. (2009) *J. Phys. Condens. Matter*, **21**, 464125.
78. Hyväluoma, J., Koponen, A., Raiskinmäki, P., and Timonen, J. (2007) *Eur. Phys. J. E*, **23**, 289.
79. Mognetti, B.M., and Yeomans, J. M. (2010) *Langmuir,*, **26**, 18162.
80. (a) Brandon, S., Wachs, A., and Marmur, A. (1997) *J. Colloid Interface Sci.*, **191**, 110;(b) Gau, H., Herminghaus, S., Lenz, P., and Lipowsky, R. (1999) *Science*, **283**, 46;(c) Darhuber, A.A., Troian, S.M., Davis, J.M., Miller, S.M., and Wagner, S. (2000) *J. Appl. Phys.*, **88**, 5119;(d) Brinkmann, M. and Lipowsky, R. (2002) *J. Appl. Phys.*, **92**, 4296;(e) Brandon, S.,

Haimovich, N., Yeger, E., and Marmur, A. (2003) *J. Colloid Interface Sci.*, **263**, 237.

81. Ruiz-Cabello, F.J.M., Kusumaatmaja, H., Rodríguez-Valverde, M.A., Yeomans, J., and Cabrerizo-Vílchez, M.A. (2009) *Langmuir*, **25**, 8357.
82. (a) Dorrer, C. and Rühe, J. (2007) *Langmuir*, **23**, 3820; (b) Moulinet, S. and Bartolo, D. (2007) *Eur. Phys. J. E*, **24**, 251.
83. Choi, W., Tuteja, A., Mabry, J.M., Cohen, R.E., and McKinley, G.H. (2009) *J. Colloid Interface Sci.*, **339**, 208.
84. (a) Bienia, M., Quilliet, C., and Vallade, M. (2003) *Langmuir*, **19**, 9328; b Klingner, A., Buehrle, J., and Mugele, F. (2004) *Langmuir*, **20**, 6770.
85. (a) Jin, W., Koplik, J., and Banavar, J.R. (1997) *Phys. Rev. Lett.*, **78**, 1520; (b) Collet, P., De Coninck, J., Dunlop, F., and Regnard, A. (1997) *Phys. Rev. Lett.*, **79**, 3704; (c) Lundgren, M., Allan, N.L., and Cosgrove, T. (2006) *Langmuir*, **23**, 1187.
86. (a) Luo, K., Kuittu, M.-P., Tong, C., Majaniemi, S., and Ala-Nissila, T. (2005) *J. Chem. Phys.*, **123**, 194702; (b) Vedantam, S. and Panchagnula, M.V. (2007) *Phys. Rev. Lett.*, **99**, 176102; (c) Vedantam, S. and Panchagnula, M.V. (2008) *J. Colloid Interface Sci.*, **321**, 393.
87. (a) Do-Quang, M. and Amberg, G. (2010) *Math. Comput. Simul.*, **80**, 1664; (b) Do-Quang, M. and Amberg, G. (2009) *Phys. Fluids*, **21**, 022102.
88. (a) Anderson, D.M., McFadden, G.B., and Wheeler, A.A. (1998) *Annu. Rev. Fluid Mech.*, **30**, 139; (b) Jacqmin, D. (2000) *J. Fluid Mech.*, **402**, 57; (c) Nguyen, H.-B. and Chen, J.-C. (2010) *Phys. Fluids*, **22**, 062102; (d) Scardovelli, R. and Zaleski, S. (1999) *Annu. Rev. Fluid Mech.*, **31**, 567; (e) Das, A.K. and Das, P.K. (2010) *Langmuir*, **26**, 9547.
89. Fakhari, A. and Rahimian, M.H. (2010) *Phys. Rev. E*, **81**, 036707.
90. Kusumaatmaja, H. and Yeomans, J.M. (2010) in *Simulating Complex Systems by Cellular Automata* (eds A.G. Hoekstra, J. Kroc, and P.M.A. Sloot), Springer-Verlag, Berlin, p. 241.
91. Kuksenok, O., Jasnow, D., Yeomans, J., and Balazs, A.C. (2003) *Phys. Rev. Lett.*, **91**, 108303.
92. Kusumaatmaja, H., Blow, M.L., Dupuis, A., and Yeomans, J.M. (2008) *Europhys. Lett.*, **81**, 36003.
93. Mognetti, B.M. and Yeomans, J.M. (2009) *Phys. Rev. E*, **80**, 056309.
94. Hu, D.L. and Bush, J.W.M. (2005) *Nature*, **437**, 733.
95. Dressick, W.J., Kondracki, L.M., Chen, M.-S., Brandow, S.L., Matijević, E., and Calvert, J.M. (1996) *Colloids Surf., A*, **108**, 101.

13
Membrane Modification by CVD Polymers

Rong Yang

Membranes, organic or inorganic, are porous materials that enable the separation of components in a fluid based on their size and chemical or physical affinities. Membrane-based separation processes are efficient, effective, and environmentally friendly. The application of membranes spans almost every aspect of the society, ranging from the production of energy and fresh water to implantable medical devices.

The major limitations of the current membrane separation technologies are low selectivity and severe fouling. These limitations are rooted in the size-exclusion separation mechanism, which serves as the most prevalent mechanism for industrial separation processes. For a size-exclusion separation, it is extremely challenging and costly to fabricate membranes with monodispersed pore sizes (i.e., membranes with 100% size selectivity), and therefore the separation selectivity is compromised. Fouling refers to the nonspecific attachment of organic matters, biomolecules, and microorganisms on a submerged surface. During a membrane separation process, the rejected large molecules and microorganisms tend to deposit on feed side of the membrane, form a biofilm, and can develop into a dense cake in severe cases. Fouling blocks the selective pores, contaminates the permeate stream, and even degrades certain polymeric membranes. Therefore, the significant economic and environmental penalties demand novel material designs that lead to highly selective and fouling-resistant membranes.

Surface modification of a membrane decouples the bulk properties and surface properties and enables their respective independent optimization, which can lead to highly specific and efficient separation and reduced fouling propensity. In addition, novel separation mechanisms can be realized via surface modification. Instead of size-exclusion membranes, ones that separate constituents based on their chemical, physical, and molecular characteristics have been fabricated, allowing the recognition and separation of molecules with comparable sizes and thus very specific selectivity.

A surface modification method that is compatible with porous membrane substrates should have the following characteristics:

CVD Polymers: Fabrication of Organic Surfaces and Devices, First Edition. Edited by Karen K. Gleason.

1) conformal, that is, coatings with uniform thickness should be formed on membrane top and inner pore surfaces;
2) benign, that is, a method should not damage membrane substrates during the modification;
3) scalable, that is, a method should be compatible with industrial infrastructures.

A vapor-phase method prevents the undesirable surface tension effects and dewetting that can lead to an uneven coating or partial coverage on porous substrates as well as solvent-induced degradation or swelling of membrane substrates. However, application of high-energy plasma in certain chemical vapor deposition (CVD) methods can damage delicate membrane substrates. Initiated CVD (iCVD) and photoinitiated CVD (piCVD) keep the substrates at room temperature and moderate vacuum during the surface modification, and thus are more appropriate for delicate membrane substrates, such as reverse osmosis (RO) membranes used in seawater desalination. In addition, the ability of iCVD to synthesize controllably cross-linked thin films in a single step and the excellent control of coating thickness are two important factors in regulating the permeability and thus selectivity of various fluid constituents.

The rich library of organic functionalities that are compatible with CVD has enabled numerous novel membrane functions, including antifouling, self-cleaning, antimicrobial, and stimuli-responsive properties, as well as highly efficient separations based on specific molecular interactions. A selection of the functional monomers and copolymers used in novel membrane design and fabrication is shown in Figure 13.1.

(Hydroxyethyl)methacrylate (HEMA) polymerizes into a hydrophilic polymer that has been used to mitigate fouling [1–6]. 2-(Dimethylamino)ethyl methacrylate (DMAEMA) is a precursor monomer to synthesize an antifouling zwitterionic copolymer (Figure 13.1) to prevent fouling [7, 8]. Ethylene glycol dimethacrylate (EGDMA) is the cross-linker incorporated in the zwitterionic copolymer to avoid dissolution of the extremely hydrophilic zwitterionic polymer [7, 8]. 4-Vinylpyridine (4VP) and divinylbenzene (DVB) have been copolymerized similarly to synthesize a very stable and antifouling coating for desalination membranes [9]. 1*H*, 1*H*, 2*H*, 2*H*-perfluorodecyl acrylate (PFDA) represents a typical hydrophobic monomer, which has been used to fabricate membranes with unprecedented capability of separating molecules base on their hydrophobicity [10]. Dimethylaminomethylstyrene (DMAMS) is a antimicrobial monomer that can mitigate bacterial growth on membrane surfaces [11, 12].

In this chapter, we start by discussing two different membrane modification strategies using iCVD, specifically, the modification of membrane top surface and internal pores (Section 13.1). The goal of many membrane surface modifications can be categorized into one of the following two types: surface energy control, which will be the focus of Section 13.2, and mitigation of fouling, which will be discussed in Section 13.3. Finally, we end with an overview of the thin-film-coated membranes for sustainability, specifically for water purification and energy production and storage.

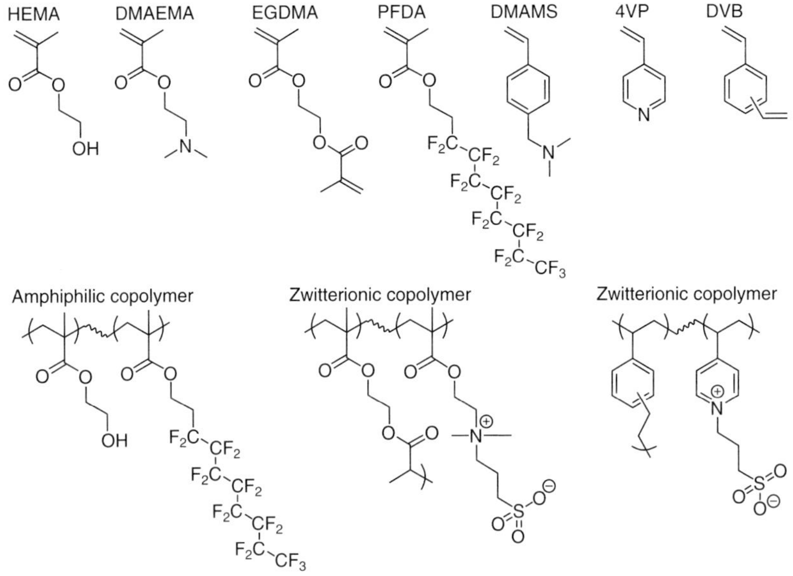

Figure 13.1 A selection of functional monomers and copolymers for novel membrane design and fabrication via CVD methods.

13.1 Modification of Membrane Surface and Internal Pores

13.1.1 Conformal Coatings for Membrane Surface Modification

The uniform coverage of the internal pores (Figure 13.2a) enabled by excellent coating conformality is highly desirable for the simultaneous control of pore size and membrane surface chemistry. As we discussed in Chapter 5, a monomer with small sticking probability allows the initiator radical to travel further down the pore by bouncing off the side walls without "sticking" (i.e., reacting with the vinyl bonds), and therefore leads to more uniform coverage along the side walls. The SEM image in Figure 13.2a demonstrates the conformal coverage achieved on microtrenches cut into a silicon wafer. The step coverage (i.e., ratio of coating thickness on top and bottom of the trench) is close to 1 in this SEM image. Such conformal coatings allow for the tuning of the membrane pore sizes and surface chemistries simultaneously.

In general, conformal coatings can be achieved virtually with any functional monomer by optimizing the deposition conditions (the reader is directed to Chapter 5 for detailed discussion on coating conformality control). Monomers with small propagation constants tend to have low sticking probabilities, and thus more conformal and even coverage. For instance, PFDA (Figure 13.1) is a relatively reactive acrylate monomer, and poly-(1*H*, 1*H*, 2*H*, 2*H*-perfluorodecylacrylate)

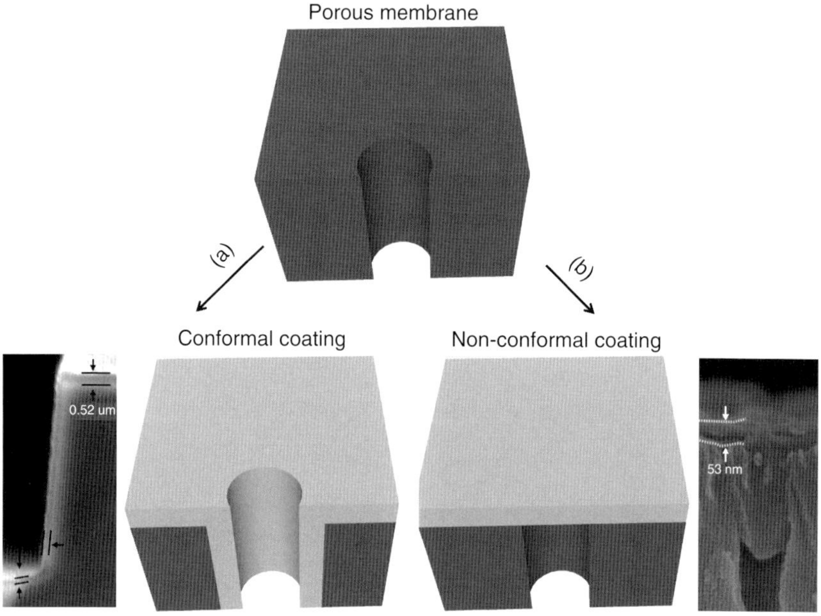

Figure 13.2 Two different membrane modification strategies using iCVD. (a) A conformal coating covers the membrane top surface and internal pores, allowing the simultaneous control over surface chemistry and membrane pore size. The step coverage is close to 1 in the SEM image. (b) A nonconformal coating covers the membrane top surface only, where the membrane surface becomes nonporous and membrane selectivity is dominated by the coating layer.

(pPFDA) coating may decrease in thickness proceeding into the pore (Figure 13.3, top right), resulting in a bottleneck cross section. However, upon optimizing the deposition conditions, pPFDA coating can become very conformal and may penetrate to the bottom of very small pores with aspect ratios as high as 4000 [10, 13]. Alternatively, a double-layer fabrication method can be used to achieve exceptionally conformal coatings (Figure 13.3), where cross-linked film of poly(divinyl benzene) (pDVB) (Figure 13.1) can be deposited first to narrow the pores in a controlled manner. Once the pore size is narrowed to a desirable diameter, a second coating of pPFDA can be deposited to render the inner pores hydrophobic. Cylindrical cross sections and hydrophobic pores with precise diameters on a nanometer scale can be achieved with this scheme. As-fabricated membranes can separate pairs of small molecules of similar size, but of different hydrophobicity. The separation mechanism is conceptually different from the common size-based separation membranes in that the selectivity arises from solute–pore wall interactions emphasized by the nano-confinement, which is only possible via the simultaneous and precise control of pore size and size wall chemistry.

The capability to modify the inner walls of membrane pores has enabled numerous novel membrane designs and beyond. A “smart” membrane has been fabricated using stimuli-responsive polymers, where autonomous control of protein

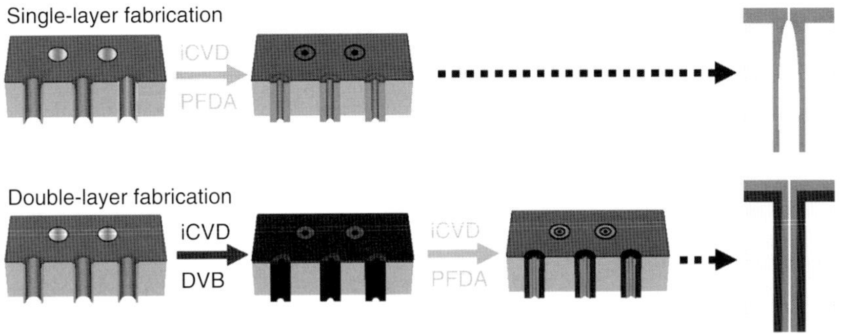

Figure 13.3 Two distinct fabrication routes to a hydrophobic membrane for the separation of pairs of small molecules of similar size but different hydrophobicity. Single-layer fabrication involves a single deposition of pPFDA, which results in a bottleneck cross-section due to the high sticking probability of the PFDA monomer. The double-layer fabrication involves two sequential depositions of pDVB and pPFDA. The first layer (i.e., pDVB) narrows the pore size to the desirable level, and a second layer (i.e., pPFDA) renders the inner walls hydrophobic. The independent control of pore sizes and inner wall chemistry can therefore be realized.

flow through the membrane upon external temperature stimuli is realized [14]. Furthermore, membrane pores can be used as a template, where the conformal polymeric films grow to form cylindrical tubes [15]. The membrane is later dissolved to extract the polymer tubes to be used in drug delivery [16], bioimaging [16], and sensing [17]. The reader is directed to Chapter 6 for detailed discussion on templating and micromolding in iCVD using membrane substrates.

13.1.2
Nonconformal Coatings for Membrane Surface Modification

Deposition conditions can be tuned so that the coating covers merely the top surface without going into the pores, where a thin-film composite (TFC) membrane can be fabricated (Figure 13.2b). The unique nonconformal coverage was achieved by increasing the surface concentrations of the monomer precursors. During a nonconformal deposition, monomers consume initiator radical rapidly on the topmost membrane surface, form a blanket layer that bridges the pore openings, blocks the access of initiator radicals to the pores, and prevents further polymerization on the inner walls of pores. A rich library of polymers can be used as the skin layers for specific separations, whereas the membrane substrates act merely as a mechanical support for the thin films. Very rapid permeation rates are possible in this case due to the thin or ultra-thin nature of the coating layer. For size-selective membranes fabricated in this manner, selectivity can be tuned easily by changing the cross-linking density of the skin layer. For instance, prototype TFC membranes made of permeable, size-selective hydrogel thin films on top of porous alumina membrane supports have been fabricated (Figure 13.2b) [18].

Sub-100 nm thickness of the hydrogel top layer allowed the rapid passage of glucose while excluding bovine serum albumin (BSA).

Flux, selectivity, and durability are the three major performance parameters one should consider for the design of TFC membranes. To achieve a high membrane *flux,* the coating layer has to be thin or ultra-thin while maintaining the blanket coverage for good selectivity. For an iCVD system, film thickness can be monitored through *in situ* interferometry system and controlled with the precision of tens of nanometers. In the example in Figure 13.2b, smooth hydrogel films with sub-100 nm thicknesses were readily synthesized via iCVD in the absence of micrometer-sized pinholes. Ultra-thin (30 nm) coatings have been deposited on commercial RO membranes and water flux of the coated membrane was only reduced by 14% [9], whereas a 100-nm coating resulted in ~50% flux reduction [7]. Lower cross-linking density can also lead to greater flux [7]. However, the selectivity and durability of the skin layer can be adversely impacted upon reducing the cross-linking density. Membrane *selectivity* relies on a pinhole-free top layer, and the selective permeation of specific molecules can be controlled by incorporating new functionalities or tuning the cross-linking density of the skin layer. The top selective layer in Figure 13.2b is a terpolymer, poly[maleic anhydride-*co*-dimethylacrylamide-*co*-di(ethylene glycol) di(vinyl ether)] where maleic anhydride is incorporated to enhance ionic swelling, dimethylacrylamide acts as a water absorbent, and di(ethylene glycol) di(vinyl ether) renders the terpolymer insoluble in water. *Durability* increases with higher cross-linking density, however, at the expense of possibly losing the high flux or selectivity of a membrane. For the case of hydrogel (Figure 13.2b), swelling is necessary for the membrane selectivity, but has to be controlled within limits by the incorporation of cross-linkers because swelling leads to compromised mechanical properties and polymer durability. Delamination of the skin layer from the underlying membrane supports is another common failure mechanism. Some common grafting methods to covalently link the skin layer to the membrane substrates include silane treatment of silicon and alumina membranes [18], passivation of silicon membranes [19], plasma treatment, and organic functionalization of polymeric membranes [7]. The reader is directed to Chapter 4 for detailed discussion on various grafting methods and mechanisms that are compatible with the CVD process.

The intermediate case between conformal (Figure 13.2a) and nonconformal (Figure 13.2b) coatings, where the coating penetrates only the top part of the pores, but still leaves the pores open, can also be a very useful membrane configuration. For example, membrane distillation (MD) for seawater desalination requires a membrane that allows the permeation of water vapor, but not liquid water and thus is able to desalinate feed using low-grade thermal energy sources. Such membranes can be fabricated by coating the inner pores of a membrane with hydrophobic materials, but only to a certain length along the pores. The coated portion should have a diameter that is narrow enough to block liquid water but wide enough to allow rapid transport of water vapor. In addition, the length of the coated portion should be short enough to reduce the transport resistance of water vapor. A hydrophobic bottle neck at the entrance of the pore (Figure 13.3) can meet these conflicting needs very well.

13.2 Membrane Surface Energy Control Via Thin-Film Coatings

Surface energy is a critical and sometimes dominating factor for membrane selectivity and lifetime. The hydrophilicity/hydrophobicity of a specific surface is determined by its chemistry and roughness. For membranes, the material, membrane porosity, pore size, and roughness can all have significant impacts on the overall hydrophilicity/hydrophobicity of a membrane surface. The application of a thin coating is a versatile method to control membrane surface interactions by deconvoluting surface energy from the other membrane properties, and thus represents a powerful tool in the fundamental membrane research. In this section, we discuss hydrophobic and hydrophilic surface modifications separately, and end with a brief discussion on stimuli-responsive coatings.

13.2.1 Hydrophobic Thin-Film Coatings for Membranes

Hydrophobic or superhydrophobic membranes are in great demand due to their water repellency, self-cleaning capability, and potential of realizing molecular property-based separations. However, the application of hydrophobic materials to porous substrates is very challenging for solution-based methods because of the insolubility of fluorinated molecules and difficulty in processing and forming uniform layers inside the membrane pores.

Hydrophobic polymer pPFDA (Figure 13.1) has been one of the most prevalent chemistry for the modification of membranes to achieve a low-energy surface. It has been applied to a variety of substrates (Figure 13.4), including high aspect ratio (80 : 1) capillary pore membranes (Figure 13.4a), with the coating thickness ranging from 10 to 150 nm [13]. Presence of the polymer coating along the pore wall was evidenced by X-ray photoelectron spectroscopy (XPS), scanning electron microscopy (SEM), and contact angle measurements (Figure 13.4a, inset). The greatest water contact angle was achieved with a membrane of 50% porosity. After the iCVD surface modification, water contact angles were ~151° on the top side and ~152° on the bottom side of membrane, which represents an initial example of utilizing surface energy and roughness simultaneously to achieve greater hydrophobicity.

Systematic studies have been enabled by the substrate-independent surface modification technologies to understand the complementary roles of surface energy and roughness, where membranes with a variety of pore sizes and porosities were coated with fluoropolymers. A composite membrane made of nanoscale fibers of poly(caprolactone) (pCL) and micron-scale beads have been modified with a thin layer (70 – 100 nm) of poly(perfluoroalkyl ethyl methacrylate) (pPFEMA, Figure 13.4b) [20]. With a water contact angle of 119° on planar

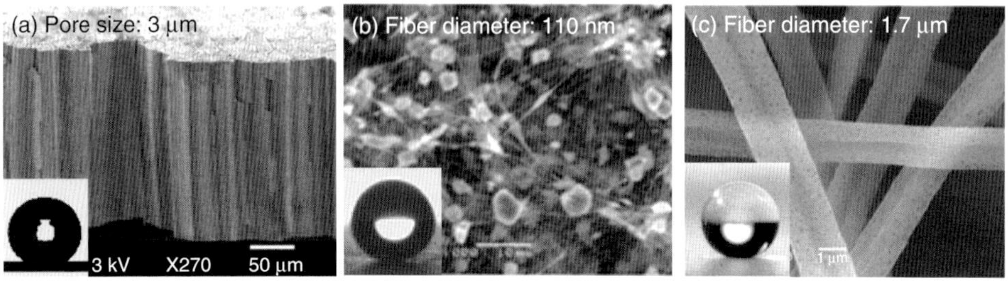

Figure 13.4 Micrographs of surface-modified membranes with very low surface energies. (a) Capillary pore membranes modified with pPFDA coatings, whose water contact angles were ~151° (inset) on the membrane top side and ~152° on the bottom side. (b) An iCVD-modified composite membrane made of nanoscale fibers of poly(caprolactone) (PCL) and micron-scale beads. The pPFEMA coating resulted in a superhydrophobic membrane surface with a contact angle of 175° (inset). (c) iCVD-modified hierarchically structured membranes. After the iCVD modification with the pPFEMA coating, the advancing and receding contact angles were increased by 163° (inset) and 160°, respectively.

substrates, the pPFEMA achieved a superhydrophobic surface with a contact angle of 175° on the membrane surface with the twofold surface structures (Figure 13.4b, inset). In this example, the first level of roughness (associated with the fibers) and the second level of roughness (associated with the beads) were essentially mixed rather than hierarchical. Hierarchical structures can be fabricated directly on the membrane surface to further increase hydrophobicity. In one such example, nanometer-scale pore structures were introduced directly onto fibers, which were then woven into a membrane, and coated with pPFEMA via iCVD [21]. The decorated membrane had an advancing water contact angle of 147° and a receding contact angle of 74°; whereas after the iCVD modification, the advancing and receding contact angles were increased by 163° (Figure 13.4c, inset) and 160°, respectively. Therefore, high water contact angles and low contact angle hysteresis, both of which are highly desirable for the next-generation membrane design with self-cleaning properties, can be obtained simultaneously via the surface modification with fluoropolymers.

The effects of iCVD fluoropolymer coatings are summarized in Table 13.1, where the advancing water contact angles before and after the surface modifications are listed. The complementary effects of roughness and surface energy are illustrated by the "critical length" column in Table 13.1. The "critical length" refers to the pore size for the capillary pore membrane (Figure 13.4a); whereas for the fiber mats, the "critical length" represents the diameters of fibers (Figure 13.4b,c).

13.2.2
Hydrophilic Thin-Film Coatings for Membranes

At the other end of the surface energy spectrum, hydrophilic or superhydrophilic surfaces are also highly desirable in various membrane-related applications, including biomedical implants [22], tissue engineering [23], drug delivery [24],

Table 13.1 Water contact angles of membranes before and after the iCVD modification.

Substrate	Water contact angle before modification (°)	Water contact angle after modification (°)	Critical length (μm)
Fabrics with beads	139	175	0.11
Fiber mats	147	163	1.7
Capillary pore membranes	105	151	3

and biosensor fabrication [25] due to their good antifouling performance and extraordinary biocompatibility. Hydrophilic nanostructures (e.g., hydrogel nanotubes) with large surface area to volume ratio are desired for tissue engineering, drug release, cell culture, and fundamental studies on protein adsorption mechanisms, and can be fabricated conveniently and economically via membrane templates [23, 26]. The reader is directed to Chapter 6 for detailed discussion on templating and micromolding in iCVD using membrane substrates.

Modification of surfaces with a thin hydrophilic layer has been a common method to limit nonspecific adsorption of biomacromolecules (e.g., protein, polysaccharide) and microorganisms. Hydrophilic surfaces tend to have less fouling because the enthalpic driving force for the adsorption of the protein at the interface is lower. By applying hydrophilic thin films via vapor-phase treatment, antifouling and wetting behavior of virtually any membrane substrate can be changed without affecting bulk properties. Poly(hydroxyethyl methacrylate) (pHEMA) (Figure 13.1) and polyzwitterion (Figure 13.1) are the two most prevalent hydrophilic materials for membranes.

Cross-linking is frequently introduced to pHEMA films to control the mechanical properties, even though linear pHEMA chains do not completely dissolve in water [27]. Plasma enhanced chemical vapor deposition (PECVD) methods have long been used to form cross-linked pHEMA coatings via high-energy plasma excitation [28–31], and cross-linking density can be tuned with the power of plasma. However, surface composition of as-deposited pHEMA coatings deviated from solution-phase synthesized pHEMA due to the loss of hydroxyl groups during the high-energy PECVD process [29]. pHEMA deposited on nonwoven poly(butyleneterephtalate) filters using a discharge power as low as 40 W resulted in a water contact angle of $59 \pm 3°$ [29]. The water contact angles of the plasma-coated filters increased slightly with the higher discharge power. Thin films of pHEMA deposited using iCVD [6, 32, 33] and piCVD yields 100% retention of the hydroxyl functionality at deposition rates greater than 100 nm min^{-1} [34, 35]. The mechanical properties can be systematically controlled through the metered addition of cross-linkers (e.g., ethylene glycol diacetate (EGDA)) during the deposition.

Zwitterionic materials represent a new category of hydrophilic or superhydrophilic chemistry with contact angles close to 0°. The strong electrostatic hydration of the zwitterionic surface is considered to be the origin of its superhydrophilicity [36, 37]. Poly(sulfobetaine) (pSB)-based zwitterionic coatings have been synthesized using iCVD and grafted onto RO membranes through a vapor-phase reaction [7]. Antifouling properties of as-fabricated zwitterionic coating were tested with static bacterial adhesion, and the coated RO membrane exhibited no adhesion of *Escherichia coli*. In this example, the sulfobetaine moieties were afforded by a methacrylate monomer, which is prone to degradation under oxidative conditions. The methacrylate monomer was later substituted by a pyridine monomer, which demonstrated very good resistance to oxidative damages [9]. These coatings serve as excellent inert surfaces to render fouling-resistant membranes, which is discussed in greater details in Section 13.3.

A final category of coatings for controlling surface energies of membrane substrates is made of stimuli-responsive polymers, which can lead to dynamic surfaces or internal pores with switchable surface energies upon external stimuli, such as change in temperature, pH values, humidity, electric current, or feed composition. Many stimuli-responsive polymers have been synthesized via iCVD, and the reader is directed to Chapter 9 for detailed discussion on stimuli-responsive polymer chemistry that is compatible with the solventless method. Membranes that respond to external stimuli can serve as a smart valve to control the flow rate or pore size (i.e., selectivity of a membrane) in response to parameters such as pH or concentration of specific species [18]. Stimuli can also serve as a tool to regenerate and/or clean membranes by attaching and detaching specific molecules upon stimuli [38, 39].

13.3 Antifouling and Antimicrobial Coatings for Membranes

Biofouling is a universal problem for membrane separation and membrane chromatography, which can cause reduction in permeate flux, selectivity, and lifetime of a membrane. Biofouling refers to the undesirable settlement of organic matters, biomolecules, and growth of microorganisms, which occurs immediately when a clean surface is immersed in a nonsterile aqueous solution. Membrane separation processes are particularly susceptible to the fouling problem because it leads to blockage of selective pores, formation of a thick biofilm on top of the membrane, contamination of the permeate stream, and degradation of the membrane material. A number of approaches have been proposed or adopted in research and industry to mitigate the fouling problem. In industrial processes, fouling can be reduced or delayed by pretreating the feed stream or periodic cleaning of the membrane. However, both methods are costly, time consuming, and pose serious environmental concerns. Therefore, the design and fabrication of nonfouling membranes become one of the most important topics in membrane research.

The design rules for an inert (i.e., antifouling) surface have been generalized to three molecular-level characteristics: [40, 41]

1) the surface is hydrophilic;
2) the surface contains hydrogen bond acceptor groups but not hydrogen bond donor groups;
3) the net charge of the surface is neutral.

This set of properties was identified by screening self-assembled monolayers (SAMs) with approximately 60 different surface chemistries. Conformational flexibility was the characteristic of many of these surface chemistries, but it was found nonessential [40]. These rules have also been observed in other antifouling chemistries discovered later on. For example, the ultra-low fouling zwitterionic coatings via SAMs [41], atom transfer radical polymerization (ATRP) [37], and vapor-phase method [7] exhibited the same traits. It is worth noting that the rules apply to most, but not all of the antifouling surfaces. For instance, amphiphilic surfaces with compositional heterogeneities on the length scale of a foulant of interest may discourage the thermodynamically favorable interactions between the foulant and the surface, which in turn would limit adsorption events [42–44].

With the rapidly growing number of antifouling surface chemistries, come the challenges of applying such functionalities to a much wider range of membrane substrates and more sophisticated membrane structures. SAMs, solution polymerization and solvent evaporation, and ATRP are the most common methods for membrane surface modifications. However, these methods require specific surface functionalities, which are not possessed by all membrane substrates or could not be possessed through pretreatment. In addition, they generally involve harsh process conditions such as organic solvents, which may cause damage to delicate membrane substrates. For example, RO membranes have shown an 8- to 10-fold increase in Knudsen diffusion-based gas permeance after soaking in ethanol for merely 5 min, which was attributed to an increase in the number or size of membrane defects generated by the ethanol treatment [45]. The nonconformal nature resulted from the surface tension effects of solvents renders various solution-phase synthesis methods inadequate to achieve complete coverage of porous membranes. Solution polymerization and solvent evaporation may also lead to high surface roughness, which is undesirable for the purpose of antifouling.

Using a vapor-phase deposition method avoids the potential for the damages by solvents. A variety of antifouling coatings have been synthesized via iCVD [46], among which amphiphilic and zwitterionic coatings give rise to the best fouling resistance (Table 13.2). The reader is directed to reference [46] for detailed comparison of the amphiphilic and zwitterionic surfaces with other hydrophilic coatings synthesized via iCVD, such as pHEMA.

Amphiphilic surfaces are smooth surfaces with compositional heterogeneities at a microorganism or molecular length scale to disrupt surface–microorganism or surface–protein interactions. Such surfaces can be formed by the self-assembly of amphiphilic macromolecules via microphase separation [41, 47–49]. However, in general, the preparation of such films is challenging because it requires the copolymerization of hydrophilic and hydrophobic monomers, which have

Table 13.2 Antifouling capabilities of various iCVD antifouling thin films.

Foulant	Fouling reduction (%)		
	Zwitterionic polymethacrylate	Zwitterionic polypyridine	Amphiphilic copolymer
BSA	93	>99	88
Sodium alginate	>99	>99	51
E. coli	>99	n.a.	98
V. cyclitrophicus	n.a.	99	n.a.

n.a., not applicable

highly contrasting solubility and lack a common solvent. CVD-based patterning offers a promising alternative in creating amphiphilic coatings with alternating hydrophilic and hydrophobic regions of customizable size and shape, as it allows the use of a multitude of patterning techniques [50–54]. For example, iCVD was used in the preparation of micropatterned surfaces with independently customizable topographic and chemical contrast [50]. A blanket film of pHEMA was deposited first, and then etched with plasma under a mask such as a transmission electron microscopy (TEM) grid. A hydrophobic film such as pPFDA was deposited in the second step and the surface become topologically and chemically patterned with amphiphilic materials. A similar fabrication approach was used in conjunction with pulsed plasma CVD although without the same topological control. The high-energy plasma can render the fabrication incompatible with delicate membrane substrates [54].

Preventing protein adsorption with this approach, however, presents a greater challenge, because the chemical heterogeneities need to be at a scale that is smaller than the size of a protein. Despite the area dimensions of 10–1000 nm^2 for typical protein molecules, the surface interacting area has been calculated to be only 1–2 nm^2 for a variety of proteins [55], indicating that the scale of chemical patterning needs to be on the order of ~1 nm. iCVD has successfully synthesized and applied random copolymers of HEMA and PFDA with the heterogeneous domain size of 1.4–1.75 nm^2 directly to commercial RO membranes as a result of its solvent-free nature [44]. The modified membranes demonstrated low surface adsorption for a large variety of foulants (Table 13.2) [4, 44, 3, 56]. Quartz crystal microbalance (QCM) experiments showed that films that contain approximately 40% PFDA adsorbed less protein than either of the homopolymer films, including the highly hydrophilic pHEMA film, demonstrating the use of molecular heterogeneities to prevent protein adsorption. The casting of similar polymers into thin films by solution methods is difficult and yields rough and uneven coatings [57].

Zwitterionic coatings synthesized via iCVD are among the best antifouling materials for membrane applications [7–9]. Zwitterionic materials are composed of positive and negative charges mixed homogeneously on a molecular level, and thus bind water very strongly through electrostatic force, which leads to their exceptional fouling resistance. Zwitterionic materials synthesized via ATRP have been studied extensively [37], and the recent development in the fabrication of zwitterionic thin film through solvent-free methods has facilitated its application to membrane substrates significantly [7–9]. A two-step, vapor-phase synthesis scheme has been developed to form cross-linked zwitterionic coatings directly on delicate membrane substrates. In the first step, a tertiary amine or pyridine-containing polymer was formed, which was converted to zwitterionic groups (Figure 13.1) in the second step via diffusion-limited vapor-phase reaction with 1,3-propanesultone. The zwitterionic groups are most concentrated on the topmost surface due to the diffusion-limited reaction conditions [8], precisely where the antifouling properties are exploited. The pyridine-based zwitterionic polymers are resistant to the oxidative damages by chlorine (i.e., the active ingredient in bleach and the most prevalent disinfection reagent in water purification industry), which leads to the discovery of an unprecedented synergistic effects between surface coating and solution disinfection reagent.

The iCVD zwitterionic surfaces have also been tested against various foulants for a range of applications, a selection of which is listed in Table 13.2. BSA is a commonly used protein to evaluate the attachment of biomacromolecules. Sodium alginate is a major component in the extracellular matrix, which leads to severe fouling in marine applications. *Vibrio cyclitrophicus* (Figure 13.5a,b) and *E. coli* (Figure 13.5c,d) are among the most prevalent species in costal water, where the fouling problem is most severe. Bare glass and zwitterionic polypyridine-coated glass surfaces shown in Figure 13.5a,b were treated with concentrated culture of *V. cyclitrophicus* (~200 times of the bacterial concentration in the ocean) for 5 h. Each bright dot corresponds to a bacteria colony and the black background indicates a clean surface that resists fouling. In Figure 13.5c,d, uncoated and zwitterionic polymethacrylate-coated RO membranes were exposed to concentrated

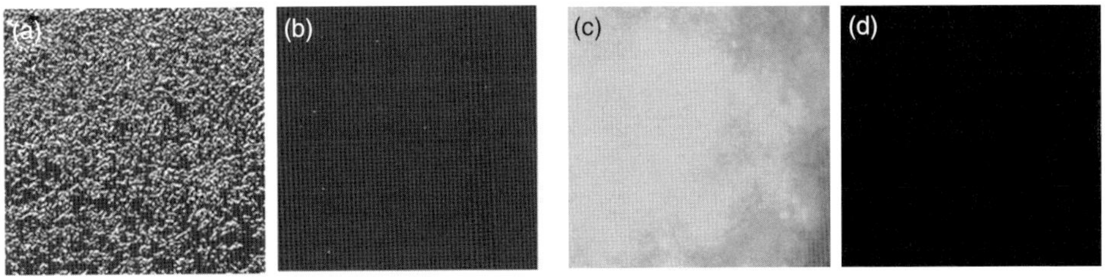

Figure 13.5 Bacterial adhesion on iCVD zwitterionic coatings. (a) Micrograph of a glass surface after the treatment with concentrated culture of *V. cyclitrophicus* (~200 times of the bacterial concentration in the ocean) for 5 h. (b) Micrograph of zwitterionic polypyridine-coated surface after the treatment with concentrated culture of *V. cyclitrophicus* for 5 h. (c) Fluorescent micrograph of uncoated RO membrane after being exposed to concentrated *E. coli* culture for 2 h. (d) Fluorescent micrograph of zwitterionic polymethacrylate-coated RO membrane after being exposed to concentrated *E. coli* culture for 2 h.

E. coli cultures for 2 h before the fluorescent micrographs were captured. The excellent resistance to both microorganisms suggests that the iCVD zwitterionic coatings are effective fouling-mitigating materials.

Zwitterionic chemistry is only one example in the rich library of antifouling chemistries that are compatible with iCVD [14, 33, 44, 46, 58]. Poly(ethylene oxide) (PEO), which is considered the gold standard for antifouling chemistry, has been synthesized using iCVD via a unique ring-opening cationic polymerization mechanism [58]. The PEO brushes were grafted and patterned onto amine-terminated surfaces, and good fouling resistance was evidenced by the sharp contrast of patterns when the surface is subject to BSA treatment.

In addition to antifouling coatings, active antimicrobial surfaces have been fabricated to not only resist the attachment of living organisms, but effectively disinfect the surrounding media by disrupting the cells of microorganisms. Such surfaces were first fabricated from DMAMS (Figure 13.1), a polymer that contains tertiary amine side chains [11]. The tertiary amines possess positive charge at neutral pH and can therefore insert themselves into negatively charged bacteria membranes and break the cell. Such coatings have been applied to a nylon membrane to extract the intracellular lipid directly from wet microalgae in an energy-efficient manner (Figure 13.6) [12]. The poly(dimethylaminomethylstyrene) (pDMAMS)-coated membrane demonstrated a high disruption yield of 25.6 ± 2.18% upon co-culturing and agitating with a microalgal culture of *Aurantiochytrium* sp. *KRS101*.

In conclusion, iCVD represents a promising method to mitigate fouling in membrane applications. The conformal and benign nature of the iCVD process gives rise to uniform coverage on porous substrates without damaging the delicate membranes. Antifouling surfaces, which resist the attachment of various fouling species, and antimicrobial surfaces, which disinfect the surface and surrounding

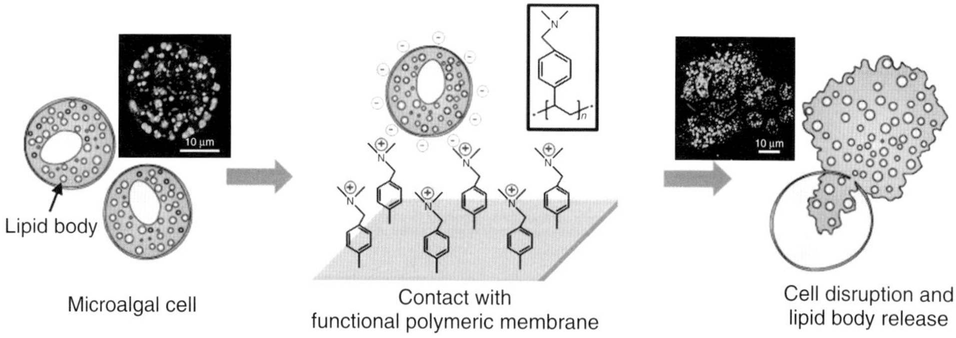

Figure 13.6 Mechanism and effects of the iCVD antimicrobial coating pDMAMS. Intact microalgal cells with negatively charged phospholipid bilayers (inset: confocal microscopic image of intact *Aurantiochytrium* cells) interact with the pDMAMS surface via electrostatic forces. The lipid bilayers of the microalgae are ruptured due to the perturbation of local electrostatic equilibrium of the cell membrane, which is induced upon the direct contact with the pDMAMS surface (inset: molecular structure of a pDMAMS surface). The lipids are released upon cell rupture (inset: confocal microscopic image of *Aurantiochytrium* cells after contacting pDMAMS surface for 24 h).

media, have been fabricated via iCVD and have demonstrated great antifouling potency.

13.4 Membrane Modification for Sustainability

Membranes are an efficient, effective, and green separation method. Membrane separation is most prevalent in the industry of water treatment, and has become a critical component in energy generation and storage because of the recent development in fuel cells and batteries.

The production of fresh water is at the core of sustainability due to its essential role in the development of agriculture, industry, and health care. Fresh water can be produced by removing salts and marine organisms from seawater, a process termed seawater desalination. However, the cost to produce fresh water via desalination is higher than that for natural fresh water. Reducing energy consumption has been one of the focuses for lowering the cost of desalted water. The breakthrough of seawater desalination process occurred in early 1960s with the invention of RO technology. It has become the dominating technology among recently constructed and future-planned desalination facilities. In a RO process, seawater is pressurized against a semipermeable membrane that selectively allows the transport of water across membrane but retains salts. RO technology has improved drastically as its invention and new designs of membrane modules and stage configurations can bring the energy consumption of an RO process close to the thermodynamic minimum energy of 1.06 kWh m^{-3} [59]. However, the RO process suffers considerably from fouling, which drives up the energy consumption, reduces salt rejection efficiency, and degrades the RO membranes. Fouling originates from the surface properties of the state-of-the-art RO membranes, namely TFC membranes. Biofouling occurs immediately when a membrane module is installed and has to be tolerated until the flux is reduced to 25% of the initial value [60]. The desalination process has to be interrupted for periodic cleaning, which reduces productivity and shortens the lifetime of the TFC membranes. Currently, fouling is mitigated by feed pretreatment, which represents the most energy-intensive (responsible for >36% of total plant energy consumption) and chemical-intensive step in a desalination plant, and one that poses environmental risks to marine organisms when treated water is discharged back into the ocean.

A fouling-resistant membrane would bring major improvements in energy usage, process reliability, and lower the environmental impact of seawater desalination. However, the direct surface modification of commercial TFC membrane is challenging due to its delicate nature. The salt rejection of TFC membranes relies solely on a 100–200-nm-thick polyamide layer, which is prone to damage when treated with solvents or at elevated temperature. In addition, the flux of water across membrane decreases rapidly with additional layers of coating. Therefore, an ideal surface modification method should be solvent-free, at mild temperature, and produces ultra-thin (<100 nm) antifouling coatings.

In spite of the challenges, the modification of delicate TFC membranes has been achieved with the benign iCVD process and the resulting membranes resist biofouling very effectively [7, 8, 61]. The long-term stability of the surface-modified membranes was enhanced in two ways. First, covalent chemical bonds were formed to tether the antifouling coating directly to the surface of a commercial membrane [7]. Second, the methacrylate-based zwitterionic structure was later re-designed to resist the oxidative damages by chlorine [61]. The chlorine-resistance enabled an interesting synergistic interaction between surface zwitterionic coating and solution-dissolved hypochlorite, which improved the fouling resistance by 10 times compared to the case where multiplicative effects between chlorination and coating were assumed (Figure 13.7a) [61]. The coated surface was subject to natural seawater for up to 100 h and no fouling was observed (Figure 13.7b). Desalination performance of the modified membranes remains unchanged, which indicates that iCVD is able to render membrane surface antifouling without damaging the delicate underlying membrane substrate.

MD is a separation process where the volatile species in a solution is evaporated and transported through a hydrophobic and microporous membrane driven by the partial pressure difference resulted from temperature difference between the feed and permeate side. MD process has many attractive features including

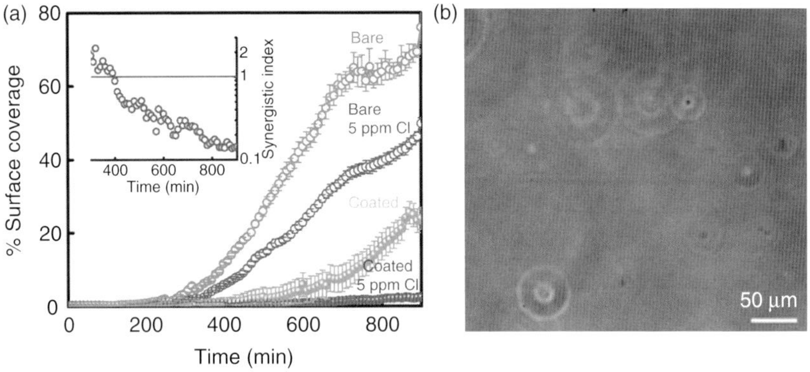

Figure 13.7 Antifouling coatings with excellent long-term stability in natural seawater. (a) Surface coverage by *Vibrio cyclitrophicus* bacteria under different conditions. The synergistic treatment – integrating zwitterionic polypyridine coating with low-level (5 ppm) chlorination – shows exceptional long-term antifouling activity even under accelerated fouling conditions, when each method in isolation begins to fail. Inset: time series of the synergistic index (*S*), quantifying synergistic effect of the two antifouling strategies. Values of $S < 1$ indicate a positive synergy between the two treatments. The monotonic decrease in *S*, with on signs of saturation over 15 h, demonstrates the importance of synergistic effect on long-term fouling resistance. (b) Optical micrograph of a zwitterionic polypyridine-coated glass after the exposure to natural seawater for 100 h. No surface fouling or coating degradation was observed.

but not limited to low-operating temperature and pressure [62], ability to utilize sustainable and alternative energy sources [63], and excellent salt rejection factor [62]. So far, most MD studies focus on theoretical modeling of transport process in MD and experimental studies on effects of MD operating condition. In most of these studies, commercial hydrophobic membrane fabricated from polytetrafluoroethylene (PTFE), polypropylene (PP), or polyvinylidenedifluoride (PVDF) have been used without further processing. Few studies have attempted to design and synthesize membranes [64], and even fewer have exploited surface modification of membranes in vapor phase. Kong *et al.* and Wu *et al.* modified hydrophilic cellulose nitrate membranes by plasma polymerizations of octafluorocyclobutane (OFCB) and vinyltrimethylsilicon (VTMS)/carbon tetrafluoride (CF_4) respectively [65, 66]. Song *et al.* employed plasma polymerization of fluorosilicone on outside surface of porous PP hollow fibers, and tested for MD performance [67]. Silane chemistry was also investigated as vapor-phase surface modification for MD membrane [68, 69]. 1*H*, 1*H*, 2*H*, 2*H*-perfluoro decyltriethoxysilane (PFS) and trichloromethylsilane (TCS) were applied as surface modifier for inorganic porous membranes [69]. The modified membrane gives sufficiently high hydrophobicity and liquid entry pressure for application in MD [69].

iCVD has shown great potential in this application. The $-CF_3$ terminated side groups of fluorinated acrylate and methacrylate are completely retained after iCVD polymerization and form comb-like structures, which give rise to desired hydrophobicity [70]. Microporous membranes formed from electrospun mats have been conformally modified by iCVD of (2-(perfluorooctyl)ethyl methacrylate) (PFOEMA), and exhibited superhydrophobicity [20]. iCVD pPFDA has been used to fabricate nanoporous membrane for hydrophobicity-based separation (Figure 13.3) [10]. The pore walls displayed hydrophobicity after iCVD modification and achieved high selectivity when tested with pairs of small molecules of similar size, but of different hydrophobicity [10]. The hydrophobic membranes for MD require larger pore sizes than reported in this work, which can be fabricated using a similar methodology.

Interestingly, free-standing membranes have been fabricated with iCVD recently [71], which promises to modify the surface morphology and chemistry simultaneously, or even enable the one-pot fabrication of antifouling and high-flux next-generation separation membranes.

An important type of membranes in sustainable energy production, among other applications, is ion-selective membranes. One of the greatest material challenges for fuel cells is the development of an effective proton-exchange membrane to separate the electrodes of the cell. Ion-selective membrane is also a critical component in the purification of heavily contaminated water (such as produced water) via electrodialysis (ED) [72] or ion concentration polarization (ICP). The current commercial standard membrane is coated with Nafion®, which is impermeable to water and costly, with limited long-term stability. The technological challenges represent great opportunities for CVD polymer coatings as hydrophobic fluoropolymers are difficult to produce by solution-phase methods. In fact, ultra-thin and conformal fluoropolymer coatings have been

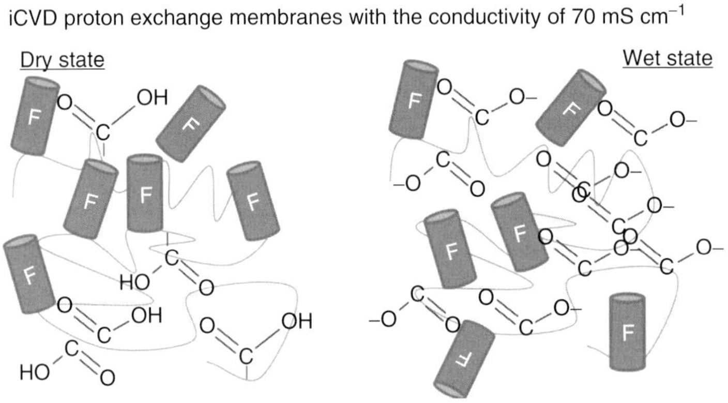

Figure 13.8 First proton-exchange membrane enabled by iCVD with comparable performance as Nafion®. In air (i.e., in dry state), the hydrophobic fluorinated chains oriented outward at the interface while the hydrophilic carboxylic acid groups oriented inward to minimize the interface free energy. After soaking in water at 80 °C, the film underwent significant chain reconfiguration. The surface concentrations of CF_2 and CF_3 components decreased substantially and a proton transport pathway formed by simple water soaking. The cylinders labeled with "F" indicate fluorinated side chains.

synthesized successfully with the iCVD technique (Figure 13.8) [73–76], and the small quantity of fluoromonomers required in the iCVD process can reduce the cost of membranes significantly. The good conformality guarantees uniform coverage over porous membrane substrates, and durability of the membrane can be increased considerably by enhancing the mechanical strength via cross-linking [76] or by grafting the fluoropolymer covalently to the substrate [74]. A proton-exchange membrane has been enabled by copolymerizing a highly hydrophobic monomer fluoroacrylate (1*H*, 1*H*, 2*H*, 2*H*-perfluorodecyl acrylate) and a highly hydrophilic monomer bearing a carboxylic acid moiety (methacrylic acid) via iCVD [70]. The resulting copolymer has comparable ion conductivity ($70\ mS\ cm^{-1}$) as commercial Nafion®, and is a promising membrane alternative for electrical-based desalination methods.

References

1. Matin, A. *et al* (2014) Surface modification of seawater desalination reverse osmosis membranes: characterization studies & performance evaluation. *Desalination*, **343**, 128–139. doi: 10.1016/j.desal.2013.10.023.
2. Matin, A. *et al* (2011) Random copolymer films as potential antifouling coatings for reverse osmosis membranes. *Desalin. Water Treat.*, **34**, 100–105. doi: 10.5004/dwt.2011.2804.
3. Matin, A. *et al* (2014) Surface-modified reverse osmosis membranes applying a copolymer film to reduce adhesion of bacteria as a strategy for biofouling control. *Sep. Purif. Technol.*, **124**, 117–123. doi: 10.1016/j.seppur.2013.12.032.
4. Ozaydin-Ince, G., Matin, A., Khan, Z., Zaidi, S.M.J., and Gleason, K.K. (2013) Surface modification of reverse osmosis desalination membranes by thin-film coatings deposited by initiated chemical

vapor deposition. *Thin Solid Films*, **539**, 181–187. doi: 10.1016/j.tsf.2013.04.133.

5. Amadei, C.A., Yang, R., Chiesa, M., Gleason, K.K., and Santos, S. (2014) Revealing amphiphilic nanodomains of anti-biofouling polymer coatings. *ACS Appl. Mater. Interfaces*, **6**, 4705–4712. doi: 10.1021/am405159f.
6. Chan, K. and Gleason, K.K. (2005) Initiated chemical vapor deposition of linear and cross-linked poly(2-hydroxyethyl methacrylate) for use as thin-film hydrogels. *Langmuir*, **21**, 8930–8939.
7. Yang, R., Xu, J., Ozaydin-Ince, G., Wong, S.Y., and Gleason, K.K. (2011) Surface-tethered zwitterionic ultrathin antifouling coatings on reverse osmosis membranes by initiated chemical vapor deposition. *Chem. Mater.*, **23**, 1263–1272.
8. Yang, R. and Gleason, K.K. (2012) Ultrathin antifouling coatings with stable surface zwitterionic functionality by initiated chemical vapor deposition (iCVD). *Langmuir*, **28**, 12266–12274. doi: 10.1021/la302059s
9. Yang, R., Jang, H., Stocker, R., and Gleason, K.K. (2014) Synergistic prevention of biofouling in seawater desalination by zwitterionic surfaces and low-level chlorination. *Adv. Mater.*, **26**, 1711–1718. doi: 10.1002/adma.201304386.
10. Asatekin, A. and Gleason, K.K. (2011) Polymeric nanopore membranes for hydrophobicity-based separations by conformal initiated chemical vapor deposition. *Nano Lett.*, **11**, 677–686. doi: 10.1021/nl103799d.
11. Martin, T.P., Kooi, S.E., Chang, S.H., Sedransk, K.L., and Gleason, K.K. (2007) Initiated chemical vapor deposition of antimicrobial polymer coatings. *Biomaterials*, **28**, 909–915.
12. Yoo, G. *et al* (2014) An effective, cost-efficient extraction method of biomass from wet microalgae with a functional polymeric membrane. *Green Chem.*, **16**, 312–319. doi: 10.1039/c3gc41695j.
13. Gupta, M., Kapur, V., Pinkerton, N.M., and Gleason, K.K. (2008) Initiated chemical vapor deposition (iCVD) of conformal polymeric nanocoatings for the surface modification of high-aspect-ratio pores. *Chem. Mater.*, **20**, 1646–1651.
14. Ye, Y. and Mao, Y. (2011) Vapor-based synthesis of ultrathin hydrogel coatings for thermo-responsive nanovalves. *J. Mater. Chem.*, **21**, 7946–7952. doi: 10.1039/c1jm10713e.
15. Ozaydin, I.G., Demirel, G., Gleason, K.K., and Demirel, M.C. (2010) Highly swellable free-standing hydrogel nanotube forests. *Soft Matter*, **6**, 1635–1639.
16. Ozaydin-Ince, G., Dubach, J.M., Gleason, K.K., and Clark, H.A. (2011) Microworm optode sensors limit particle diffusion to enable in vivo measurements. *Proc. Natl. Acad. Sci. U.S.A.*, **108**, 2656–2661. doi: 10.1073/pnas.1015544108.
17. Bhattacharyya, D., Senecal, K., Marek, P., Senecal, A., and Gleason, K.K. (2011) High surface area flexible chemiresistive biosensor by oxidative chemical vapor deposition. *Adv. Funct. Mater.*, **21**, 4328–4337. doi: 10.1002/adfm.201101071.
18. Tenhaeff, W.E. and Gleason, K.K. (2009) Surface-tethered pH-responsive hydrogel thin films as size-selective layers on nanoporous asymmetric membranes. *Chem. Mater.*, **21**, 4323–4331.
19. Yang, R., Buonassisi, T., and Gleason, K.K. (2013) Organic vapor passivation of silicon at room temperature. *Adv. Mater.*, **25**, 2078–2083. doi: 10.1002/adma.201204382.
20. Ma, M., Mao, Y., Gupta, M., Gleason, K.K., and Rutledge, G.C. (2005) Superhydrophobic fabrics produced by electrospinning and chemical vapor deposition. *Macromolecules*, **38**, 9742–9748.
21. Ma, M. *et al* (2007) Decorated electrospun fibers exhibiting superhydrophobicity. *Adv. Mater.*, **19**, 255–259. doi: 10.1002/adma.200601449.
22. Yammine, P., Pavon-Djavid, G., Helary, G., and Migonney, V. (2005) Surface modification of silicone intraocular implants to inhibit cell proliferation. *Biomacromolecules*, **6**, 2630–2637.
23. Stevens, M.M. and George, J.H. (2005) Exploring and engineering the cell surface interface. *Science*, **310**, 1135–1138. doi: 10.1126/science.1106587.

24. Hsiue, G.-H., Guu, J.-A., and Cheng, C.-C. (2001) Poly(2-hydroxyethyl methacrylate) film as a drug delivery system for pilocarpine. *Biomaterials*, **22**, 1763–1769.
25. Gavalas, V., Berrocal, M., and Bachas, L. (2006) Enhancing the blood compatibility of ion-selective electrodes. *Anal. Bioanal. Chem.*, **384**, 65–72.
26. Ko, S. and Jang, J. (2007) Protein immobilization on aminated poly(glycidyl methacrylate) nanofibers as polymeric carriers. *Biomacromolecules*, **8**, 1400–1403.
27. Alf, M.E. *et al* (2010) Chemical vapor deposition of conformal, functional, and responsive polymer films. *Adv. Mater.*, **22**, 1993–2027.
28. Lopez, G.P., Ratner, B.D., Rapoza, R.J., and Horbett, T.A. (1993) Plasma deposition of ultrathin films of poly(2-hydroxyethyl methacrylate): surface analysis and protein adsorption measurements. *Macromolecules*, **26**, 3247–3253.
29. Morra, M. and Cassinelli, C. (1995) Surface field of forces and protein adsorption behavior of poly(hydroxyethyl methacrylate) films deposited from plasma. *J. Biomed. Mater. Res.*, **29**, 39–45.
30. Tarducci, C., Schofield, W.C.E., Badyal, J.P.S., Brewer, S.A., and Willis, C. (2002) Monomolecular functionalization of pulsed plasma deposited poly(2-hydroxyethyl methacrylate) surfaces. *Chem. Mater.*, **14**, 2541–2545.
31. Pfluger, C.A., Carrier, R.L., Sun, B., Ziemer, K.S., and Burkey, D.D. (2009) Cross-linking and degradation properties of plasma enhanced chemical vapor deposited poly(2-hydroxyethyl methacrylate). *Macromol. Rapid Commun.*, **30**, 126–132.
32. Marí-Buyé, N. *et al* (2009) Functionalized, swellable hydrogel layers as a platform for cell studies. *Adv. Funct. Mater.*, **19**, 1276–1286.
33. Bose, R.K. and Lau, K.K.S. (2011) Mechanical properties of ultrahigh molecular weight PHEMA hydrogels synthesized using initiated chemical vapor deposition. *Biomacromolecules*, **11**, 2116–2122.
34. Baxamusa, S.H. *et al* (2008) Protection of sensors for biological applications by photoinitiated chemical vapor deposition of hydrogel thin films. *Biomacromolecules*, **9**, 2857–2862.
35. Montero, L., Baxamusa, S.H., Borros, S., and Gleason, K.K. (2008) Thin hydrogel films with nanoconfined surface reactivity by photoinitiated chemical vapor deposition. *Chem. Mater.*, **21**, 399–403.
36. Tegoulia, V.A., Rao, W., Kalambur, A.T., Rabolt, J.F., and Cooper, S.L. (2001) Surface properties, fibrinogen adsorption, and cellular interactions of a novel phosphorylcholine-containing self-assembled monolayer on gold. *Langmuir*, **17**, 4396–4404.
37. Jiang, S. and Cao, Z. (2010) Ultralow-fouling, functionalizable, and hydrolyzable zwitterionic materials and their derivatives for biological applications. *Adv. Mater.*, **22**, 920–932.
38. Alf, M.E., Godfrin, P.D., Hatton, T.A., and Gleason, K.K. (2010) Sharp hydrophilicity switching and conformality on nanostructured surfaces prepared via initiated chemical vapor deposition (iCVD) of a novel thermally responsive copolymer. *Macromol. Rapid Commun.*, **31**, 2166–2172.
39. Alf, M.E., Hatton, T.A., and Gleason, K.K. (2011) Novel N-isopropylacrylamide based polymer architecture for faster LCST transition kinetics. *Polymer*, **52**, 4429–4434. doi: 10.1016/j.polymer.2011.07.051.
40. Prime, K.L. and Whitesides, G.M. (1993) Adsorption of proteins onto surfaces containing end-attached oligo(ethylene oxide): a model system using self-assembled monolayers. *J. Am. Chem. Soc.*, **115**, 10714–10721.
41. Banerjee, I., Pangule, R.C., and Kane, R.S. (2010) Antifouling coatings: recent developments in the design of surfaces that prevent fouling by proteins, bacteria, and marine organisms. *Adv. Mater.*, **23**, 690–718.
42. Finlay, J.A. *et al* (2008) Settlement of ulva zoospores on patterned fluorinated and PEGylated monolayer surfaces. *Langmuir*, **24**, 503–510. doi: 10.1021/la702275g.

43. Krishnan, S. *et al* (2006) Anti-biofouling properties of comblike block copolymers with amphiphilic side chains. *Langmuir*, **22**, 5075–5086.
44. Baxamusa, S.H. and Gleason, K.K. (2009) Random copolymer films with molecular-scale compositional heterogeneities that interfere with protein adsorption. *Adv. Funct. Mater.*, **19**, 3489–3496.
45. Louie, J.S., Pinnau, I., and Reinhard, M. (2011) Effects of surface coating process conditions on the water permeation and salt rejection properties of composite polyamide reverse osmosis membranes. *J. Membr. Sci.*, **367**, 249–255.
46. Yang, R., Asatekin, A., and Gleason, K.K. (2012) Design of conformal, substrate-independent surface modification for controlled protein adsorption by chemical vapor deposition (CVD). *Soft Matter*, **8**, 31–43.
47. Krishnan, S., Weinman, C.J., and Ober, C.K. (2008) Advances in polymers for anti-biofouling surfaces. *J. Mater. Chem.*, **18**, 3405–3413.
48. Jackson, A.M., Myerson, J.W., and Stellacci, F. (2004) Spontaneous assembly of subnanometre-ordered domains in the ligand shell of monolayer-protected nanoparticles. *Nat. Mater.*, **3**, 330–336.
49. Centrone, A. *et al* (2008) The role of nanostructure in the wetting behavior of mixed-monolayer-protected metal nanoparticles. *Proc. Natl. Acad. Sci. U.S.A.*, **105**, 9886–9891. doi: 10.1073/pnas.0803929105.
50. Baxamusa, S.H., Montero, L., Borros, S., and Gleason, K.K. (2010) Self-aligned micropatterns of bifunctional polymer surfaces with independent chemical and topographical contrast. *Macromol. Rapid Commun.*, **31**, 735–739.
51. Ross, A.M., Zhang, D., Deng, X., Chang, S.L., and Lahann, J. (2011) Chemical-vapor-deposition-based polymer substrates for spatially resolved analysis of protein binding by imaging ellipsometry. *Anal. Chem.*, **83**, 874–880.
52. Chen, H.-Y. and Lahann, J. (2005) Fabrication of discontinuous surface patterns within microfluidic channels using photodefinable vapor-based polymer coatings. *Anal. Chem.*, **77**, 6909–6914.
53. Suh, K.Y., Langer, R., and Lahann, J. (2004) A novel photodefinable reactive polymer coating and its use for microfabrication of hydrogel elements. *Adv. Mater.*, **16**, 1401–1405.
54. Malkov, G.S. *et al* (2008) Pulsed-plasma-induced micropatterning with alternating hydrophilic and hydrophobic surface chemistries. *Plasma Processes Polym.*, **5**, 129–145.
55. Macritchie, F. (1978) Proteins at interfaces. *Adv. Protein Chem.*, **32**, 283–326.
56. Matin, A., Shafi, H.Z., Khan, Z., Khaled, M., Yang, R., Gleason, K. and Rehman, F. (2014) Surface modification of seawater desalination reverse osmosis membranes: characterization studies & performance evaluation. *Desalination*, **343**, 128–139. doi: 10.1016/j.desal.2013.10.023.
57. Grignard, B., Jerome, C., Calberg, C., Detrembleur, C., and Jerome, R. (2007) Controlled synthesis of carboxylic acid end-capped poly(heptadecafluorodecyl acrylate) and copolymers with 2-hydroxyethyl acrylate. *J. Polym. Sci., Part A: Polym. Chem.*, **45**, 1499–1506. doi: 10.1002/pola.21920.
58. Bose, R.K., Nejati, S., Stufflet, D.R., and Lau, K.K.S. (2012) Graft polymerization of anti-fouling PEO surfaces by liquid-free initiated chemical vapor deposition. *Macromolecules*, **45**, 6915–6922. doi: 10.1021/ma301234z.
59. Elimelech, M. and Phillip, W.A. (2011) The future of seawater desalination: energy, technology, and the environment. *Science*, **333**, 712–717. doi: 10.1126/science.1200488.
60. Fritzmann, C., Löwenberg, J., Wintgens, T., and Melin, T. (2007) State-of-the-art of reverse osmosis desalination. *Desalination*, **216**, 1–76. doi: 10.1016/j.desal.2006.12.009.
61. Yang, R., Jang, H., Stocker, R., and Gleason, K.K. (2014) Synergistic prevention of biofouling in seawater desalination by zwitterionic surfaces and low-level chlorination. *Adv. Mater.*, **26**, 1711–1718. doi: 10.1002/adma.201304386.

62. Alkhudhiri, A., Darwish, N., and Hilal, N. (2012) Membrane distillation: a comprehensive review. *Desalination*, **287**, 2–18. doi: 10.1016/j.desal.2011.08.027.
63. Galvez, J.B., Garcia-Rodriguez, L., and Martin-Mateos, I. (2009) Seawater desalination by an innovative solar-powered membrane distillation system: the MEDESOL project. *Desalination*, **246**, 567–576. doi: 10.1016/j.desal.2008.12.005.
64. Suk, D.E., Matsuura, T., Park, H.B., and Lee, Y.M. (2010) Development of novel surface modified phase inversion membranes having hydrophobic surface-modifying macromolecule (nSMM) for vacuum membrane distillation. *Desalination*, **261**, 300–312. doi: 10.1016/j.desa1.2010.06.058.
65. Kong, Y., Lin, X., Wu, Y., Chen, J., and Xu, J. (1992) Plasma polymerization of octafluorocyclobutane and hydrophobic microporous composite membranes for membrane distillation. *J. Appl. Polym. Sci.*, **46**, 191–199.
66. Wu, Y.L., Kong, Y., Lin, X., Liu, W.H., and Xu, J.P. (1992) Surface-modified hydrophilic membranes in membrane distillation. *J. Membr. Sci.*, **72**, 189–196. doi: 10.1016/0376-7388(92)80199-t.
67. Song, L., Li, B., Sirkar, K.K., and Gilron, J.L. (2007) Direct contact membrane distillation-based desalination: novel membranes, devices, larger-scale studies, and a model. *Ind. Eng. Chem. Res.*, **46**, 2307–2323.
68. Krajewski, S., Kujawski, W., Bukowska, M., Picard, C., and Larbot, A. (2006) Application of fluoroalkylsilanes (FAS) grafted ceramic membranes in membrane distillation process of NaCl solutions. *J. Membr. Sci.*, **281**, 253–259. doi: 10.1016/j.memsci.2006.03.039.
69. Hendren, Z.D., Brant, J., and Wiesner, M.R. (2009) Surface modification of nanostructured ceramic membranes for direct contact membrane distillation. *J. Membr. Sci.*, **331**, 1–10. doi: 10.1016/j.memsci.2008.11.038.
70. Coclite, A.M., Lund, P., Di Mundo, R., and Palumbo, F. (2013) Novel hybrid fluoro-carboxylated copolymers deposited by initiated chemical vapor deposition as protonic membranes. *Polymer*, **54**, 24–30.
71. Seidel, S., Kwong, P., and Gupta, M. (2013) Simultaneous polymerization and solid monomer deposition for the fabrication of polymer membranes with dual-scale porosity. *Macromolecules*, **46**, 2976–2983. doi: 10.1021/ma302607c.
72. Xu, T. and Huang, C. (2008) Electrodialysis-based separation technologies: a critical review. *AIChE J.*, **54**, 3147–3159. doi: 10.1002/aic.11643.
73. Coclite, A.M., Shi, Y., and Gleason, K.K. (2012) Controlling the degree of crystallinity and preferred crystallographic orientation in polyperfluorodecylacrylate thin films by initiated chemical vapor deposition. *Adv. Funct. Mater.*, **22**, 2167–2176. doi: 10.1002/adfm.201103035.
74. Coclite, A.M., Shi, Y., and Gleason, K.K. (2012) Grafted crystalline polyperfluoroacrylate structures for superhydrophobic and oleophobic functional coatings. *Adv. Mater.*, **24**, 4534–4539. doi: 10.1002/adma.201200682.
75. Gupta, M. and Gleason, K.K. (2009) Surface modification of high aspect ratio structures with fluoropolymer coatings using chemical vapor deposition. *Thin Solid Films*, **517**, 3547–3550. doi: 10.1016/j.tsf.2009.01.037.
76. Yagüe, J.L. and Gleason, K.K. (2013) Enhanced cross-linked density by annealing on fluorinated polymers synthesized via initiated chemical vapor deposition to prevent surface reconstruction. *Macromolecules*, **46**, 6548–6554. doi: 10.1021/ma4010633.

14
CVD Polymer Surfaces for Biotechnology and Biomedicine

Anna Maria Coclite

14.1
Introduction

Recent developments in biotechnology and biomedicine have been boosted by the rational design of devices with biomimetic functions, specific cellular responses, or tissue-mediated biomolecular recognition [1]. Polymer properties can be easily engineered from the nanoscale to the microscale: a large variety of functionalized end groups are available for biomolecule immobilization, and different levels of stiffness and mechanical properties can be obtained. This versatility makes them a leading tool for future progress in delivery of drugs and genes as well as in regenerative medicine. Chemical composition, hardness, elasticity, hydrophobicity, morphology, biodegradability, and biocompatibility can have a profound impact on biological response regulation, cell proliferation, and tissue engineering [2–4]. The tissues of the human body are characterized by different mechanical properties ranging from soft (brain, about 0.5 kPa) to moderately stiff (skin and muscles, around 100 kPa) and stiff (bone, >100 MPa). The differences in elastic moduli create mechanically defined microenvironments that effectively affect cell proliferation.

CVD polymers, in particular, are also easily patterned and shaped, so they can be readily integrated on different types of substrates and spatially resolved biodevices, at room temperature, without altering the mechanical properties of the bulk materials.

Solvents entrained in the polymers are a major cause for irritation and implant rejection in implantable devices. CVD being a completely dry technique (i.e., does not require use of solvents) overcomes this problem. The resulting polymers are of high purity and free of leachable and plasticizer.

The main goal of biomedical research has been how to combine fine-tuning of the surface chemistry with precise tailoring of the other material physical properties. Biocompatibility and biopassivation properties are achieved when fluorocarbon polymers retain the composition of Teflon, while hydrophilicity and swelling properties are obtained by retention of the hydroxyl groups, in hydrogels, for example. The immobilization of enzymes, peptides, proteins, polysaccharides,

CVD Polymers: Fabrication of Organic Surfaces and Devices, First Edition. Edited by Karen K. Gleason.

and other bioactive molecules on polymer surfaces is realized through specific functional groups that generate the covalent bonding of the desired biomolecule on the surface. For this purpose, the retention of carboxylic or amine groups is fundamental.

CVD has been proven as durable and efficient biomaterial modification method; indeed it typically results in polymer thin films, which are well adherent, chemically inert, insoluble, mechanically tough, and thermally stable. Stability over extended periods of time and resistance to solvent treatments, multiple exposures to heat, humidity, and atmosphere and mechanical wear are fundamental requirements for CVD polymers to be successfully used for biomaterials modification.

The translation of the capabilities of CVD polymers finds direct application in device fabrication (e.g., biosensor, drug delivery system, bioMEMS), tissue engineering, and biopassivating and antimicrobial coating deposition on clinically used biomaterials. CVD surface modification techniques currently applied in biomedicine and biotechnology include plasma-enhanced CVD (PECVD) [5], vapor-phase polymerization (VPP) [6], initiated CVD (iCVD) [7], oxidative CVD (oCVD) [8], and parylene [9].

14.2 Biosensors

According to the IUPAC definition, a biosensor is "A device that uses specific biochemical reactions mediated by isolated enzymes, immunosystems, tissues, organelles or whole cells to detect chemical compounds usually by electrical, thermal or optical signals." Biosensors are widely used as medical diagnostic tools. CVD polymers are an ideal platform for biosensors because they can be easily functionalized with the aforementioned biomolecules.

The main component of a biosensor is the bioactive material, which is sensitive to the analyte to be detected. When the analyte is present into the sensor's environment, it induces some structural changes in the active material. A schematic of a biosensor is presented in Figure 14.1a. In this case, the active material is represented by a conductive CVD polymer, which is functionalized with an analyte-specific molecule, avidin. When the analyte is in the environment, in this case biotin, it binds to the analyte-specific molecule, triggering some changes in the resistance of the conductive polymer, measured by the electrical contacts that are deposited on top and bottom of the active layer.

A good sensor requires short response time (i.e., time between the introduction of the analyte and the consequent structural changes in the polymer) and high sensitivity (low level of analyte should induce significant response). A high density of bonding sites of the analyte-specific molecule and high surface area play a fundamental role on the sensitivity of the sensors. For this reason, a good control over the surface composition is crucial, as well as the possibility of patterning and shaping the active material. A common approach to improve the performance

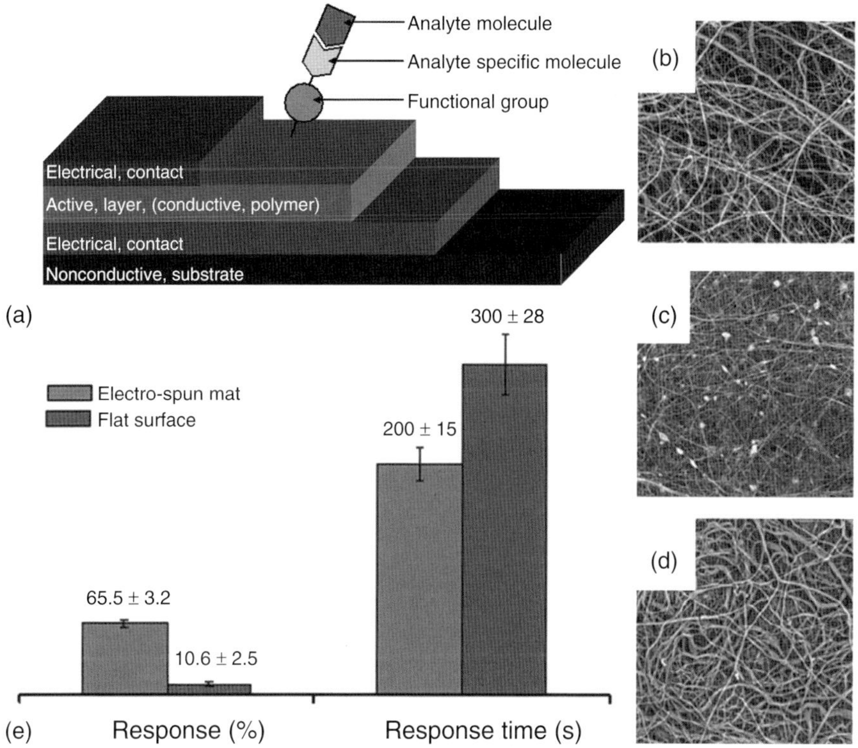

Figure 14.1 (a) Schematic of a biosensor, in which the active layer is a conductive polymer, whose resistivity changes when the analyte molecule binds with the analyte-specific molecule, immobilized on the surface. The active layer is sandwiched between two electrical contacts for the detection of sensor response. Scanning electron micrographs of the (b) bare electrospun nylon mat, (c) mat coated with the conductive copolymer, and (d) after immobilization with avidin (scale: 100 μm). (e) Comparison between response and response time of the same biosensor deposited on the electrospun mat and on flat substrate. The response increases and the response time decreases on the mat due to higher surface area. ((a–e) Adapted with permission from Ref. [3]. Copyright 2011, Wiley.)

of these sensors is to immobilize biomolecules on nanorods, nanowires, nanoribbons, or nanofibers in order to obtain much higher surface area as compared to flat substrates.

The active layer of the biosensor represented in Figure 14.1a was a copolymer of 3,4-ethylenedioxythiophene (EDOT) and 3-thiopheneethanol (3-TE), conformally deposited on electrospun mats by oCVD [3]. The EDOT monomer was used to create conductive units in the copolymer, while the 3-TE was used to provide the –OH functional groups for the successive immobilization of avidin, as the analyte-specific molecule. The deposition mechanism of the oCVD process allows a complete retention of the monomer's functionalities. In this case, this directly translated into high density of -OH groups for covalent bonding of avidin.

The copolymer was deposited on electrospun mat to obtain a high surface to volume ratio for the device and thus increase the sensitivity of the sensor. Figure 14.1b shows the bare nylon fibers. These were still visible after the oCVD deposition (Figure 14.1c) and after the avidin immobilization (Figure 14.1d). Other deposition methods were not so successful in retaining the electrospun mat structure: solution coating resulted into aggregation of particulates without coating the fibers while evaporation produced a continuous layer over the mat, eliminating the high surface area. Figure 14.1e compares the response and the response time for the same device deposited on flat substrates and electrospun mat. The biosensor performed well in terms of high sensitivity and low response time due to the high functional group retention and high conformality of the oCVD method used to synthetize the conductive polymer. The high surface area of the nanofiber mats improved the sensor response sixfold and significantly lowered the response time.

Other types of biosensors were fabricated using the swelling properties of hydrogel materials in water. The hydrogel response to aqueous environment results in changes in mechanical properties [10], protein adsorption capabilities [11], and hydrophilicity of the polymer [12]. All these changes can be monitored and used in biosensors. In the swollen state, hydrogels can have the same water content as living tissues, this makes them highly biocompatible [13]. The swelling properties of the hydrogels are strongly dependent on the monomer structure retention, the hydrophilicity, the lack of damage created by UV irradiation, and/or ion bombardment. The degree of crosslinking has also a key role on swellability and analyte passage through the mesh of the hydrogel. The iCVD method has been successfully used for the conformal deposition of polyhydroxyethylmethacryate (PHEMA). The possibility to systematically control the crosslinking degree, the thickness, and the molecular weight of the PHEMA deposited by iCVD has boosted a lot of study on biosensors made with an active iCVD PHEMA layer [14, 15]. The deposition of crosslinked polymer is difficult to achieve by other techniques, for example, spin coating, because of the insolubility of such polymers. The precise control over the crosslinking is obtained during the iCVD process by varying the degree of incorporation of a crosslinking monomer, for example, ethylene glycol diacrylate (EGDA). Yague and Gleason [16] have demonstrated that the mesh size of the hydrogel systematically changes from 0.4 to 2.0 nm with the ratio EGDA/HEMA used in the deposition process, as shown in Figure 14.2a. The sensor selectivity was very much improved by the control over the mesh size. Figure 14.2b shows a significant change in intensity of the UV emission at 490 nm of the fluorescein released through a hydrogel mesh size of 0.5 and 2 nm.

PHEMA iCVD hydrogels were also used for encapsulating multiple types of biosensors. In this case, the role of the thin hydrogel coatings was to prevent the agglomeration or shape deformation of the sensors without affecting the response and the response time of the encapsulated sensor. An optode sensor for *in vivo* detection of sodium was encapsulated into PHEMA microworm structure and subcutaneously injected [17]. The microworm shape was obtained by conformally coating the high aspect ratio pores of Anodized Aluminum Oxide (AAO) membranes. The conformal nature of the iCVD deposition process perfectly

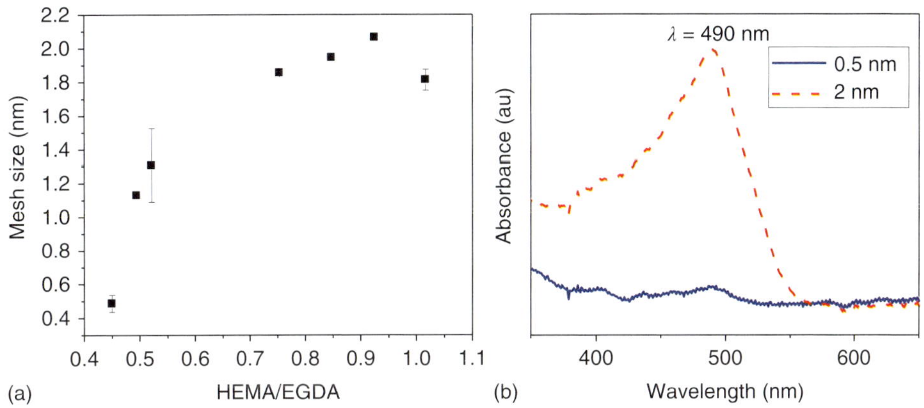

Figure 14.2 (a) Plot of the mesh size of a PHEMA hydrogel layer, when varying the HEMA/EGDA (i.e., monomer/crosslinker) ratio. The mesh size varies from 0.4 to 2 nm. (b) Plot of the characteristic UV emission peak of fluorescein at 490 nm. Depending on the mesh size, the fluorescein is retained in the hydrogel structure or eluted. ((a, b) Reproduced from Ref. [11] with permission from the Centre National de la Recherche Scientifique (CNRS) and The Royal Society of Chemistry.)

enabled this unique template fabrication of microworms. The cylindrical shape of the biosensor enhanced the response time (for its large surface-area-to-volume ratio) and minimized the diffusion away from the injection site, contrarily to spherical sensors.

Encapsulating microelectrode-based biosensors with an ultrathin PHEMA hydrogel was demonstrated to enhance the sensor response by reducing the surface impedance [18]. The mesh size of the hydrogel was optimized to allow the permeation of the electrolyte, while excluding larger molecules, in order to have optimal contact at the electrode–electrolyte interface, with the potential to greatly improve the performance of implantable electrodes for short- and long-term studies.

The conformal nature of ultrathin iCVD hydrogel layer was also used to obtain surface-imprinted polymeric nanotube sensors for the detection of immunoglobulin G (IgG) [19]. First, a nanoporous membrane with covalently bonded IgG was coated by iCVD with a hydrogel layer. Then the membrane with the IgG was etched away, leaving imprinted the reverse form of the biomolecule on the hydrogel. The shape factor was used for the selective detection of IgG. Some key elements made of iCVD form the ideal candidate technique for achieving such structure: (i) the gentleness of the process permitted the deposition a hydrogel layer on IgG molecule without causing degradation and (ii) the conformality of the process allowed to obtain a high density of imprinted recognition sites.

Very recently, a new protein detection platform was obtained on an iCVD vinyl functionalized surface [20]. The solvent-free aspect and the gentleness of the iCVD process translated into a surface with spatially resolved protein-bonded sites on disposable, low-cost substrates, such as plastic and paper.

14.3 Controlled Drug Release

Drug-eluting coatings are widely studied nowadays to achieve intelligent delivery of pharmaceuticals in different parts of the body [21, 22]. The controlled release is due to the encapsulation of the pharmaceutical agent in a synthetic polymer, which acts as a protection shell and allows the site-specific delivery [23, 24].

The polymers that can act as smart encapsulants for controlled drug release are so-called stimuli-responsive, that is, their properties dynamically switch depending on external stimuli. The external stimuli can come from a different pH environment or with a difference in temperature. pH-responsive polymers are, for example, used for enteric release. In this case, the polymer has to be stable at low pH, when the drug travels through the stomach, while at high pH it has to dissolve or swell to release the drug in the lower gastrointestinal tract. Such timed release of the drugs only in the lower part of the intestine is important for enabling new therapies based on noninvasive methods, such as oral administration, while the efficiency of the drug is not affected by any degradation during the transit to the specific target.

The requirements for a good encapsulation process are (i) the gentle deposition on the drug to avoid degradation and (ii) the retention of the monomer functionalities, which allow a timed response of the encapsulant. CVD polymerizations do not require harmful solvents or high temperatures; therefore, they are highly suitable for such applications.

Copolymers of methacrylic acid (MAA) are well known for enteric release: they show a considerable transition in swelling properties at pH between 5 and 6. In acidic conditions, the swelling is very limited, while it is significantly higher at neutral or even higher pH, when the MAA units are deprotonated. The swelling of the polymer allows the release of the drugs encapsulated, while the unswollen state acts as a diffusion barrier.

One example of controlled drug release device was built on a biodegradable p-Si support [25]. The pores of the support were loaded with the drug and capped with a pH-responsive copolymer of MAA and a crosslinking monomer, ethylene dimethacrylate (EDMA). The mechanism of drug release is simplified in Figure 14.3a. At high pH, the polymer swells and the drug is released four times faster than at low pH. The copolymer was deposited by iCVD, which offered the key advantage of being independent on the surface chemistry and pore size of the nanoporous matrix.

The same copolymer was also deposited directly on fine ibuprofen microcrystals below 100 mm [26]. The release of uncoated and coated ibuprofen at pH 1.2 and 7.4 is compared in Figure 14.3b. At pH of 7.4, the release of the drug was relatively fast for both encapsulated and bare drugs, at low pH, instead the release through the encapsulation was significantly delayed.

The swelling properties of PHEMA and the thermoresponsiveness of a shape memory polymer were combined to achieve burst release in response to an increase in the environmental temperature [27]. Specifically, the shape memory

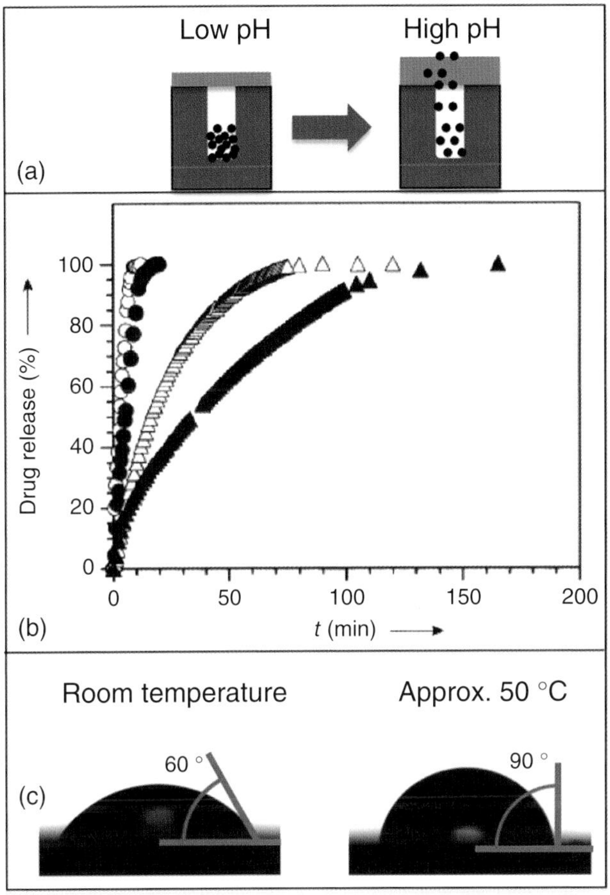

Figure 14.3 (a) Schematic of drug release depending on the swelling of the polymer layer. The polymer does not allow the elution of the drug when it is in the unswollen state. Instead the drug is completely released when the polymer swells. (b) Time release profiles of ibuprofen, without coating (open symbols) and with the iCVD coating (filled symbols), at pH 1.2 (triangle) and pH 7.4 (circle). (c) Optical micrographs of a water droplet deposited on pNIPAAm layer at two different temperatures. The thermoresponsivity of the pNIPAAm layer induces different wetting behaviors. ((b) Reproduced with permission from Ref. [21]. Copyright 2007, Wiley. (c) Adapted with permission from Ref. [7]. Copyright 2011, Elsevier.)

polymer was crosslinked poly(tert-butyl acrylate). This polymer was conformally deposited by iCVD in the pores of an AAO membrane, and subsequently PHEMA was deposited inside the same pores to form coaxial nanotubes with a hydrogel core and a shape memory shell. The release of fluorescein was demonstrated as a model, but the same technology can be easily applied to the delivery of actual drugs. The burst release of the fluorescent dye occurred due to the stress applied by the shape memory outer layer when activated at elevated temperature.

The possibility to systematically control the PHEMA mesh size, discussed in Section 14.2, is extremely important also for drug delivery to enable the selective diffusion of species through the polymer network [11].

Another thermoresponsive polymer, which is receiving increasing attention, is the poly(*N*-isopropylacrylamide) (pNIPAAm). This polymer has an interesting lower critical solution temperature (LCST), approximately 32 °C, very close to body temperature. Depending on the external temperature, the pNIPAAm surface can be hydrophilic or hydrophobic, as shown in Figure 14.3c [7, 28]. Below the LCST (i.e., at room temperature), the hydrophilic amide groups of NIPAAm are exposed to the surface, and therefore the contact angle is low. It has been demonstrated that below the LCST, the polymer's chains assume a random coil configuration in water, corresponding to a swollen state [29]. Above the LCST, the amide groups are interchain and intrachain bonded. As a result, only the hydrophobic isopropyl groups interact with water and the chains evolve into a globular unswollen state.

The ability of the iCVD method to control coating chemistry was used to deposit the pNIPAAm with a graded composition: the crosslinking degree was gradually decreased to the surface, resulting in a NIPAAm-rich upper layer. This improved the swelling and deswelling kinetics in response to small temperature changes compared to homogeneously crosslinked films [30].

14.4 Tissue Engineering

Tissue engineering is a key area of biotechnology and regenerative medicine. The basic idea is to use biodegradable polymers as a template for tissue regeneration and repair. The polymer acts as a suitable microenvironment for the desired cells to attach and grow and degrades upon formation of the neotissue. It has been demonstrated that 3D environments, such as scaffolds or microgrooves, are ideal for tissue formation [31]. In this case, chemistry and topography play a complementary role [3, 32].

Covalent grafting of proteins, peptides, and other biomolecules on the surface of the scaffold promotes cell adhesion, as an imitation of the extracellular matrix (ECM), the natural environment where cells proliferate in living tissues. Often the biomolecule anchors are added through the conformal deposition of a polymer coating with the proper functional groups, within the scaffold pores. CVD surface modification offers a method to incorporate functional groups without altering the bulk properties of the substrate material, such as shape and biodegradability. Other techniques, wet chemical processes, for example, can lead to a faster degradation rate due to the presence of solvents.

Plasma [33, 34] and iCVD [2] polymers have been largely tested as biocompatible platforms for cell adhesion and proliferation. PECVD of Hexamethyldisiloxane/HMDSO) polymers showed high retention of fibrinogen as well as increased cell proliferation [35]. Many other plasma polymers have showed biocompatibility and cell adhesion, for example, fluorocarbon [36] and allylamine [37–39]. Pulsed plasma conditions or atmospheric plasmas were used to maintain a sufficient number of functional groups to promote the cell adhesion [40]. Parylene was also functionalized with thiol molecules to obtain cell adhesion and controlled biological responses [4].

The ability of the iCVD technique to homogeneously incorporate 100% of functionality at high deposition rates and high conformality is ideally suited for tissue engineering applications. The thermoresponsive pNIPAAm, introduced in Section 14.3, was conformally deposited by iCVD in the microgrooves of a polydimethylsilane (PDMS) substrate. The microgrooves offered the 3D template for the formation of longitudinal tissue from NIH-3T3 fibroblasts; one of these constructs is shown in Figure 14.4a [41]. The temperature-dependent swelling and change in hydrophilicity of the pNIPAAm allowed to form and easily retrieve the tissue constructs, making these templates of high potential application in tissue engineering. The ideal temperature for cell culture, that is, 37 °C, is above the LCST of pNIPAAm; therefore, the polymer is in its unswollen and hydrophobic state that promotes cell adhesion. Bringing the device back at room temperature, that is, below the LCST, causes the swelling of the polymer and cell detachment from the culture surface without disturbing cell–cell and cell–ECM interactions. As it can be observed in Figure 14.4b, both the PDMS and the

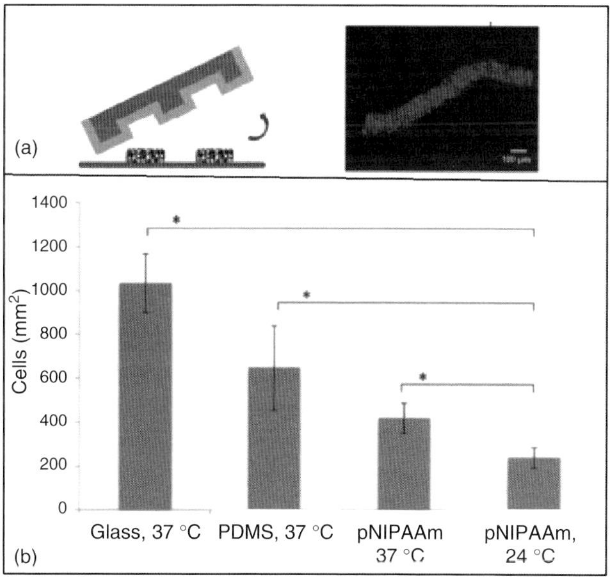

Figure 14.4 (a) Detachment of microgrooves from glass substrate and fluorescent images of tissue constructs. (b) Cell adhesion on 300 nm of pNIPAAm surfaces incubated for 2 h at 37 °C and at 24 °C. Bare PDMS and glass surfaces were used as control. The bars indicated with * shows a statistically significant difference in variance ($p < 0.05$). ((a,b) Adapted with permission from Ref. [36]. Copyright 2011, American Chemical Society.)

pNIPAAm are biocompatible and promote cell adhesion. A glass substrate was also immersed in a solution of NIH-3T3 fibroblasts as control test. Cells adhesion on the pNIPAAm surface at 24 °C was significantly less than that on the same surface kept at 37 °C. The principal reasons for such lower degree of cell adhesion and cell detachment are the swelling of the polymer at lower temperatures and its transition to the hydrophilic state.

PHEMA in a pure or functionalized state was also tested for cell adhesion and aggregation. Although pure PHEMA is not considered an ideal substrate for cell culture, it has been shown that the high molecular weight [15] or the highly crosslinked plasma-deposited form [42] allows for the attachment and culture of certain cell lines. Figure 14.5a shows the comparison between the Human Dermal Fibroblasts (HDF) cells morphology and distribution between a control well (top figure) and on the iCVD PHEMA surface (bottom figure) [43]. The comparability of the two images demonstrates the good cell adhesion and proliferation capabilities of pure, free-standing PHEMA.

A successful functionalization of PHEMA was done by confining pentafluorophenylmethacrylate (PFM) functional groups on the top surface, Figure 14.5b [44]. This allowed keeping intact the swelling properties of the pure PHEMA, while the PFM groups were used to immobilize peptides on the surface and enhance human umbilical vein endothelial cells (HUVEC) growth.

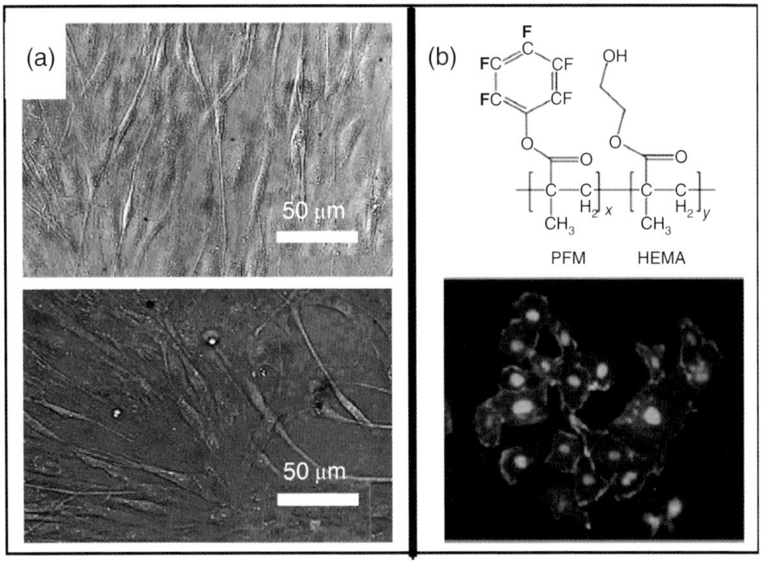

Figure 14.5 Optical micrographs of (a) HDF cell adhesion at 48 h on control well (top) and iCVD PHEMA (bottom). The density and morphology of the HDF cells on iCVD PHEMA films appear similar to the control. (b) HUVEC cells on the surface of the copolymer of PFM and HEMA (structure on top) functionalized with peptide. ((a) Adapted with permission from Ref. [39]. Copyright 2011, Elsevier. (b) Adapted with permission from Ref. [40]. Copyright 2009, Wiley.)

14.5 Bio-MEMS

Biological microelectrochemical systems (BioMEMS) are micro-/nanofabricated devices in which biological molecules are an integral part. The biological material could be the active component, for example, microcantilever biosensors in which a biochemical reaction produces a mechanical deformation of the device; or the microdevice can be used for the construction of biological entities, drug delivery or tissue engineering, for example, microfluidics or lab on a chip. Microfluidic arrays on the cellular scale or smaller ensure superior control overmixing compared to batch reactions, therefore, can create a microenvironment for cell growth in parallel, allowing the analysis of multiple (sub)cellular phenomena and selective biochemical treatment in the microchannel. Polymers such as PDMS are very attractive for the fabrication of BioMEMS devices because they enable soft lithography and have high biocompatibility. Conformal coatings that limit the lateral diffusion in the PDMS without occluding the microchannels are often required to avoid chemicals intermixing. The iCVD method highly conforms to these requests. iCVD polymers were used to bond PDMS microchannels together: one side of the device was coated with poly(glycidyl methacrylate) p(GMA) and the other side with poly(4-aminostyrene) p(AS) [45]. The epoxy groups of p(GMA) were reacted with the amino groups of the p(AS) to assemble the device.

One very interesting application of iCVD for BioMEMS was for the fabrication of an implantable microfluidic lung-assisted device [46]. Figure 14.6a shows a cross section of a lung vascular channel. The pulmonary vasculature is made for selective permeation of carbon dioxide and oxygen between the blood in the vascular channels and the air in the adjacent alveolar chamber. In lung-assist devices, this is achieved with a branched microfluidic gas exchange membrane designed to allow physiologic permeation of oxygen and carbon dioxide while acting as a liquid barrier to prevent the permeation of blood out of the vascular vessels. Figure 14.6b shows such branched microfluidic membrane made of a free-standing membrane of crosslinked maleic anhydride, synthetized by iCVD, and bonded onto a microfluidic device. No liquid lateral leakage of the dyed water pressurized in the channels was observed, while the O_2 and CO_2 permeance was 1.3 times higher than the control membrane. This capability was achieved also on composite membranes, obtained by deposition of poly(1,3,5-trivinyltrimethylcyclotrisiloxane) by iCVD on Polytetrafluoroethylene (PTFE) [47]. The iCVD polymer formed a so-called skin layer on the membrane (Figure 14.6c), which prevented liquid permeation, as demonstrated in Figure 14.6d: the microfluidic channels all preserved their defined shape after burst pressure tests.

14.6 Biopassivating Coatings

Implantable medical devices require biopassivating coatings for protection, electrical insulation, and maximization of the implant biocompatibility. Examples of

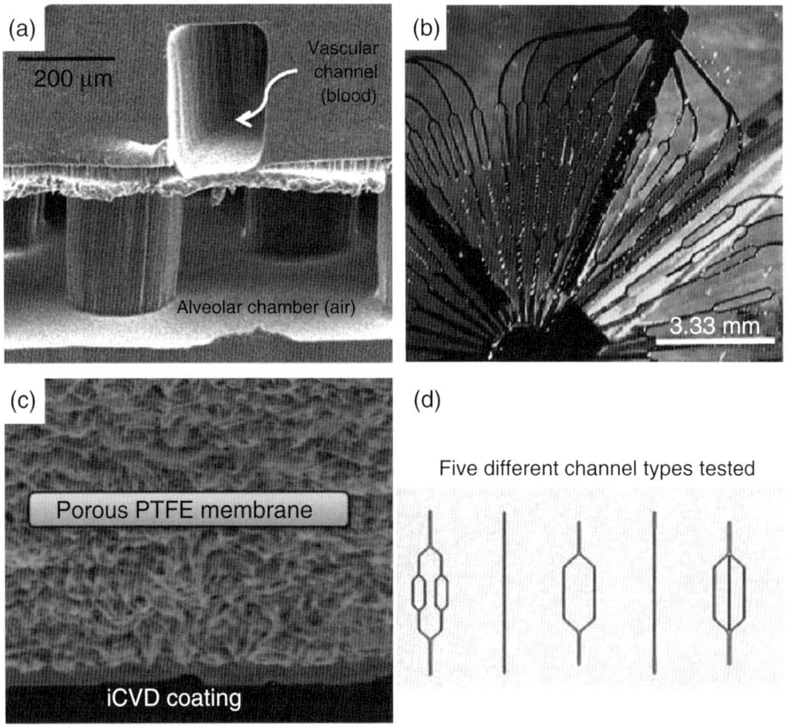

Figure 14.6 (a) Scanning electron microscopy image in cross section of a vascular channel. (b) Free-standing iCVD membrane bonded to a microfluidic device withstanding perfusion pressure of dyed water, which appears dark in the channels. (c) Cross-sectional scanning electron microscopy image of the iCVD coating on PTFE membrane. The iCVD layer acts as a skin, preventing liquid diffusion in the membrane but allowing selective passage of gas. (d) Five different microfluidic channel types tested to demonstrate that iCVD layer withstands liquid pressure. ((b) Reproduced with permission from Ref. [15]. Copyright 2011, Elsevier. (c–d) Adapted with permission from Ref. [42]. Copyright 2012, Springer.)

this type of coatings are insulating coatings on pacemaker leads or antithrombotic coatings on arterial stents. Protection from biofouling when the implant is placed in contact with biological or physiological fluids is also of extreme importance. Biofouling takes place when proteins or whole organisms attach to a surface, degrading the performance of biomedical implants, for example, occluding biomedical tubing. A complete review on CVD polymers that resist biofouling has been recently published [48]. High density of functional groups has been shown to result in excellent antifouling properties, especially ether groups or compositional gradients on the molecular scale of the proteins. The details on this are given in Chapter 13.

Specific coating design is required for neural probes. They indeed require electrical insulation, keeping intact the interaction with the surrounding biological environment. Electrical continuity between active site of the probe and neurons in the brain needs to be maintained over long periods (>6 months). The probes are made of high aspect ratio shanks; therefore, they require to be coated with smooth, even film, adding as little thickness increase as possible. Finally, also some mechanical flexibility is required so that the coating does not crack when the probe array and its lead wires are bent.

Durable, conformal, and adherent fluorocarbon and organosilicon CVD coatings have been investigated for the insulation of neural probes. It was demonstrated that the most flexible fluorocarbon films were the ones having a structure very similar to the Teflon, repeating units of CF_2, which was achieved at low deposition rates with pulsed plasma deposition and with much higher speed from HFCVD [49].

Organosilicon polymers of poly(trivinyltrimethylcyclotrisiloxane) (poly(V3D3)) were deposited by iCVD on gold wires (diameter 50 μm) [50–52]. As it can be appreciated from Figure 14.7a, the coating keeps its integrity even when the wire is bent to 250 μm, indeed no cracking or buckling is visible even at higher magnification. Such high flexibility and conformality is a unique characteristic of the iCVD method. Figure 14.7b shows that the coating retains its electrical properties for almost 3 years in a simulated biological environment.

14.7 Antimicrobial Coatings

Infections of medical implants or surgical wounds are nowadays a major cause of health threat. Biomaterial technology can offer strategic preventive solutions to fight bacterial colonization. As bacterial adhesion and biofilm formation are very superficial processes, the best solution is to engineer the biomaterial so that it has two active components: the bulk material with the desired properties of flexibility and specific topography, and a thin adherent antibacterial film coating on top. Engineering a coating with bacterial resistance and high adhesion to the soft and deformable material of the implants is not an easy task.

Silver, and specifically Ag^+, has long been known to have antimicrobial activity: it disrupts the membranes of the bacteria, so they die. Ag-containing coatings, obtained from plasma polymerization and cosputtering from a silver target, have been largely documented in the literature [53, 54]. Strongly bonded polymer layers have to be used to ensure enough stability of these coatings and avoid thickness losses due to a percolation-like effect during silver release. Controlled Ag^+ release was achieved with a bioactive multilayer, consisting of two layers of plasma-polymerized maleic anhydride [55]. Ag nanoparticles were trapped between the two polymer layers, and the top layer was opportunely cracked to create diffusive channels for Ag delivery. The aperture of the cracks and the consequent Ag delivery were controlled by the mechanical stimulation coming naturally from the implant location.

Nevertheless, the leaching of Ag from the surface leads to environmental, health, and safety concerns. Nonleaching antimicrobial polymers offer a better

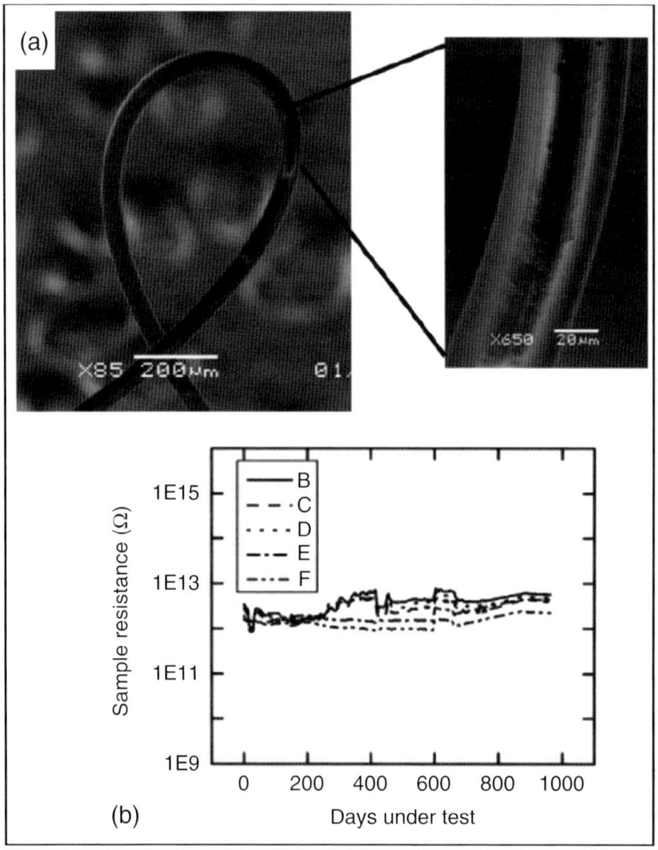

Figure 14.7 (a) Scanning electron micrograph of 50-μm-diameter gold wire coated with iCVD passivating layer. The wire is bent into a 250–μm-diameter loop without evidence of film cracking or buckling. Enlargement shows e-beam ablation of the coating as well as a lack of cracking or buckling at higher magnification. (b) Electrical resistance of the biopassivating coating in different deposition conditions (B–F) as a function of aging time. ((a,b) Reprinted with permission from Ref. [45]. Copyright 2007, American Chemical Society.)

solution, in this sense. Polymers with covalently immobilized polycationic groups create permanent nonleachable antibacterial surfaces, for example, polymers bearing quaternary ammonium groups. Atmospheric plasma in mild deposition conditions was utilized to polymerize 3-(trimethoxysilyl)-propyldimethyloctadecylammonium, retaining the monomer structure and a high density of ammonium groups [56]. The antibacterial activity showed 99.9999% reduction of *Escherichia coli* after 24 h of incubation.

The elevated chemical group retention, conformality, and flexibility granted by the iCVD method also offered significant advantages for the realization of antimicrobial coatings. Combinatorial iCVD was used to study the deposition of polydimethylaminomethylstyrene (DMAMS) [57]. The term combinatorial iCVD refers to the examination of various iCVD-processing parameters at the same time in order to optimize the deposition rate, while retaining the functional groups believed responsible for antimicrobial activity. Figure 14.8a shows the coated nylon fabric: no damages to the structure can be seen on the fibers due to the high conformality of the iCVD process [58]. Antimicrobial tests against *E. coli* and *Bacillus subtilis* bacteria showed the 99.9999% reduction rate, and the leaching tests demonstrated that the coating does not dissolve off the fabric to kill bacteria.

The DMAMS was also copolymerized with EGDA to obtain tertiary amine groups in a crosslinked polymer network [59]. At neutral pH, these groups get partially protonated. The *E. coli* bacteria killing efficiency was tested depending on the crosslinking degree of the polymer, that is, in terms of molar ratio between DMAMS and EGDA units. Figure 14.8b shows the bacteria reduction rate as a function of the molar ratio n_{DMAMS}/n_{EGDA}. The increase in bacteria killing efficiency with the number of DMAMS units is remarkable. The number of cationic units increases in the polymer network when the ratio DMAMS/EGDA is high. These charges penetrate into the bacterial membrane causing the loss of integrity. It is also important to notice that the 99.99% reduction of bacteria is obtained for copolymer coatings with a 22% crosslinking degree. A certain degree of crosslinking is important for the stability of such coatings; therefore,

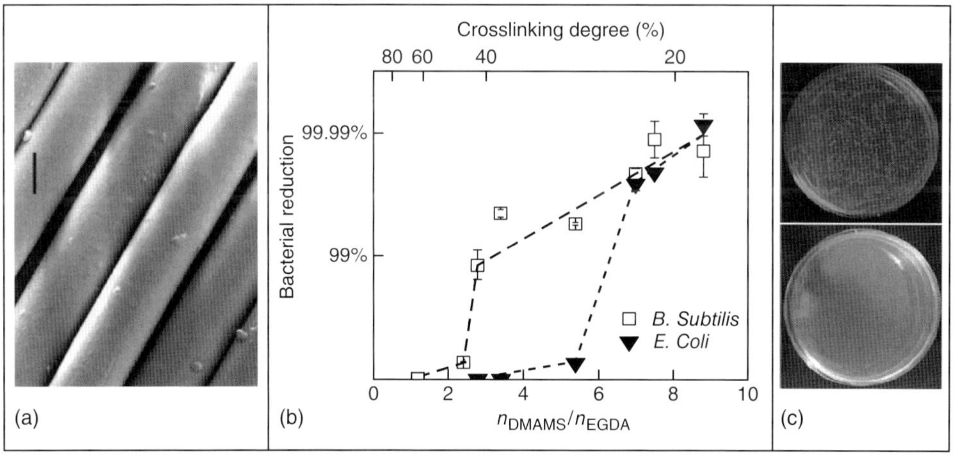

Figure 14.8 (a) Scanning electron micrograph of antimicrobial coating deposited on nylon fibers. The coating follows conformally the fiber structure. Scale bar is 10 μm. (b) Bacteria reduction rate of iCVD coating as a function of the DMAMS/EGDA molar ratio (i.e., monomer/crosslinker). The coating with 22% of crosslinking showed a reduction rate of 99.99% against *B. subtilis* and *E. coli*. (c) Optical microscopy images of bacterial colonies formed on the agar plates from antibacterial tests of control (top) and coating with 22% crosslinking (bottom). No bacteria are observable in the coated well. ((a) Adapted with permission from Ref. [52]. Copyright 2006, Elsevier. (b,c) Adapted with permission from Ref. [54]. Copyright 2011, The Royal Society of Chemistry.)

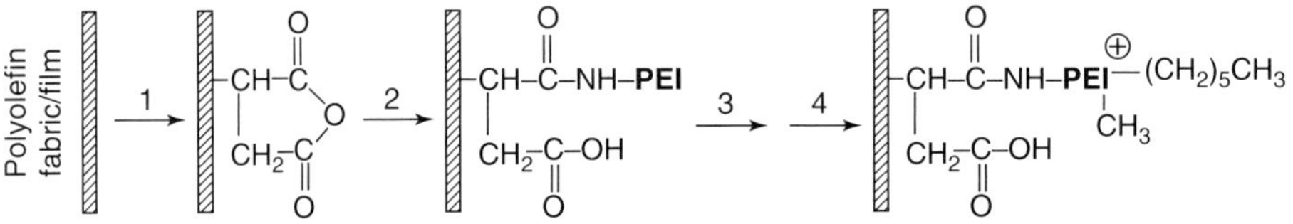

Figure 14.9 Functionalization process of polyolefin fabric to make them antimicrobial and hydrophobic. (Reproduced with permission from Ref. [55]. Copyright 2003, Springer.)

it is remarkable that a strong bactericidal activity was observed even with a significant presence of nonbiocidal EGDA components. Figure 14.8c shows the optical micrographs of the bacterial colonies formed on the agar plates, used as control for the antibacterial tests and on the surface coated with the 22% crosslinked polymer. No colony-forming unit is observable on the plate with the iCVD coating.

A multistep process to form cationic species on fabric was investigated by Lin *et al.* [60]. Figure 14.9 shows that first the polyolefin fabric was grafted with maleic anhydride to covalently bind polyethylenimine (PEI) and then subsequently N-alkylated and N-methylated. The bactericidal and fungicidal action was reached, thanks to the formation of long, moderately hydrophobic, and positively charged "nano-whiskers," capable of fatally damaging the microbial cell membranes. The same process was repeated on two polyolefin fabrics: polypropylene and polyethylene. The ability of the fabrics to kill on contact airborne human pathogenic microorganisms is reported in Table 14.1. Four bacterial representatives (*Staphylococcus aureus, S. epidermidis, Pseudomonas aeruginosa*, and *E. coli*) and the fungus, *Candida albicans*, were tested and reduction rates up to 99% were obtained. A subsequent hot filament CVD (HFCVD) deposition of a superhydrophobic coating on the top of the PEI-functionalized surfaces made the fabric also water repellent, keeping intact the bactericidal properties.

Table 14.1 Reduction rate of four bacteria and one fungus when exposed to the PEI-functionalized surface.

Microorganism	Microbial activity (%)	
	Polypropylene fabric	Polyethylene fabric
Staphylococcus aureus	90 ± 1	98 ± 1
Staphylococcus epidermidis	92 ± 4	97 ± 1
Pseudomonas aeruginosa	94 ± 1	99 ± 1
Escherichia coli	96 ± 1	99 ± 1
Candida albicans	80 ± 6	81 ± 3

Source: Reproduced with permission from Ref. [60]. Copyright 2003, Springer.
Two types of fabric were PEI functionalized: polypropylene and polyethylene.

14.8
Significance and Future Directions

Biotechnology and biomedicine are rapidly growing fields that largely benefit from advances in surface engineering and surface modification methods. The biological response depends on many different properties, for example, surface chemical composition or surface roughness and also bulk properties of the entire device as shape, elasticity, and stiffness. CVD allows to finely tailoring the surface properties, keeping intact the bulk properties of the biomaterial, opening the door to new therapies and new preventing solutions.

CVD polymerization incorporates the advantages related to a vacuum deposition method (e.g., dryness, conformality) to the versatility of organic synthesis to deposit functional and responsive materials. The polymerization methods discussed in this chapter, that is, iCVD, oCVD, PECVD, VPP, and parylene, are all based on low substrate temperature, thus allowing the coating of labile materials as fabrics, paper, or plastic. The polymers are synthetized with high purity, as required by biological applications, because the initial monomers are highly pure and the whole process happens in controlled vacuum environments without the use of solvents, leachable, plasticizers, or catalyzers.

The full degree organic functional group retention is highly desirable to address the high specificity of biological interactions. High density of functional groups is required to covalently bond biomolecules to the surface in order to induce cell attachment and growth on the surface in tissue regenerative medicine. Also the specificity and the timeliness of the biosensors' response are strictly connected to the surface chemical composition. iCVD and oCVD have shown retention of 100% of functionalities at high deposition rates (about 100 nm/min).

Conformality to nonplanar material objects is another key requirement for successful surface modification of biomaterials. Very often, the latter has tubular shape or pores and other kinds of nonplanar geometries, which are very difficult to coat by typical solution processing.

Most of the cell – biomaterial interaction happens in aqueous media, and, in general, biomaterials are subject to mechanical stress. The adhesion and robustness of the coating is essential for the integration of the surface modification step into the biodevice fabrication. Delamination of biopassivating layers leads inevitably to implant failures, for example. CVD has demonstrated to be effective also in this field: organosilicon coatings were tested for over 3 years, showing constant performances.

In conclusion, many applications can benefit from the CVD polymer properties, including, but not limited to, biosensors, controlled release, tissue engineering, and bioMEMS based on membranes, microfluidic channels. The outstanding potentiality of CVD is especially appreciable for the applications for which wet chemistry methods are not possible. The diversity of applications for CVD polymers is anticipated to expand for the new generation of functional surfaces and devices.

References

1. Lendlein, A., Rehahn, M., Buchmeiser, M.R., and Haag, R. (2010) Polymers in biomedicine and electronics. *Macromol. Rapid Commun.*, **31**, 1487–1491.
2. Mitragotri, S. and Lahann, J. (2009) Physical approaches to biomaterial design. *Nat. Mater.*, **8**, 15–23.
3. Rehfeldt, F., Engler, A.J., Eckhardt, A., Ahmed, F., and Discher, D.E. (2007) Cell responses to the mechanochemical microenvironment—Implications for regenerative medicine and drug delivery. *Adv. Drug Deliv. Rev.*, **59**, 1329–1339.
4. Kong, H.J., Liu, J., Riddle, K., Matsumoto, T., Leach, K., and Mooney, D.J. (2005) Non-viral gene delivery regulated by stiffness of cell adhesion substrates. *Nat. Mater.*, **4**, 460–464.
5. Vasudev, M.C., Anderson, K.D., Bunning, T.J., Tsukruk, V.V., and Naik, R.R. (2013) Exploration of plasma-1enhanced chemical vapor deposition as a method for thin-film fabrication with biological applications. *ACS Appl. Mater. Interfaces*, **5**, 3983–3994.
6. Chen, H.Y. and Lahann, J. (2011) Designable biointerfaces using vapor-based reactive polymers. *Langmuir*, **27**, 34–48.
7. Yagüe, J.L., Coclite, A.M., Petruczok, C., and Gleason, K.K. (2013) Chemical vapor deposition for solvent-free polymerization at surfaces. *Macromol. Chem. Phys.*, **214**, 302–312.
8. Bhattacharyya, D., Senecal, K., Marek, P., Senecal, A., and Gleason, K.K. (2011) High surface area flexible chemiresistive biosensor by oxidative chemical vapor deposition. *Adv. Funct. Mater.*, **21**, 4328–4337.
9. Sun, H.-Y., Fang, C.-Y., Lin, T.-J., Chen, Y.-C., Lin, C.-Y., Ho, H.-Y., Chen, M.H.-C., Yu, J., Lee, D.-J., Chang, C.-H., and Chen, H.-Y. (2014) Thiol-reactive parylenes as a robust coating for biomedical materials. *Adv. Mater. Interfaces*. doi: 10.1002/admi.201400093.
10. Refojo, M.F. and Leong, F.-L. (1981) Poly(methyl acrylate-co-hydroxyethyl acrylate) hydrogel implant material of strength and softness. *J. Biomed. Mater. Res.*, **15**, 497–509.
11. Baxamusa, S.H. and Gleason, K.K. (2009) Random copolymer films with molecular-scale compositional heterogeneities that interfere with protein adsorption. *Adv. Funct. Mater.*, **19**, 3489–3496.
12. Alf, M.E., Hatton, T.A., and Gleason, K.K. (2011) Initiated chemical vapor deposition of responsive polymeric surfaces. *Thin Solid Films*, **519**, 4412–4414.
13. Baxamusa, S.H., Montero, L., Dubach, J.M., Clark, H.A., Borros, S., and Gleason, K.K. (2008) Protection of sensors for biological applications by photoinitiated chemical vapor deposition of hydrogel thin films. *Biomacromolecules*, **9**, 2857–2862.
14. Bose, R.K. and Lau, K.K.S. (2007) Initiated chemical vapor deposition of poly(2-hydroxyethyl methacrylate) hydrogels. *Macromol. Biosci.*, **7**, 429–434.
15. Bose, R.K. and Lau, K.K.S. (2010) Mechanical properties of ultrahigh molecular weight PHEMA hydrogels synthesized using initiated chemical vapor deposition. *Biomacromolecules*, **11**, 2116–2122.
16. Yague, J.L. and Gleason, K.K. (2012) Systematic control of mesh size in hydrogels by initiated chemical vapor deposition. *Soft Matter*, **8**, 2890–2894.
17. Ozaydin-Ince, G., Dubach, J.M., Gleason, K.K., and Clark, H.A. (2011) Microworm optode sensors limit particle diffusion to enable in vivo measurements. *Proc. Natl. Acad. Sci. U.S.A.*, **108**, 2656–2661.
18. Montero, L., Gabriel, G., Guimerà, A., Villa, R., Gleason, K.K., and Borrós, S. (2012) Increasing biosensor response through hydrogel thin film deposition: Influence of hydrogel thickness. *Vacuum*, **86**, 2102–2104.
19. Ozaydin-Ince, G., Armagan, E., Erdogan, H., Buyukserin, F., Uzun, L., and Demirel, G. (2013) One-dimensional surface-imprinted polymeric nanotubes for specific biorecognition by initiated chemical vapor deposition (iCVD). *ACS Appl. Mater. Interfaces*, **5**, 6447–6452.
20. Jeong, G.M., Seong, H., Kim, Y.S., Im, S.G., and Jeong, K.J. (2014) Site-specific

immobilization of proteins on nonconventional substrates via solvent-free initiated chemical vapour deposition (iCVD) process. *Polym. Chem.*, **5**, 4459–4465.
21. Jeong, B., Bae, Y.H., Lee, D.S., and Kim, S.W. (1997) Biodegradable block copolymers as injectable drug-delivery systems. *Nature*, **388**, 860–862.
22. Coclite, A.M. (2013) Smart surfaces by initiated chemical vapor deposition. *Surf. Innovations*, **1**, 6–14.
23. Langer, R. (1998) Drug delivery and targeting. *Nature*, **392**, 5–10.
24. Lin, C.-C. and Anseth, K.S. (2009) PEG hydrogels for the controlled release of biomolecules in regenerative medicine. *Pharm. Res.*, **26**, 631–643.
25. McInnes, S.J.P., Szili, E.J., Al-Bataineh, S.A., Xu, J., Alf, M.E., Gleason, K.K., Short, R.D., and Voelcker, N.H. (2012) Combination of iCVD and porous silicon for the development of a controlled drug delivery system. *ACS Appl. Mater. Interfaces*, **4**, 3566–3574.
26. Lau, K.K.S. and Gleason, K.K. (2007) All-dry synthesis and coating of methacrylic acid copolymers for controlled release. *Macromol. Biosci.*, **7**, 429–434.
27. Ozaydin-Ince, G., Gleason, K.K., and Demirel, M.C. (2011) A stimuli-responsive coaxial nanofilm for burst release. *Soft Matter*, **7**, 638.
28. Alf, M.E., Godfrin, P.D., Hatton, T.A., and Gleason, K.K. (2010) Sharp hydrophilicity switching and conformality on nanostructured surfaces prepared via initiated chemical vapor deposition (iCVD) of a novel thermally responsive copolymer. *Macromol. Rapid Commun.*, **31**, 2166–2172.
29. Alf, M.E., Hatton, T.A., and Gleason, K.K. (2011) Insights into thin, thermally responsive polymer layers through quartz crystal microbalance with dissipation. *Langmuir*, **27**, 10691.
30. Alf, M.E., Hatton, T.A., and Gleason, K.K. (2011) Novel N-isopropylacrylamide based polymer architecture for faster LCST transition kinetics. *Polymer*, **52**, 4429.
31. Dhandayuthapani, B., Yoshida, Y., Maekawa, T., and Kumar, D.S. (2011) Polymeric scaffolds in tissue engineering application: a review. *Int. J. Polym. Sci.*, **2011**, Article ID290602, 1–19
32. Rashidi, H., Yang, J., and Shakesheff, K.M. (2014) Surface engineering of synthetic polymer materials for tissue engineering and regenerative medicine applications. *Biomater. Sci.* doi: 10.1039/c3bm60330j.
33. Foerch, R., Zhang, Z., and Knoll, W. (2005) Soft plasma treated surfaces: tailoring of structure and properties for biomaterial applications. *Plasma Processes Polym.*, **2**, 351–372.
34. Desmet, T., Morent, R., De Geyter, N., Leys, C., Schacht, E., and Dubruel, P. (2009) Nonthermal plasma technology as a versatile strategy for polymeric biomaterials surface modification: a review. *Biomacromolecules*, **10**, 2351–2378.
35. Gandhiraman, R.P., Muniyappa, M.K., Dudek, M., Coyle, C., Volcke, C., Killard, A.J., Burham, P., Daniels, S., Barron, N., Clynes, M., and Cameron, D.C. (2010) Interaction of plasma deposited HMDSO-based coatings with fibrinogen and human blood plasma: the correlation between bulk plasma, surface characteristics and biomolecule interaction. *Plasma Processes Polym.*, **7**, 411–421.
36. Rosso, F., Marino, G., Muscariello, L., Cafiero, G., Favia, P., D'Aloia, E., d'Agostino, R., and Barbarisi, A. (2006) Adhesion and proliferation of fibroblasts on RF plasma-deposited nanostructured fluorocarbon coatings: evidence of FAK activation. *J. Cell. Physiol.*, **207**, 636–643.
37. Detomaso, L., Gristina, R., Senesi, G.S., d'Agostino, R., and Favia, P. (2005) Stable plasma-deposited acrylic acid surfaces for cell culture applications. *Biomaterials*, **26**, 3831–3841.
38. Harsch, A., Calderon, J., Timmons, R.B., and Gross, G.W. (2000) Pulsed plasma deposition of allylamine on polysiloxane: a stable surface for neuronal cell adhesion. *J. Neurosci. Methods*, **98**, 135–144.
39. Zelzer, M., Albutt, D., Alexander, M.R., and Russell, N.A. (2012) The role of albumin and fibronectin in the adhesion of fibroblasts to plasma polymer

surfaces. *Plasma Processes Polym.*, **9**, 149–156.

40. Da Ponte, G., Sardella, E., Fanelli, F., d'Agostino, R., and Favia, P. (2011) Trends in surface engineering of biomaterials: atmospheric pressure plasma deposition of coatings for biomedical applications. *Eur. Phys. J. Appl. Phys.*, **56**, 24023.
41. Tekin, H., Ozaydin-Ince, G., Tsinman, T., Gleason, K.K., Langer, R., Khademhosseini, A., and Demirel, M.C. (2011) Responsive microgrooves for the formation of harvestable tissue constructs. *Langmuir*, **27**, 5671–5679.
42. Pfluger, C.A., Burkey, D.D., Wang, L., Sun, B., Ziemer, K.S., and Carrier, R.L. (2010) Biocompatibility of plasma enhanced chemical vapor deposited poly(2-hydroxyethyl methacrylate) films for biomimetic replication of the intestinal basement membrane. *Biomacromolecules*, **11**, 1579–1584.
43. Bose, R.K. and Lau, K.K.S. (2011) Initiated chemical vapor deposition of poly(2-hydroxyethyl methacrylate) hydrogels. *Thin Solid Films*, **519**, 4415–4417.
44. Marı-Buye, N., O'Shaughnessy, S., Colominas, C., Semino, C.E., Gleason, K.K., and Borros, S. (2009) Functionalized, swellable hydrogel layers as a platform for cell studies. *Adv. Funct. Mater.*, **19**, 1276–1286.
45. Bong, K.W., Xu, J.J., Kim, J.H., Chapin, S.C., Strano, M.S., Gleason, K.K., and Doyle, P.S. (2012) Non-polydimethylsiloxane devices for oxygen-free flow lithography. *Nat. Commun.*, **3**, 805.
46. Sreenivasan, R., Bassett, E.K., Hoganson, D.M., Vacanti, J.P., and Gleason, K.K. (2011) Ultra-thin, gas permeable free-standing and composite membranes for microfluidic lung assist devices. *Biomaterials*, **32**, 3883–3889.
47. Sreenivasan, R., Bassett, E.K., Cervantes, T.M., Hoganson, D.M., Vacanti, J.P., and Gleason, K.K. (2012) Solvent-free surface modification by initiated chemical vapor deposition to render plasma bonding capabilities to surfaces. *Microfluid. Nanofluid.*, **12**, 835–839.
48. Yang, R., Asatekin, A., and Gleason, K.K. (2012) Design of conformal, substrate-independent surface modification for controlled protein adsorption by chemical vapor deposition (CVD). *Soft Matter*, **8**, 31–43.
49. Lau, K.K.S., Caulfield, J.A., and Gleason, K.K. (2000) Structure and morphology of fluorocarbon films grown by hot filament chemical vapor deposition. *Chem. Mater.*, **12**, 3032.
50. O'Shaughnessy, W.S., Murthy, S.K., Edell, D.J., and Gleason, K.K. (2007) Stable biopassive insulation synthesized by initiated chemical vapor deposition of poly(1,3,5-trivinyltrimethylcyclotrisiloxane). *Biomacromolecules*, **8**, 2564–2570.
51. O'Shaughnessy, W.S., Edell, D.J., and Gleason, K.K. (2008) Initiated chemical vapor deposition of biopassivation coatings. *Thin Solid Films*, **516**, 684–686.
52. O'Shaughnessy, W.S., Edell, D.J., and Gleason, K.K. (2009) Initiated chemical vapor deposition of a siloxane coating for insulation of neural probes. *Thin Solid Films*, **517**, 3612–3614.
53. Zanna, S., Saulou, C., Mercier-Bonin, M., Despax, B., Raynaud, P., Seyeux, A., and Marcus, P. (2010) Ageing of plasma-mediated coatings with embedded silver nanoparticles on stainless steel: an XPS and ToF-SIMS investigation. *Appl. Surf. Sci.*, **256**, 6499.
54. Beier, O., Pfuch, A., Horn, K., Weisser, J., Schnabelrauch, M., and Schimanski, A. (2013) Low temperature deposition of antibacterially active silicon oxide layers containing silver nanoparticles, prepared by atmospheric pressure plasma chemical vapour deposition. *Plasma Processes Polym.*, **10**, 77.
55. Kulaga, E., Ploux, L., Balan, L., Schrodj, G., and Roucoules, V. (2014) Mechanically responsive antibacterial plasma polymer coatings for textile biomaterials. *Plasma Processes Polym.*, **11**, 63–79.
56. Sarghini, S., Paulussen, S., and Terryn, H. (2011) Atmospheric pressure plasma technology: a straightforward deposition of antibacterial coatings. *Plasma Processes Polym.*, **8**, 59–69.

57. Martin, T.P. and Gleason, K.K. (2008) Combinatorial initiated CVD for polymeric thin films. *Chem. Vap. Deposition*, **12**, 685–691.
58. Martin, T.P., Kooi, S.E., Chang, S.H., Sedransk, K.L., and Gleason, K.K. (2007) Initiated chemical vapor deposition of antimicrobial polymer coatings. *Biomaterials*, **28**, 909–911.
59. Ye, Y., Song, Q., and Mao, Y. (2011) Single-step fabrication of non-leaching antibacterial surfaces using vapor crosslinking. *J. Mater. Chem.*, **21**, 257–262.
60. Lin, J., Murthy, S.K., Olsen, B.D., Gleason, K.K., and Klibanov, A.M. (2003) Making thin polymeric materials, including fabrics, microbicidal and also water-repellent. *Biotechnol. Lett.*, **25**, 1661–1665.

15
Encapsulation, Templating, and Patterning with Functional Polymers

Gozde Ozaydin Ince

15.1
Introduction

The utilization of vapor-deposited polymers in fields that so far have been dominated by solution-synthesized polymers has significantly increased in the recent years. In addition to the elimination of hazardous solvents or protection of delicate substrates, improved conformality was among the benefits that prompted the research groups to adopt the vapor deposition techniques.

Conformality that defines how well the polymer coating follows the contour of a surface is a critical parameter for applications that involve high aspect ratio substrates. The need to preserve the thickness and the composition of the coatings over the surface stems from high performance requirements and standardization. Challenges in attaining high degrees of conformality via solution polymerization techniques due to capillary action could be overcome by using polymerization methods that utilize the vapors of the precursors. Controlling the amount of precursor molecules that adsorb on the surface or the reaction rate of polymerization on the surface are the ways to tune the conformality of the coatings. In Chapter 5, the active surface mechanisms during deposition that affect the conformality degree of the coating are discussed, and the relations between the surface monomer concentration, the polymerization rate, and conformality are derived.

In this chapter, the applications of conformal coatings deposited by chemical vapor deposition (CVD) methods are discussed. In the first part, conformal coatings of functional polymers for the encapsulation of 1D and 2D structures with an emphasis on carbon nanotubes and micro/nanostructures are introduced. Chemical and topographical patterning of surfaces via selectively coating or etching the surface using various masks is presented in the second part. In the final part, studies on the utilization of CVD techniques in the synthesis of polymeric micro/nanostructures, which involve conformal coating of sacrificial templates, are discussed.

CVD Polymers: Fabrication of Organic Surfaces and Devices, First Edition. Edited by Karen K. Gleason.

15.2
Encapsulation of 1D and 2D Structures with Functional Polymers

Encapsulation of micro/nanostructures with polymer coatings significantly widens the application areas of these structures due to new functionalities attained. Through encapsulation with functional polymers, surfaces of the inorganic structures can be chemically modified to control their interactions with the surrounding. These strong interactions enable the immobilization of the micro/nanostructures on surfaces with the potential use in sensor applications. Furthermore, diffusion of molecules out of and into these structures can be controlled by tuning the physical or chemical properties of the polymer coating. Interactions of the functional coatings can also be used to guide the structures to specific locations for targeted drug delivery applications.

A vast number of encapsulation methods exist depending on the coating material and the application, including emulsion polymerization, dip coating, spraying, or extrusion methods. Most of the conventional methods used for particle encapsulation require use of solvents that poses challenges due to interactions between the solvent and particle or agglomeration of the particles in the solution. Recently, vapor-phase polymer-coating techniques have been gaining attention as alternative encapsulation methods. The absence of solvents in the process and high degrees of conformality achieved are major advantages of these techniques.

15.2.1
Encapsulation of Carbon Nanotubes (CNTs)

The interest in CNTs (carbon nanotubes) has significantly increased since their discovery in 1991 [1], thanks to their high electrical and thermal conductivities and mechanical strength. However, for their integration in devices such as sensors or composite membranes, additional functionalization of the CNT surfaces is needed. Traditionally, surface modification of the CNTs is done by chemical functionalization, involving oxidative processes generating defects on the walls as attachment sites or chemical polymerization directly on the CNT surface. These processes generally involve solvents or oxidants that reduce the electrical conductivity of the CNTs, and preserving the vertical alignment can be challenging. Conformal nature of the CVD polymerization enables encapsulation of high aspect ratio CNTs, eliminating the problems encountered with the wet processing techniques. Operating under the conformal regime where the monomer surface concentration and the surface reaction rate are low, full encapsulation of the CNTs with functional polymers can be achieved.

Ye *et al.* [2] utilized initiated chemical vapor deposition (iCVD) technique to conformally coat VACNTs (vertically aligned carbon nanotubes) with a pH-responsive hydrogel poly(methacrylic acid-*co*-ethylene glycol diacrylate) (P(MAA-*co*-EGDA)). The absence of solvents in the coating process enabled to preserve the alignment of the CNTs and to control the thickness of the polymer encapsulating the CNTs. Figure 15.1a shows the VACNTs before and after

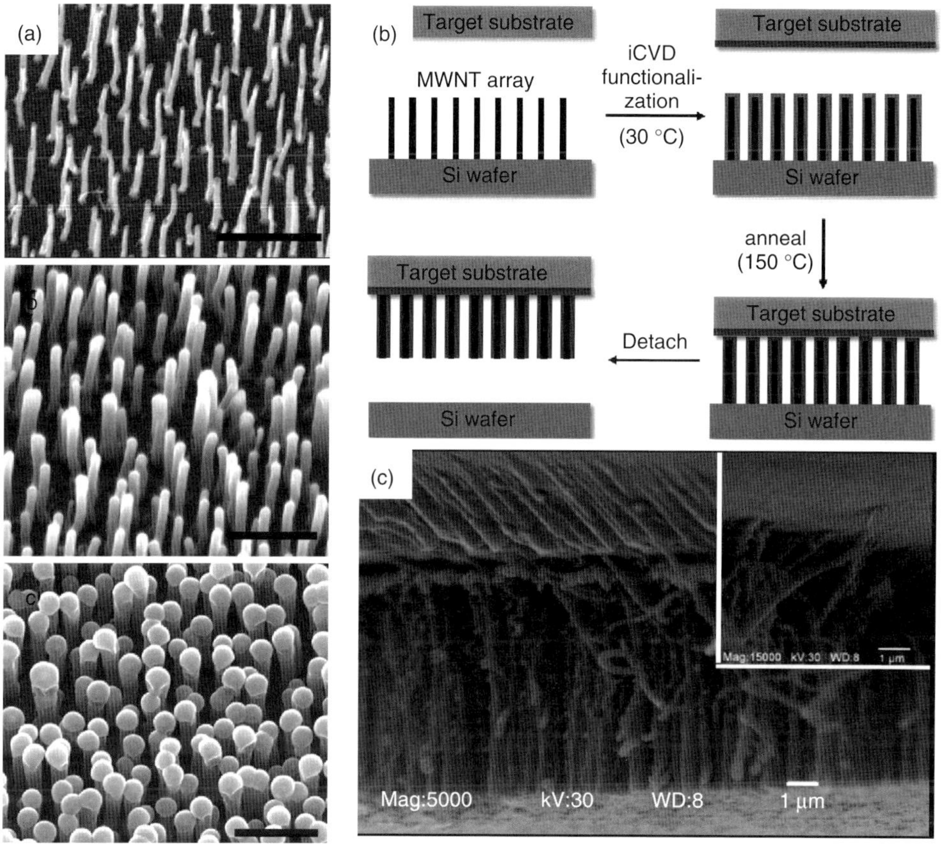

Figure 15.1 (a) SEM images of pristine (top image) VACNTs and VACNTs coated with hydrogels of thicknesses 50 and 140 nm (middle and bottom images). (b) Schematic of the transfer process of MWNTs (multiwalled nanotubes). Functionalization of both the substrate and MWCNT surfaces with iCVD PGMA polymer promotes adhesion upon contact of the surfaces at high temperature. Strong bonding between the target substrate and the MWCNTs due to the self-crosslinking of PGMA enables removal of the MWCNTs from Si wafer. (c) SEM images of cross section of the transferred MWNTs. References: (a) [2], (b) [3], (c) [3].

the P(MAA-*co*-EGDA) deposition. Conformal encapsulation of the CNTs was achieved at 50 nm and the vertical alignment of the nanotubes was preserved. On the other hand, at coating thickness of 140 nm mushroom-like structures formed on top of the CNTs due to accumulation of the polymer coating as a result of the difference in the mass transport of the reactants.

The pH-responsive behavior of the polymer coating was maintained and the surface wettability could be tuned by controlling the swelling ratio at different pH values. At pH 2, the pristine, hydrophobic VACNT film had a contact angle of 112° while the hydrogel-coated VACNT films had significantly lower contact angles of 18° and 24° at coating thickness values of 50 and 140 nm, respectively. The enhanced wettability obtained with the hydrogel-coated CNTs confirmed the successful encapsulation of the CNTs and the retention of the functional groups of the polymer.

In a different study, Ye *et al.* [3] conformally coated MWCNTs (multiwalled carbon nanotubes) with poly(glycidyl methacrylate) (PGMA) to modify the nanotube surfaces with epoxy functionality for transfer purposes. The PGMA-coated MWCNTs were brought into contact with a planar target substrate with PGMA coating. Upon applying heat, crosslinking reactions between the epoxy groups of the two surfaces occurred, enabling the transfer of the MWCNTs to the target substrate (Figure 15.1b). The transferred CNTs were well aligned and firmly attached to the substrate (Figure 15.1c).

The transferred MWCNTs had improved mechanical properties as a result of stronger crosslinked PGMA support and improved adhesion between the CNTs and the substrate. The elastic modulus of the encapsulated MWCNTs was measured to increase more than 650% after the transfer process. Furthermore, encapsulation and transfer improved the wetting stability of the nanotubes. While the pristine nanotubes collapsed upon wetting and subsequent drying, encapsulated nanotubes, although formed clusters due to capillary force, could resist collapse.

15.2.2
Encapsulation of Micro/Nanostructures

Conformal CVD techniques can be utilized in the surface functionalization of a wide range of nano/microparticles with polymers. Lau and Gleason [4], in a custom-built rotary particle bed reactor, coated glass microspheres with PGMA using iCVD technique. The precursor vapors were fed into the rotating chamber where the heated filaments initiated the polymerization. The rotary reactor enabled conformal coating of the microspheres at low coating thickness of about 135 nm without any agglomeration of the particles (Figure 15.2a). The uniformity and stability of the polymer coatings on the particles were demonstrated by immobilizing the FTSC fluorescent dye on the particle surfaces. The uniform fluorescent ring observed around the particles confirmed the stability and conformality of the PGMA coating (Figure 15.2b).

Using a similar iCVD rotary reactor, Parker *et al.* [5] encapsulated NaCl particles with PGMA to act as a barrier layer. The dissolution of the coated NaCl particles in water was observed to be 13 times slower than the bare NaCl particles. Furthermore, dissolution of the Al-coated NaCl particles encapsulated by PGMA layer was 72 times slower compared to that of Al-coated NaCl particles. The delay in the dissolution of PGMA-coated NaCl particles confirmed the improved barrier properties of the conformal PGMA coating. In the case of Al-coated NaCl particles, the conformal nature of the iCVD process also enabled to fill in the pinholes of the Al layer with PGMA, further slowing the dissolution rate.

Thickness and chemical structure of the coatings, encapsulating molecules for release studies, are the dominant factors that affect the release performance.

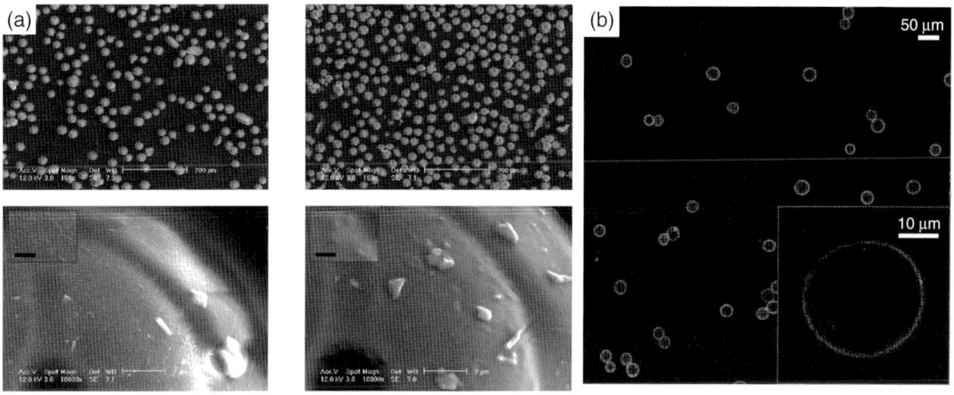

Figure 15.2 (a) SEM images of uncoated glass microspheres (top-left, bottom-left) and microspheres encapsulated with iCVD-coated PGMA (top-right, bottom-right). (b) CLSM image of PGMA-coated microspheres reacted with an amine containing fluorescent label. References: (a), (b) [4].

The iCVD technique that offers separate control of the film thickness and the chemical composition was successfully used for controlled release applications. Bose *et al.* [6] studied the effects of iCVD coating thickness and the crosslinking of two different polymer systems on the controlled release of a water-soluble crop protection compound. Among the two different polymers, PGMA and poly(cyclohexyl methacrylate) (PCHMA) studied, PCHMA had slower release rates due to its more hydrophobic nature. Crosslinking of the studied polymers was successfully carried out to further reduce the release rate. The impact of the process parameters on the coated films was also tested by operating the filament in continuous and pulsed modes. The thicker films obtained in the pulsed mode resulted in slower release and thus better control of the process.

A good control of the coating chemistry, in addition to improved release rates, also enables tuning of the permeation through the coating. Baxamusa *et al.* [7] used photoinitiated CVD (piCVD) to encapsulate biosensors with a selectively permeable coating that would allow passage of analytes but not the proteins from the medium to the sensor. This requires tuning of the polymer mesh size in the swollen state that is dependent on the crosslinking ratio. UV irradiation of the hydroxyethyl methacrylate monomer produced crosslinked coatings with average mesh sizes between 2 and 8 nm. Although this mesh size was large enough to allow the passage of the analytes, passage of proteins would be prevented, protecting the sensor surface.

Encapsulation can also be used to improve the mechanical robustness and stability of the structures in order to resist harsh treatments. This is especially important in the textile industry where the coatings are exposed to harsh treatments on a daily basis. Yoo *et al.* [8] coated a polyester mesh with a stacked polymer film composed of poly(1,3,5,7-tetravinyl-1,3,5,7-tetramethylcyclotetrasiloxane) (p(V4D4)) and poly(1*H*,1*H*,2*H*,2*H* perfluorodecylacrylate) (p(PFDA)) layers

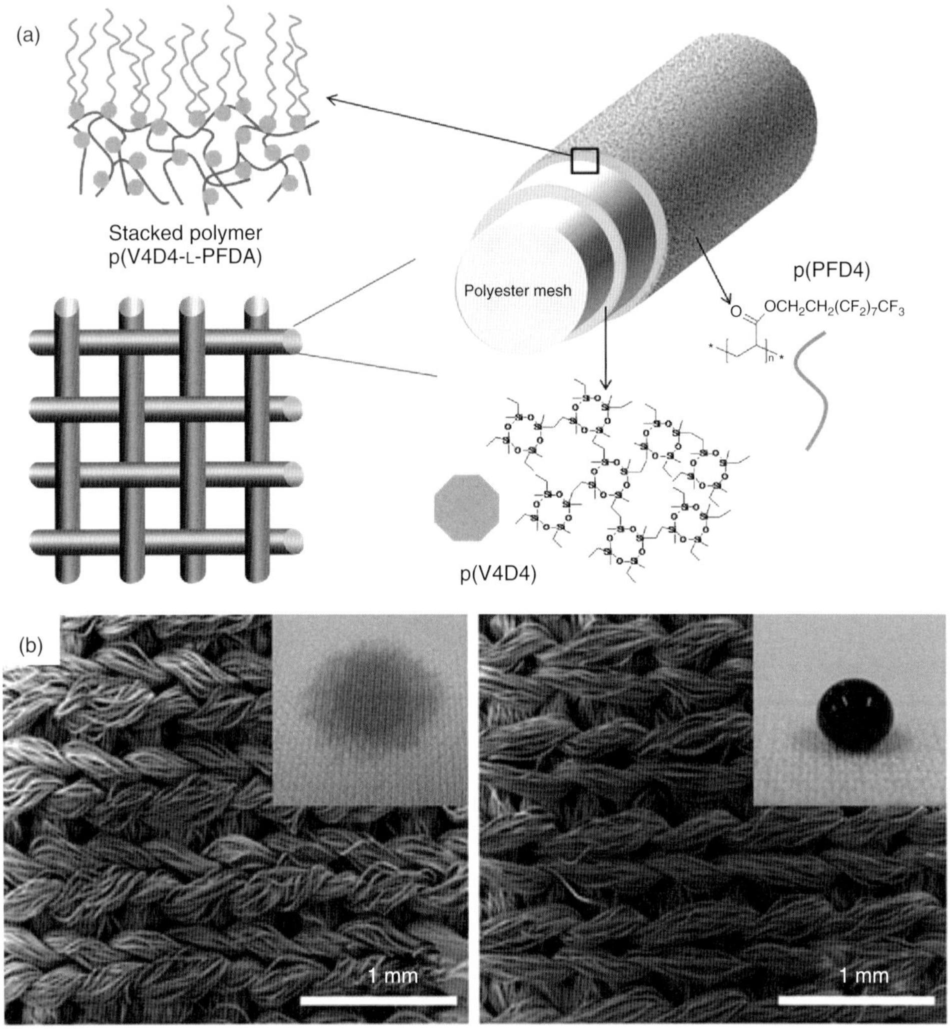

Figure 15.3 (a) Structure of a fabric fiber coated with a stacked polymer composed of p(V4D4) and p(PFDA) layers. (b) SEM images of a bare (left) and hydrophobic p((V4D4)-l-p(PFDA))-coated polyester fabric. References: (a), (b) [8].

(Figure 15.3a). The low substrate temperatures and the absence of solvents in the iCVD process enabled coating of the fabric. The mechanical robustness of the coating was enhanced by the covalent bonding between the polymer layers, which was achieved by continuously delivering the radicals in the reactor between subsequent depositions of the layers. This maintained the chain propagation reactions to continue with the new monomer molecules. The hydrophobic nature of the polymer layers when combined with the rough textile surface resulted in the formation of a superhydrophobic coating (Figure 15.3b).

Modification of the surface chemistry of the nanostructures may also be used to form self-assembled structures with improved stability. Chen *et al.* [9] coated elastomeric pillars with poly(hydroxyethyl methacrylate) and poly(methacrylic acid) (PMAA) coatings to make the hydrophobic pillar surface hydrophilic for easier water penetration. Treating the surface with water resulted in the collapse of the pillars due to capillary and adhesive forces, creating self-assembled structures. Depending on the pillar size and the polymer chemistry, dominant forces can be tuned to control the pattern formation. Furthermore, encapsulating the pillars with different monomers would make the structures sensitive to different solvents for potential applications in drug delivery or microfluidics.

Conformal coatings of microchannels with functional polymers for microfluidic applications were also reported. Kaya *et al.* [10] coated inner surfaces of metallic microtubes using iCVD with p(HEMA) (poly(hydroxyethylmethacrylate)) coatings for improved heat transfer. Microtubes with inner diameters in the range 200–1000 μm were coated with 30-nm-thick p(HEMA) films. The surface modification of the inner walls led to an increase in the heat transfer surface area, improved liquid replenishment following bubble departure, and extended critical heat flux, achieving an increase by 109% in the heat transfer coefficients compared to bare microtubes. In a similar study, Çıkım *et al.* [11] systematically modified the thickness of the p(HEMA) coatings in the range 50–150 nm and observed that increasing the thickness of the polymer coatings resulted in higher heat transfer coefficients and further improvements in critical heat flux values, due to increased surface roughness acting as active nucleation sites. Modification of the inner walls of the microtubes with functional polymers, by changing the coating thickness or chemical composition of the polymer, therefore, allowed to control the heat transfer capacities of the tubes for improved performance.

15.3 Patterning of Surfaces

Surface patterning is generally defined as changing the chemical or topographical properties of a surface in a specific spatial order. The contrast in the properties of neighboring sites on a patterned surface will lead to different interactions between the surface atoms and external molecules. The types and strength of these interactions can be detected or manipulated, prompting the usage of patterned surfaces in different applications. Detection of these interactions finds use in chemical sensors, molecular diagnostics, or cell behavior studies, whereas surfaces with tunable patterns can be used as optical gratings, information storage devices, or in microelectronics circuitry.

Until recently, top-down lithographic techniques have been the conventional methods to pattern surfaces. However, the high cost and low throughput of these processes increased the interest in other patterning techniques. Furthermore, as the device sizes decreased to nanometer scale, the resolution limit of the lithographic techniques was reached, increasing the need to develop higher resolution patterning methods. Some of these high-resolution, low-cost techniques include bottom-up self-assembly methods, strain-induced wrinkling, or selective chemical etching of coatings.

15.3.1 Patterning of Multifunctional Surfaces

Applications, such as sensors or molecular diagnostics that are based on differentiating between molecules that interact with the surface, require use of surfaces with multiple functionalities. By integrating different functionalities in a pattern, surfaces can be made responsive to different stimuli with spatial control of the surface properties. Fabrication of multifunctional surfaces requires selective etching or deposition of different polymers with functionalities. Generally, a mask is used to transfer the pattern onto the surface by shielding some regions of the surface from etchants or deposited polymers. Studies in this field mostly focus on adding multiple functionalities in a more controlled way using fewer steps. Developing low-cost masks for better resolution and improved coupling with the surface has also been a topic of interest.

Baxamusa *et al.* [12] using a TEM copper grid as a mask created patterned bifunctional polymer surfaces. The mask was used to etch a pattern on the first polymer layer, followed by the conformal coating of the second layer using iCVD. The use of mask during deposition ensured the coating of the exposed regions only (Figure 15.4a). The dual-patterned surface with different functionalities was obtained with the removal of the mask. The polymer film thickness and etching times determined the topography of the surface. By using two different polymers, p(HEMA) and poly(perfluorodecylacrylate) p(PFA) of hydrophilic

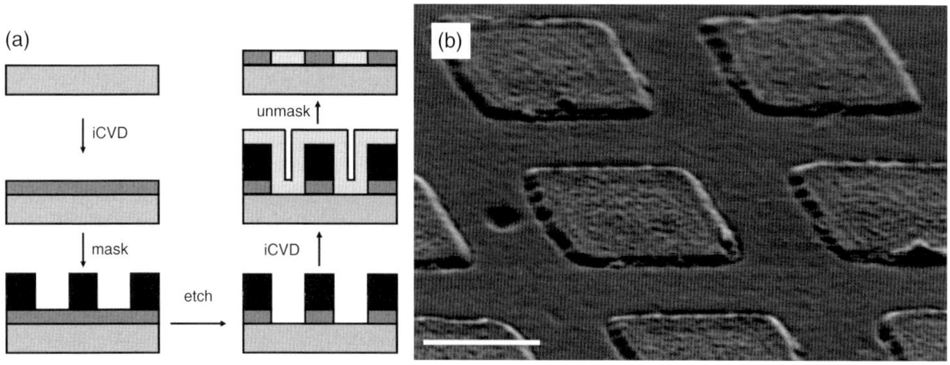

Figure 15.4 (a) Schematic of the patterning process to create patterned bifunctional surfaces. A mask is used to etch the exposed areas of an iCVD polymer film. iCVD deposition of another polymer layer after the etching process followed by removal of the mask reveals the patterned bifunctional surfaces. (b) SEM image of the patterned surface with p(PFA) in the squares and p(HEMA) surrounding the squares. Reference: [12].

and hydrophobic nature, respectively, they could create bifunctional surfaces for controlled wetting (Figure 15.4b). The microcondensation experiments they performed showed that the water droplets preferentially condensed on the hydrophilic regions of the pattern. While the functionalities on the surface could be controlled by the polymers chosen, the pattern was set by the mask, enabling the separate controls of the surface chemistry and topography.

In order to improve the resolution limit of the traditional masks, nanostructures can be used where the pattern resolution is determined by the particle size. Lee *et al.* [13] used CNTs as masks to create patterned surfaces for covalent tethering of quantum dots (QDs). They spin coated the CNTs on a poly(styrene-*alt*-maleic anhydride) (PSMa) thin film deposited on Si wafer via iCVD as a first step of patterning. As the next step, oxygen plasma was applied on the surface, etching the polymer layer, whereas the areas underneath the CNTs were shielded. Upon removal of the CNTs from the surface, the underlying polymer had the pattern of CNT shapes in slightly smaller scale due to shielding effects. The QDs were then attached to this patterned polymer, forming QD assemblies. Using this inexpensive method pattern sizes as low as 20 nm could be achieved, with the ability to spatially control the QD assemblies.

On the other hand, colloidal lithography that involves self-assembly of monodisperse colloidal particles into 2D hexagonal arrays is another simple and low-cost alternative to conventional lithography techniques. The hexagonal array of particles acts as a mask in sputtering or evaporation processes that are used to fill in the interstices to create patterned surfaces. In order to fill in the interstices, utilization of a conformal coating technique is crucial. Trujillo *et al.* [14] used conformal iCVD technique to create patterned surfaces via colloidal lithography. They first stabilized the colloidal particles on the surface, followed by treating the surface with a silane coupling agent to graft the deposited film for improved adhesion (Figure 15.5a). After the deposition of a polymer film the colloidal particles were removed, revealing a patterned surface with bowl-shaped nanostructures. Using this technique they created nanopatterned functional surfaces from organic polymers, fluoropolymers, and organosilicones with different surface properties ranging from hydrophilic to water repellent (Figure 15.5b).

Conventional lithography techniques to create patterns generally involve spin coating a polymer resist on the surface, selective crosslinking of this layer by exposing to light/e-beam through a mask, followed by development to remove the unreacted polymer. The low resolution of the patterns created using resins, the wet spin coating process and the challenges of spin coating on nonplanar surfaces, prompted the search on new resist polymers. Mao *et al.* [15] used hot filament CVD (HFCVD) to deposit PGMA thin film as a resist to improve sensitivity and resolution. The deposited PGMA had a linear polymeric structure, and all the epoxy functional groups remained intact after the CVD process. PGMA films had negative-tone feature so that only the e-beam exposed part remained after development. Furthermore, sensitivity of the films, which is the e-beam dosage required to retain 50% of the original film thickness, was high, with values around 27 $\mu C\ cm^{-2}$.

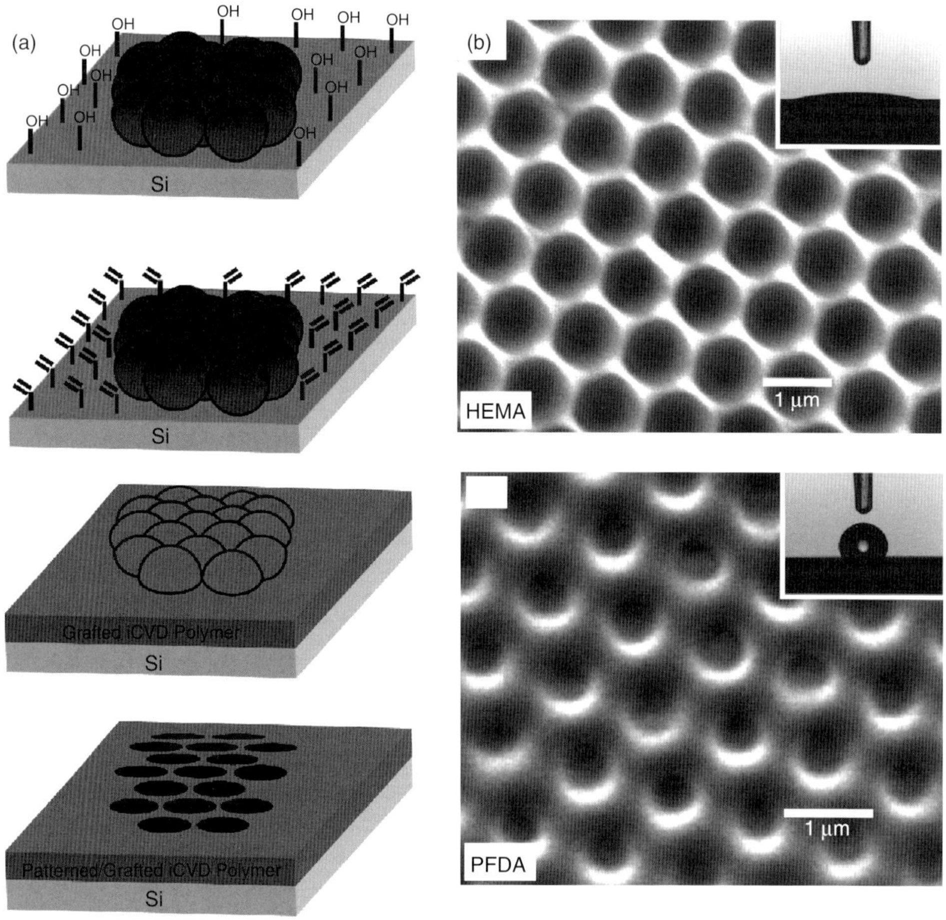

Figure 15.5 (a) Schematic of the patterning process using colloidal lithography. 2D assembly of colloid nanoparticles is deposited on a substrate that is treated with silane coupling agent for improved adhesion of a functional polymer that is subsequently deposited by iCVD. Removal of the colloids reveals the bowl-shaped patterns. (b) SEM images of the hydrophilic p(HEMA) (top-right) and hydrophobic p(PFDA) (bottom-right) surfaces obtained after the removal of the colloid template. Reference: [14].

The sensitivity and resolution of the patterns depended on the molecular weights of the PGMA films, which were regulated by changing the monomer-to-initiator ratio. The sensitivity decreased as the molecular weight of the films decreased due to larger number of chains that require high e-beam dosages to reach a certain crosslink density. On the other hand, the resolution increased as the molecular weight decreased; features as small as 80 nm could be obtained using PGMA films with molecular weight of 4700 g mol^{-1}. For development of the coatings, supercritical CO_2 was used as an alternative to conventional wet chemistry methods.

In a similar study [16], copolymers of methacrylic acid (MAA) and perfluoroalkyl ethyl methacrylate (PFEMA) were synthesized via iCVD, to improve solubility in supercritical CO_2, due to the CF_3 end groups in PFEMA. Using this all-dry lithographic process, positive tone contrast with features as small as 300 nm could be obtained. The resolution of the positive-tone features could be improved further using terpolymers of methyl *R*-chloroacrylate (MCA), MAA, and methacrylic anhydride (MAH) [17]. This improvement in the resolution was attributed to intramolecular cyclization that prevents crosslinking but leads to MAH anhydride formation. With this new polymer system, positive-tone pattern of feature sizes of 60 nm could be created.

Yoshida *et al.* [18] also studied iCVD-deposited PGMA as a lithographic polymer and investigated the effects of deposition conditions on the smoothness and the conformality of the films. Achieving conformal coatings of the PGMA polymer enables the resist to be applied on 3D structures, which is especially important for MEMS applications. They observed that GMA partial pressure/total pressure ratio was a factor in obtaining smooth films. Conformality of the depositions that is critical for patterning on 3D surfaces could be improved by operating at high substrate temperatures or low precursor partial pressures that reduced the amount of precursor molecules on the surface. The results agreed with the conformality studies reported by Baxamusa and Gleason [19].

Patterning nonplanar substrates with functional polymers has a wide range of potential applications, especially in biomimetics where curved surfaces are commonly used. However, one of the major challenges in patterning nonplanar surfaces stems from the difficulty in applying resists on a curved surface. Petruczok and Gleason [20] used iCVD to conformally coat thin polymer films on curved substrates for patterning purposes. They utilized iCVD to coat glass rods with poly(4-vinylpyridine) (P4VP) polymer and then functionalized the surface with photoactive 10,12-tricosadiynoic acid molecule (TDA). Conformality of the polymer coating on the curved glass rod was essential for the uniformity of the TDA coating; adhesion of which was via the hydrogen bonds between the acid moieties of TDA and nitrogen atoms in the pyridine rings. Therefore, conformality of the P4VP layer ensured conformal functionalization. Exposure at 245 nm through a mask and subsequent development led to the creation of negative-tone pattern. Stability of the exposed P4VP-TDA film during development was the result of hydrogen bonding between the acid moieties and the pyridine rings, as well as the photochemical crosslinking of P4VP. The same masking process was also used to create bifunctional surfaces that involved alternating layers of hydrophilic P4VP-TDA and hydrophobic p(PFDA) polymers, achieved first by coating the surface with p(PFDA) layer.

The robust P4VP-TDA layer formed after functionalization and exposure was also used as a mask to pattern curved metal surfaces. Evaporation of thin metal films onto the masked surface, followed by the etching of the polymer mask, revealed the patterned metal microstructures. This method also ensured that the mask was securely fixed on the curved surface, preventing the metal undercutting the mask.

Selective crosslinking through a mask was also used to create reversibly swelling patterns [21]. Poly(vinyl cinnamate) (PVC), a light-responsive polymer, which undergoes crosslinking reactions upon exposure to UV at 254 nm, was deposited by iCVD to create selectively crosslinked surfaces by exposing the surface to UV through a patterned mask. Using this process, patterns with low and high crosslinked regions could be created. Copolymerizing vinyl cinnamate with the thermoresponsive *N*-isopropylacrylamide monomer enabled to create functionalized patterned surfaces that were responsive to temperature changes. P(VCin-*co*-NIPAAm) copolymer was exposed through a mask that led to crosslinking of the exposed regions. PNIPAAm exhibits hydrophilic–hydrophobic transition around LCST, and the degree of the swelling depends on the crosslinking. Therefore, the exposed areas that were highly crosslinked swelled less than the unexposed regions. By drying and rewetting the sample at 20 °C, they could observe reversible swelling of the pattern (Figure 15.6).

In addition to nonplanar surfaces, porous substrates could also be successfully patterned with functional polymers. Haller *et al.* [22] used iCVD to deposit a thin layer of photoresponsive poly(*o*-nitrobenzyl methacrylate) (P*o*NBMA) on porous chromatography paper. The hydrophilic fibers of the paper were conformally coated with the hydrophobic polymer through the entire depth of the paper.

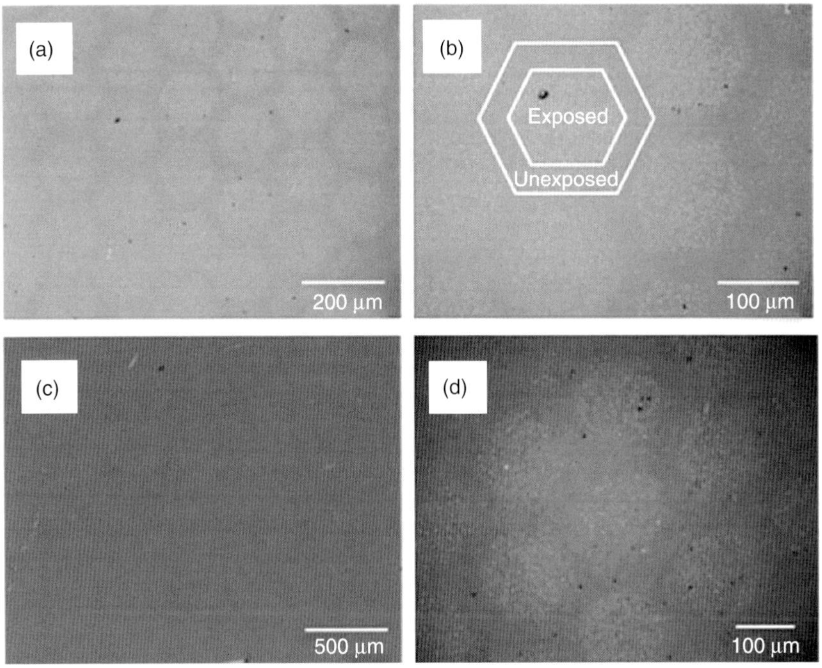

Figure 15.6 Optical images of patterned p(VCin-*co*-NIPAAm) surfaces. (a, b) Submersion in 20 °C water reveals the patterns due to the swelling of the selectively noncrosslinked regions. Patterning is reversible; disappears when dried (c) and appears when resubmersed in water (d). Reference: [21].

Upon exposure to UV, the nitrobenzyl groups of P*o*NBMA were cleaved, leaving behind the carboxylic acid groups, thus converting P*o*NBMA to PMAA, which is pH responsive. Electrostatic repulsion of the ionized carboxylic acid groups at high pH values caused PMAA to swell significantly. The surface was exposed to UV through a mask, rendering the exposed regions to become soluble at high pH. The coating in the exposed regions was then removed by washing the surface with pH 8 buffer, while the coating of the unexposed regions remained intact, resulting in the functional patterning of the surface. Using this method, the authors patterned hydrophilic channels into porous chromatography paper for paper-based diagnostics applications.

Another method to create patterned functional surfaces was demonstrated by Kwong *et al.* [23], which, as opposed to selective removal of coating, involved directed deposition of the polymer. In this process, a metal salt solution was first applied in a pattern on the surface, and then a functional polymer was deposited via iCVD. The metal salts acted as inhibitors, preventing polymerization where the solution was applied. The method was tested on porous chromatography paper, using different functional polymers such as hydrophobic PPFA, click-active poly(pentafluorophenyl methacrylate), and light-responsive P*o*NBMA. Among various metal salts tested, $CuCl_2$ and $Cu(NO_3)_2$ were the most effective in inhibiting the polymerization. This method of controlling the location of polymerization could be applied to different polymer systems and substrates and would reduce the processing steps in the fabrication of patterned surfaces.

15.3.2
Surface Wrinkling

Surface patterning methods that eliminate the use of masks are preferred for industrial applications as they reduce the cost and number of steps in the process, making them easier to scale up. One of these spontaneous pattern formation techniques is based on surface wrinkling as a result of induced strain. In this method, first a stiff coating is deposited on a stretched elastomeric substrate, followed by slow release of the strain that leads to the spontaneous formation of ordered patterns. In the small deformation limit, the wavelength of the wrinkles that form on the surface is given by

$$\lambda_c = 2\pi h \left[\frac{\bar{E}_f}{3\bar{E}_s} \right]^{1/3} \tag{15.1}$$

where h is the thickness of the stiff coating, $\bar{E}_s$ and $\bar{E}_f$ are the plane strain moduli of the substrate and the coating, respectively.

Evaporated metal thin films have generally been used as the coating because of the large elastic moduli difference with the substrate layer. However, with the advances in polymer thin-film deposition techniques, functional polymers could also be used as the coating layer for the fabrication of functional patterned surfaces.

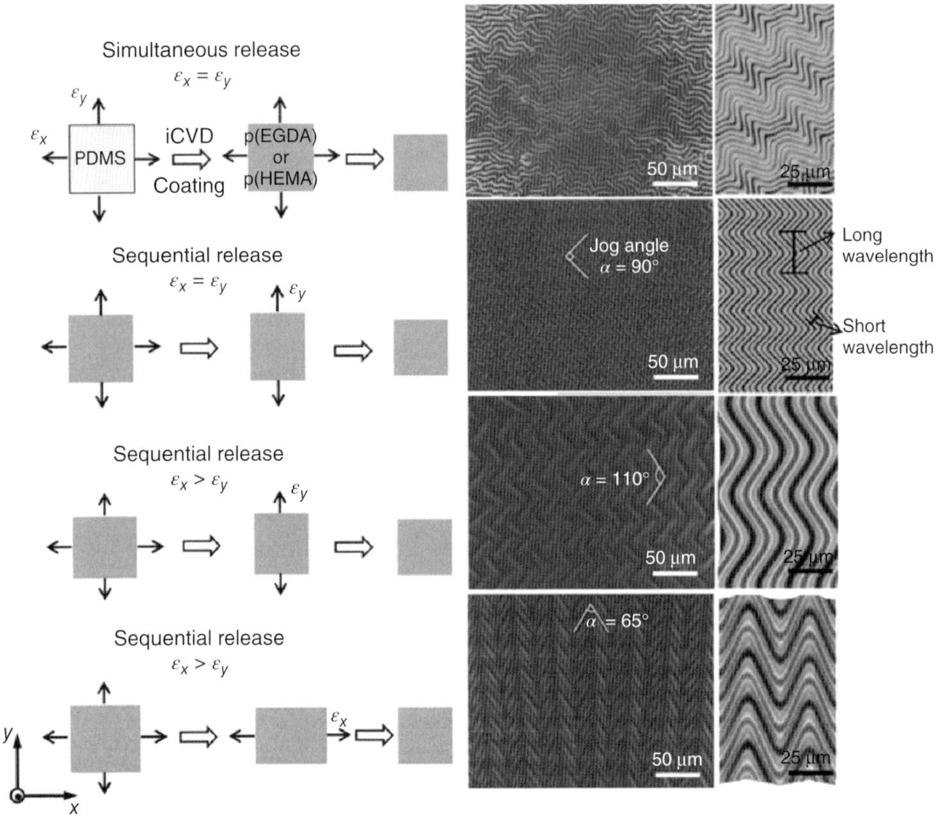

Figure 15.7 Schematics of surface wrinkling (left column). Biaxially stretched PDMS substrate is first coated with a copolymer film using iCVD and then the strain is released to form the surface patterns. Disordered patterns form upon simultaneous release of the strains whereas sequential release leads to formation of highly ordered herringbone patterns. The jog angles of the patterns depend on the order of the release of nonequibiaxial prestrains. Simulated surfaces are shown on the right column. Reference: [24].

Yin *et al.* [24] used the iCVD technique to graft p(EGDA) and p(HEMA-EGDA) polymer thin films on PDMS substrate to obtain wrinkled topologies. Upon applying the iCVD coating to a biaxially stretched PDMS substrate, the strains were released simultaneously or sequentially, producing wrinkled patterns on the polymer surface (Figure 15.7). The grafting is essential for preventing delamination. In the case of simultaneous release, energetically favorable disordered labyrinth patterns formed on the surface. On the other hand, sequential release of the strain resulted in ordered herringbone patterns with jog angles that depended on the amount of strain. Jog angle of 90° was obtained for the sequential release of an equibiaxial strain. By changing the relative amounts of the nonequibiaxial prestrain and the order of the strain release in different directions, the jog angle could be controlled. Furthermore, the angle was observed to depend only on the ratio of the applied biaxial strain.

The sequential release led to the formation of wrinkles at two different wavelengths, both increasing with the coating thickness. They concluded that the ratio between these two wavelengths depended only on the modulus ratio between the coating and substrate and is given as

$$\frac{\lambda_1}{\lambda_m} \approx 1.03(1-\nu_f^2)^{1/4}\left[\frac{\overline{E}_f}{3\overline{E}_s}\right]^{1/6} \tag{15.2}$$

This relation could be used to obtain the coating properties without measuring the film thickness.

Sequential release that led to transition from 1D to 2D patterns could also be reversed as reported by Yagüe *et al.* [25]. Sequential restretching of a herringbone pattern on a p(EGDA) coating was tested. Initial restretching in one axis resulted in the formation of 1D wrinkles. By subsequently restretching in the other axis, the initial flat surface was obtained. Herringbone patterns and flat surfaces could be obtained alternatively in cyclic sequential release and restretch processes [26], whereas this reversible behavior was not observed on the simultaneously released patterns. This facile and low-cost technique of reversible pattern formation could therefore be used to create surfaces with tunable functionalities.

15.4 Synthesis of Polymeric Micro/Nanostructures

Polymeric micro/nanostructures with applications ranging from drug delivery devices to adhesives and paints have attracted considerable interest in the recent years, prompting the development of different fabrication methods. Until recently, conventional techniques used for the synthesis of polymeric micro/nano structures followed top-down approaches, focusing mainly on lithographic methods. These methods generally involve use of masks to define the structures and etchants to remove the sacrificial layers. With the advances in probe microscopy technology, direct fabrication of micro/nano structures could be achieved via carving or etching a substrate. Although structures with fine-tuned shapes and resolution in the nanoscales can be fabricated with these methods, the long process times and low throughput are significant drawbacks. Low-cost and high-throughput techniques, such as stamping or molding, are alternative methods that have successfully produced high aspect ratio of polymeric micro/nanostructures.

Most of the conventional techniques used to synthesize polymeric structures require dissolving the polymer in a compatible solvent and evaporation of the solvent after the casting or molding process. The main drawback of these techniques lies in dissolving the polymer for processing; especially finding solvents for responsive polymers that are highly crosslinked poses a big challenge. Employing techniques that involve *in situ* polymerization during synthesis helps to overcome this challenge. CVD techniques, therefore, help to eliminate solvent-related problems, enabling fabrication of micro/nanostructures of crosslinked polymers.

15.4.1 Templating Using Porous Membranes

Fabrication of polymeric micro/nanostructures with controlled shapes and sizes can be achieved using templated CVD techniques. Generally, porous track-etched anodic aluminum oxide (AAO) or polycarbonate (PC) membranes are used as templates due to their wide range of pore sizes and high solubility in compatible solvents.

In the templated iCVD technique, the template is used as a substrate, and the polymer film growth takes place inside the pores of the template. Operating in the adsorption-limited regime, the monomer molecules can diffuse into the pores and polymerization takes place on the pore walls. After the polymer deposition via iCVD, the template membranes are dissolved in NaOH or dilute HCl solution. Damage to the polymer layer can be prevented by using a solution that selectively etches the template. The polymer layer that is deposited on the walls of the pores remains and takes the shape of the cylindrical pores, resulting in nanotubular structures. The polymer maintains this cylindrical shape even after the removal of the template.

The overall diameter and the length of the fabricated tubular structures depend on the aspect ratios of the membrane pores and, thus, can be controlled by using membranes with different pore dimensions. The wall thickness of the polymeric structures, on the other hand, depends on the deposited polymer film thickness and can be controlled by changing the deposition time or rate. This allows synthesis of structures ranging from hollow tubes to rods or fibers.

One important factor in the synthesis of micro/nanostructures with well-defined shapes is the conformality of the iCVD process. The ability to conformally coat the walls of the pores throughout the pore depth prevents clogging of the pore openings and formation of keyhole structures. Furthermore, conformality of the deposition ensures constant micro/nanotube wall thickness along the length of the tubular structures. Conformal deposition via iCVD technique is achieved by operating under low P_m/P_{sat} conditions, ensuring low monomer concentration on the surface, thus reducing the deposition rate. Low deposition rates enable the monomer molecules to diffuse inside the pores, resulting in a homogenous coating. Using iCVD technique, pores with aspect ratios ranging from 40 : 1 to 4000 : 1 can be conformally coated with functional polymers for nanotube synthesis [27] or membrane pore functionalization [28].

The ability of iCVD method to deposit stimuli-responsive polymers with protected functional groups enables fabrication of “smart micro/nanostructures” when combined with templates. Smart micro/nanostructures with tunable response to external stimuli have a wide range of potential applications. Ozaydin-Ince *et al.* [29] fabricated crosslinked p(HEMA) nanotubes using the templated iCVD method. Operating in the conformal regime, AAO membrane pores with aspect ratios of 400 : 1 were successfully coated with the polymer. By systematically changing the crosslink ratio of the deposited polymer, the swelling degrees of the synthesized nanotubes could be tuned. The protein adsorption studies showed that as the degree of crosslinking increases, the amount of adsorbed protein on the nanotube samples decreases. Therefore, protein adsorption could be controlled by tuning the swelling properties of the nanotubes, making them potential templates for biomaterials.

The hydrogel p(HEMA) nanotubes due to their biocompatible nature and high surface-to-volume ratios enable their integration in sensors for *in vivo* measurements. "Microworm" sensors with optode cores were fabricated by filling the pores of the polymer-coated AAO membranes [30]. An extra step of iCVD polymer coating was performed prior to etching to cap the openings of the pores to confine the optode inside (Figure 15.8a). The tubular shapes of the nanotubes significantly decreased the diffusion coefficient, immobilizing the sensor microworms at the

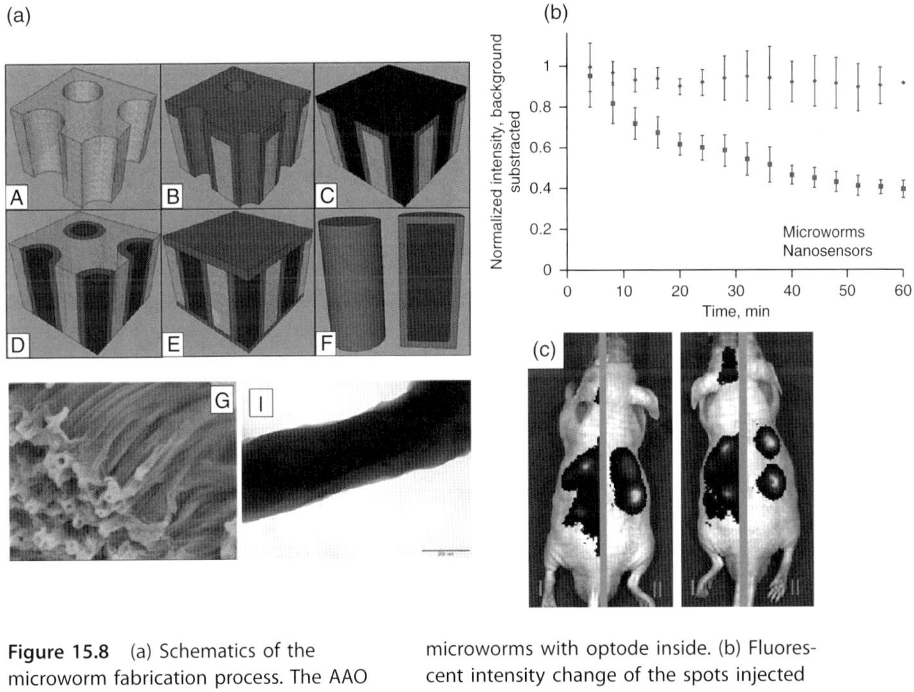

Figure 15.8 (a) Schematics of the microworm fabrication process. The AAO templates are first coated with p(HEMA) via iCVD and filled with the optode solution. The top and bottom parts of the templates are then coated with p(HEMA) to cap the optode. The templates are then etched away to release the microworms. SEM (bottom-left) and TEM (bottom-right) images show the fabricated microworms with optode inside. (b) Fluorescent intensity change of the spots injected with microworms and nanosensors over time. Diffusion of the nanosensors away from the target site leads to a decrease in the fluorescence intensity, whereas no diffusion of microworms is observed as confirmed by the stable fluorescence intensity. (c) Microworm sensors can successfully be imaged *in vivo*. Reference: [30].

site of injection for longer measurement times. Spherical nanosensors fabricated for sodium sensing lost 60% of their fluorescence intensity in 1 h because of their diffusion away from the injection spot, whereas the microworm sensors maintained their efficiency, confirming their immobility (Figure 15.8b). Furthermore, microworm sensors could successfully be imaged *in vivo* (Figure 15.8c).

The mesh size of the polymer encapsulating layer is critical for the performance of the sensors. The polymer should allow the diffusion of sodium ions but prevent permeation of larger macromolecules. Yagüe *et al.* [31] calculated the mesh size, ξ, of the p(HEMA) polymers deposited on flat Si surfaces using the formulation:

$$\xi = (1/Q)^{-1/3} l \left(\frac{2\overline{M}_c}{M_m} \right)^{1/2} C_n^{1/2} \tag{15.3}$$

where Q is the swelling ratio, l is the C-C bond length, M_m is the molecular weight, and C_n is the characteristic ratio for HEMA. $\overline{M}_c$ is the average molecular weight between crosslinks obtained using modified version of Flory's equation. They systematically varied the crosslinker, ethylene glycol diacrylate (EGDA), amount in the polymer and observed that as the crosslinker amount was increased in the films the mesh size decreased from around 2.0 to 0.5 nm for HEMA/EGDA ratios of approximately 0.93–0.45, respectively. The diffusion of the fluorescein molecules with an average diameter of 1 nm through the p(HEMA) microworms of different compositions was monitored to study the effect of mesh size. The fluorescein molecules could diffuse out of the 2-nm mesh size microworms but not through the 0.5 nm mesh size. The ability to control the mesh size of the nano/microstructures, therefore, would enable their utilization as drug delivery devices.

For delivery applications, the onset of release through the nano/microstructures needs to be well controlled. Ozaydin-Ince *et al.* [32] fabricated coaxial nanotubes with an outer shape memory polymer (SMP) layer surrounding the inner hydrogel layer (Figure 15.9a) using the templated iCVD method. The SMP poly(*tert*-butyl acrylate-*co*-diethylene glycol divinyl ether) can be shaped above its T_g of 60 °C and subsequent cooling of the polymer temporarily fixes this shape. Further heating of the polymer above its T_g helps the system to recover its original shape. In the study reported, the swelling of the inner hydrogel layer in a water medium at high temperatures enlarged the outer SMP layer of the coaxial nanotubes, and the subsequent cooling of the system fixed the nanotubes at the enlarged diameter of 350 ± 50 nm. Heating the nanotubes a second time triggered the SMP layer to shrink back to its original diameter, compressing the inner hydrogel layer. The fluorescein dye molecules that were uptaken by the nanotubes during initial swelling were released instantly with the second heating. At room temperature, the nanotubes held the dye molecules for up to 100 h before the burst release was triggered by heating (Figure 15.9b). The sensitive control of the release onset could, therefore, be achieved using coaxial nanotubes with layers of different functional polymers.

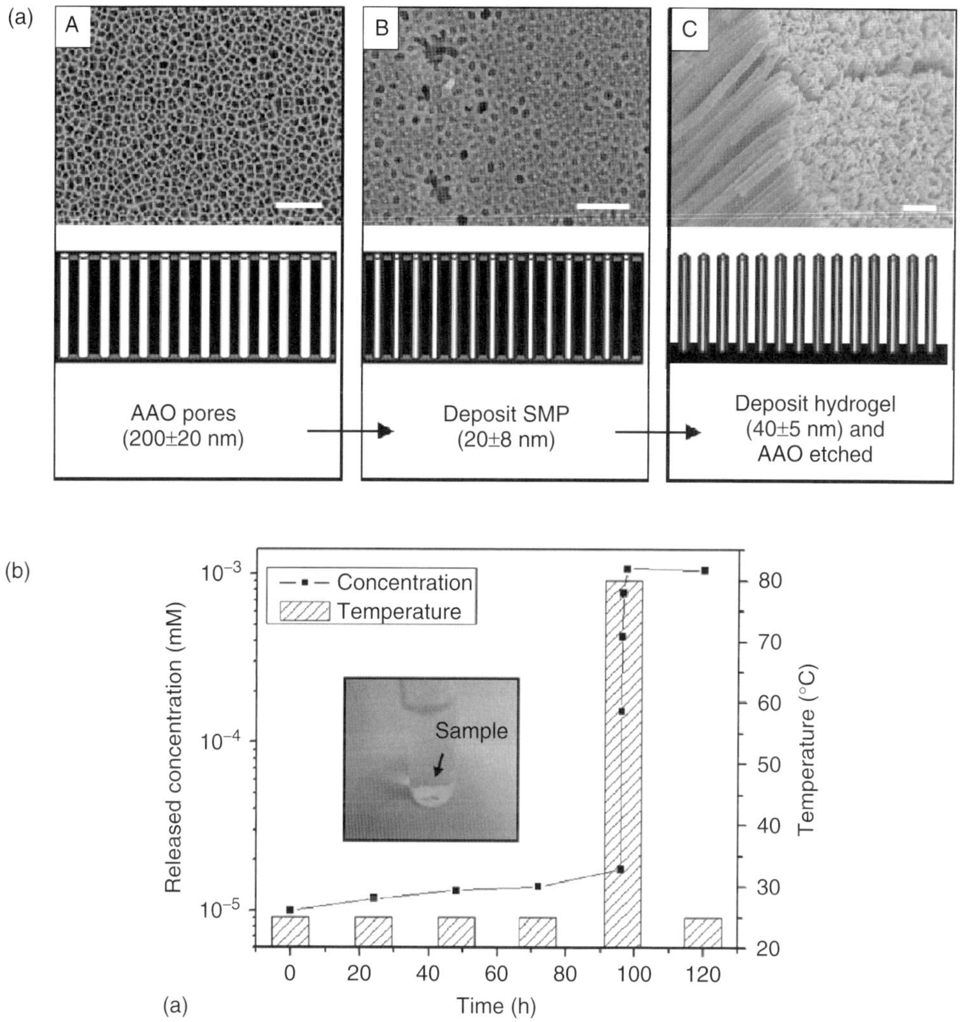

Figure 15.9 (a) Fabrication process of the coaxial nanotubes. The bare AAO templates are first coated with SMP polymer using iCVD, followed by iCVD deposition of the hydrogel layer. The template is etched to release the coaxial nanotubes. (b) The release of the dye at the end of 4 days is triggered by heating the nanotubes above the T_g of the SMP, which results in the shrinking of the outer layer. The outer SMP layer then compresses the inner hydrogel layer, causing the stress-induced burst release of the dye molecules. Reference: [32].

Furthermore, coaxial nanotubes with multilayers of functional polymers can be used in sensor applications where adjacent layers detect different molecules, improving the selectivity of the device. The application areas can be widened depending on the functionalities of the polymer layers, which are protected during fabrication via iCVD technique.

15.4.2 Micromolding

Transfer of patterns on polymer surfaces or fabrication of natural surface replicas from synthetic polymers can be easily achieved by using molding techniques. In micromolding, an elastomeric template is used as the mold to pattern polymer surfaces, making micromolding a low-cost, easy-to-apply soft-lithography technique. Combining micromolding with conformal CVD enables transfer of high-resolution patterns on crosslinked functional polymers.

Karaman *et al.* [33] fabricated a self-supporting, two-layered polymer surfaces with rose petal structures. Polyvinyl alcohol (PVA) molds of inverse petal structures were first created by pouring the PVA solution on rose petal and drying the solution to remove from the petal surface. The fabricated PVA mold was then conformally coated with hydrophobic p(PFDA) and PGMA using iCVD. In the final step, removal of the PVA mold in water resulted in superhydrophobic polymer sheets with rose petal patterns that have contact angles of $152 \pm 3°$. The conformality of the coating enabled inverse replication of the petal pattern.

Im *et al.* [34] used micromolding to form periodic nanostructures on a polymeric surface. A PDMS or poly(urethaneacrylate) (PUA) mold was pressed on the surface of a multilayered polymer film and heated above the T_g of the polymer. The top poly(propargyl methacrylate) (PPMA) layer was deposited by iCVD and was noncrosslinked, whereas the PECVD deposited bottom layer of poly(allylamine) (PAAm) was highly crosslinked, preventing its deformation above T_g temperatures. The lack of crosslinking enabled deformation of the iCVD-deposited layer above T_g and pressing the mold to the surface led to capillary rise, resulting in the nanopattern formation on the polymer surface. Patterning by micromolding exposed the bottom layer, and thus a dual functionalized nanopatterned surface was obtained. The dual functionality was tested using two different fluorescent dyes that assembled on the patterns self-sorting according to the functionality of the polymer surface. The control over the crosslinking of individual layers was critical to be able to deform one layer while keeping the other layer intact even above T_g. Furthermore, solventless CVD methods enabled processing of individual layers without damaging the adjacent polymer layer.

An important challenge in patterning via micromolding is caused by diffusion of solvents or oxygen through the mold that inhibits polymerization or damages the mold. Permeation through the mold surface can be hindered by coating the mold surface with a barrier layer, thus minimizing the interaction between the mold and the surface to be patterned. High conformality of the coating without any cracks or delamination is required for an effective barrier performance. Jeong *et al.* [35] coated the PDMS mold surface using iCVD with a hydrophobic p(PFDA) thin layer that acts as a barrier layer. The coated mold was used to fabricate polymer microparticles by pouring a photo-curable monomer on the mold, followed by photopolymerization activated by UV exposure (Figure 15.10a). The polymerized microparticles could be detached easily from the mold due the low surface energy of the barrier layer. As the microparticles retained the shape

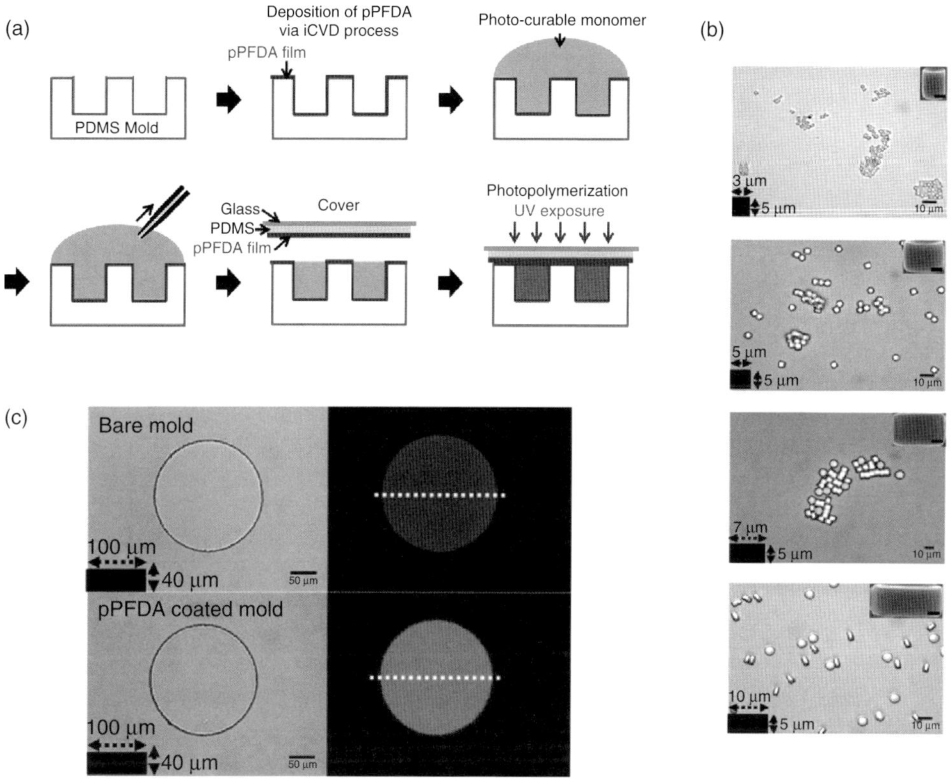

Figure 15.10 (a) Schematic of the PEG-DA microparticle fabrication process. iCVD p(PFDA) coating on the PDMS mold acts as a barrier to prevent diffusion of the photo-curable monomer. In order to prevent oxygen penetration, top side is covered and at the final step, the monomer mixture is exposed to UV for polymerization. (b) Optical microscope images of PEG-DA microparticles with diameters of 3, 5, 7, and 10 μm. (c) Fluorescence intensities of the PEG-DA microparticles loaded with FITC. Particles synthesized on bare molds had lower intensities compared to the particles synthesized on p(PFDA)-coated molds due to increased diffusion of FITC in the absence of barrier polymer layer. Reference: [35].

of the mold, microparticles with different shapes and sizes could be synthesized (Figure 15.10b). The effect of the barrier layer against diffusion through the mold was tested by pouring a monomer solution mixed with a fluorescent dye (FITC) on the mold and fabricating FITC loaded polyethylene glycol diacrylate (PEG-DA) microparticles. The higher fluorescent intensity of the microparticles synthesized on iCVD-coated molds confirmed that in the presence of the barrier layer, FITC in the prepolymer solution did not diffuse into the bulk of the mold (Figure 15.10c). The results of this study showed that the p(PFDA) coating, acting as the barrier layer could be conformally coated on the high aspect ratio mold pattern using the vapor-phase iCVD method.

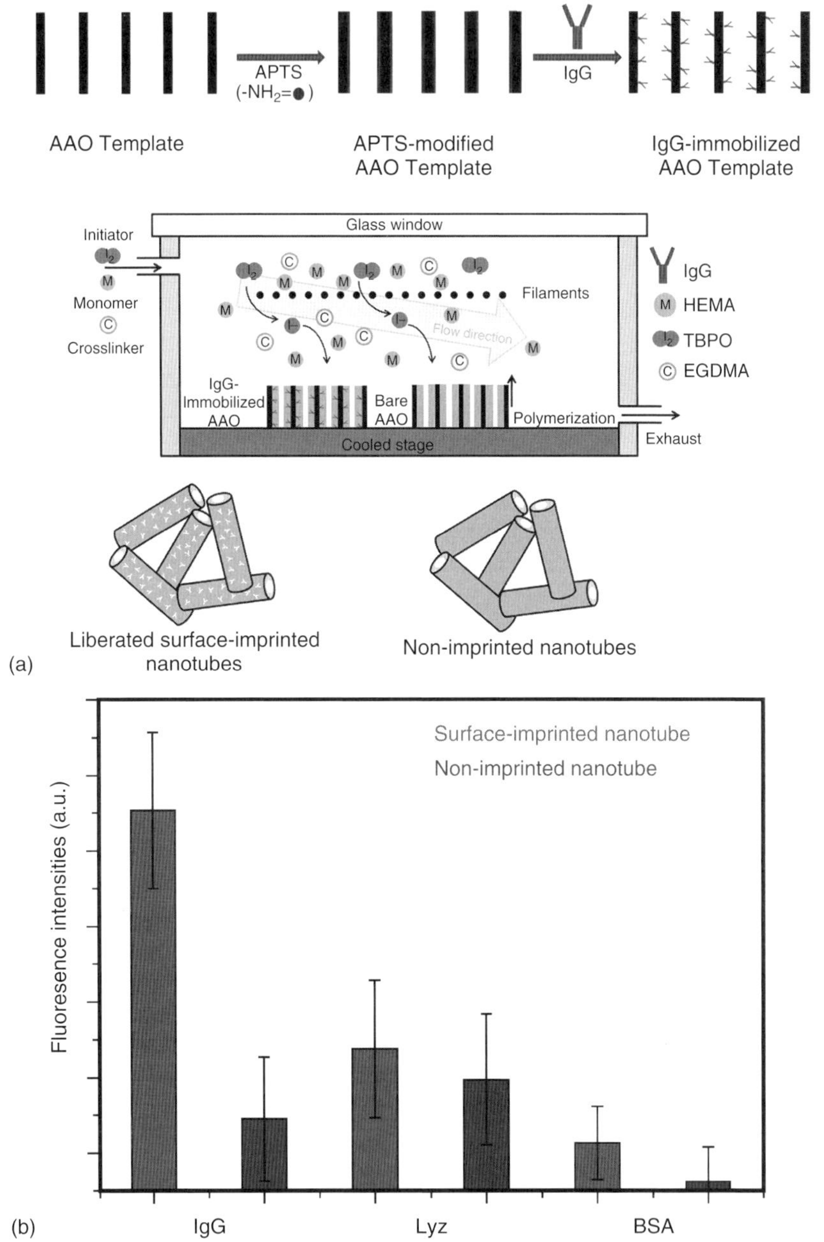

Figure 15.11 (a) Schematic of the imprinted nanotube fabrication. IgG-conjugated AAO templates are coated with p(HEMA) using iCVD. Etching of templates releases the imprinted hydrogel nanotubes. (b) Fluorescent intensities of the imprinted and nonimprinted nanotubes loaded with IgG, Lyz, and BSA. Greatest affinity of the imprinted nanotubes for IgG confirms the selectivity of the specific binding of the target molecules. Reference: [36].

15.4.3 Surface-Imprinted Micro/Nanostructures

Molecular imprinting involves creating voids with shapes of specific molecules on a surface to be used for recognition of the target molecules. The target molecules are first attached on a surface, which is then coated with a polymer layer, covering also the target molecule. The molecules are later removed from the surface, leaving behind the polymer coating with cavities in the shapes of target molecules. The imprinted surface can then be used in the recognition of the target molecules by selectively binding molecules with the correct shape that fits the cavities.

Fabrication of surface-imprinted polymeric micro/nanostructures combines the benefits of large surface area with recognition capability. Ozaydin-Ince *et al.* [36] used templated iCVD to fabricate surface-imprinted functional p(HEMA) nanotubes for biorecognition of immunoglobulin G (IgG). IgG-conjugated AAO templates were prepared by treating the aminosilane-coupled AAO membranes in a IgG solution. The AAO templates were then coated with p(HEMA) polymer using iCVD in the conformal coating regime to obtain high aspect ratio nanotubes. Removal of the templates by immersing in the HCl solution also removed the IgG molecules, thus p(HEMA) nanotubes with surface cavities in the shapes of IgG molecules were obtained (Figure 15.11a).

The IgG binding capacity of the imprinted nanotubes was studied by immersing the nanotubes in FITC-IgG solution and measuring the changes of the fluorescence intensities. Amount of IgG adsorbed on surface-imprinted nanotubes was significantly larger than the nonimprinted nanotubes. Similar studies were performed using FITC labeled Lyz and BSA to study the selectivity of the imprinted nanotubes. The result showed that while the surface-imprinted nanotubes had higher uptake ratios compared to nonimprinted nanotubes for all molecules tested, the uptake of the IgG molecules by the imprinted nanotubes was significantly higher than Lyz or BSA, confirming the specific binding of target molecules (Figure 15.11b). The results showed that conformal coating of the pore walls and encapsulation of the target molecules inside the pores could successfully be carried out using iCVD method.

15.5 Summary

Conformal depositions of functional polymers on patterned surfaces are achieved using vapor-phase CVD techniques, which enable keeping the functional groups intact and protect delicate substrates due to low process temperatures and absence of solvent. The good control over the thickness and chemical composition of the coatings widens the application areas to include surface patterning and fabrication of 1D or 2D structures.

Micro/nanostructures can be encapsulated with functional polymers to improve mechanical properties or to add new functionalities to control surface reactions. Ability to control the swelling properties of the coatings by tuning the chemical composition increases the use of CVD coatings as encapsulants for controlled release applications with improved release rates.

Templates can be used to selectively coat the surfaces for chemical patterning to obtain bifunctional surfaces. By changing the mask size and shape, surfaces with patterns of different characteristic dimensions can be formed. On the other hand, maskless patterning of surfaces through wrinkling is achieved by tuning the mechanical properties of the polymer coatings.

Polymeric structures in the micro or nanoscales can be synthesized using sacrificial templates that are coated conformally with functional polymers using CVD techniques. Use of templates provides control over the shapes and dimensions of the structures, which is critical for their integration in devices. Finally, the vapor-phase CVD techniques enable controlling the response of these structures to stimuli by tuning the polymer composition, widening the application areas of these responsive structures.

References

1. Iijima, S. (1991) Helical microtubules of graphitic carbon. *Nature*, **354**, 56–58.
2. Ye, Y., Mao, Y., Wang, H., and Ren, Z. (2012) Hybrid structure of pH-responsive hydrogel and carbon nanotube array with superwettability. *J. Mater. Chem.*, **22**, 2449–2455.
3. Ye, Y., Mao, Y., Wang, F., Lu, H., Qu, L., and Dai, L. (2011) Solvent-free functionalization and transfer of aligned carbon nanotubes with vapor-deposited polymer nanocoatings. *J. Mater. Chem.*, **21**, 837–842.
4. Lau, K.K.S. and Gleason, K.K. (2006) Particle surface design using an all-dry encapsulation method. *Adv. Mater.*, **18**, 1972–1977.
5. Parker, T.C., Baechle, D., and Demaree, J.D. (2011) Polymeric barrier coatings via initiated chemical vapor deposition. *Surf. Coat. Technol.*, **206**, 1680–1683.
6. Bose, R.K., Heming, A.M., and Lau, K.K.S. (2012) Microencapsulation of a crop protection compound by initiated chemical vapor deposition. *Macromol. Rapid Commun.*, **33**, 1375–1380.
7. Baxamusa, S.H., Montero, L., Dubach, J.M., Clark, H.A., Borros, S., and Gleason, K.K. (2008) Protection of sensors for biological applications by photoinitiated chemical vapor deposition of hydrogel thin films. *Biomacromolecules*, **9**, 2857–2862.
8. Yoo, Y., You, J.B., Choi, W., and Im, S.G. (2013) A stacked polymer film for robust superhydrophobic fabrics. *Polym. Chem.*, **4**, 1664–1671.
9. Chen, B., Seidel, S., Hori, H., and Gupta, M. (2011) Self-assembly of pillars modified with vapor deposited polymer coatings. *ACS Appl. Mater. Interfaces*, **3**, 4201–4205.
10. Kaya, A., Demiryurek, R., Armagan, E., Ozaydin-Ince, G., Sezen, M., and Kosar, A. (2013) Boiling heat transfer enhancement in mini/microtubes via polyhydroxyethylmethacrylate (pHEMA) coatings on inner microtube walls at high mass fluxes. *J. Micromech. Microeng.*, **23**, 115017.
11. Çıkım, T., Armagan, E., Ozaydin Ince, G., and Koşar, A. (2014) Flow boiling enhancement in microtubes with crosslinked pHEMA coatings and the effect of coating thickness. *J. Heat Transfer*, **136**, 081504.
12. Baxamusa, S.H., Montero, L., Borros, S., and Gleason, K.K. (2010) Self-aligned micropatterns of bifunctional polymer surfaces with independent chemical and topographical contrast. *Macromol. Rapid Commun.*, **31**, 735–739.

13. Lee, C.-H., Tenhaeff, W., and Gleason, K.K. (2009) Nano-patterning of an iCVD polymer, followed by covalent attachment of QDs. *Thin Solid Films*, **517**, 3619–3621.
14. Trujillo, N.J., Baxamusa, S.H., and Gleason, K.K. (2009) Grafted functional polymer nanostructures patterned bottom-up by colloidal lithography and initiated chemical vapor deposition (iCVD). *Chem. Mater.*, **21**, 742–750.
15. Mao, Y., Felix, N.M., Nguyen, P.T., Ober, C.K., and Gleason, K.K. (2004) Towards all-dry lithography: electron-beam patternable poly(glycidyl methacrylate) thin films from hot filament chemical vapor deposition. *J. Vac. Sci. Technol. B*, **22**, 2473–2478.
16. Mao, Y., Felix, N.M., Nguyen, P.T., Ober, C.K., and Gleason, K.K. (2006) Positive- and negative-tone CVD polyacrylic electron-beam resists developable by supercritical CO2. *Chem. Vap. Deposition*, **12**, 259–262.
17. Mao, Y. and Gleason, K.K. (2006) Positive-tone nanopatterning of chemical vapor deposited polyacrylic thin films. *Langmuir*, **22**, 1795–1799.
18. Yoshida, S., Kobayashi, T., Kumano, M., and Esashi, M. (2012) Conformal coating of poly-glycidyl methacrylate as lithographic polymer via initiated chemical vapor deposition. *J. Micro/Nanolithogr. MEMS MOEMS*, **11**, 023001.
19. Baxamusa, S.H. and Gleason, K.K. (2008) Thin polymer films with high step coverage in microtrenches by initiated CVD. *Chem. Vap. Deposition*, **14**, 313–318.
20. Petruczok, C.D. and Gleason, K.K. (2012) Initiated chemical vapor deposition-based method for patterning polymer and metal microstructures on curved substrates. *Adv. Mater.*, **24**, 6445–6450.
21. Petruczok, C.D., Armagan, E., Ozaydin Ince, G., and Gleason, K.K. (2014) Initiated chemical vapor deposition and light-responsive cross-linking of poly(vinyl cinnamate) thin films. *Macromol. Rapid Commun.*, **35**, 1345–1350.
22. Haller, P.D., Flowers, C.A., and Gupta, M. (2011) Three-dimensional patterning of porous materials using vapor phase polymerization. *Soft Matter*, **7**, 2428–2432.
23. Kwong, P., Flowers, C.A., and Gupta, M. (2011) Directed deposition of functional polymers onto porous substrates using metal salt inhibitors. *Langmuir*, **27**, 10634–10641.
24. Yin, J., Yagüe, J.L., Eggenspieler, D., Gleason, K.K., and Boyce, M.C. (2012) Deterministic order in surface micro-topologies through sequential wrinkling. *Adv. Mater.*, **24**, 5441–5446.
25. Yagüe, J.L., Yin, J., Boyce, M.C., and Gleason, K.K. (2013) Design of ordered wrinkled patterns with dynamically tuned properties. *Physics Procedia*, **46**, 40–45.
26. Yin, J., Yague, J.L., Boyce, M.C., and Gleason, K.K. (2014) Biaxially mechanical tuning of 2-D reversible and irreversible surface topologies through simultaneous and sequential wrinkling. *ACS Appl. Mater. Interfaces*, **6**, 2850–2857.
27. Ozaydin Ince, G. and Gleason, K.K. (2010) Tunable conformality of polymer coatings on high aspect ratio features. *Chem. Vap. Deposition*, **16**, 100–105.
28. Asatekin, A. and Gleason, K.K. (2011) Polymeric nanopore membranes for hydrophobicity-based separations by conformal initiated chemical vapor deposition. *Nano Lett.*, **11**, 677–686.
29. Ozaydin-Ince, G., Demirel, G., Gleason, K.K., and Demirel, M.C. (2010) Highly swellable free-standing hydrogel nanotube forests. *Soft Matter*, **6**, 1635–1639.
30. Ozaydin-Ince, G., Dubach, J.M., Gleason, K.K., and Clark, H.A. (2011) Microworm optode sensors limit particle diffusionto enable in vivo measurements. *Proc. Natl. Acad. Sci. U.S.A.*, **108**, 2656–2661.
31. Yague, J.L. and Gleason, K.K. (2012) Systematic control of mesh size in hydrogels by initiated chemical vapor deposition. *Soft Matter*, **8**, 2890–2894.
32. Ozaydin-Ince, G., Gleason, K.K., and Demirel, M.C. (2011) A stimuli-responsive coaxial nanofilm for burst release. *Soft Matter*, **7**, 638–643.
33. Karaman, M., Cabuk, N., Ozyurt, D., and Köysüren, O. (2012) Self-supporting superhydrophobic thin polymer sheets

that mimic the nature's petal effect. *Appl. Surf. Sci.*, **259**, 542–546.

34. Im, S.G., Bong, K.W., Kim, B.-S., Baxamusa, S.H., Hammond, P.T., Doyle, P.S., and Gleason, K.K. (2008) Patterning nanodomains with orthogonal functionalities: solventless synthesis of self-sorting surfaces. *J. Am. Chem. Soc.*, **130**, 14424–14425.
35. Jeong, J.-M., Oh, M.S., Kim, B.J., Choi, C.-H., Lee, B., Lee, C.-S., and Im, S.G. (2013) Reliable synthesis of monodisperse microparticles: prevention of oxygen diffusion and organic solvents using conformal polymeric coating onto poly(dimethylsiloxane) micromold. *Langmuir*, **29**, 3474–3481.
36. Ozaydin-Ince, G., Armagan, E., Erdogan, H., Buyukserin, F., Uzun, L., and Demirel, G. (2013) One-dimensional surface-imprinted polymeric nanotubes for specific biorecognition by initiated chemical vapor deposition (iCVD). *ACS Appl. Mater. Interfaces*, **5**, 6447–6452.

16
Deposition of Polymers onto New Substrates

Malancha Gupta

The initiated chemical vapor deposition (iCVD) process is typically used to deposit thin conformal coatings onto solid substrates at room temperatures. This chapter focuses on expanding the capabilities of the iCVD technique by depositing polymers onto new substrates. The first part of this chapter focuses on patterning polymer coatings onto porous paper-based microfluidic devices. These devices can be used for diagnostic assays for healthcare applications. The iCVD process is ideal for coating porous materials because the precursors diffuse throughout the 3D matrix. Patterning the location of polymer growth within the porous matrix can be difficult. In this chapter, several methods to pattern polymers onto paper-based devices will be discussed including light-based techniques and chemical inhibitors. Although the focus is on applying these coatings to paper-based microfluidic devices, these patterning techniques can also be used to apply coatings to other porous substrates including membranes and textiles. The second part of this chapter deals with coating elastomeric substrates that are patterned with micron-scaled features. Large arrays of pillars can be coated with hydrophilic polymers and then self-assembled via capillary forces into hierarchical microstructures. Elastomeric microfluidic devices can be uniformly coated to increase the durability and function of these devices.

The third part of this chapter involves deposition of polymers onto liquid substrates. These liquids have extremely low vapor pressures and are therefore stable under vacuum conditions. Polymerization can occur at both the liquid surface and within the liquid. At the liquid surface, the surface tension interaction between the polymer and the liquid can lead to the formation of new polymer architectures such as nanoparticles and free-standing polymer films. If the precursors are soluble within the liquid, polymerization can also occur within the liquid, which can lead to the formation of gels. The final part of this chapter focuses on deposition onto low-temperature substrates. Lowering the temperature of the substrate below the freezing point of the monomer leads to the formation of macroporous polymer membranes. These membranes can be deposited onto porous substrates to create hierarchical porous-on-porous structures that enable improved filtration for water purification and sensor applications.

CVD Polymers: Fabrication of Organic Surfaces and Devices, First Edition. Edited by Karen K. Gleason.

16.1
Paper-Based Microfluidic Devices

The iCVD process is ideal for uniformly coating porous substrates because the monomer and initiator radicals diffuse throughout the depth of the porous matrix. Patterning the location of the coating is therefore very difficult. Although iCVD polymerization has been patterned using techniques such as colloidal lithography [1], electron-beam lithography [2, 3], and capillary force lithography [4], these types of patterning methods are limited to flat substrates and cannot be applied to porous substrates.

The ability to pattern polymers onto porous substrates is useful for applications such as filtration, optics, and biosensing. For example, new techniques for patterning polymers onto porous substrates could be useful for making next-generation paper-based microfluidic devices. These devices are made by patterning hydrophobic wax barriers into hydrophilic porous chromatography paper [5]. The wax ink is printed onto the chromatography paper and then melted through the entire thickness of the paper to form the hydrophobic barriers. The channels are typically several centimeters long and several millimeters wide. Paper-based microfluidic devices are a good platform for diagnostic assays because they are easy to use and because diagnostic fluids are able to wick into the channels using natural capillary forces instead of syringe pumps [6, 7]. The iCVD process is ideal for modifying paper-based microfluidic devices because the polymer coating wraps around each individual paper fiber, and therefore the porosity of the chromatography paper is maintained to allow for capillary-driven flow. The ability to pattern the location of polymer growth on these devices is important for making on-chip actuators, switches, and filters, which would improve the capabilities of these devices.

One method to pattern the location of polymer growth on paper-based microfluidics is to deposit a light-responsive polymer throughout the porous matrix and then selectively remove it. Haller *et al.* patterned regions of hydrophobicity and hydrophilicity through the depth of porous chromatography paper by depositing poly(*ortho*-nitrobenzyl methacrylate) (P*o*NBMA) onto porous chromatography using iCVD and then using UV light to selectively remove the polymer [8]. In the first step, the iCVD process was used to conformally deposit hydrophobic P*o*NBMA onto the entire chromatography paper. Scanning electron microscopy (SEM) was used to confirm that the P*o*NBMA coating wraps around the individual paper fibers, and therefore the porosity of the cellulose paper remains the same before and after the iCVD coating process. X-ray photoelectron spectroscopy and contact angle measurements were used to confirm that the coating penetrated the entire depth of the paper, and therefore both the front and back of the paper were coated. In the second step, the coated paper was exposed to UV light through a transparency. The UV light cleaved the nitrobenzyl moieties, and the exposed polymer was then able to be removed by dissolving in pH 8 buffer. The areas without polymer regained the hydrophilicity of the cellulose paper. Fluid was able to be retained with the hydrophilic channels that

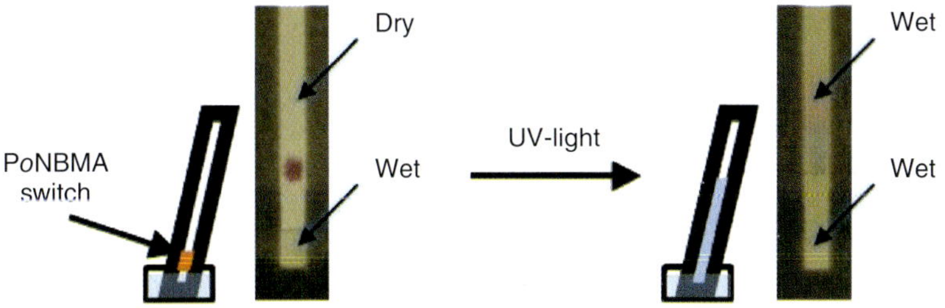

Figure 16.1 Poly(methacrylic acid) (PMAA) and poly(*o*-nitrobenzyl methacrylate) (PoNBMA) coatings were sequentially deposited onto a paper-based microfluidic device to make a light-responsive activation switch and to separate charged molecules (toluidine blue O and ponceau S). (Adapted with permission from Ref. [9]. Copyright 2012 American Chemical Society.)

confirmed the ability of this technique to pattern regions of P*o*NBMA within the depth of the paper. This patterning process does not require any organic solvents during polymerization or patterning and is therefore environmentally friendly relative to other patterning techniques.

Another reason that the iCVD process is ideal for modifying paper-based microfluidic devices is that multiple polymers can be sequentially deposited onto the chromatography paper in order to add several capabilities to the devices. For example, Kwong *et al.* sequentially deposited poly(methacrylic acid) (PMAA) and P*o*NBMA coatings to fabricate a device that had both a light-responsive activation switch and could separate charged molecules [9]. In the first step, PMAA was deposited onto the chromatography paper to serve as a coating to separate anionic and cationic analytes, and then P*o*NBMA was sequentially deposited and patterned onto the top of the PMAA layer to make a photo-responsive switch. Exposing the P*o*NBMA switch to UV light allowed aqueous fluid to travel up the PMAA-coated channel where the analytes (toluidine blue O and ponceau S) were then subsequently separated (Figure 16.1).

The earlier patterning technique requires a light-responsive moiety on the polymer. It is useful to have a more general patterning technique that can be applied to all iCVD polymers. One method to pattern other iCVD polymers involves the use of transition metal salt inhibitors to selectively inhibit polymerization in certain regions. Kwong *et al.* used a combinatorial screening approach to determine the effects of different transition metal salts and reaction parameters on the inhibition of several iCVD polymers [10]. Contact angle measurements were used to determine whether polymerization was inhibited. Three polymers were tested: poly(1*H*,1*H*,2*H*,2*H*-perfluorodecyl acrylate) (PPFDA), poly(pentafluorophenyl methacrylate) (PPFM), and P*o*NBMA. If the transition metal salt did not inhibit, polymerization occurred on the chromatography paper and the surface became hydrophobic. If the salt did inhibit, polymerization did not occur and the chromatography paper remained hydrophilic. It was found that copper chloride ($CuCl_2$) and copper nitrate ($Cu(NO_3)_2$) were effective at uniformly inhibiting the deposition of all three polymers through the depth of the porous chromatography paper. The inhibition process lost uniformity after very long deposition times, which is likely due to the passivation of the inhibiting salt by the accumulation of deactivated precursor molecules, however only short deposition times are required for practical applications.

The ability to use transition metal salts to inhibit polymerization is useful in fabricating paper-based microfluidic devices that are compatible with organic solvents. Traditional paper-based microfluidic devices are only compatible with aqueous-based solutions because organic solvents dissolve the wax barriers. Chen *et al.* replaced the wax barriers with fluoropolymer barriers because fluoropolymers do not dissolve in organic solvents [11]. The fluoropolymer barriers were patterned onto the chromatography paper by first applying $CuCl_2$ as the inhibiting salt and then depositing PPFDA onto the paper via iCVD. The PPFDA only deposited in the areas without the $CuCl_2$ salt. The salt was washed away after the deposition (Figure 16.2a). The PPFDA barriers were able to contain a wide variety of organic solvents including hexane, ethyl acetate, acetone, and methanol (Figure 16.2b). The ability to use organic solvents in paper-based microfluidic devices expands the potential applications of these devices to detect water-insoluble compounds such as pharmaceutical drugs and pesticides.

The patterning techniques discussed in this section were used to pattern functional polymers onto paper-based microfluidic devices. These patterning techniques can also be used to pattern functional polymers onto other porous substrates including membranes and textiles.

16.2 Elastomeric Substrates

Elastomeric poly(dimethylsiloxane) (PDMS) substrates are very common because they can be molded into different shapes and structures using techniques such as soft lithography. In the soft lithography process, a template is typically made using photolithography. PDMS solution is then poured onto this template and subsequently cured. The template can be reused dozens of times to make reproducible PDMS structures that allows for systematic studies. Many types of microfluidic devices are made using the soft lithography process. In addition, surfaces with microstructured periodic features can also be made using the soft lithography process. Patterning surfaces with periodic structures can lead to interesting adhesive, optical, and hydrophobic properties. There are several examples in Nature such as the Lotus leaf and the Gecko, which have shown that microstructured features on a surface can lead to improved material properties such as self-cleaning surfaces and reversible adhesion.

Chen *et al.* demonstrated that the iCVD process can be used to make PDMS surfaces with periodic microstructures [12]. Hydrophobic PDMS pillar arrays were coated with hydrophilic poly(hydroxyethyl methacrylate) (PHEMA) and PMAA coatings via iCVD. The iCVD process is ideal for coating PDMS micropillar arrays because the precursor molecules diffuse isotropically and conformally coat all the

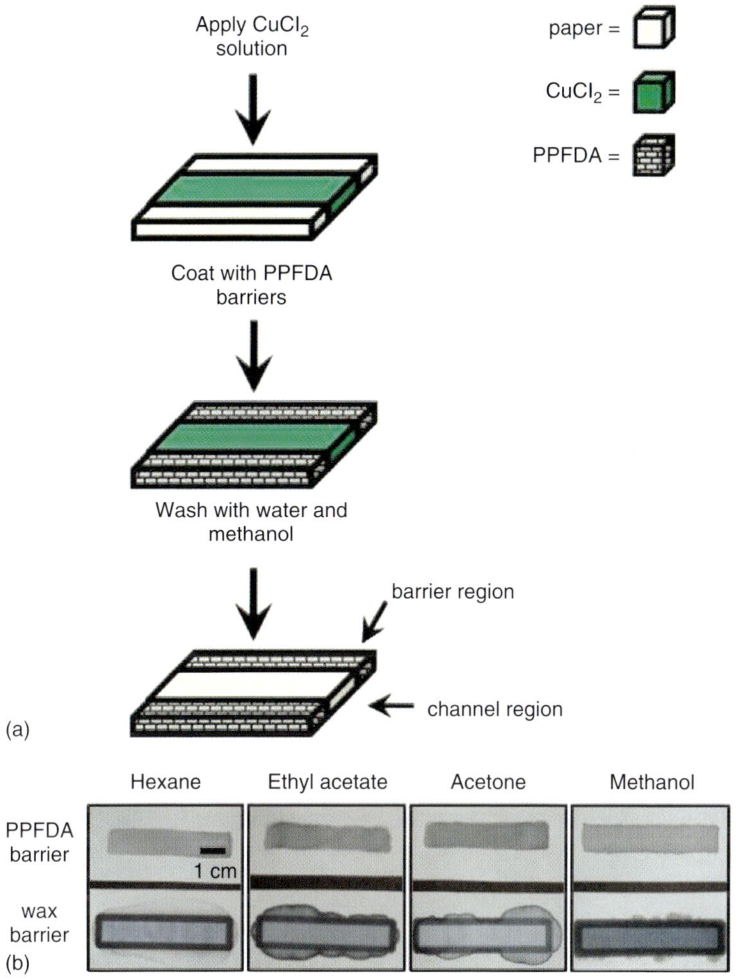

Figure 16.2 (a) $CuCl_2$ was used to pattern poly(1*H*,1*H*,2*H*,2*H*-perfluorodecyl acrylate) (PPFDA) through porous cellulose to make paper-based microfluidic devices that are compatible with organic solvents. (b) The PPFDA barriers are able to contain a wide variety of organic solvents, whereas traditional wax barriers dissolve. (Adapted with permission from Ref. [11]. Copyright 2013 American Chemical Society.)

sidewalls and features. Capillary forces were then used to self-assemble the coated pillars into microstructures by adding water to the pillars and allowing the water to evaporate. The aspect ratio of the pillars affected the self-assembly process. PDMS pillars with a diameter of 22 μm, edge-to-edge spacing of 18 μm, and heights of 34, 54, and 65 μm were tested. In 34-μm tall pillars, the capillary force was not great enough to bend the pillars to cause physical contact. In contrast, the 54-μm pillars made contact and formed perfect clusters of four pillars (Figure 16.3a). However, when the pillar height was increased to 65 μm, several clusters of six pillars were

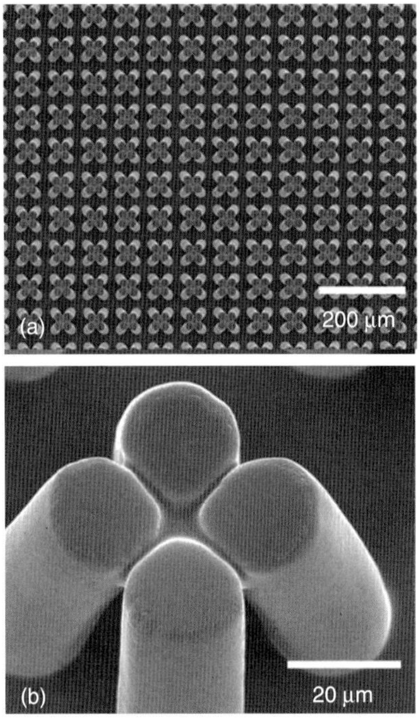

Figure 16.3 (a) SEM image of microstructures that were formed by capillary-forced self-assembly of 54-μm tall PDMS pillars coated with 250 nm of PMAA. (b) A close-up view of the microstructures shows that polymer welds form between adjacent pillars and within the center of the microstructure. (Adapted with permission from Ref. [12]. Copyright 2011 American Chemical Society.)

formed in addition to clusters of four pillars. During the self-assembly process, the water mobilizes the hydrophilic PHEMA and PMAA polymer chains. The chains interdiffuse when the pillars make contact leading to the formation of polymer welds between the pillars. These welds are important because they stabilize the microstructures by providing an adhesive force. The adhesive force prevents the collapsed pillars from reverting back to their upright position after evaporation. For thin coatings, polymer welds are formed between adjacent pillars, whereas polymer welds are formed between adjacent pillars and within the center of the microstructures for thicker coatings (Figure 16.3b).

A major benefit for using the iCVD process to modify pillar arrays is the ability to stack several layers of functional polymers. Stacking layers using solution-phase polymerization techniques is difficult because solvents might dissolve underlying polymer layers. In contrast, the iCVD technique does not require solvents, and therefore the dissolution of the underlying layer is not an issue. Polymer layers can be stacked to make regions of hydrophobicity and hydrophilicity, which can guide self-assemble. For example, the location of self-assembly can be controlled by first depositing PHEMA onto the PDMS pillars and then depositing a layer of light-responsive P*o*NBMA onto the PHEMA layer. The P*o*NBMA can then be selectively removed by exposure to UV light through a mask. The regions with the P*o*NBMA are hydrophobic while the regions with the exposed PHEMA underlayer are hydrophilic. Self-assembly by capillary forces only occurs in the hydrophilic regions leading to patterned regions of microstructures (Figure 16.4a).

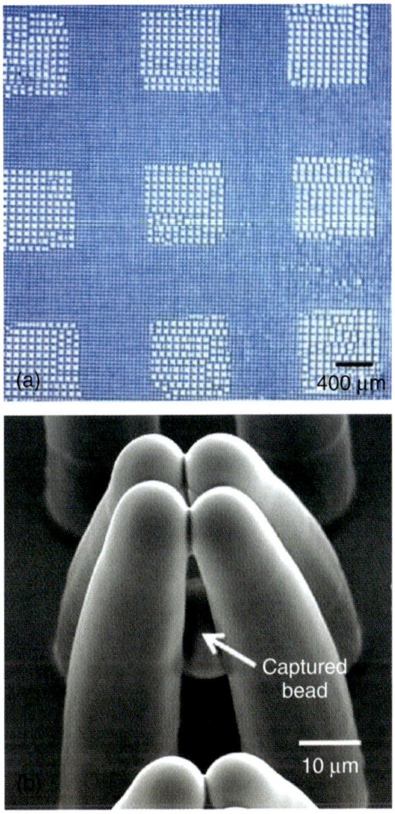

Figure 16.4 (a) Stereoscope image showing that the location of self-assembly can be controlled by patterning regions of hydrophilicity and hydrophobicity using poly(*o*-nitrobenzyl methacrylate) (PoNBMA). (b) SEM image showing that beads can be captured with the microstructures. (Adapted with permission from Ref. [12]. Copyright 2011 American Chemical Society.)

The polymer welds allow the microstructures to be reversibly opened and closed, which have applications in tunable adhesion and optics. For example, the microstructures can be opened and closed using different solvents. The microstructures remain closed if the polymer coating is incompatible with the solvent and the microstructures remain open if the solvent is compatible. The time required for the microstructures to open can be tuned by mixing solvents or changing the thickness of the polymer coating. The microstructures have potential applications in drug delivery and biomedical applications because they can be fabricated using biocompatible materials such as PDMS and PHEMA. Beads can also be captured within the microstructures (Figure 16.4b). The release of the beads can be controlled using different solvents for potential applications in drug delivery, catalysis, and microfluidics.

Soft lithography is commonly used to make PDMS microfluidic devices for use in medical diagnostics and chemical synthesis. The main advantage for using PDMS for microfluidic applications is that soft lithography is a cheap, fast, and reproducible manufacturing technique that does not require a clean room. However, the main disadvantages for using PDMS are that organic solvents can absorb into it and swell it and small molecules can diffuse into it. In order to create robust PDMS microfluidic devices that can be operated for long periods of time, it is important to prevent absorption and diffusion. Riche *et al.* demonstrated that the interior surfaces of pre-assembled PDMS channels can be modified with fluoropolymer coatings using the iCVD process [13]. The channels had a height of 450 μm and widths of either 200 or 1000 μm. The coatings were composed of 1*H*,1*H*,2*H*,2*H*-perfluorodecyl acrylate (PFDA) cross-linked with ethylene glycol diacrylate (EGDA). The cross-linking density was systematically varied, and it was found that there is an optimal cross-linking density where the surface energy is very low and the polymer coating is stable. At this ideal cross-linking density, the coating prevented the diffusion of Rhodamine B into PDMS for 3 h and prevented the absorption of hexane. Coatings with higher cross-linking densities were not as effective at preventing diffusion and swelling because of the higher surface energy. The optimal cross-linked fluoropolymer coatings have been recently applied to PDMS devices that are being used to synthesize nanoparticles [14, 15].

16.3 Liquids Substrates

The iCVD process is typically used to deposit polymers onto solid substrates. Deposition onto liquid substrates can lead to new polymer morphologies because the polymer chains are more mobile on liquid surfaces. In addition, the monomer and initiator molecules can absorb into the liquid substrate, leading to polymerization at both the liquid surface and within the liquid. The main requirement for liquid substrates is that they must have negligible vapor pressures in order to remain stable under vacuum conditions. Ionic liquids (ILs) can serve as liquid substrates because these salts are liquids at room temperature and have extremely low vapor pressures. Silicone oil, Krytox, and glycerol also have extremely low vapor pressures and are therefore stable under vacuum conditions.

Haller *et al.* showed that performing iCVD depositions onto liquid substrates leads to the formation of polymer particles and polymer films depending on the surface tension interaction between the polymer and the liquid [16, 17]. A systematic study involving the deposition of six different polymers onto seven different liquids showed that the spreading coefficient, S, could be used to predict whether films or particles will form for a given polymer and liquid system. The spreading coefficient measures the free energy required for the polymer to spread over the surface of the liquid and it can be written in terms of the liquid–vapor surface tension (γ_{LV}), the polymer–vapor surface tension (γ_{PV}), and the advancing contact angle of the liquid on the polymer (θ) [18]:

$$S = \gamma_{LV}{}^{*}(1 + \cos\theta) - 2\gamma_{PV}$$

Table 16.1 shows that polymer films form when it is energetically favorable for the polymer to spread on the liquid surface (positive spreading coefficients) and polymer particles form when it is energetically favorable for the polymer to minimize its area of contact with the liquid (negative spreading coefficients).

Table 16.1 Shaded entries show polymer–liquid systems that formed particles and unshaded entries show polymer–liquid systems that formed films for 5 min depositions at a rate of 10 nm min^{-1}.

Liquid (γ_{LV} in mN m^{-1})	Polymer (γ_{PV} in mN m^{-1})					
	PPFDA (13.6)	PPFM (25.7)	PnBA (35.1)	PnBMA (35.4)	PHEMA (48.5)	P4VP (57.5)
Glycerol (63.4)	116° **8.2**	100° **1.4**	80° **4.3**	82° **1.8**	62° **−4.0**	46° **−11.6**
[emim][BF$_4$] (55.6)	107° **11.8**	83° **10.8**	71° **3.1**	77° **−3.0**	62° **−14.9**	61° **−32.0**
[bmim][BF$_4$] (45.6)	102° **8.8**	80° **1.9**	63° **−4.2**	64° **−5.4**	46° **−20.0**	55° **−43.4**
[bmim][PF$_6$] (45.1)	104° **7.0**	86° **−2.9**	55° **0.8**	56° **−0.4**	48° **−21.7**	55° **−44.0**
Squalene (31.5)	86° **6.4**	47° **1.7**	54° **−15.5**	30° **−12.0**	28° **−37.6**	22° **−54.3**
[dmim][BF$_4$] (30.3)	88° **4.0**	65° **−8.2**	43° **−21.9**	37° **−16.3**	31° **−40.7**	46° **−63.7**
Silicone oil (22.8)	74° **2.0**	31° **−9.1**	50° **−32.8**	40° **−30.4**	24° **−53.3**	25° **−71.6**

The estimated spreading coefficients are in bold within each entry (mN m^{-1}), and the measured advancing contact angles are listed. PPFDA = poly(1*H*,1*H*,2*H*,2*H*-perfluorodecyl acrylate), PPFM = poly(pentafluorophenyl methacrylate), PnBA = poly(*n*-butyl acrylate), PnBMA = poly(*n*-butyl methacrylate), PHEMA = poly(2-hydroxyethyl methacrylate), P4VP = poly(4-vinylpyridine), [emim][BF$_4$] = 1-ethyl-3-methylimidazolium tetrafluoroborate, [bmim][BF$_4$] = 1-butyl-3-methylimidazolium tetrafluoroborate, [bmim][PF$_6$] = 1-butyl-3-methylimidazolium hexafluorophosphate, [dmim][BF$_4$] = 1-decyl-3-methylimidazolium tetrafluoroborate.
Source: Adapted with permission from Ref. [17]. Copyright 2013 American Chemical Society.

The polymer PPFDA has a low surface tension and formed films on all the liquids that were tested, whereas the polymer poly(4-vinylpyridine) (P4VP) has a high surface tension and formed particles on all the liquids that were tested (Figure 16.5). The spreading coefficient determines the thermodynamically favorable polymer morphology, and therefore all the PPFDA films that formed on all the tested liquids were stable because of the positive spreading coefficients. The deposition rate could be increased to form films in systems with negative spreading coefficients, but these films are not thermodynamically stable and eventually de-wet. For example, poly(*n*-butyl acrylate) (PnBA) formed particles on 1-butyl-3-methylimidazolium tetrafluoroborate at deposition rates of 10, 23, and 29 nm min^{-1}; however, films were formed at 43 and 51 nm min^{-1}. Since the spreading coefficient is negative for this system ($S = -4.2$ mN m^{-1}), the films that formed at the higher deposition rates were not stable and eventually de-wetted.

The ability to tune the polymer morphology on different liquids can be used to make free-standing polymer films for a variety of applications in tissue engineering, optics, sensing, and separations. Frank-Finney *et al.* made free-standing

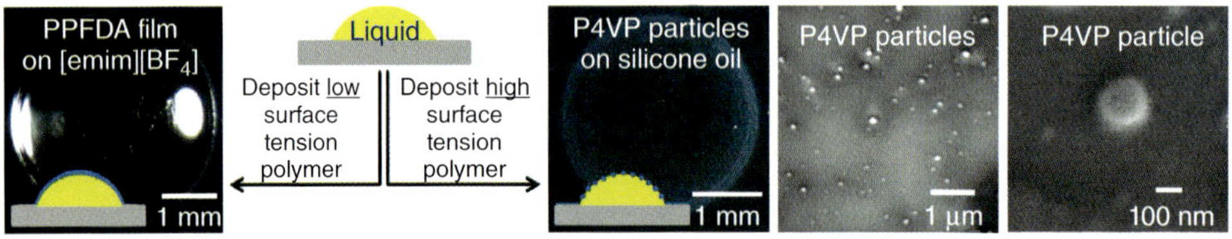

Figure 16.5 Low surface tension polymers such as poly(1*H*,1*H*,2*H*,2*H*-perfluorodecyl acrylate) (PPFDA) form polymer films on liquid substrates, whereas high surface tension polymers such as poly(4-vinylpyridine) (P4VP) form polymer particles on liquid substrates. The P4VP particles were removed from the silicone oil and drop-casted onto a silicon wafer for SEM imaging. (Adapted with permission from Ref. [17]. Copyright 2013 American Chemical Society.)

films by depositing polymer onto patterned liquid substrates [19]. Shaped free-standing PHEMA and poly(*N*-isopropylacrylamide) (PNIPAAm) films were made by depositing the polymer onto shaped regions of ionic liquid surrounded by silicone oil. The patterned liquid substrate was made by drawing the outline of a shape onto a silicon wafer using wax. The IL was then confined within the shape, and the IL was then surrounded by silicone oil. Polymer films formed on the IL surface, whereas polymer particles formed on the silicone oil surface. The films could be easily lifted off because they were not attached to the underlying substrate. The benefits of this technique for making free-standing films are that the shape of the films can be easily controlled with the wax, and the thickness of the films can be varied from tens of nanometers to microns by changing the deposition time.

In the earlier examples, the liquid substrates served as a template and were not incorporated into the polymer. There are several applications in which it would be useful to incorporate the liquid within the polymer, particularly for ILs because ILs have unique properties such as high thermal stability, high conductivity, and wide electrochemical window that make them useful for a variety of applications such as chemical synthesis, gas absorption, and electrochemical applications. Bradley *et al.* demonstrated that IL droplets can be encapsulated within polymer shells composed of PFDA cross-linked with EGDA [20]. The IL droplets were first rolled into marbles using micron-sized poly(tetrafluoroethylene) (PTFE) particles. These marbles were then placed on a bed of PTFE particles in a Petri dish and coated via iCVD (Figure 16.6a). The bed of PTFE particles ensured that the coated marbles did not adhere to the Petri dish and, therefore, could be easily removed without puncturing (Figure 16.6b). SEM images showed that the PTFE particles that surround the IL marbles become integrated into the polymer shell (Figure 16.6c). This PTFE integration adds to the mechanical integrity of the polymer shell.

Another method to incorporate IL within the polymer matrix is to promote polymerization within the bulk liquid, which leads to the formation of polymer–IL gels. These polymer–IL gels have potential applications as fuel cell membranes and polymer actuators. Polymerization can occur within the bulk liquid if the monomer is soluble within the liquid. For example, 2-hydroxyethyl methacrylate (HEMA) monomer is soluble in many ILs and will polymerize in both the bulk liquid and on the liquid surface, whereas PFDA monomer is not soluble in most

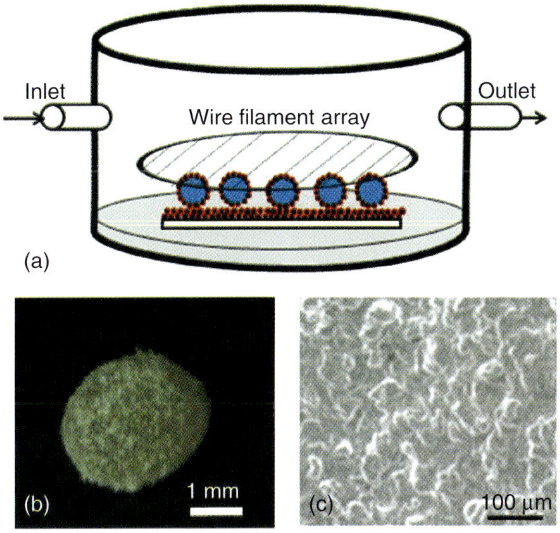

Figure 16.6 (a) Schematic showing the process for encapsulating ionic liquid droplets within polymer shells. PTFE particles are used to stabilize the droplets into spheres. (b) A [emim][BF4] droplet was encapsulated in a P(PFDA-*co*-EGDA) polymer shell. (c) SEM image of the polymer shell showing that the PTFE particles are incorporated within the polymer. (Adapted with permission from Ref. [20]. Copyright 2012 American Chemical Society.)

ILs and will only polymerize on the liquid surface. Frank-Finney *et al.* made a polymer–IL gel by polymerizing HEMA on thin layers of [emim][BF_4] [21]. As more polymerization occurs within the IL layer, there is a transition from a viscous liquid to a solid-like gel (Figure 16.7a). The gels are robust enough to be handled with tweezers (Figure 16.7b,c). Fourier transform infrared spectroscopy (FTIR) showed that the gels consisted of both the IL and the polymer (Figure 16.7d). It has been shown that the molecular weight of polymer formed in ILs is much larger than the molecular weight of polymer formed in organic solvents likely due to increased propagation rates and decreased termination rates in ILs [22, 23]. Gel permeation chromatography experiments show that the polymer formed within the polymer–IL gel has two distinct molecular weight distributions, reflecting polymerization at both the liquid surface and within the bulk IL. The molecular weight of the polymer formed within the IL (10^7 Da) is several orders of magnitude higher than the polymer formed at the liquid surface (10^4 Da), which is consistent with observations of higher molecular weights in IL solvents. The benefit of using the iCVD technique to make polymer–IL gels is that gels with high concentrations of IL can be formed in deposition times less than an hour. The polymerization of soluble monomers within liquid substrates also enables the fabrication of heterogeneous stacked polymer films. The order that soluble and insoluble monomers are introduced into the iCVD chamber can be systematically varied in order to produce stacked polymer films with different chemical compositions [24].

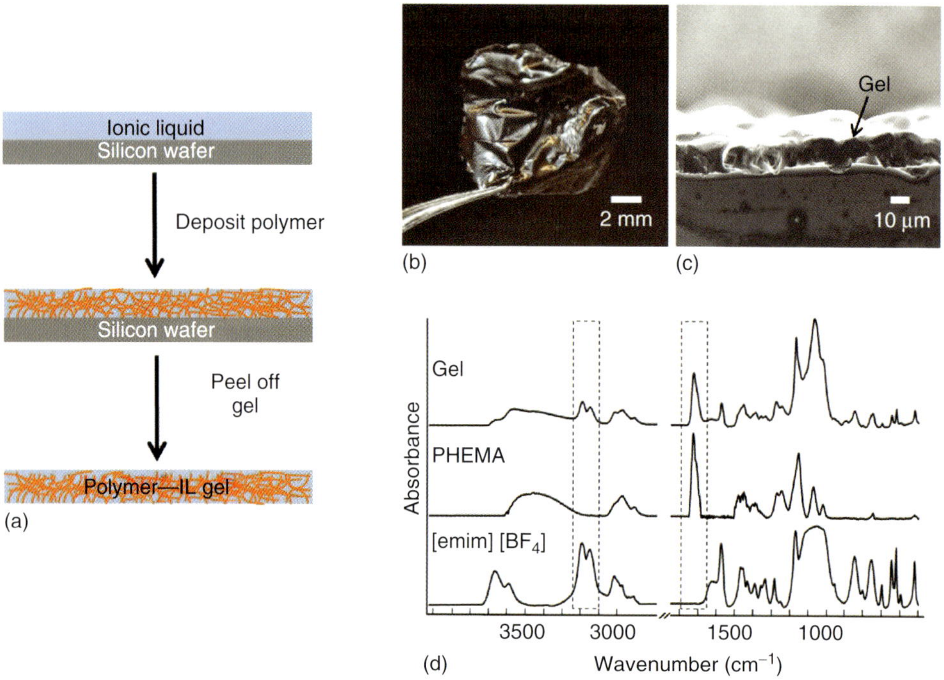

Figure 16.7 (a) Schematic of the process to make a polymer-ionic liquid gel. (b) PHEMA-[emim][BF_4] gels are robust enough to be handled. (c) Cross-sectional image of a 43 wt% PHEMA-[emim][BF_4] gel on a silicon wafer. (d) FTIR spectra showing that the gel is composed of both PHEMA and [emim][BF_4]. The dashed lines show the carbonyl stretching of PHEMA and the aromatic C-H symmetric stretching of [emim][BF_4]. (Adapted with permission from Ref. [21]. Copyright 2013 American Chemical Society.)

Future work in this area includes studying particle formation on the liquid surface. The ability to make nanoparticles via iCVD deposition can lead to a new high-throughput manufacturing process for fabricating functional polymer particles for photonic, sensing, and drug delivery applications. Liquid substrates are also interesting because they can be manipulated *in situ* to create waves and patterns. Future work will include determining whether magnetic and electrical forces can be used to shape the fluid surface to form new polymer shapes.

16.4 Low-Temperature Substrates

In the iCVD process, the substrate is typically at room temperature and the monomer partial pressure is below its saturation pressure. Seidel *et al.* has shown that decreasing the substrate temperature below the freezing point of the monomer and increasing the monomer partial pressure above the saturation

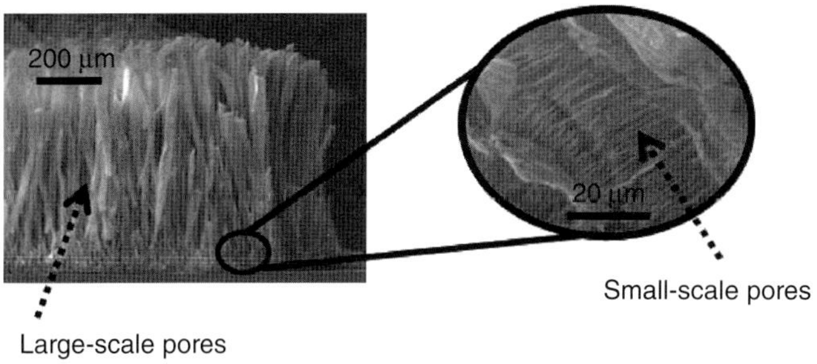

Figure 16.8 PMAA membranes made using low substrate temperatures have two pore sizes: large-scale pores between the microstructures and small-scale pores within the microstructures. (Adapted with permission from Ref. [25]. Copyright 2013 American Chemical Society.)

pressure lead to the formation of porous polymer membranes [25]. These membranes have several potential applications in separations, tissue engineering, and biosensing. Although several liquid-based methods exist for making polymer membranes, the advantages of using iCVD is that solubility is not an issue and copolymer and cross-linked membranes can therefore be made. In addition, the iCVD method is a bottom-up growth process; therefore, the membranes can be deposited onto porous substrates to make porous-on-porous materials that have better separation capabilities.

In order to achieve very low substrate temperatures, the iCVD system was modified with a thermoelectric cooler. When the substrate temperature is below the freezing point of the monomer, the monomer simultaneously freezes as a solid and polymerizes. The solid monomer deposits as macroscale structures and can be sublimated after the deposition process. FTIR spectroscopy of the membranes before and after sublimation showed that the sample is mostly monomer before sublimation and is pure polymer after sublimation. The resulting macroporous polymer membranes have dual-scale porosity: large-scale pores are formed by the void spaces between the macroscale structures formed by the solid monomer, and small-scale pores are formed by the void spaces within the macroscale structures after the solid monomer is sublimated (Figure 16.8). The deposition rate of the membranes can be varied between 10 and 40 $\mu m\ min^{-1}$ by varying the monomer partial pressure and the substrate temperature. These deposition rates are several orders of magnitude greater than traditional iCVD deposition rates (10–100 $nm\ min^{-1}$).

The generality of using low substrate temperatures to make membranes was shown by depositing a variety of different polymers. Stimuli-responsive PMAA and PNIPAAm membranes were fabricated (Figure 16.9), and robust copolymer membranes were made using ethylene glycol diacrylate as a cross-linking agent (Figure 16.10). The solid monomer provides the overall template for the polymerization, and therefore the structure of the membrane can be tuned by controlling the deposition of the solid monomer. For example, in the case of methacrylic acid, the solid monomer grows as 3D pillar-like microstructures

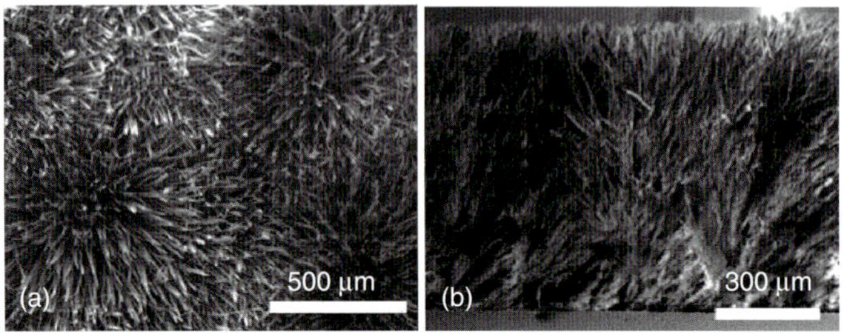

Figure 16.9 (a) Top-down and (b) cross-sectional SEM images of a PNIPAAm membrane. (Adapted with permission from Ref. [25]. Copyright 2013 American Chemical Society.)

Figure 16.10 Photograph of a free-standing P(MAA-*co*-EGDA) membrane. (Adapted with permission from Ref. [25]. Copyright 2013 American Chemical Society.)

at low substrate temperatures (−20 °C), whereas the growth is more 2D as the substrate temperature is increased. The resulting PMAA polymer membranes are therefore thicker and more porous at lower substrate temperatures [26]. The membranes can be deposited onto porous substrates to make porous-on-porous materials. For example, Kwong *et al.* deposited a circular membrane composed of poly(methacrylic acid-*co*-ethylene glycol diacrylate) at the inlet of a paper-based microfluidic channel in order to separate cationic and anionic dyes [27] (Figure 16.11).

Future work in this area includes further studying the nucleation and growth of these membranes at different reactor conditions in order to control the porosity of the membranes. In addition, preliminary work shows that a dense capping layer can be formed on the membranes that would increase the applications for these membranes. Further research is necessary to understand which parameters control the formation of these dense polymer layers on the porous structures.

Acknowledgments

This work was supported, in part, by the National Science Foundation Division of Civil, Mechanical, and Manufacturing Innovation Award Number 1069328, the National Science Foundation CAREER Award CMMI-1252651, and the Donors of the American Chemical Society Petroleum Research Fund.

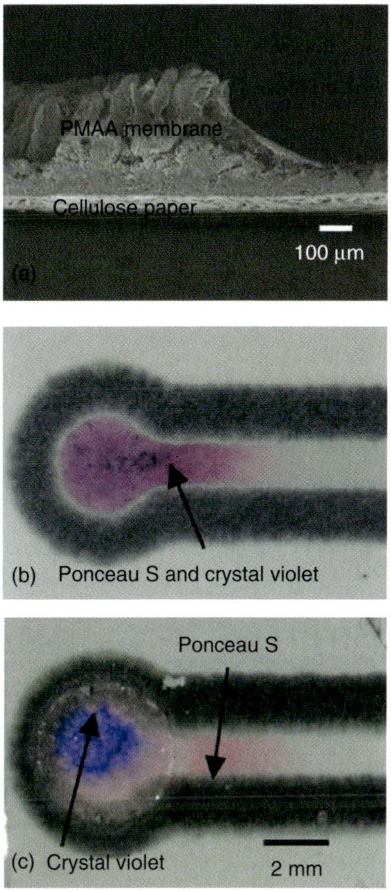

Figure 16.11 (a) Cross-sectional SEM image of a PMAA membrane deposited at the inlet of a cellulose paper-based microfluidic device. (b) Without the membrane, a mixture of crystal violet and ponceau S flow together through the channel. (c) The addition of the membrane allows for the separation of the analytes. (Adapted with permission from Ref. [27]. Copyright 2013 American Chemical Society.)

References

1. Trujillo, N.J., Baxamusa, S., and Gleason, K.K. (2009) *Thin Solid Films*, **517**, 3615.
2. Im, S.G., Kim, B.S., Lee, L.H., Tenhaeff, W.E., Hammond, P.T., and Gleason, K.K. (2008) *Macromol. Rapid Commun.*, **29**, 1648.
3. Mao, Y., Felix, N.M., Nguyen, P.T., Ober, C.K., and Gleason, K.K. (2004) *J. Vac. Sci. Technol. B*, **22**, 2473.
4. Im, S.G., Bong, K.W., Kim, B.-S., Baxamusa, S.H., Hammond, P.T., Doyle, P.S., and Gleason, K.K. (2008) *J. Am. Chem. Soc.*, **130**, 14424.
5. Carrilho, E., Martinez, A.W., and Whitesides, G.M. (2009) *Anal. Chem.*, **81**, 7091.
6. Martinez, A.W., Phillips, S.T., Butte, M.J., and Whitesides, G.M. (2007) *Angew. Chem. Int. Ed.*, **46**, 1318.
7. Lu, Y., Shi, W., Jiang, L., Qin, J., and Lin, B. (2009) *Electrophoresis*, **30**, 1497.
8. Haller, P.D., Flowers, C.A., and Gupta, M. (2011) *Soft Matter*, **7**, 2428.
9. Kwong, P. and Gupta, M. (2012) *Anal. Chem.*, **84**, 10129.
10. Kwong, P., Flowers, C.A., and Gupta, M. (2011) *Langmuir*, **27**, 10634.
11. Chen, B., Kwong, P., and Gupta, M. (2013) *ACS Appl. Mater. Interfaces*, **5**, 12701.
12. Chen, B., Seidel, S., Hori, H., and Gupta, M. (2011) *ACS Appl. Mater. Interfaces*, **3**, 4201.

13. Riche, C.T., Marin, B.C., Malmstadt, N., and Gupta, M. (2011) *Lab Chip*, **11**, 3049.
14. Lazarus, L.L., Riche, C.T., Marin, B.C., Gupta, M., Malmstadt, N., and Brutchey, R.L. (2012) *ACS Appl. Mater. Interfaces*, **4**, 3077.
15. Riche, C.T., Zhang, C., Gupta, M., and Malmstadt, N. (2014) *Lab Chip*, **14**, 1834.
16. Haller, P.D., Frank-Finney, R.J., and Gupta, M. (2011) *Macromolecules*, **44**, 2653.
17. Haller, P.D., Bradley, L.C., and Gupta, M. (2013) *Langmuir*, **29**, 11640.
18. Adamson, A.W. and Gast, A.P. (1997) *Physical Chemistry of Surfaces*, 6th edn, John Wiley & Sons, Inc., New York.
19. Frank-Finney, R.J., Haller, P.D., and Gupta, M. (2012) *Macromolecules*, **45**, 165.
20. Bradley, L.C. and Gupta, M. (2012) *Langmuir*, **28**, 10276.
21. Frank-Finney, R.J., Bradley, L.C., and Gupta, M. (2013) *Macromolecules*, **46**, 6852.
22. Zhang, H., Hong, K., and Mays, J.W. (2002) *Macromolecules*, **35**, 5738.
23. Strehmel, V., Laschewsky, A., Wetzel, H., and Görnitz, E. (2006) *Macromolecules*, **39**, 923.
24. Bradley, L.C. and Gupta, M. (2013) *Langmuir*, **29**, 10448.
25. Seidel, S., Kwong, P., and Gupta, M. (2013) *Macromolecules*, **46**, 2976.
26. Seidel, S. and Gupta, M. (2014) *J. Vac. Sci. Technol., A*, **32**, 041514.
27. Kwong, P., Seidel, S., and Gupta, M. (2013) *ACS Appl. Mater. Interfaces*, **5**, 9714.

17
Organic Device Fabrication and Integration with CVD Polymers

Hyejeong Seong, Bong Jun Kim, Jae Bem You, Youngmin Yoo, and Sung Gap Im

17.1
Introduction

Polymer thin films are essential components of various device applications such as electronic devices (e.g., energy devices, optical devices), microfluidic devices, and sensors. In the field of electronic devices, conducting polymers, semiconducting polymers, and other functional polymers receive attention owing to their flexibility and light weight [1]. To achieve high device performance, it is important to secure the method to deposit polymer thin film with high quality.

The devices mentioned earlier are usually composed of a stack of many thin organic or inorganic layers. The interface between the layers is one of the significant factors that determines the overall device performance as well as the coating process [2]. Therefore, designing and controlling the surface of the polymers and choosing proper deposition process must be conducted thoughtfully. In this regard, the initiated chemical vapor deposition and oxidative chemical vapor deposition (iCVD and oCVD, respectively) processes can be powerful tools for device applications [3]. These processes require low substrate temperature and dry condition during the whole procedure, giving no damage to the underlying layer or devices. Also, a variety of functional and conducting polymers can be deposited using these versatile deposition methods. Furthermore, both processes are easily scalable to achieve high-throughput production, making them suitable for mass production equipment such as roll-to-roll process.

In this chapter, we discuss the application of iCVD and oCVD processes for various device applications. Recent works with oCVD, especially on the energy devices, will be briefly reviewed. Since numerous functional polymers deposited via iCVD have been reported, it is difficult to overview every application of the process in this chapter. Some recent applications of iCVD functional polymers to optical devices, adhesives, microfluidics, and electronics devices are examined in this chapter, and merging of the iCVD with inorganic deposition methods are also discussed.

CVD Polymers: Fabrication of Organic Surfaces and Devices, First Edition. Edited by Karen K. Gleason.

17.2
Energy Devices

17.2.1
Organic Photovoltaics (OPVs)

17.2.1.1 Application of oCVD PEDOT to Cathode Buffer Layer

The controllability of work function (WF) of conducting polymer film for electronic devices is of extreme importance. The WF determines the height of energy barrier for the charge injection, which is directly related to the device efficiency [4, 5]. Since the poly(3,4-ethylenedioxythiophene) (PEDOT) has been widely used for the electrode for organic photovoltaics (OPVs), various investigations aimed to control the WF of PEDOT were attempted to achieve highly efficient OPVs [5, 6]. Recently, controlling of the WF of PEDOT deposited via oCVD, namely oCVD PEDOT, was achieved by introducing various kinds of proper dopants into the PEDOT layer as post-treatments. On the as-deposited PEDOT layers, a chemical treatment was applied with tetrakis(dimethylamino)ethylene (TDAE) or cesium carbonate (Cs_2CO_3) (in Figure 17.1a) [6]. The oCVD PEDOT treated with liquid TDAE showed about 0.8 eV of decrease in WF (5.35 ± 0.04 eV to 4.53 ± 0.05 eV). Similarly, a thin layer (0.5 nm) of Cs_2CO_3 was evaporated on the oCVD PEDOT surface and about 0.3 eV of decrease in WF (5.35 ± 0.04 eV to 5.03 ± 0.03 eV) was observed. The doping of oCVD PEDOT with TDAE or Cs_2CO_3 at the cathode in inverted tetraphenyldibenzoperiflanthene (DBP)/fullerene (C_{60}) heterojunction OPVs could significantly reduce the electron-limiting Schottky barrier height at the cathode (in Figure 17.1c). Thus, improved fill factor (FF) and open-circuit voltage (V_{oc}) could be accomplished compared to a device with unbuffered cathode interface. With this cathode of oCVD PEDOT on indium tin oxide (ITO) with lowered WF, improved FF (0.62) and V_{oc} (0.90 V) could be achieved, which is comparable to the conventional Ag cathode device (FF = 0.62 and V_{oc} = 0.91 V) (in Figure 17.1d,e). The range of controllability of WF of oCVD PEDOT was up to 1 eV, which enables the adaptation of oCVD PEDOT to cathode, as well as anode in OPV device fabrication. The controllability of WF also allows a wide degree of freedom in the designing of device architecture for optimizing the OPV performance.

17.2.1.2 Application of oCVD PEDOT to Top-Contact Anode

Top-illuminated OPV configuration is highly advantageous compared to the conventional orientation in that various kinds of non-transparent substrates can also be used. It is also useful in tandem device structures. The freedom to use non-transparent substrate in OPV fabrication can enable its deployment on various surfaces that we encounter in our daily life such as walls, apparel, and papers. This can be a strong candidate technology for powering flexible, embeddable electronic devices. Conventional PEDOT:poly(styrene sulfonate) (PSS) must be processed in aqueous solution. The solution-based process can dissolve or damage the underlying organic semiconductor layers, which limits

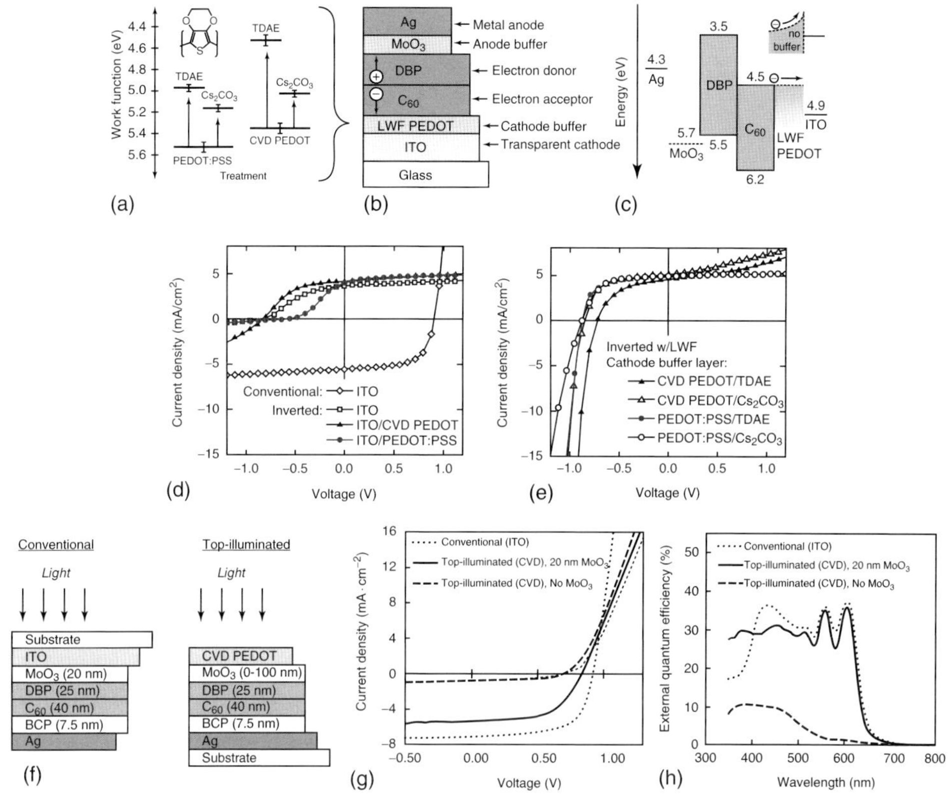

Figure 17.1 (a) WF of both PEDOT:PSS and oCVD PEDOT treated with TDAE or Cs_2CO_3. (b) Schematic of the inverted device architecture with transparent ITO cathode and low WF PEDOT buffer layer inserted between ITO/C_{60} interfaces. (c) Flat band energy level diagram for the inverted device architecture. The top-right inset shows the proposed electron-limiting Schottky barrier formed at an unbuffered ITO/C_{60} interface. (d) Comparison of the conventional device orientation (ITO/MoO_3 (20 nm)/DBP (25 nm)/C_{60} (40 nm)/Ag) (open diamonds) with the same device stack inverted (ITO cathode) with no buffer layer on the ITO (open squares) and with untreated PEDOT layers on the ITO (filled triangles – CVD PEDOT, filled circles – PEDOT:PSS). (e) The same inverted device structure incorporating a low WF PEDOT buffer layer on the ITO cathode: CVD PEDOT/TDAE (filled triangles), CVD PEDOT/Cs_2CO_3 (open triangles), PEDOT:PSS/TDAE (filled circles), and PEDOT:PSS/Cs_2CO_3 (open circles). *J*–*V* curves were obtained under AM1.5G simulated solar illumination. (f) Conventional orientation OPV device with transparent ITO anode (device is illuminated from the substrate side) and top-illuminated orientation OPV device with transparent oCVD PEDOT anode (device is illuminated from the device side). (g) Representative *J*–*V* performance curves measured under 1.1 sun illumination and (h) external quantum efficiency spectra, for the conventional device with ITO anode (dotted) and top-illuminated devices with oCVD PEDOT anode, with (solid) and without (dashed) MoO_3 as a buffer layer. All devices are on silver-coated glass substrates. ((a–c) from Ref. [6]. Copyright © 2012 American Institute of Physics. (d–h) from Ref. [7].)

its application as top electrode with multilayered device configurations. On the other hand, oCVD is a solvent-free process and the damage to the underlying layer can be avoided. A top-illuminated OPV could be fabricated with an oCVD PEDOT as a transparent anode on top of a small-molecule organic heterojunction based on vacuum-evaporated DBP and C_{60}. To minimize the damage to the underlying layers from the vaporized oxidants used during the oCVD process, a buffer layer of molybdenum trioxide (MoO_3) was deposited prior to oCVD process. This resulted in an increase in the OPV photocurrent nearly 10 times compared to the oCVD PEDOT electrode without the oxide buffer layer. The power conversion efficiencies (η_p) also increased by 2.8% for these top-illuminated configurations with ITO-free electrode, which was about 75% of η_p from ITO-based OPV with conventional orientation ($\eta_p = 3.7\%$) [7]. Inverted OPV with oCVD PEDOT top electrode also showed decreased short-circuit current, $J_{sc} = 4.7 \pm 1.6$ mA cm^{-2}, $V_{oc} = 0.84 \pm 0.01$ V, FF = 0.58 ± 0.01, and $\eta_p = 2.1 \pm 0.6\%$. The decrease in the device performance can be ascribed to small absorptive losses in the less transparent oCVD PEDOT and the increased series resistance of oCVD PEDOT compared with ITO.

17.2.1.3 **Application of oCVD PEDOT to Hole Injection Layer (HIL)**

PEDOT has been widely used as a hole injection layer (HIL) for p-type semiconductor [5], and the suitability of oCVD PEDOT as a HIL is investigated for its application to flexible electronics. oCVD PEDOT has unique advantageous characteristics compared to solution-casted PEDOT, including high electrical conductivity, controllability of WF, and formation of a good interface with active layers [8]. Owing to these characteristics, OPV with oCVD PEDOT HIL showed enhanced device performance compared to the OPV with solution-casted PEDOT HIL [9].

A thin layer (50 nm) of oCVD PEDOT was deposited on ITO surface and incorporated into OPV devices with the device structure of copper(II) phthalocyanine (CuPc)/C_{60} (or 3,4,9,10-perylenetetracarboxylic bisbenzimidazole, PTCBI)/ bathocuproine (BCP)/Ag molecular organic heterojunction architecture. As control devices, bare ITO and ITO coated with solution-cast PEDOT:PSS with same device architecture were also fabricated. Both oCVD PEDOT and PEDOT:PSS devices showed improved performances in an V_{oc} (0.48 V) compared to those on bare ITO (0.41 V). The electrical conductivity of oCVD PEDOT thin films (300–1000 S cm^{-1}) was about two orders of magnitude greater than that of the PEDOT:PSS buffer layer (<10 S cm^{-1}), which resulted in an improved higher FF (>0.6) in the devices with ITO/oCVD PEDOT electrodes as shown in Figure 17.2a.

The surface-independent coatability of oCVD PEDOT film could be utilized to modify the electronic surface properties of other electrode materials such as graphene. Recently, many works have reported the excellent electronic properties of graphene such as superior electron and hole mobility (up to 200 000 cm^2 V^{-1}s^{-1}) [11, 12], high current-carrying capability (up to 3×10^8 A cm^{-2}) [13], and great mechanical robustness. However, the WF of bare graphene is 4.27 eV, which is far lower than that of many electron-donor layers

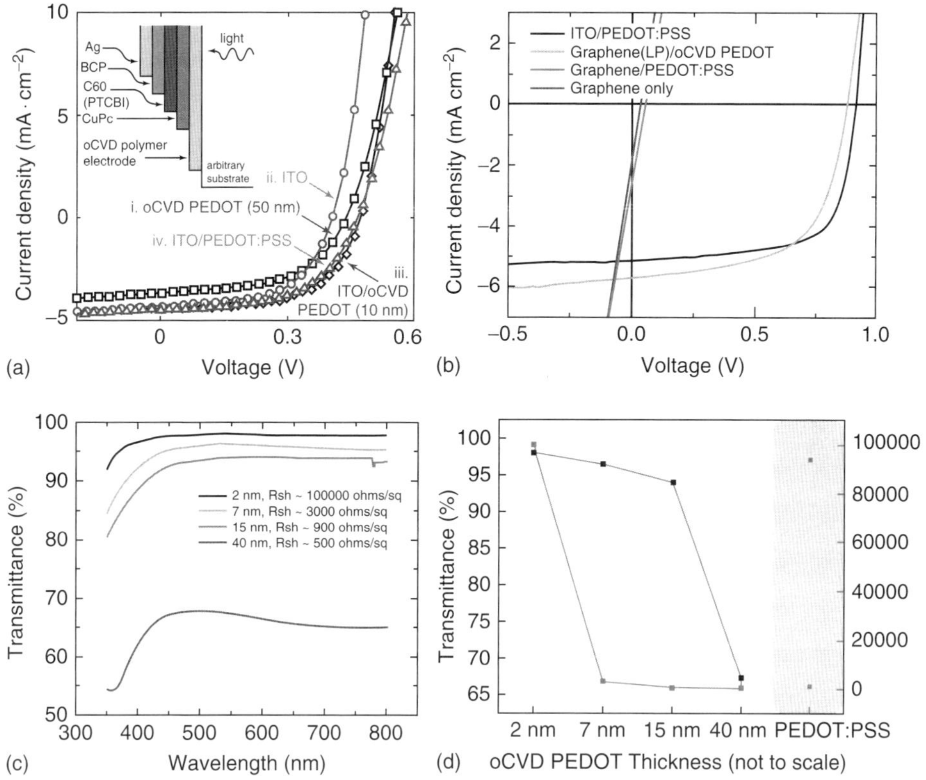

Figure 17.2 (a) Current density–voltage characteristics under illumination (AM1.5, 100 mW cm^{-2}) for oCVD PEDOT PVs on glass differing only in anode structure. The inset shows the device structure used in the work. (b) Current density–voltage characteristics under illumination (AM1.5, 100 mW cm^{-2}) for ITO coated with PEDOT:PSS, bare graphene, graphene coated with oCVD PEDOT and PEDOT:PSS, respectively. (c) Transmittance data for the oCVD PEDOT hole transportation layers (HTLs), measured using ultraviolet–visible (UV–vis) spectroscopy over wavelengths from 350 to 800 nm. (d) Sheet resistance values for each thickness and the transmittance at 550 nm. The oCVD PEDOT R_{sh} was measured using a four-point probe (taking the average of 10 measurements). Transmittance and R_{sh} values of PEDOT:PSS are also shown for comparison. ((a) From Ref. [9]. (b–d) From Ref. [10]. Copyright © 2012 American Chemical Society.)

such as Au or ITO (4.7–5.2 eV) [10]. In spite of many hole injection layers available to modify the surface of anode, most of them could not be applied directly onto the graphene surface due to its inherent water-repellent characteristics. oCVD process can coat conducting polymer film on graphene layer without the concern of de-wetting. The oCVD PEDOT-modified graphene electrode was integrated in the OPV devices, and the device performance was compared with the control device made of ITO electrode (Figure 17.2b). The obtained performance was quite similar with that of PEDOT:PSS-coated ITO electrode, the reference device. The thickness of the oCVD PEDOT was further optimized to achieve both higher transmittance (generally >90%) and lower sheet resistance (R_{sh}). The optimal thickness of the PEDOT layer was found to be within the range of 7–40 nm for both optical transparency and electrical conductivity, as shown in Figure 17.2c,d.

17.2.1.4 Application of oCVD PEDOT Electrode to Flexible, Cost-Effective Photovoltaics

- *oCVD PEDOT on plastic substrates:* The oCVD PEDOT electrodes deposited on flexible plastic substrates and the OPV devices made thereby showed excellent electrical stability upon applying mechanical deformations, which is of critical importance for flexible electronics and also highly desirable for low-cost, roll-to-roll processing. A set of 1000-time compressive, flexing cycles at the radius of less than 5 mm did not show any apparent degradation of electrical conductivity of the oCVD PEDOT deposited on flexible poly(ethyleneterephthalate) (PET). In contrast, only a 20-cycle flexing caused an abrupt decrease in the conductivity of ITO-coated PET substrates. Moreover, even after over 100 compressive flexing cycles, the device performance of OPV device with oCVD PEDOT electrodes on PET remained practically unchanged, as shown in Figure 17.3 [9].
- *Conformal coverage and mechanical robustness of oCVD PEDOT on paper substrates:* The oCVD PEDOT electrode showed excellent conformal coverage; no large variation in thickness was observed at the top, at the side wall, and at the bottom, without any blanketing, or welling-up of the coating [9, 14, 15]. As shown in Figure 17.4a,b, the oCVD PEDOT could fully fill in features of varying shapes and sizes as low as 50 nm. The conductivity of the oCVD PEDOT

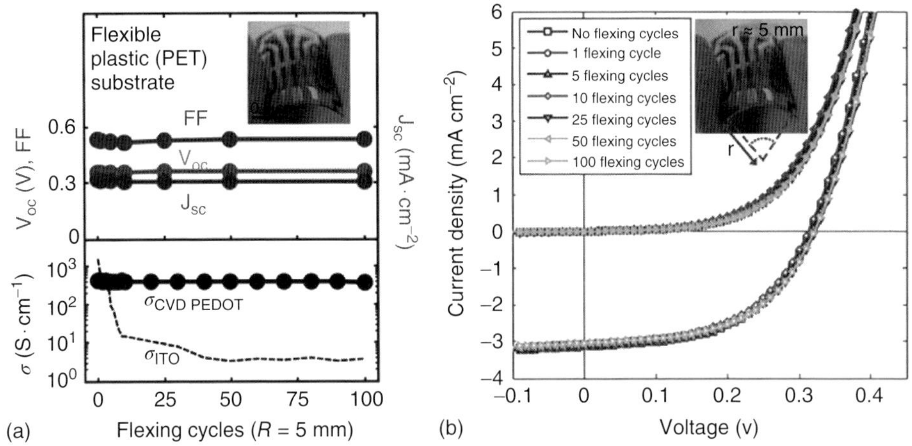

Figure 17.3 OPVs with oCVD PEDOT on flexible PET substrate. (a) Device characteristics under illumination (AM1.5, 100 mW cm^{-2}) and electrode conductivity for the OPVs with oCVD PEDOT on PET (5-mm thick) after repeated flexes to 5-mm radius. The dashed black line shows the conductivity of ITO/PET for reference. (b) *J–V* characteristics of an ITO-free OPV on PET under illumination (AM1.5, 100 mW cm^{-2}). The thickness of oCVD PEDOT was 50 nm. The inset photograph illustrates the orientation of flexing for an array of these OPVs. (From Ref. [9].)

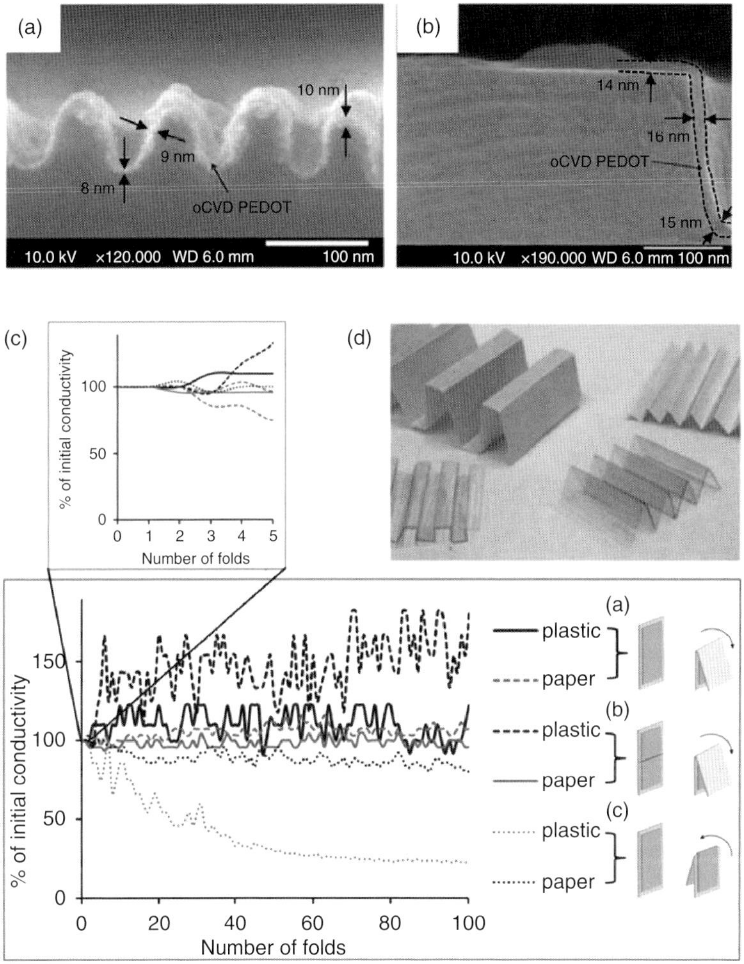

Figure 17.4 Scanning electron microscope (SEM) images of oCVD PEDOT on (a) nanowedges textured PMMA, and (b) square trench textured PMMA. The sputtered gold layer is visible on the top surface of the images and may affect the appearance of the morphology. Dashed yellow lines have been added to (b) to denote where the oCVD PEDOT layer is located. Arrows have been added to show film thickness at various locations. (c) Change in conductivity of PEDOT layers on plastic and paper substrates after multiple folding cycles, (i) regular folding, with PEDOT interior to fold, of initially flat substrate, (ii) regular folding of PEDOT deposited on a pre-folded substrate, and (iii) back folding, with PEDOT exterior to fold, of initially flat substrate. Inset shows values for first fivefolds and (d) PEDOT deposited on folded polycarbonate and paper substrates. (Republished from Ref. [15]. Copyright © 2013 Elsevier.)

on paper substrates decreased by only less than 5% of the initial value even after 100-folding cycles, as shown in Figure 17.4c. In case of the back-folded samples, larger decrease in conductivities to 80% and 25% was observed for the paper and plastic samples, respectively. The mechanical robustness of the oCVD conducting polymer coatings allows for a wide variety of manipulation of the structure of flexible substrates.

- *Bottom-contact and top-contact oCVD PEDOT electrode on paper substrates:* OPVs were fabricated directly on various as-purchased paper substrates enabled by the oCVD PEDOT electrodes due to its excellent conformal coverage. Figure 17.5a shows completed OPVs on tracing paper, tissue paper, and copy paper. The ultra-light weight paper-based OPV devices was fully operational even after a complete folding of the device [9]. Characteristic $J-V$ performance curves clearly showed a significant power generation capability from the paper-based devices. Furthermore, oCVD PEDOT could be used as a transparent electrode in both conventional structures, where the electrode is deposited onto a transparent substrate, and in inverted, top-illuminated structures, where the transparent electrode is deposited onto a device stack with the option of using an opaque substrate [7]. Various kinds of opaque substrates, commonly available from our daily life including photo paper, magazine print, US first-class stamp, and plastic food packaging were demonstrated to be used to fabricate inverted OPVs with oCVD PEDOT electrode (Figure 17.5b). The $J-V$ performance curves with the inverted orientation of illumination displayed J_{sc} close to that of the OPVs deposited on the transparent glass substrate, despite the low substrate transparency. The comparatively low η_p and V_{oc} as compared to the standard glass-based devices is likely due to the increased current pathways through the non-conformal photoactive layers that allow facile charge transport opposite the photo-generated current.
- *Large-area applicability of oCVD PEDOT electrode on paper substrates:* The expansion of these paper-thin photovoltaics to a larger scale was demonstrated by utilizing the vapor-patterned oCVD PEDOT electrodes by shadow mask to monolithically integrate the paper solar cell into a large area of serially connected photovoltaic arrays directly on paper and glass [9]. The current–voltage characteristics for 250-cell ($0.1 \times 0.3\,cm^2$ each) monolithic arrays on paper and glass substrates showed high V_{oc} of 50 and 67 V, respectively (Figure 17.5c, left). The obtained values correspond closely to the summation of voltages from all working cells in each series (minus losses from shorted devices). Spatial distribution maps of the individual solar cell performance in the series-integrated arrays were recorded to gain insight into overall cell statistics and the morphological impacts of the substrates. The cell voltage distributions on both glass and paper substrates (Figure 17.5c, right) show that the device on each substrate achieved similar maxima (~0.4 V), with higher variance across the rougher paper, due to the inherent non-homogeneity of the paper. This all-dry fabrication and integration strategy will enable the design and implementation of new, low-cost photovoltaic and optoelectronic systems without the substrate limitation.

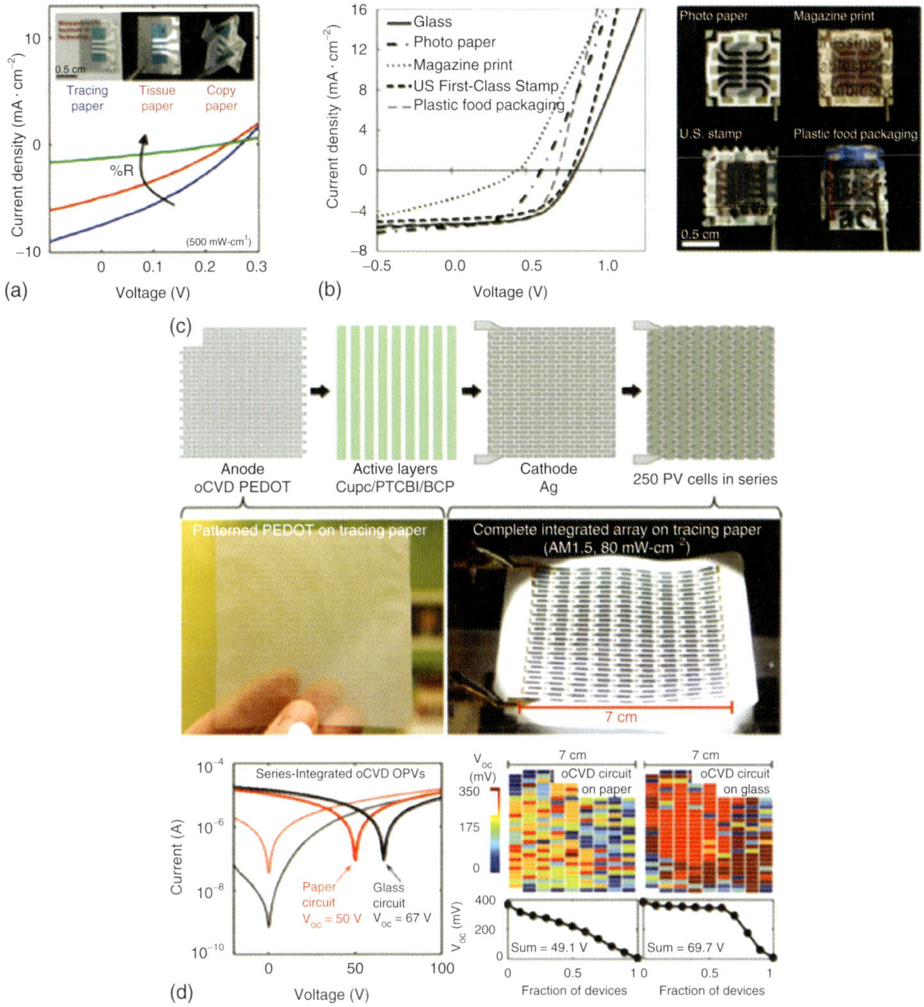

Figure 17.5 (a) Current density–voltage characteristics under illumination (AM1.5, 500 mW cm^{-2}) for oCVD PVs vapor printed on as-purchased tracing paper, tissue paper (~40-μm thick), and copy paper (~120-μm thick). (b) *J–V* curves for top-illuminated OPVs fabricated on the top side of some common opaque substrates under 1.1 sun illumination, including photo paper, magazine print, a US first-class stamp, plastic food packaging, and glass for reference. Photographs of completed 10 device arrays are also shown. (c) Printing schematic for 250-cell, series-integrated large-area monolithic photovoltaic arrays. The photographs show the printed PEDOT (~50-nm thick) pattern (left) and a completed array (right) on tracing paper. (d) Current–voltage performance curves for series-integrated photovoltaic arrays with vapor-patterned oCVD electrodes on paper (red) and glass (black) under illumination (AM1.5, 80 mW cm^{-2}) (bold) and in the dark (thin) (left). Spatial map of individual-cell open-circuit voltages across the respective ~50 cm^2 arrays. The lower insets show the cumulative fraction of devices producing at or below a given voltage (right). ((a, c, d) From Ref. [9]. (b) From Ref. [7].)

17.2.2
iCVD Polymer for Dye-Sensitized Solar Cell (DSSC)

Dye-sensitized solar cell (DSSC) has attracted a great attention for harvesting sunlight due to its low cost and high efficiency [16]. In the DSSC, mesoporous titanium dioxide (TiO_2) structure is covered with nano-sized dye to increase the surface area of photosensitizer for improved η_p [17]. The dye releases electron with the light absorption, and the released electrons are in turn transferred to transparent electrode through the electrolyte. Therefore, increasing the ionic conductivity of the electrolyte is highly desirable. For this purpose, liquid electrolyte has been widely used for its high ion conductivity. However, constant leakage or evaporation of electrolyte has occurred due to its incomplete sealing. Therefore, many research efforts are being made to replace liquid electrolyte with solid or gel electrolyte. For improved device performance, it is critical to fill up the pores of mesoporous TiO_2 completely with solid electrolyte so that the contact of dye with electrolyte is maximized. To completely fill in the pore, poly(2-hydroxyethyl methacrylate) (PHEMA) was introduced via iCVD [18]. The ester and hydroxyl groups in the PHEMA play an important role in facilitating ion transport. Also, PHEMA can form a stable gel electrolyte meeting with ethylene and propylene carbonate. Consequently, the DSSC made of PHEMA electrolyte showed higher cell efficiency and V_{oc} compared to liquid electrolyte DSSC because of the effect of pore filling achieved by iCVD process, inhibiting the electron recombination at the electrolyte-electrode interface (Figure 17.6).

17.2.3
oCVD PEDOT for Supercapacitor

To maximize the power and energy efficiency, it is significant to raise the cell voltage of supercapacitors, which is dependent on the properties of electrolyte and the electrode materials. Therefore, for active electrodes of supercapacitors, carbon-based materials such as activated carbon, graphene, and carbon nanotubes (CNTs) have been extensively investigated owing to their high surface area and excellent electrical conductivity. However, since the charge accumulation

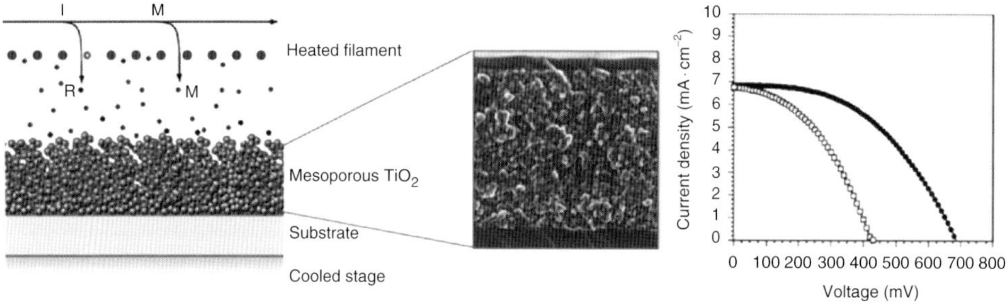

Figure 17.6 Scheme of the PHEMA depositing process via iCVD for infiltrating mesoporous TiO_2 layer and its SEM image. (From Ref. [18]. Copyright © 2011 American Chemical Society.)

takes place only on the surface of CNTs, increasing the charge density of CNTs in the application of supercapacitors is highly challenging and specific capacitance is limited. To improve the storage capacitance of the CNT-based supercapacitor device, highly ordered, aligned carbon nanotubes (A-CNTs) were conformally covered with oCVD PEDOT (Figure 17.7b) [19]. The electrochemical window of oCVD PEDOT is higher than that of many other conducting polymers

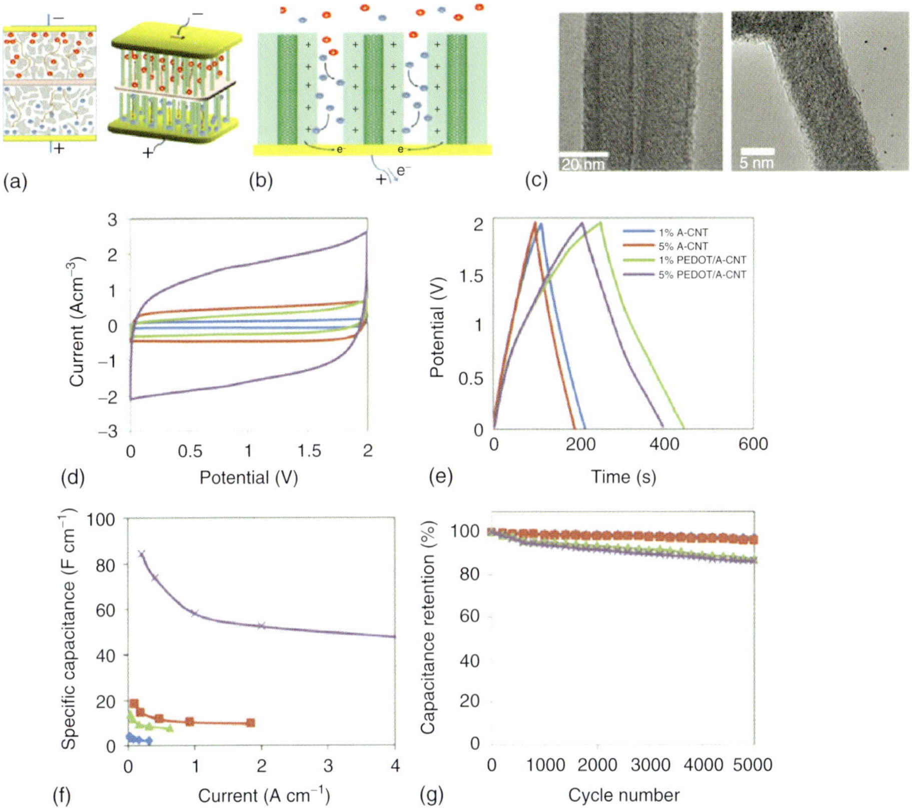

Figure 17.7 (a) Schematic comparison of the tortuous ion transport paths in nanoporous electrodes formed, for example, from activated carbons (left) and parallel ion pathways in the A-CNTs (right). (b) Schematic representation of conducting polymer-coated A-CNT used as supercapacitor electrode. Cations and anions along with electrons and holes are shown in the figure. (c) Transmission electron microscopy (TEM) images of A-CNT conformally coated with PEDOT. The contrast in the image on the right shows the uniform coating of PEDOT on A-CNT in consistence with the image on the left, showing the detailed morphology of CNT used in this work. (d) Cyclic voltammetry curves for the pure A-CNT and PEDOT/A-CNT electrodes at 100 mV s^{-1} scan rate. Volumetric current density as a function of applied voltage. (e) Galvanostatic charge/discharge curves for A-CNT electrodes at 0.5 A g^{-1} discharge rate. (f) Volumetric specific capacitance obtained from galvanostatic cycles as functions of discharge rate. (g) Capacitance retention for the A-CNT electrodes during 5000 charge/discharge cycles at 10 A g^{-1} discharge rate. (From Ref. [19]. Copyright © 2013 Elsevier.)

and synergistic effects between the two components of A-CNT and PEDOT, with complementary properties leading to a dramatic increase in capacitive performance for the PEDOT/A-CNT electrodes, relative to non-coated A-CNTs. Moreover, significantly enhanced charge storage capability was also exhibited. The conformally covered oCVD PEDOT layer also strongly adhered to A-CNT increasing its mechanical robustness. This led to higher retention of the capacitance compared to the electrodes of the conducting polymers deposited on randomly packed CNT networks. A slight reduction of the capacitance with the charge/discharge was observed in oCVD PEDOT/A-CNT electrode due to the volume change in oCVD PEDOT during the oxidation/reduction cycles compared with the non-coated A-CNTs. The synergistic effect of independently optimized the carbon-based electrode materials will enable further improvement of performance of the various types of energy devices [20].

17.3
Optical Devices

17.3.1
Bragg Mirror

Bragg mirror (dielectric mirror) consists of multiple dielectric layers, where the difference between the refractive indices of two layers is large [21]. Usually, a thin-film Bragg mirror is manufactured on a rigid substrate with the precisely controlled film thickness, achieved by vapor-phase processes. In order to achieve high reflectivity from all-organic Bragg mirror, it is inevitable to stack large number of layers because the difference of refractive indices of common organic materials is generally quite small. Moreover, the adjacent polymer layer must be cross-linked to avoid the damage to the other layer during the wet coating process. Overall, the aforementioned requirements increase the total thickness of the Bragg mirror.

Rapid and reversible responsive dielectric mirror was fabricated by alternating deposition of inorganic TiO_2 and organic PHEMA layer [22]. The capability of precise control of film thickness by the iCVD process was demonstrated (Figure 17.8a). The fabricated hybrid dielectric mirror showed unique characteristics of changing its reflectance bands in the visible range. It was due to the hygroscopic nature of PHEMA, leading to the change in PHEMA thickness. The change of the reflectance bands was reversible. The mild and dry nature of the iCVD process enabled the application of Bragg mirror to various delicate substrates such as papers and plastics (Figure 17.8b–e). Also, the fabricated Bragg mirror showed quick response to the water vapor. A seven-layered structure of alternating TiO_2 and PHEMA was coated on quartz cell, and its color changed from green to red by contact with 1 mol% water vapor in N_2 within 0.3 s. However, immediately after purging the water vapor, the color was returned back to its original green color within 0.3 s (Figure 17.8f).

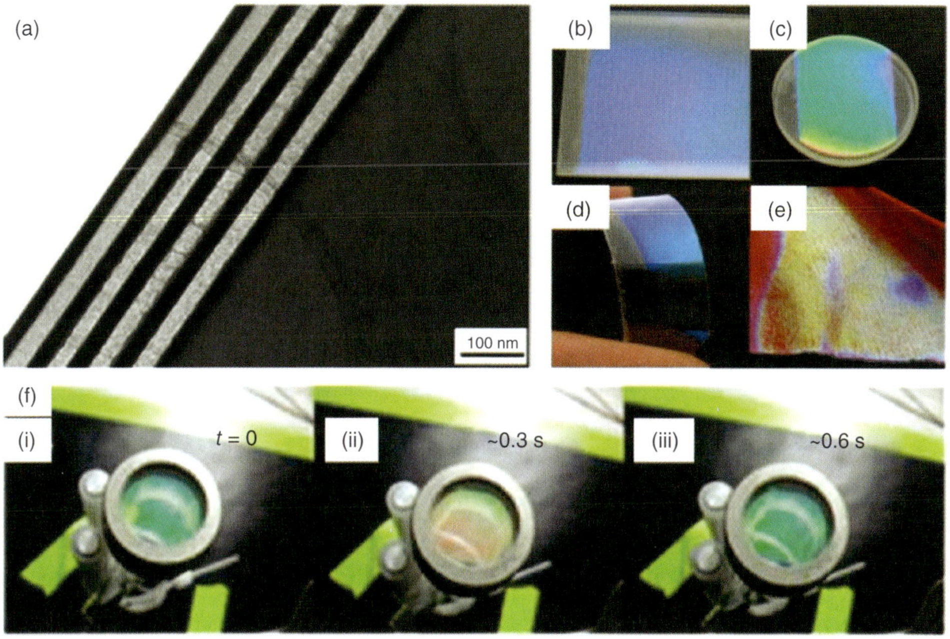

Figure 17.8 (a) TEM image of TiO_2 (black)/PHEMA (white) multilayer. Hybrid Bragg structure on various substrates: (b) microscope glass, (c) quartz, (d) polycarbonate, and (e) red paper. (f) Color change of TiO_2/PHEMA multilayer (i) in bare state, (ii) after contact with water vapor, and (iii) after purging the water vapor. (From Ref. [22]. Copyright © 2008 American Chemical Society.)

17.3.2 Electrochromic Devices

In an electrochromic device, the color of the material changes reversibly by applying a burst of charge to the material [23]. In case of PEDOT, applied voltage can control its electronic structure that alters the conductivity and color. When conductive, PEDOT has sky blue color because it is oxidized, and when insulating, it becomes nearly opaque because of reduction. When PEDOT is oxidized by the injected charge, a cation is generated in every two or three EDOT segments [24], which is electrically neutralized with a counter-anion. In this oxidized state, the unpaired electrons mostly absorb low-energy, near-infrared light that makes it transparent to the visible light. On the other hand, in the reduced state, the electrons are stabilized in the conjugated alternating single-bond/double-bond configuration. The conjugated electrons are at a lower energy than unpaired electrons, and absorption is shifted into the higher energy visible region resulting in the dark color in the reduced state. In oCVD PEDOT, the oxidized PEDOT is charge-balanced with Cl^- ions.

For the application of oCVD PEDOT films to electrochromic device, about 100-nm-thick oCVD PEDOT was deposited on ITO/glass, and electrochemical power was applied to the oCVD PEDOT layer. Cyclic voltammetry scan shows that reduction peaks appeared at −430, −450, and −460 mV and oxidation peaks appeared at 22, 40, and 62 mV. It indicates that a potential of at least ~0.5 V is necessary to electrochemically switch between the oxidized and reduced states of PEDOT [25]. The reduction peak is broader than oxidation peak because higher potential is required for less conductive PEDOT in reduced state. On the contrary, the oxidation peak is sharp because a conducting reaction front facilitates electron transfer. The color contrast was 45% at 566 nm and is 85% stable over 150 redox cycles. A maximum contrast for this film was found to be 25.8% at 566 nm as the potential was gradually stepped from its oxidized state at 400 mV to its reduced state at −600 mV. A cycle step voltages of −600 and 400 mV were applied to the oCVD PEDOT layer with a cycle time of 1 min. After the first 15 cycles, a slight degradation in the contrast was observed: the contrast fell from 45 to 42.5%. The contrast at 566 nm was still maintained over 85% of the initial value even after 150 cycles (Figure 17.9).

17.4 Nano-Adhesives

17.4.1 iCVD Polymer as Nano-Adhesives

Adhesives are substances that allow the bonding between two surfaces via a two-step mechanism: wetting and curing into a cohesively strong solid [26]. Currently, various types of adhesives including ultraviolet (UV)-curable adhesives and pressure-sensitive adhesives are widely used in our everyday life [27]. To achieve high-performance adhesives with improved bond strength, the interfacial adhesion between microstructured and nanostructured surfaces should be considered. As a result, there has been a growing demand for nanoscale adhesives that can adhere two surfaces with micro- and nanostructured topography without damaging them, which were burst into the research on "nano-adhesives." These "nano-adhesives" are useful in many areas such as sealing of nano- or microanalytical systems [28], or increasing the adhesion between electroless deposited copper and plastic substrates that are used as printed circuit boards (CBs) [29, 30]. However, the currently used liquid-phase adhesives are not useful in bonding substrates with micro- and nanostructures on the surface, because these adhesives are generally highly viscous and the thickness of the applied adhesive is no less than few micrometers. Therefore, it is too difficult to deposit the adhesive that is thin enough to preserve the detailed nanomorphologies of the surface.

The iCVD process serves as an effective tool to circumvent the issues of current adhesive systems and provides a nano-adhesive layer to achieve strong bonding between the surfaces. The conformal characteristic of iCVD process allows the preservation of micro- and nanostructures on the surfaces of the substrates. Since the thickness of the nano-adhesive layer is on the order of a few hundred nanometers, it retains the flexibility even after complete curing. Also, the nano-adhesive

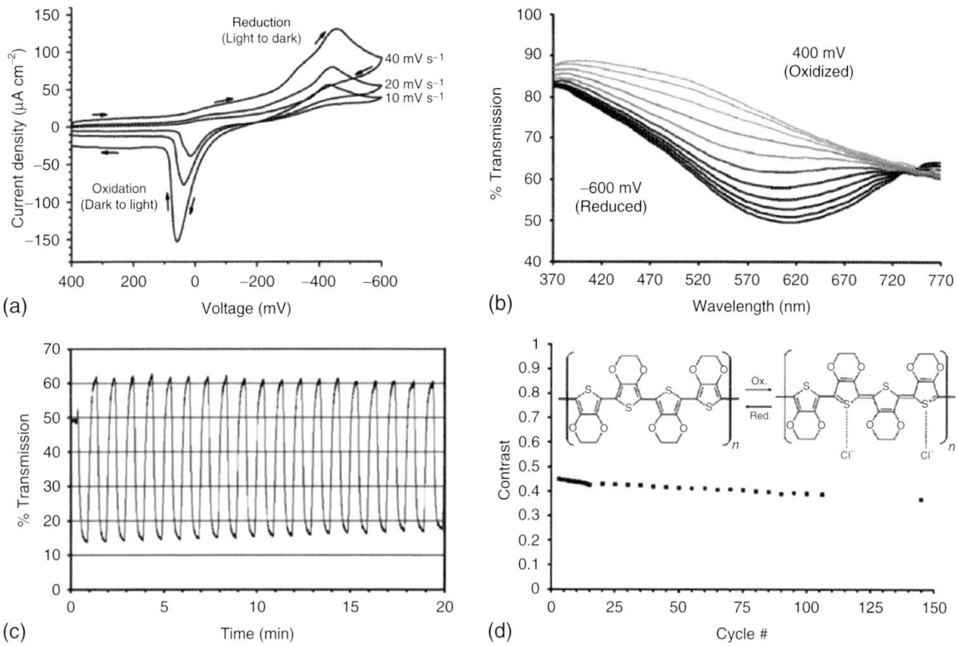

Figure 17.9 (a) Cyclic voltammetry indicates that PEDOT is reduced gradually, but oxidizes more suddenly as a result of the conductivity swing during redox cycles. The optical properties of the two states are shifted as well. (b) The maximum optical contrast achieved with oCVD PEDOT is 45% at 566 nm. The film shown here throughout a full redox cycle had a maximum contrast of 25.8%, also at 566 nm. (c) Average switching speeds for the first five cycles (based on 80% of full response) were 13 s for light-to-dark and 8.5 s for dark-to-light transitions. (d) oCVD PEDOT exhibited greater than 85% optical contrast stability after 150 redox cycles. Neutral PEDOT (left, inset) is oxidized to form a conducting polycation (right, inset) in the presence of charge-balancing anions. Oxidized PEDOT has a transparent sky blue color that turns dark purple upon reduction. (From Ref. [25]. Copyright © 2007 Elsevier.)

can be applicable to various substrates such as paper or fabric that are vulnerable to conventional liquid-phase adhesives. Recently, several conformal dry adhesive layers with nanometer-scale thickness were developed via the iCVD process. Figure 17.10 shows the basic scheme for the nano-adhesive layers that have been developed so far using the iCVD process.

The first nano-adhesive layer (Gen I) developed via iCVD consisted of depositing an epoxide-containing polymer, poly(glycidyl methacrylate) (PGMA), on one substrate via iCVD and amine-containing poly(allylamine) (PAAm) on the other substrate via plasma polymerization (PP) [31]. The highly nucleophilic nature of the primary amines attacks the epoxide ring, which is highly unstable owing to the ring strain arising from the low bond angle. Once the ring opens by the attack of the amine group, hydroxyl group is formed and a new C-N covalent bond is created. The reaction scheme is shown in Scheme 17.1.

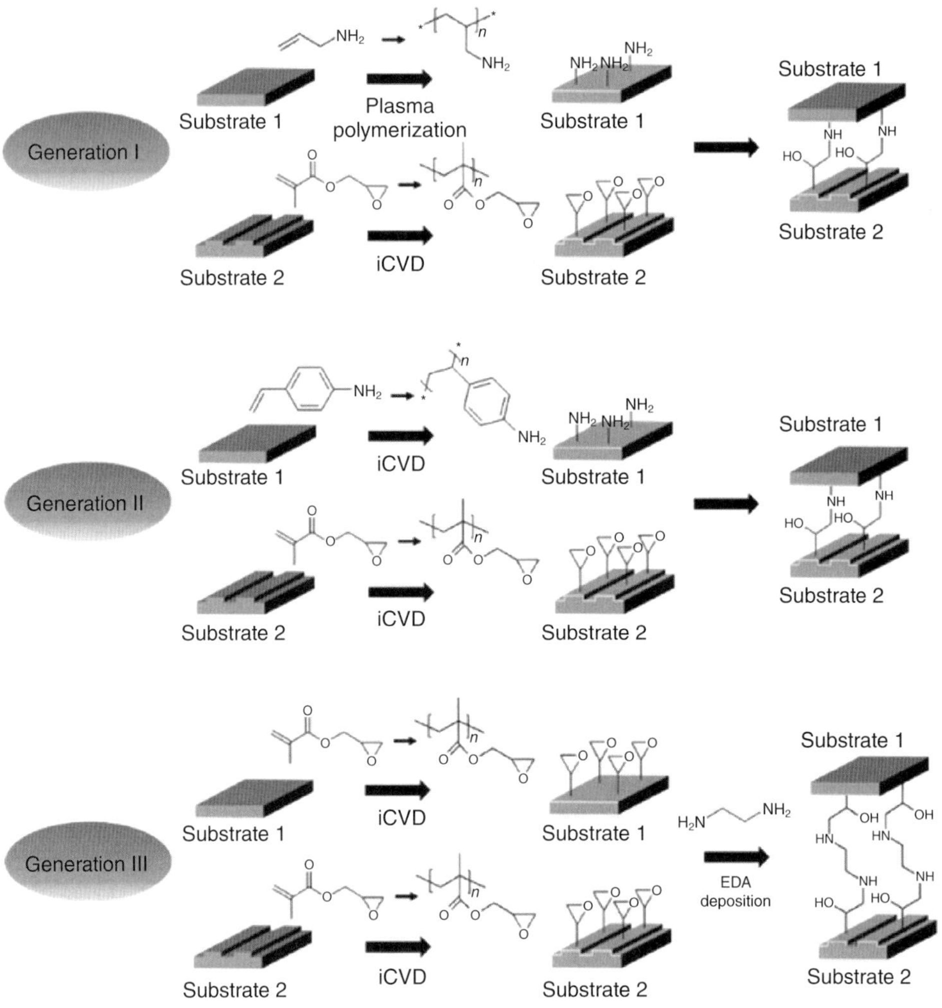

Figure 17.10 Reaction scheme of Generations I, II, and III Nano-adhesive deposited via iCVD process.

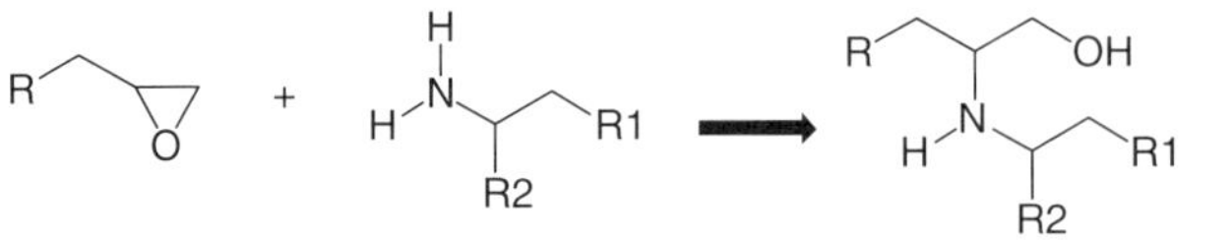

Scheme 17.1 Scheme of epoxy-amine ring opening curing reaction.

The epoxy-amine ring opening curing reaction is an ideal chemistry for developing the nano-adhesive because it is fast and does not produce any by-product. In fact, reactions that form by-product is undesired for the application of nano-adhesive because the volatile reaction by-product such as water or methanol vapors can produce surface defects and lower the bonding strength.

The Gen I nano-adhesive deposited via iCVD and PP was able to bond different substrates such as poly(dimethylsiloxane) (PDMS), Si wafer, poly(tetrafluoroethylene) (PTFE), poly(styrene) (PS) Petri dish, poly(carbonate) (PC), and PET films. The bonded substrates were able to endure at least 50 psia of pressure.

An improved version of Gen I nano-adhesive was developed shortly after it was reported. Instead of using PAAm as its amine polymer, poly(4-aminostyrene) (PAS) was used to enhance the performance of its predecessor [32]. The Gen II nano-adhesive was advantageous over its previous version in that both polymers were deposited using the iCVD process. The Gen II nano-adhesive also successfully bonded different substrates with a higher adhesion strength. The bonded substrates were able to withstand more than 150 psi of pressure for all the substrate combinations attributed to the higher density of amine functionality compared to that of plasma-polymerized PAAm with improved stability in water environment.

Although Gen I and II nano-adhesive layers are able to adhere two substrates while preserving the surface morphology, they involve two separate depositions that increase the process time. Also, since the chemical reaction in both adhesives takes places only at the interface between the two, stability in various chemical solvents that could dissolve either one of the layers is not ensured. To address this issue, a doubly cross-linked nano-adhesive (DCNA) (Gen III) was developed in which the adhesive layer was fully cross-linked at the interface as well as within the layer itself [33].

In Gen III nano-adhesive, both substrates are coated with PGMA and one of the substrates is exposed to an amine-based cross-linker, ethylenediamine (EDA) vapor at room temperature and ambient pressure. The substrates are then placed face-to-face to achieve a full curing between both PGMA layers catalyzed by EDA molecules. Upon curing, the epoxy-amine reaction takes place as in the case of Gen I and II nano-adhesives. However, in this case, the reaction yields free secondary amines that are available for cross-linking with other free epoxide groups. Moreover, the tertiary amines formed by the reaction of secondary amine and an epoxide group can catalyze the self-cross-linking between epoxide groups yielding a highly cross-linked adhesive layer. Different from Gen I and Gen II nano-adhesives where the reaction only occurs at the interface, Gen III nano-adhesive layer is cross-linked at the interface as well as within the adhesive itself.

Apart from these three nano-adhesives, an adhesion promoter layer was also developed to promote adhesion of metal to various organic and inorganic substrates such as PC, PET, PS, epoxy-based CB substrate, glass, and Si wafer [29]. The interfacial adhesion of metal, especially copper (Cu), and organic substrates is crucial for developing low aspect ratio CB substrates, following the continuous miniaturization in device size [34]. To achieve this, metal-chelating poly(4-vinylpyridine) (P4VP) was deposited as nano-adhesive layer via iCVD process to strongly bind Pd^{2+} ion and reduce it to Pd^0, which act as a reducing agent during electroless deposition of Cu. As a result, electroless deposited Cu was strongly bound to epoxy-based substrate with a maximum strength of 1.6 kgf cm^{-1}.

17.4.2
Application of iCVD Nano-Adhesives to Microfluidic Devices

Microfluidics has attracted much attention from researchers owing to its broad applications such as biological and analytical tools, microreactors, as well as tools to study the behavior of complex fluids [35]. So far, PDMS has been widely used as the material of choice for fabricating the microfluidic devices. However, the demand for the use of other materials as a substrate such as PET, poly(imide) (PI), poly(methyl methacrylate) (PMMA), and cyclic olefin copolymer (COC) are increasing owing to the wide range of applications that microfluidics can offer. However, the sealing methods of these substrates were either specific to the substrate materials or prone to damage to the substrates.

The Gen I, II, and III nano-adhesive layers ensure the bonding with high adhesion between various types of substrates that are advantageous for fabricating microfluidic devices with unconventional materials. The nano-adhesive layers via iCVD can provide an easy and effective sealing method for these substrates. Microfluidic devices sealing with the nano-adhesive layer were able to withstand at least 2 MPa of pressure with the maximum operating pressure to reach 11 MPa for PET and glass combination (Figure 17.11a). Microfluidic devices that can operate at high pressures are highly desired for applications where operation at high flow rate is needed such as inertial mixing, particle focusing, and separation [37]. Not only that, devices that are able to work at high flow rates can be used in turbulent microfluidics for large throughput chemical synthesis [38].

Apart from the high adhesion strength, the nano-adhesive layers are also highly stable in water (for Gen II) and in other chemicals such as acetone, toluene, and tetrahydrofuran (THF) (for Gen III). Specifically, the lap shear strength between glass substrates bonded with Gen III nano-adhesive remained stable even after a 10-day exposure to various organic solvents (Figure 17.11b). The nano-adhesive layer was able to maintain the flexibility of the bonded substrates. A 200-nm pattern can easily be bonded without any structural damage owing to the low thickness of the nano-adhesive (Figure 17.11c). The thin nature of the nano-adhesive permits effective sealing of flexible microfluidic devices with high adhesion strength (Figure 17.11d).

The nano-adhesive deposited via the iCVD process was recently used in fabricating non-PDMS microfluidic devices that allow oxygen-free flow lithography (FL) for the synthesis of various anisotropic particles [36]. FL enables reliable fabrication of various particles by exposing a portion of polymer solution flow through the microfluidic channel. However, PDMS is disadvantageous in that the solvents and/or monomers used in FL can swell the PDMS wall [39]. Using the iCVD, a novel non-PDMS-based microfluidic device for oxygen-free FL was highly beneficial in this regard because iCVD is capable of both inhibiting

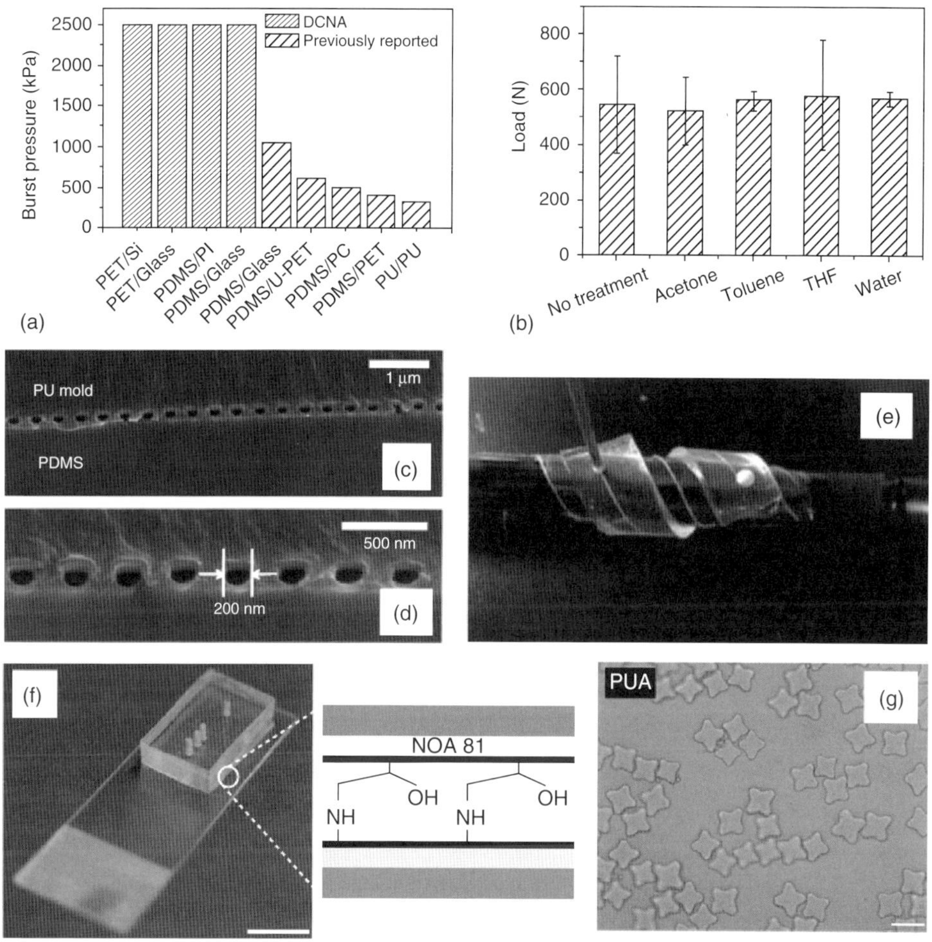

Figure 17.11 Characteristics of iCVD nano-adhesives. (a) Results obtained from burst pressure measurements. Burst pressures obtained from previously reported methods are also shown for comparison. (b) Lap shear strength measurement results of glass–glass substrates exposed to different solvents for a period of 10 days. (c–d) A cured iCVD PGMA film on pre-patterned PDMS with a flat PET film coated with PAAm with different magnification. The cross-sectional SEM image of (c) and (d) clearly shows that 200-nm channels at the interface between PDMS and poly(urethane) (PU) film are retained. (e) PDMS/PET microfluidic device was fabricated and rolled around a pen. (f) Image of fully assembled channel mounted on a glass slide. The inserted schematic describes the bonding chemistry between channel interfaces. (g) Images of particles synthesized from poly(urethane acrylate) (PUA). Scale bars, 50 μm. ((a, b, e) From Ref. [33]. Copyright © 2013 The Royal Society of Chemistry. (c, d) From Ref. [31]. Copyright © 2009 The Royal Society of Chemistry. Parts (f, g) From Ref. [36]. Copyright © 2012 Macmillan Publishers Limited.)

swelling and increasing adhesion strength [40]. UV-curable Norland Optical Adhesive 81 (NOA81), which is highly stable in strong solvents (e.g., acetone, hydrochloric acid, toluene) and gas impermeable, was used as the substrate material to replace the PDMS. The microfluidic channel patterned on PDMS-backed NOA81 substrate was strongly bonded to glass-backed blank NOA81 substrate via the nano-adhesive bonding process using the iCVD (Figure 17.11f). The fabricated microfluidic device enabled completely solvent-based synthesis of complex anisotropic particles. Also, particles were synthesized using water-insoluble monomers such as polyurethane acrylate using toluene as the solvent (Figure 17.11e).

17.5 Encapsulation of Electronic Devices

17.5.1 Thin-Film Barrier for Encapsulation of Electronic Devices

With the elevating expectations and demands for future electronics, organic electronic devices are acknowledged as one of the candidate technologies because they are highly flexible, or even stretchable. However, most of the organic materials in electronic devices degrade rapidly upon an exposure to water vapor and/or oxygen, which hinders the wide application of organic electronics [41]. Therefore, organic electronic devices must be encapsulated by barriers to obstruct the penetration of water vapor and oxygen to elongate the device lifetime.

The barrier requirement is generally guided by water vapor transmission rate (WVTR) and for organic light-emitting diode (OLED), the guideline is set as below $10^{-6}\,g\,m^{-2}\,day^{-1}$ [42]. Relatively mild requirement is demanded for OPV and organic thin-film transistors (OTFTs) as the required WVTR below 10^{-3} g m^{-2} day^{-1} [43]. To satisfy the barrier requirement, glass or metal lid was attached above the device using epoxy glue. This approach proves to be effective in the viewpoint of the barrier performance [44]. However, it cannot be used for flexible applications, eliminating one of the most advantageous features of the organic electronic devices. To solve this matter, thin-film barrier composed of alternating organic and inorganic layers was intensively investigated to encapsulate organic electronic devices. Figure 17.12a illustrates the basic structure of the organic/inorganic hybrid multilayer, which is also known as "Vitex^{TM} structure." In the Vitex^{TM} structure, inorganic layer provides the barrier property because it has dense structure that can effectively block the penetration of vapor molecules. The organic layer is generally acknowledged to provide mechanical flexibility. The organic layer also serves to decouple the defects in adjacent inorganic layers, and thus increases the tortuous path that the penetrating vapors travel. The defects in the inorganic layer can be filled and planarized by the organic layer deposition [47].

For the deposition of inorganic layer, the most common approach is using plasma-enhanced chemical vapor deposition (PECVD) or atomic layer deposition

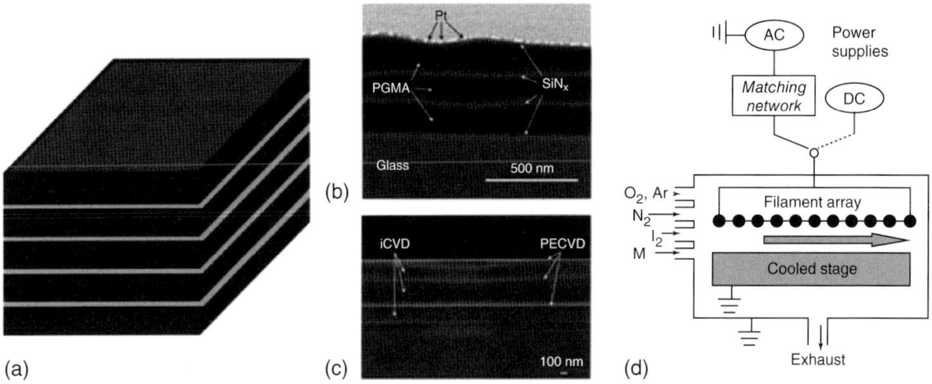

Figure 17.12 (a) Basic scheme of the structure of the organic/inorganic hybrid multilayer used for thin-film barrier. (b) Thin-film barrier developed using iCVD and HWCVD. (c) Thin-film barrier developed within single chamber that enables both iCVD and PECVD processes. (d) The scheme of the single chamber that enables both iCVD and PECVD for the fabrication of thin-film barrier. ((b) From Ref. [45]. Copyright © 2012 Elsevier. (c, d) From Ref. [46].)

(ALD). The film quality of inorganic layer is the most critical factor that decides the barrier property of the thin-film barrier. Thus, the inorganic layer must be deposited in high quality. PECVD and ALD, especially the latter, provides much finer quality compared to other deposition methods, such as sputtering. For the deposition method for organic layer, both liquid-phase deposition (e.g., flash evaporation followed by UV light curing [48], spin coating [49]) and vapor-phase deposition (e.g., parylene deposition [50], PP [51], and molecular layer deposition (MLD) [52]) had been investigated so far. The deposition methods could successfully produce organic layer; however, there are some factors that should be improved. First of all, liquid-phase deposition methods cannot be applied to vulnerable substrates and often can degrade the organic device when the solvent used is in direct contact. Although parylene deposition and MLD are free of solvent degradation, the selection of materials is highly limited in these processes because only few sources for polymerization are available to date. For the case of PP, plasma damage on devices can occur.

17.5.2 Fabrication of Multilayered Barrier Using iCVD Polymer and Inorganic Layers

iCVD has the properties that can lead this process to the optimal deposition system for the encapsulation film production owing to their dry nature, low process temperature, and availability of versatile monomer materials. It was also investigated that iCVD monomers fill the nanoporosity in inorganic layer successfully [53], and the iCVD could effectively fill the nano-defects of underlying inorganic layer, just as liquid-phase polymerized layers. iCVD polymer layers can elongate the tortuous path effectively when alternately deposited with inorganic layers.

The iCVD process is precisely suitable for organic electronic device application, and the characteristics of the iCVD polymer layer are proven to serve the expected role of organic layer within the organic/inorganic hybrid multilayer.

The characteristics of iCVD-deposited polymers were investigated to confirm that the layers can be used in combination with deposition methods for inorganic layer. iCVD-deposited PGMA was demonstrated to be thermally stable after 410 °C heat treatment for 4 h [54]. Organosilicon layer with outstanding stability was developed as well. Poly(1,3,5-trivinyl-1,3,5-trimethylcyclotrisiloxane) (PV3D3) was deposited via iCVD process and the film was highly cross-linked and stable, which indicates that it is suitable for the organic interlayer for the thin-film barrier [55]. Moreover, another highly cross-linked organosilicon layer, polyhexavinyldisiloxane (PHVDSO), was also developed via iCVD and the layer did not show cracks or delamination after a set of repeated bending cycles [56]. This mechanical robustness is a crucial factor considering that the organic electronic devices are expected to be advanced to flexible devices.

There were several approaches of applying iCVD polymer layer for the thin-film barrier fabrication. The thermally stable iCVD-deposited PGMA was combined with hot wire chemical vapor deposition (HWCVD)-deposited SiN_x to build up organic/inorganic hybrid multilayer, as shown in Figure 17.12b [45]. The multilayer demonstrated the best barrier performance when four dyads of the multilayers were deposited by alternating 200 nm of PGMA and 50 nm of SiN_x. iCVD was also capable of merging with ALD. Organosilicon layer with a similar structure to PV3D3, poly(2,4,6,8-tetravinyl-2,4,6,8-tetramethylcyclotetrasiloxane) (PV4D4), was alternated with ALD-processed Al_2O_3 layer [57]. Although the yield still remains issue, the developed multilayer exhibited sufficient barrier performance even for OLED encapsulation.

Deposition system for organic/inorganic multilayer in a single chamber also have been demonstrated, as shown in Figure 17.12c,d [46]. The filament above the cooled stage enables iCVD process while the power supplies enable PECVD process. With the iCVD process, PHVDSO was deposited. Although same precursor hexavinyldisiloxane (HVDSO) was used, with plasma assist, the deposited layer had characteristic extremely similar to SiO_x. The thickness of PHVDSO layer was around 400 nm while that of SiO_x-like layer was around 100 nm. The barrier performance was not as good as thin-film barriers obtained otherwise; however, it was for the first time that the organic/inorganic hybrid multilayer was fabricated within a single-chamber system. In addition, this work demonstrated that the iCVD polymer layer has better flattening effect than that of PECVD polymer layer.

17.6 Conclusion

This chapter reviewed the advantages of the oCVD and iCVD process for the versatile device applications, which include the performance enhancement, consequent potential lowering of the cost, mechanical flexibility, and robustness.

For oCVD, PEDOT with WF controllability and high flexibility were highlighted. Especially the PEDOT via oCVD could conformally cover various kinds of unconventional substrates including paper and plastic substrates, which could dramatically cut off the fabrication cost of OPVs. Also, owing to the excellent conformality and unique functionality of iCVD polymer, performance of electronic and microfluidic devices could be further improved.

Some current and future directions to widen the application of oCVD and iCVD for devices include the following: (i) scaling up and commercialization and (ii) continuous material development. With the promising advances in the oCVD and iCVD, there is an exciting potential for further improvements in various emerging device applications.

Acknowledgments

This work was supported by the IT R&D program of MKE/KEIT (Grant No. 10041416, the core technology development of light and space adaptable new mode display for energy saving on 7 in. and 2 W), by Grant No. EEWS-2014-N01140052 from EEWS Research Project of the KAIST EEWS Research Center (EEWS: Energy, Environment, Water and Sustainability), and in part by the Advanced Biomass R&D Center (ABC) of Global Frontier Project funded by the Ministry of Education, Science and Technology (ABC-2011-0031356).

References

1. Forrest, S.R. (2004) *Nature*, **428**, 911–918.
2. Logothetidis, S. (2008) *Mater. Sci. Eng., B*, **152**, 96–104.
3. Lock, J.P., Im, S.G., and Gleason, K.K. (2006) *Macromolecules*, **39**, 5326–5329.
4. Im, S.G. and Gleason, K.K. (2007) *Macromolecules*, **40**, 6552–6556.
5. Im, S.G., Gleason, K.K., and Olivetti, E.A. (2007) *Appl. Phys. Lett.*, **90**, 152112.
6. Barr, M.C., Carbonera, C., Po, R., Bulović, V., and Gleason, K.K. (2012) *Appl. Phys. Lett.*, **100**, 183301.
7. Barr, M.C., Howden, R.M., Lunt, R.R., Bulović, V., and Gleason, K.K. (2012) *Adv. Energy Mater.*, **2**, 1404–1409.
8. Im, S.G. and Gleason, K.K. (2011) *AIChE J.*, **57**, 276–285.
9. Barr, M.C., Rowehl, J.A., Lunt, R.R., Xu, J., Wang, A., Boyce, C.M., Im, S.G., Bulović, V., and Gleason, K.K. (2011) *Adv. Mater.*, **23**, 3500–3505.
10. Park, H., Howden, R.M., Barr, M.C., Bulović, V., Gleason, K.K., and Kong, J. (2012) *ACS Nano*, **6**, 6370–6377.
11. Berger, C., Song, Z.M., Li, X.B., Wu, X.S., Brown, N., Naud, C., Mayou, D., Li, T.B., Hass, J., Marchenkov, A.N., Conrad, E.H., First, P.N., and de Heer, W.A. (2006) *Science*, **312**, 1191–1196.
12. Bolotin, K.I., Sikes, K.J., Jiang, Z., Klima, M., Fudenberg, G., Hone, J., Kim, P., and Stormer, H.L. (2008) *Solid State Commun.*, **146**, 351–355.
13. Moser, J., Barreiro, A., and Bachtold, A. (2007) *Appl. Phys. Lett.*, **91**, 163513.
14. Im, S.G., Kusters, D., Choi, W., Baxamusa, S.H., van de Sanden, R., and Gleason, K.K. (2008) *ACS Nano*, **2**, 1959–1967.
15. Howden, R.M., Flores, E.J., Bulović, V., and Gleason, K.K. (2013) *Org. Electron.*, **14**, 2257–2268.
16. O'Regan, B. and Gratzel, M. (1991) *Nature*, **353**, 737–740.

17. Hagfeldt, A., Boschloo, G., Sun, L., Kloo, L., and Pettersson, H. (2010) *Chem. Rev.*, **110**, 6595–6663.
18. Nejati, S. and Lau, K.K.S. (2011) *Nano Lett.*, **11**, 419–423.
19. Ghaffari, M., Kosolwattana, S., Zhou, Y., Lachman, N., Lin, M., Bhattacharya, D., Gleason, K.K., Wardle, B.L., and Zhang, Q.M. (2013) *Electrochim. Acta*, **112**, 522–528.
20. Zhou, Y., Lachman, N., Ghaffari, M., Xu, H., Bhattacharya, D., Fattahi, P., Abidian, M.R., Wu, S., Gleason, K.K., Wardle, B.L., and Zhang, Q.M. (2014) *J. Mater. Chem. A*, **2**, 9964–9969.
21. Monch, W., Denhert, J., Prucker, O., Ruhe, J., and Zappe, H. (2006) *Appl. Opt.*, **45**, 4284–4290.
22. Karaman, M., Kooi, S.E., and Gleason, K.K. (2008) *Chem. Mater.*, **20**, 2262–2267.
23. Granqvist, C.G., Avendaño, E., and Azens, A. (2003) *Thin Solid Films*, **442**, 201–211.
24. Groenendaal, L., Zotti, G., Aubert, P., Waybright, S.M., and Reynolds, J.R. (2003) *Adv. Mater.*, **15**, 855–879.
25. Lock, J.P., Lutkenhaus, J.L., Zacharia, N.S., Im, S.G., Hammond, P.T., and Gleason, K.K. (2007) *Synth. Met.*, **157**, 894–898.
26. Cognard, P. (2006) *Adhesives and Sealants*, vol. 2, Elsevier.
27. Brockmann, W., Geiβ, P., Klingen, J., and Schröder, B. (2009) *Adhesive Bonding: Materials, Applications and Technology*, Wiley-VCH Verlag GmbH, Weinheim.
28. Kobayashi, M., Terada, M., and Takahara, A. (2011) *Soft Matter*, **7**, 5717–5722.
29. You, J.B., Kim, S.Y., Park, Y.J., Ko, Y.G., and Im, S.G. (2014) *Langmuir*, **30**, 916–921.
30. Matsui, J., Kubota, K., Kado, Y., and Miyashita, T. (2007) *Polym. J.*, **37**, 41–47.
31. Im, S.G., Bong, K.W., Lee, C.H., Doyle, P.S., and Gleason, K.K. (2009) *Lab Chip*, **9**, 411–416.
32. Xu, J. and Gleason, K.K. (2010) *Chem. Mater.*, **22**, 1732–1738.
33. You, J.B., Min, K.I., Lee, B., Kim, D.P., and Im, S.G. (2013) *Lab Chip*, **13**, 1266–1272.
34. Kim, H.J., Park, Y.J., Choi, J.-H., Han, H.S., and Hong, Y.T. (2009) *J. Ind. Eng. Chem.*, **15**, 23–30.
35. Sackmann, E.R., Fulton, A.L., and Beebe, D.J. (2014) *Nature*, **507**, 181–189.
36. Bong, K.W., Xu, J., Kim, J., Chapin, S.C., Strano, M.S., Gleason, K.K., and Doyle, P.S. (2012) *Nat. Commun.* doi: 10.1038/ncomms1800
37. Di Carlo, D. (2009) *Lab Chip*, **9**, 3038–3046.
38. Lim, J.M., Swami, A., Gilson, L.M., Chopra, S., Choi, S., Wu, J., Langer, R., Karnik, R., and Farokhzad, O.C. (2014) *ACS Nano*, **8**, 6056–6065.
39. Kim, M., Huang, Y., Choi, K., and Hidrovo, C.H. (2014) *Microelectron. Eng.*, **124**, 66–75.
40. Jeong, J., Oh, M.S., Kim, B.J., Choi, C., Lee, B., Lee, C., and Im, S.G. (2013) *Langmuir*, **29**, 3474–3481.
41. Burrows, P.E., Graff, G.L., Gross, M.E., Martin, P.M., Shi, M.K., Hall, M., Mast, E., Bonham, C., Bennett, W., and Sullivan, M.B. (2001) *Displays*, **22**, 65–69.
42. Dennler, G., Lungenschmied, C., Neugebauer, H., Sariciftci, N.S., Latrèche, M., Czeremuszkin, G., and Wertheimer, M.R. (2006) *Thin Solid Films*, **511-512**, 349–353.
43. Burrows, P.E., Graff, G.L., Gross, M.E., Martin, P.M., Hall, M., Mast, E., Bonham, C., Bennett, W., Michalski, L., Weaver, M., Brown, J.J., Fogarty, D., and Sapochak, L.S. (2001) *Proc. SPIE*, **4105**, 75–83.
44. Burrows, P.E., Bulović, V., Forrest, S.R., Sapochack, L.S., McCarty, D.M., and Thompson, M.E. (1994) *Appl. Phys. Lett.*, **65**, 2922–2924.
45. Spee, D.A., Schipper, M.R., van der Werf, C.H.M., Rath, J.K., and Schropp, R.E.I. (2013) *Thin Solid Films*, **532**, 84–88.
46. Coclite, A.M., Ozaydin-Ince, G., Palumbo, F., Milella, A., and Gleason, K.K. (2010) *Plasma Processes Polym.*, **7**, 561–570.
47. Graff, G.L., Williford, R.E., and Burrows, P.E. (2004) *Appl. Phys. Lett.*, **96**, 1840.
48. Park, J., Chae, H., Chung, H.K., and Lee, S.I. (2011) *Semicond. Sci. Technol.*, **26**, 034001.

49. Han, Y.C., Jang, C., Kim, K.J., Choi, K.C., Jung, K., and Bae, B. (2011) *Org. Electron.*, **12**, 609–613.
50. Ghosh, A.P., Gerenser, L.J., Jarman, C.M., and Fornalik, J.E. (2005) *Appl. Phys. Lett.*, **86**, 223503.
51. Seo, S., Chae, H., Seo, S.J., Chung, H.K., and Cho, S.M. (2013) *Appl. Phys. Lett.*, **102**, 161908.
52. Park, M., Oh, S., Kim, H., Jung, D., Choi, D., and Park, J. (2013) *Thin Solid Films*, **546**, 153–156.
53. Aresta, G., Palmans, J., van de Sanden, M.C.M., and Creatore, M. (2012) *Microporous Mesoporous Mater.*, **151**, 434–439.
54. Bakker, R., Verlaan, V., van der Werf, C.H.M., Rath, J.K., Gleason, K.K., and Schropp, R.E.I. (2007) *Surf. Coat. Technol.*, **201**, 9422–9425.
55. Aresta, G., Palmans, J., van de Sanden, M.C.M., and Creatore, M. (2012) *J. Vac. Sci. Technol., A*, **30**, 041503.
56. Coclite, A.M., Ozaydin-Ince, G., d'Agostino, R., and Gleason, K.K. (2009) *Macromolecules*, **42**, 8138–8145.
57. Kim, B.J., Kim, D.H., Kang, S.Y., Ahn, S.D., and Im, S.G. (2014) *J. Appl. Polym. Sci.*, 40974. doi: 10.1002/app.40974

18
CVD Polymers for the Semiconductor Industry

Vijay Jain Bharamaiah Jeevendra Kumar and Magnus Bergkvist

18.1
Introduction

Ever since Gordon Moore's prediction that the number of transistors on a chip would double every 2 years, the semiconductor industry has been in a relentless pursuit to keep up with the so-called *Moore's Law*. However, with transistor scaling nearing its physical limit, the industry is faced with challenges to undergo radical changes where new materials, processes, and technologies are the need of the hour. Furthermore, the demands in terms of control over material properties, ease of processing, cost of operation, and environmental compatibility have never been more important than today.

Polymers are widely used materials in semiconductor manufacturing where the application areas range from resists for lithography to underfills and encapsulants in device packaging [1, 2]. Polymeric thin films are currently obtained either by solution processing (spin coating, spray coating) or by high-energy vapor-phase techniques (plasma-enhanced chemical vapor deposition (PECVD), parylene CVD). However, for some emerging technologies, such as 3D integrated circuits (ICs), ultra-low *k* dielectrics/air-gaps, and dry-film resist lithography, traditional processing techniques are reaching their limits. Material incompatibility, poor yield, high operational costs, and environmental impact are some of the concerns [3–5]. Initiated chemical vapor deposition (iCVD) is an emerging technique that can offer an alternative to established chemical vapor deposition (CVD) processing techniques and has gained significant attention over the last decade. iCVD retains the advantages of bulk synthesized polymers while operating in a vapor-phase environment, enabling excellent control over polymer properties, conformal deposition on 3D high aspect ratio features, and ability to deposit extremely thin films [6–8]. The intention of this chapter is to highlight examples where iCVD has been investigated for semiconductor applications. Design and cost of ownership (COO) considerations for a volume manufacturing iCVD tool that could be integrated into a semiconductor fabrication line are also discussed. This chapter mainly focuses on iCVD for ICs; however, the

CVD Polymers: Fabrication of Organic Surfaces and Devices, First Edition. Edited by Karen K. Gleason.

examples are applicable to other areas with similar fabrication technology, such as microelectromechanical systems (MEMSs).

18.2
Application Areas for iCVD

18.2.1
Lithography

Lithography on the micro- and nanoscale has been one of the main driving technologies for semiconductor and IC industry over several decades. Lithography also plays an increasingly important role in the manufacture of commercial MEMS devices, such as accelerometers for cars and consumer electronics [9–13]. In large volume IC manufacturing, projection lithography is the principal method where ultraviolet (UV) light (193–436 nm, typically 193 nm for IC fabrication) is projected through a patterned mask to expose a spin-coated thin film of a polymer/photoacid generator blend (photoresist), where the exposed polymer undergo a chemical reaction altering its solubility. After exposure, baking, and solvent development, the exposed area either remains or is dissolved away depending on whether the resist is negative or positive tone, respectively. The patterned photoresist is then subsequently used as a protective layer that allows transfer of the design into the substrate either by etching or by material deposition processes. Projection lithography is very fast and can pattern features below 50 nm using immersion optics [14]. Electron-beam (e-beam) lithography is another common form of writing with higher resolution and can generate features below 10 nm [15]. However, in comparison to other writing methods, it has significantly slower throughput and is often to generate high-quality masks with small features that is used in projection lithography. The e-beam induces polymer crosslinking or polymer degradation that changes the solubility of the exposed areas and does not require additional additives to the resist formulation *per se*.

Common to all lithography is the use of polymer resists, which are typically applied via spin coating. Thus, solvent-based wafer processing is a key step in volume manufacturing, and high-throughput automated track systems are available for industrial scale operations. Organic solvent-based processing steps are frequently used and work well; even so, environmental health and safety (EHS) concerns with volatile organic carbon emissions drive efforts to reduce the amount of wet chemistry processing required for deposition/development steps in lithography processing. Also, economic concern such as cost per wafer pass is important to consider. Dry depositions of resist and development using supercritical CO_2 ($SCCO_2$) are some potential strategies that can reduce the use of organic solvents.

18.2.1.1 iCVD Polymers as Photoresists

CVD offers a dry processing alternative to conventional spin coating methods. Considering iCVD over plasma CVD, the former operate under milder deposition conditions resulting in minimal crosslinking during deposition that can help improve both sensitivity and resolution. Toward that end, Mao *et al.* evaluated iCVD poly(glycidyl methacrylate) (PGMA) as a negative e-beam resist using both conventional and $SCCO_2$ development [16]. The iCVD PGMA photoresists exhibited sensitivity at exposure doses of 27 μC cm^{-2} with 500 nm line/space pattern resolution using conventional solvent-based development. By tuning the deposition conditions to obtain different molecular weight PGMA from ~11 500 to 4600 M_n, they were able to resolve features down to 80 nm at 100 μC cm^{-2}. Using $SCCO_2$ development (with 2% acetone at 5000 psi, 45 °C), 300-nm line resolution could be realized at 140 μC cm^{-2}.

Further research into iCVD e-beam compatible photoresists has involved copolymerization of acrylates with fluorinated acrylic monomers [17]. PGMA has relatively low solubility in $SCCO_2$, where fluoropolymers, on the other hand, have better solubility but lower sensitivity to e-beam irradiation. In an effort to combine these properties, iCVD was used to copolymerize PGMA. With this copolymer resist, 300-nm line resolution (80 μC cm^{-2} dose) could be realized. This is on par with the performance of similar spin-on resists developed with $SCCO_2$. Attempts at improved resolution were hampered by the typical swelling behavior of negative resists, which causes distorted features. To overcome such issues, a positive-tone iCVD resist using methacrylic acid (MAA) copolymerized with a fluorinated methacrylate was prepared [17]. Annealing at 160 °C dehydrates the carboxylic acid and form mainly intramolecular anhydrides, which are highly electron withdrawing and increase the likelihood of polymer chain breaks upon electron irradiation [18]. With this approach, 300-nm line resolution patterns with no swelling could be realized at 100 μC cm^{-2} dose and pure $SCCO_2$ development. A similar anhydride formation approach was used to demonstrate 20 μC cm^{-2} dose sensitivity and line resolutions down to 60 nm using an iCVD copolymer photoresist prepared with MAA/methacrylate and developed in a 2-propanol/dimethylacetamide mix (Figure 18.1) [19]. Although

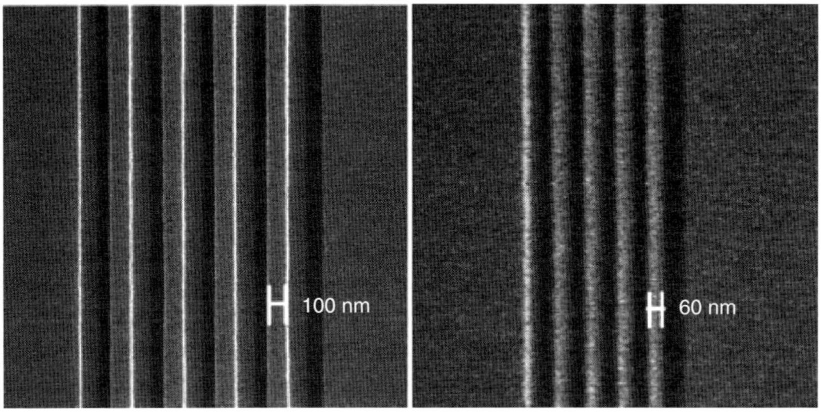

Figure 18.1 100- and 60-nm line features generated with an iCVD resist having an original MAA composition of 75% and M_w 6800. (Reprinted with permission from Ref. [19]. Copyright (2014) American Chemical Society.)

photoresists specifically formulated for e-beam lithography have demonstrated higher resolution, these results indicate the potential of positive-tone iCVD resist for high-resolution patterning at low-dose exposures.

Even though the majority of wafer processing takes place on flat "2D" substrates, there are emerging applications in MEMS and biotechnology that could benefit from a more "3D" non-conformal patterning. One attractive feature of iCVD is the ability to deposit high-quality thin polymer films on textured and curved substrates. The feasibility of using iCVD resists for lithographic patterning on curved substrates has been demonstrated by Petruczok and Gleason [20]. They deposited 50–250 nm films of poly(4-vinyl pyridine) (P4VP) on 2–6-mm diameter glass rods as a base layer. The P4VP is not very sensitive to UV light, so to realize a more photosensitive resist, the coated rods were immersed in a solution of 10,12-tricosadiynoic acid (TDA), which interact with the P4VP through hydrogen bonding. Flexible transmission electron microscopy (TEM) grids placed in direct contact with the rods were used as masks. Irradiation with 254-nm light photopolymerized the diacetylene units to produce a negative-tone patterning. Patterns from 50 μm to less than 3 μm could be resolved, having a good agreement with the mask pattern. These results indicate that iCVD-enabled photolithography on curved substrates has similar resolution as conventional contact lithography.

The iCVD resist technology demonstrates good lithographic performance and has potential advantages from an environment perspective. However, one major challenge of implementing iCVD in lithographic applications for semiconductor volume production will be wafer throughput, considering that current track-based processing systems can handle ~180 wafers/h.

18.2.2 Air Gap Dielectrics

A classic paper by Bohr pointed out that metal interconnects and their associated RC-delay presented a *"bottle-neck"* for further scaling of transistor devices [21]. Following this, copper (Cu) interconnects and low dielectric constant, amorphous carbon-doped glass dielectrics (commonly referred to as SiCOH), have been adopted into semiconductor back-end of line (BEOL) processes [22]. Dielectrics minimize "unintentional" electrical interactions between metal interconnects (crosstalk), thereby reducing device power consumption and signal delay. BEOL dielectrics are classified as inter-metal dielectrics (IMDs) and inter-layer dielectrics (ILDs), depending on their location with respect to the interconnects. With transistor scaling, it was proposed that bulk SiCOH be replaced with porous SiCOH, where studies suggesting that dielectric constant $(k) \cong 2$ could be realized [22, 23]. However, integration of porous SiCOH into standard BEOL processing has been fraught with many challenges due to poor material reliability, high cost, and manufacturing compatibility. This is reflected in *International Technology Roadmap for Semiconductors (ITRS)* reports where the target implementation of low k dielectrics have been revised since 2001 [24, 25]. One major reason behind this delay is the chemical degradation of

SiCOH under plasma conditions during processing, rendering them hydrophilic and resulting in moisture uptake. This resulted in an increase in their k values to much higher than bulk SiO_2 ($k_{SiO_2} = 4$) [26]. Furthermore, porous SiCOH layers peeled and cracked during planarization, assembly, and packaging processes due to their poor adhesion and mechanical properties. This severely compromised the yield and reliability of devices prompting for the development of a radical new dielectric material that can address these issues [27–29]. Toshiba Corporation provided a breakthrough when proposing the use of air/vacuum as IMDs by incorporating a gap between the metal interconnects [30]. They demonstrated a process flow where sacrificial materials were used to create air-gaps resulting in IMD values between 1.1 and 1.2. This attracted attention toward exploring techniques to create and integrate air-gap dielectrics in semiconductor devices, and investigations regarding their reliability and manufacturability.

18.2.2.1 Integration of Air-Gap Dielectrics

Integration of air-gap dielectrics into Cu damascene processes is considered a significant challenge facing semiconductor industry. A number of different techniques have been proposed to create these air-gaps, which fall under two broad categories: (a) non-conformal CVD technique and (b) sacrificial technique (Figure 18.2).

The first technique takes advantage of the non-conformal nature of many CVD processes to create "pinch-off" areas during deposition between Cu interconnects. Briefly, copper interconnects are formed in the dielectric material by a damascene process, after which the dielectric is either partially or completely

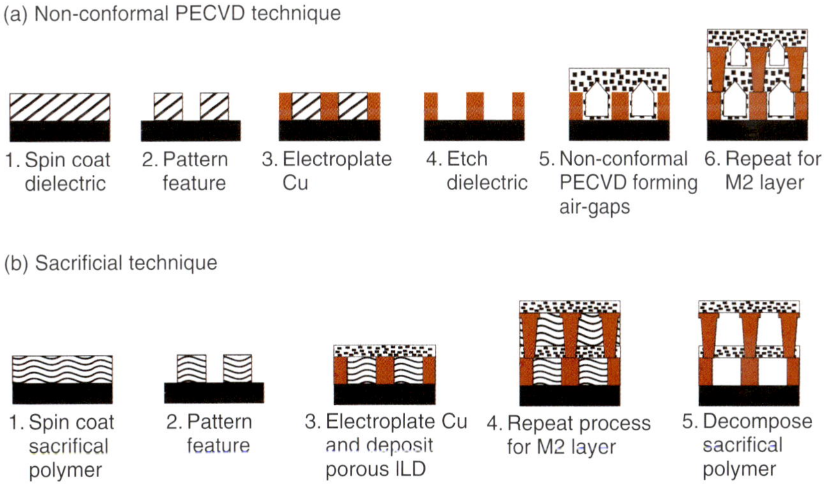

Figure 18.2 Schematics of the two techniques used to form air-gap dielectrics (M2 = second metal layer). It can clearly be seen that the non-conformal PECVD technique has additional process steps. Also, misaligned vias in PECVD technique can land in air-gaps (step 6) presenting reliability issues. (Adapted from Ref. [31] with permission from Elsevier.)

removed by wet or dry etching. Following this, the desired dielectric material is deposited non-conformally by PECVD, which pinch-off between narrow metal areas forming air-gaps. Although this technique utilizes well-established processes, it suffers from some drawbacks:

1) the non-conformal deposition technique requires a critical inter-via separation below which air-gaps are not formed. Missing air-gap structures can pose significant reliability challenges during BEOL processing.
2) misaligned copper vias can result in metal filling in the air-gaps during subsequent metal layer deposition causing significant reliability issues [31, 32]; and
3) the additional process steps and photolithography masks add to the complexity and cost of the process.

An alternative to the non-conformal CVD technique is to use a sacrificial approach [30, 33]. Here, sacrificial materials are integrated between Cu interconnects, which could then be preferentially removed to create air-gaps. Thermally degradable polymers are commonly used as "placeholders," where a porous ILD is deposited on top. This process can be repeated for all metal layers after which the sacrificial material can be thermally (UV assisted for some polymers) decomposed to create air-gaps. This technique can generate wide air-gaps while eliminating multiple process steps with significant reduction in cost. An important advantage of this approach is that metal deposition is done before gap formation, eliminating reliability issues associated with misaligned vias. One area of concern is the mechanical stability and structural integrity of wide air-gap devices. While porous ILD mechanical properties may limit global implementation of sacrificial air-gaps, recent studies have demonstrated that with careful material and process selection, reliability of devices can be significantly improved [34, 35]. Currently, there is no consensus over the technique to be employed for air-gap fabrication. Below we focus on the sacrificial technique and discuss the use of iCVD in air-gap fabrication processes.

18.2.2.2 **iCVD in Air-Gap Fabrication**

iCVD offers an alternative method to currently used spin coating for depositing sacrificial polymers. These materials should have good thermal/mechanical stability during processing and a clean degradation above their degradation temperature. iCVD has the unique ability to deposit a variety of materials, with controlled chemistry in a bottom-up manner providing excellent control over film thickness and conformity. The possibility to use the iCVD technique for integrated processing was demonstrated by Tokyo Electron Limited (TEL), where vinyl triethoxy silane (VTES) adhesion promoter, poly(ethylene glycol diacrylate) (poly(EGDA)) sacrificial polymer, and methyltriethoxy silane (MTES) porous SiCOH layer were all deposited sequentially using a single-chamber system [36]. Notably, this integration capability was demonstrated using industry-standard 300-mm silicon wafers. The structure and functionality of each material was

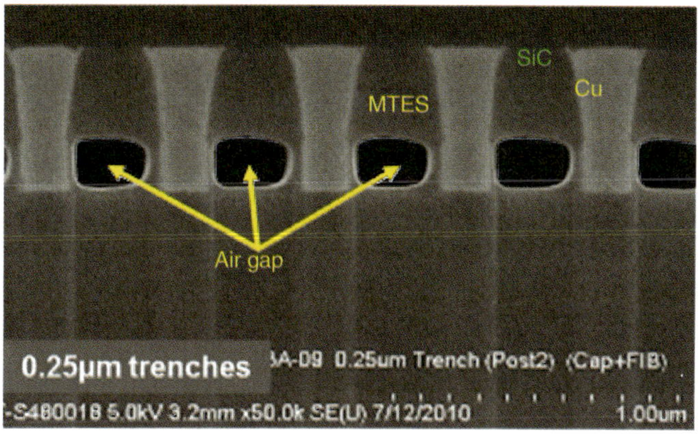

Figure 18.3 Single-layer air-gaps fabricated using iCVD p(CHMA-*co*-EGDA). (Reprinted from Ref. [36] with permission from Elsevier.)

preserved during deposition, which is critical for reliability. For example, preservation of vinyl groups on VTES molecule is critical to promote adhesion with poly(EGDA), while the porosity of MTES ILD layer facilitates clean degradation of the sacrificial iCVD polymer. Complete air-gaps were formed in 0.25-μm spaces where the poly(EGDA) films were decomposed by a combination of broadband UV irradiation and thermal anneal (Figure 18.3). These air-gaps showed a 20% decrease in capacitance values compared to baseline data [36].

Although TEL demonstrated the integration aspect of the iCVD technique, early feasibility studies to deposit sacrificial polymers for air-gap fabrication were demonstrated by the Gleason group at MIT [37]. They were interested in manipulating the chemical and thermal properties of deposited polymers to suit air-gap fabrication applications. Copolymers of poly(cyclohexyl methacrylate-*co*-ethyleneglycol diacrylate) (p(CHMA-*co*-EGDA)) were prepared via the iCVD technique, where EGDA was used as the crosslinker to manipulate the film thermal degradation onset temperature to 270 °C. Annealing the sacrificial polymer above the degradation temperature resulted in >99.7% of the material being removed [37, 38].

The iCVD technique takes advantage of an initiator to start the polymerization process, unlike PECVD or hot-wire CVD that creates initiating radicals from monomers. This allows better control over polymer formation with no observable difference between bulk synthesized and iCVD polymers. Thermal degradation can take place by a chain-unzipping mechanism where depolymerization or depropagation occurs along the backbone of the polymer chain. This is unlike PECVD polymers, where there is little control over polymer structure when random bond breakage and crosslinking can result in incomplete thermal degradation and char residues [39]. Single-layer air-gaps of 1.5 μm × 0.5 μm were demonstrated using iCVD p(CHMA-*co*-EGDA) as a sacrificial material, and conclusions were made that the technique could be adapted to produce air-gaps with much smaller feature sizes. Furthermore, they also demonstrated iCVD of 1,3,5,7-tetravinyltetramethylcyclotetrasiloxane (V4D4) dielectrics, which can potentially be incorporated as ultra-low k ILD layers above the air-gaps, which according to *the ITRS* 2003 report is essential to reduce the overall k_{eff} in a device [40]. The abovementioned results can be a step forward toward realizing dielectric materials with $k_{eff} \leq 2$ in a cost-efficient manner, where different layers such as adhesion layer, sacrificial material, and ultra-low k porous capping ILD can be deposited in a single system.

18.2.2.3 Air-Gap Dielectrics in 3D Integrated Circuit Vias

Although traditional silicon-based transistor scaling approaches its physical limits, 3D stacking of integrated chip circuits is seen as an effective way to keep up with Moore's law. The 3D circuits involve stacked functional chips that are interconnected by vertical copper through-silicon-vias (TSVs). Electrical characteristics of TSVs such as RC-delay, crosstalk, leakage current, and power consumption are of critical importance for the performance of such devices. One way to improve device performance is by reducing the capacitance between interconnects. Annular air-gaps around TSVs are proposed to reduce interconnect RC-delays based on the same principles as in 2D interconnects. Techniques such as solution pore-filling, liner deposition, and lithography patterning of resists are proposed to deposit sacrificial polymer that can then be removed to form annular air-gaps [41–44]. However, solution-based techniques commonly result in nonuniform films and void trapping in such high aspect ratio, 3D features. The iCVD process can deposit highly conformal polymer films on high aspect ratio trenches [6]; thus, iCVD can be an ideal candidate for depositing sacrificial polymer thin films for annular TSV air-gap formation. Furthermore, the bottom-up technique allows excellent control over film thickness, enabling deposition of extremely thin sacrificial films, which could help reduce concerns with mechanical reliability of air clad TSVs.

18.3 Thin-Film Adhesives

In the next generation ICs, 3D stacking is considered as a practical way to keep up with Moore's law. Wafer thinning and bonding are integral processes in 3D IC fabrication. Among the various approaches to bonding, polymer-adhesive wafer bonding has received significant attention due to its ease of application, low processing temperatures, tolerance to particles, and compatibility with CMOS technology [4, 45]. Currently, such adhesive films are obtained by solution-based techniques such as spin casting, spray coating, screen printing, and stamping [46]. However, the use of solvents poses challenges with substrate compatibility, film dewetting, and conformality [4]. Furthermore, processing of solution-based adhesive films may involve multiple bake and cure steps resulting in thermal stress induced wafer bow and warpage, rendering the substrate unusable for further processing steps [47]. In addition, there are EHS concerns with volatile organic carbon solvents to consider. Solventless, vapor-phase deposition techniques can alleviate many of these issues by forming thin functional films on a variety of substrates [48].

The concept of solventless adhesive bonding (SAB) using reactive polymer films has been explored in microfluidic applications to create leak-free polydimethylsiloxane (PDMS)-based devices using PECVD and parylene CVD [49, 50]. Xu and Gleason demonstrated that iCVD poly(aminostyrene) retained a higher degree of functional groups compared to PECVD poly(allylamine) [5]. Using substrates coated with PGMA, respectively, they assembled and covalently bonded microfluidic devices that could withstand burst pressures in excess of 150 psi after 24 h anneal at 50 °C.

18.3.1 iCVD for Wafer Bonding Applications

The potential of iCVD polymers for solventless adhesive wafer bonding in semiconductor applications have been investigated [51, 52]. Early feasibility studies demonstrated that iCVD PGMA films can be used as low-temperature adhesives for solventless bonding of silicon wafers. The bonding conditions were milder (90 °C, <30 min) compared to solution-based benzocyclobutene (>200 °C, >2 h) adhesives. Further testing has demonstrated feasibility of bonding just above the polymer's glass transition temperature (T_g ~65 °C) and at shorter bonding times (15 min is the shortest time tested). Notably, iCVD adhesives allowed void-free bonding with close to 100% bond uniformity (Figure 18.4) while exhibiting similar adhesion energies as benzocyclobutene (BCB) [4, 51, 53]. The iCVD PGMA-bonded wafers survived standard wafer thinning processes (coarse and fine chemical mechanical polishing steps) without being compromised and did not show any significant change in wafer bow or warpage, which is critical for successful integration in 3D IC fabrication technology.

Of importance is also the ability to manipulate the adhesion energy and the type of interaction the polymer has with the wafer substrate to enable both temporary and permanent bonding applications. This can be realized by varying the van der Waal's or covalent interactions between the adhesive polymer and substrate by plasma treatment or other surface chemistry. The influence of surface pre-treatment and the ability to control the adhesion behavior of iCVD PGMA through its epoxy functionality were demonstrated on various surface-modified silicon wafers [52]. Molecular vapor deposition (MVD) was used to deposit uniform monolayers of functional siloxane monolayers to manipulate the surface energy of the substrate. iCVD PGMA and siloxane-treated substrates displayed different critical adhesion energy ranging between $G_c = 4.5–35\ \mathrm{J\ m^{-2}}$ (interaction thought to be governed mainly by van der Waal type interactions). All the samples failed adhesively (at the polymer–Si interface) during mechanical testing and could be suitable for temporary bonding applications. Notably, after a 150 °C anneal, which promoted a covalent reaction between both the epoxy

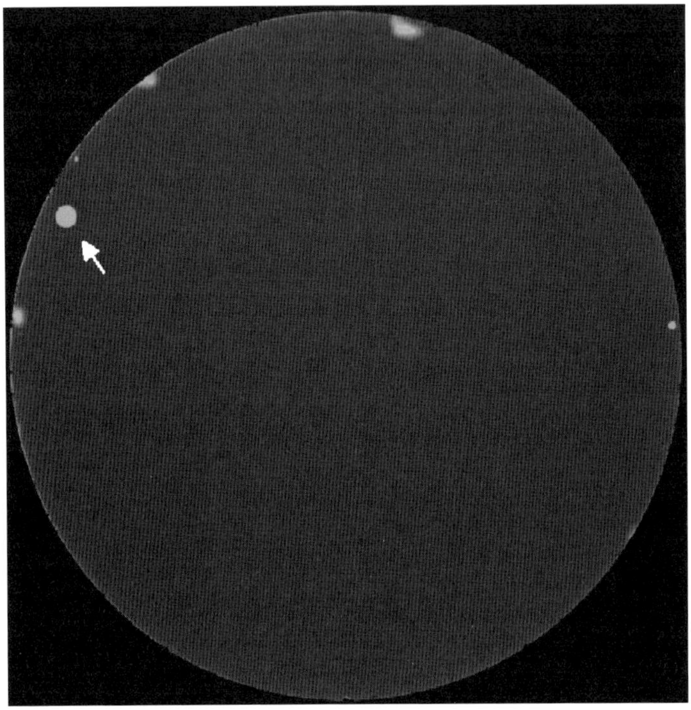

Figure 18.4 Scanning acoustic microscopy image of 300-mm Si wafers bonded using iCVD PGMA adhesive. The arrow indicates a trapped void that was attributed to dust particle introduced during manual handling of wafers [52]. (Reproduced with permission.)

groups in PGMA and the substrate surface groups (amine, epoxy, or hydroxyl), the bonding characteristics changed. These annealed films had significantly higher critical adhesion energy ($G_c \cong 100\ \mathrm{J\ m^{-2}}$) with cohesive failure (within the polymer) indicating a covalent bond. These types of covalently linked films have potential use in permanent bonding applications.

18.4 Design Considerations for iCVD Tools in Semiconductor Manufacturing

CVD enables films of nanoscale thicknesses with macroscale uniformity to be deposited, and this technology may be applied to complex geometries often found in semiconductor devices. A considerable body of research has been developed regarding the design and optimization of CVD systems for deposition of inorganic and organic materials [54–61]. For selected emerging semiconductor applications, iCVD shows unique process capability over conventional deposition processes such as spin-on or PECVD. Plasma-free (no-damage), solvent-free organic films deposited at temperatures less than 100 °C are attractive for a range

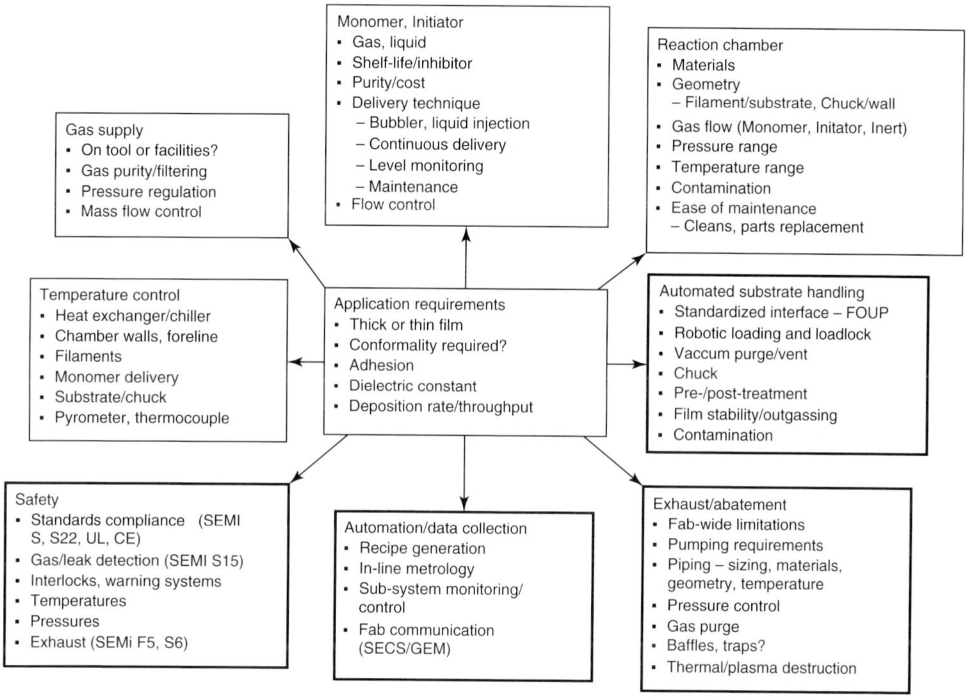

Figure 18.5 iCVD system platform engineering choices – system-wide implications. (Adapted from Ref. [57, p. 53]. Reproduced by permission of The Royal Society of Chemistry.)

of BEOL, 3D interconnect, and packaging applications. Translating this capability to production begins with a review of the application's specific requirements and the resulting process and tool requirements (Figure 18.5). For semiconductor manufacturing equipment, high-volume manufacturing (HVM) metrics such as COO, quality/defectivity, and environmental sustainability are additional drivers for many of these design decisions [62]. These sections focus on the design elements critical to iCVD processing equipment design to meet the challenges of advanced semiconductor production.

18.4.1 iCVD for Semiconductor Manufacturing

A typical HVM semiconducting wafer processing system integrates several high-level constituents:

1) system frame, computer control, and operator interface,
2) wafer handling and (optional) loadlock,
3) pre-/post-treatment chamber(s), and
4) processing chamber(s).

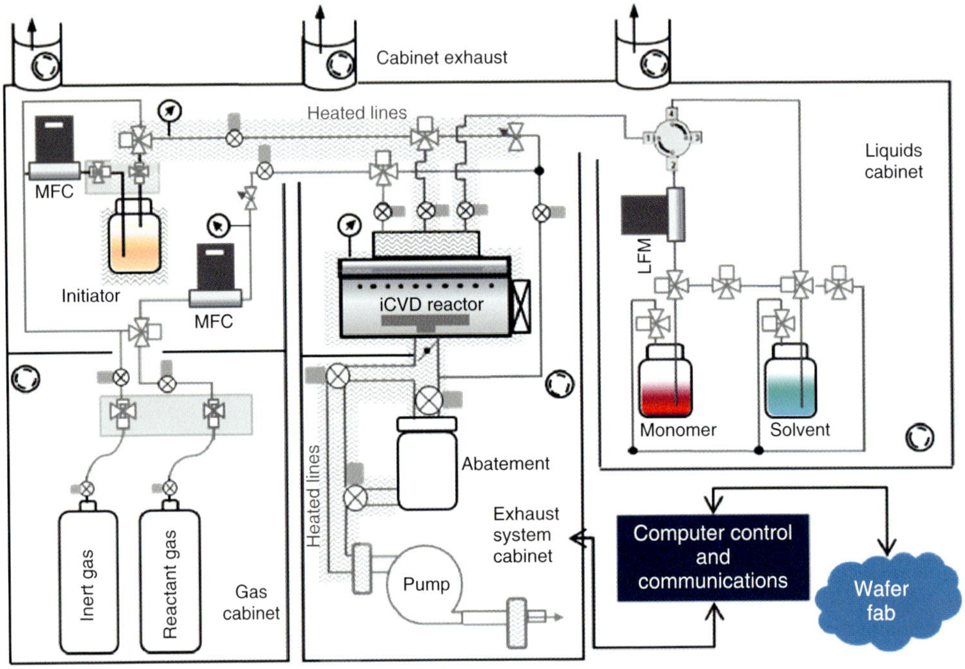

Figure 18.6 Model iCVD system.

At a minimum, a commercial iCVD system will include one or more iCVD reaction chambers with associated computer-controlled subsystems – chemical delivery; temperature-controlled reactor vessel and wafer chuck with resistively heated filament array; vacuum pumping and pressure control; process monitoring and data storage; and automated wafer handling. Although several iCVD process characterization studies have been conducted on research-grade horizontal reactors [63–65], a single-wafer vertical CVD chamber system will be used to illustrate these key subsystems and how they influence the iCVD system's overall design (see Figure 18.6).

18.4.2 iCVD Reactor Design

A necessary first step in the hardware design process is to define the range of applications to be supported by the equipment. Potential applications might include deposition of iCVD polymer adhesives for BEOL wafer processing [52] or high aspect ratio conformal iCVD polymer liner for through-silicon-vias [6, 8]. In each instance, an analysis of the application specifications and potential deposition

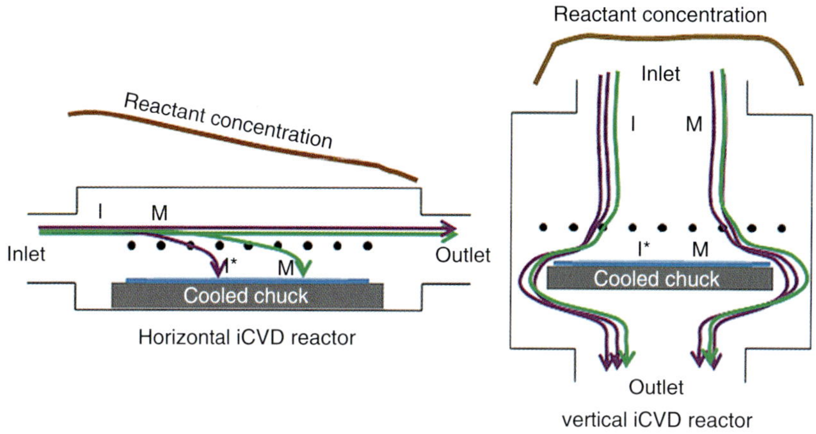

Figure 18.7 Reactant depletion comparison – horizontal versus vertical iCVD reactor.

chemistries [66] will bear directly on iCVD hardware, process conditions, and ultimately the tool's cost, performance, and safety.

For the preferred iCVD operating conditions – lower pressures (<10 Torr) and lower temperatures (<80 °C) – a vertical axisymmetric chamber configuration provides laminar gas flow and minimizes gas-phase monomer depletion between inlet and pump port, which can often result in across-wafer film non-uniformity in horizontal reactors (Figure 18.7). This class of CVD reactor has the added benefit of years of proven performance in high-volume semiconductor manufacturing.

A separate chapter would be needed to provide a comprehensive discussion of CVD hardware and process simulation methods, but a sample of this active body of research may be found in Refs. [67–72]. After the application specifications and preferred polymer chemistry reaction set have been defined, iCVD chamber geometry may be prepared for computational modeling of the thermodynamics, kinetics, and transport phenomena (heat, mass, and momentum), which occur within the reactor and on the wafer surface. Reactor scale simulators such as ANSYS Fluent and COMSOL Multiphysics that deal with different types of mass transport on non-moving grids are commercially available and are routinely used in CVD reactor design.

In addition to conventional CVD chamber design elements such as reactor dimensions, gas injection location, wafer chuck diameter, chuck-to-chamber wall distance, and exhaust port placement, one must define iCVD-specific variables, including filament-to-wafer and gas injection-to-filament distances, and the number of filaments in the array. Each of these may be considered adjustable parameters that need to be optimized for the iCVD reactor, where modeling and simulations can help guide the design. A research-scale iCVD reactor may be constructed to generate validation data for further refinements to both model and hardware [73]. Combining iCVD hardware/process modeling (see Figure 18.8)

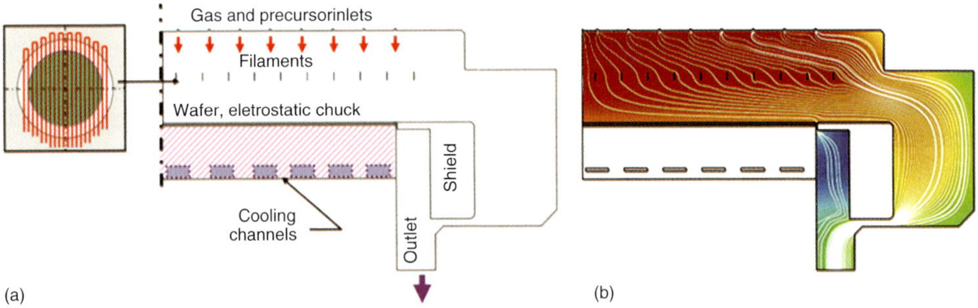

Figure 18.8 (a) 2D model geometry of an axisymmetric iCVD chamber and (b) simulation of thermal distribution and gas transport.

with experimental validation provides an iterative path for linking predicted design outputs to observed deposition characteristics [74–76].

18.4.3 iCVD Subsystem Design

18.4.3.1 Filaments

When designing a filament array for a commercial iCVD reactor, one should take into account the following parameters:

1) Filament and array support material compatibility with semiconductor devices (e.g., avoiding parts that could contribute mobile metal ions detrimental to device performance),
2) Heat load as a function of filament-to-wafer distance and chamber pressure (impacts cooling subsystem specifications and influences achievable monomer saturation pressures),
3) Monitoring and control, and
4) Ease of maintenance.

It is worth noting that filaments used in iCVD processing do not exhibit aging effects such as embrittlement or silicidation [77] observed in hot wire or catalytic CVD because of the low filament temperatures required for free radical production. This contributes to long filament lifetimes and low material consumption. Nichrome or tungsten filament wires are acceptable choices for semiconductor processing.

18.4.3.2 Chemical Delivery and Vaporization

Precursors and initiators for the iCVD process are vaporized and introduced into the reactor typically by means of direct liquid injection (DLI) or passivated stainless steel bubblers coupled with mass flow metering control [78]. Vapor draw systems can be difficult to start, require very close control of temperature and pressure, and are less efficient at generating well-controlled vapor mass flows necessary in a manufacturing environment. In addition, chemical lifetime and safety concerns may dictate a vapor draw temperature range that limits maximum flow rates for certain monomers and initiators. Purity of the precursor, control of polymerization inhibitor concentration, and duration of storage (on- and off-tool) can also influence vaporization performance. The combination of DLI with mass flow metering couples high-efficiency vaporization with microprocessor control of gas flows. To minimize the opportunity for recondensation of the vaporized monomers before introduction to the process chamber, gas delivery lines should be held to the shortest achievable lengths while minimizing temperature gradients between injector and process reactor.

Chemical delivery design should take into account materials compatibility with the monomers to be delivered and vaporized including the affinity of monomers to bond or condense on interior surfaces and orifices. Wherever possible, chemical mixing prior to process chamber vapor introduction should be limited to minimize reliability issues because of crosscontamination and/or undesired chemical reactions. The amount of material delivered in a high-volume iCVD system can lead to problems less likely to be encountered in a research-level tool. As an example, Figure 18.9 shows clogging a DLI outlet nozzle due to repeated exposure to and build up of incompatible monomers from a shared iCVD vaporization chamber.

Figure 18.9 Blocked DLI due to crosscontamination at the outlet nozzle.

Although iCVD does not require the use of solvents for film formation, the use of certain precursors may necessitate adding solvent delivery and disposal capability for injector and/or delivery line cleaning to the tool design.

18.4.3.3 Temperature Control

iCVD metrics such as deposition rate, film non-uniformity, and process repeatability are dependent on stable temperature control. This extends from the vaporization temperature(s) of the initiator and monomer(s) to the filament temperatures needed to create free radicals from the vaporized initiator to the substrate surface temperature and the corresponding monomer saturation vapor pressure (P_{sat}). Reactor modeling results should indicate various thermal loads to be considered in designing interior geometry (chamber walls, filaments) and wafer chuck cooling capacity. For example, during iCVD processing, chamber walls are typically heated to temperatures unfavorable for polymer deposition. However, this can lead to reduced deposition rate at the edge of the wafer if radiative heating effects from the walls are not considered.

Wafer surface temperature control – maintaining adequate thermal contact between the substrate and the chuck – is critical to the iCVD process as surface temperature directly influences P_{sat} and thus the corresponding surface kinetics (P_m/P_{sat}). (See other chapters in this book for additional details.) A temperature-controlled electrostatic chuck (ESC) with backside gas flow has been adopted by the semiconductor industry for lower temperature vacuum processes such as dry etch and physical vapor deposition (PVD) to provide constant, uniform, and reproducible wafer temperatures. The low-temperature iCVD operating range makes an ESC, rather than a conventional CVD susceptor, the preferred choice of wafer stage.

It is important to note that optical access for front-surface temperature monitoring possible on a research-scale system may not be practical for HVM CVD vacuum chambers. A less direct means of monitoring such as ESC temperature may be necessary. Instrumented wafers may be used to determine actual surface temperature under simulated process conditions and to calibrate results obtained from indirect temperature monitoring schemes.

18.4.3.4 Process Control

In the demanding semiconductor manufacturing environment, equipment and wafer state monitoring and control are essential for operating at or near "optimal" conditions and for minimizing process variability [79]. A number of sensors may be used as part of a process control system. *In situ* CVD equipment state monitors may include thermocouples, pressure gauges, and liquid/gas flow meters, as well as wafer state monitors such as laser interferometer, pyrometer, and quartz crystal microbalance (QCM) [80, 81]. These sensors may be integrated in a feed-forward/feedback computer control system [82].

Important iCVD process variables to control are substrate and filament temperature and chamber pressure, since they impact initiator production and P_m/P_{sat}

ratio and ultimately control deposition rate and wafer throughput. The system controller may utilize active feedback from ESC temperature monitors to control coolant bath temperature and to minimize deviations from desired ESC temperature setpoint. Similarly, chamber-mounted pressure gauge measurements relayed to the controller result in throttle valve position changes needed to maintain the desired process pressure.

As part of contamination control during iCVD polymer deposition, one can consider the case of undesired film formation on the outer edge of the ESC and on the wafer bevel. Polymer build up on this area of the ESC can lead to poor thermal contact at the wafer edge with a corresponding center-to-edge wafer temperature gradient and variation in deposition rate. Unwanted polymer is not only a potential film non-uniformity problem, but it can also result in particle formation and contamination of the front or rear surface of the wafer. Methods have been developed to minimize unwanted ESC polymer deposits [83] and should be considered in wafer chuck design and as part of an overall process control strategy.

18.4.3.5 **Exhaust/Abatement**

Careful design and construction of the exhaust system is essential for safe operation of the system. A substantial body of vacuum engineering literature exists for guidance in calculations and component selection [84–86]. Reactor operating pressure and gas flow limits are governed by the pump system speed and the exhaust line conductance. In many semiconductor manufacturing wafer fabrication facilities (fabs), minimum exhaust lengths can be measured in meters as vacuum pumps are located well away from the process reactor. Failure to take into account conductance losses due to pump location and the number of bends between the chamber and the vacuum pump can result in limited iCVD process capability and costly modifications after tool installation to expand the process window. An iCVD exhaust subsystem would include heated foreline tubing, a heated throttle, or butterfly valve to control process pressure in the reactor; a heated gate valve to isolate the reactor from the foreline; an incinerating system for safe decomposition of unreacted monomer and initiator; and a high-pumping speed system such a rotary vane/roots blower pump combination (Figure 18.10).

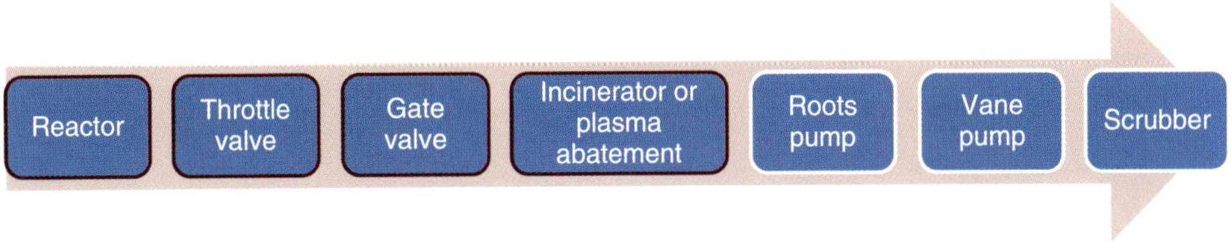

Figure 18.10 Schematic illustration of iCVD exhaust system.

To minimize deposition of unreacted monomer within foreline piping or valves, these components should be heated above monomer condensation temperature. The desired temperature range will dictate valve material compatibility and suitable monitoring and feedback control for safe operation. Additional foreline purge with inert gas can also help to minimize unwanted deposits. The throttle valve is particularly prone to monomer/polymer deposition as the first-flow restriction point in the foreline (Figure 18.11). Monitoring valve response to pressure set-point coupled with periodic inspections can help avoid catastrophic failure of this critical part and loss of reactor pressure control.

No CVD process completely utilizes all incoming species. Therefore, the resulting reactor exhaust must be suitably treated before it can be safely released to the environment. Effluent treatment options chosen for implementation must be compliant with semiconductor manufacturing safety standards (SEMI F5) [87].

18.4.3.6 Wafer Handling and Integration

Fabrication of semiconductor devices involves a number of interrelated process steps. Design of the iCVD tool and process must take into account both the expected incoming substrate state and the expected deposited film surface and bulk states. Pre- and/or post-deposition processes and reaction chambers in addition to the iCVD reactor may be required to satisfy device requirements. For example, post-deposition treatment of iCVD films to prepare polymer surfaces for the next process step or to minimize contamination effects due to evolution of unreacted monomers from the bulk film (outgassing) could be implemented to improve overall device yield. Unlike research-scale reactors typically configured for manual loading/unloading of substrates, HVM systems will have an automated wafer-handling subsystem designed to load/unload multiple wafers

(a) (b)

Figure 18.11 Polymer build-up on mixed use iCVD throttle valve after sustained service – (a) chamber-facing side and (b) pump-facing side.

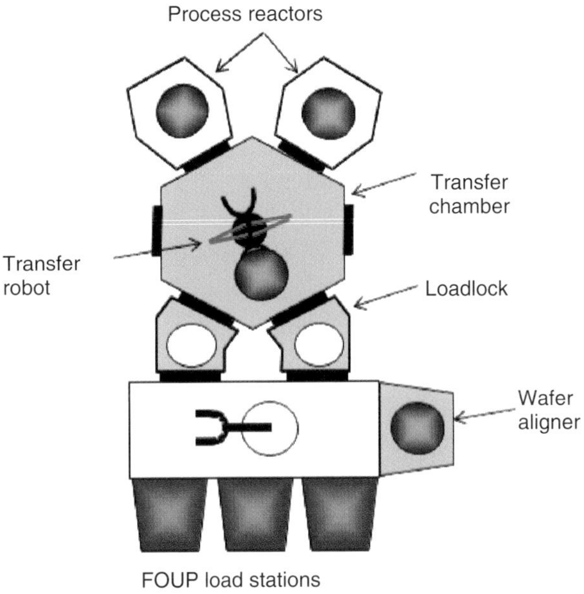

Figure 18.12 Multichamber platform.

from self-contained mini-environments known as front-opening unified pods or FOUPs (Figure 18.12).

18.4.4
Economic Considerations

The semiconductor industry has adopted COO as a standard (SEMI E35) [88] for analyzing and minimizing all costs associated with the acquisition, use, and maintenance of production equipment. Any equipment or process considered for high-volume semiconductor manufacturing will need to meet or exceed specifications for a number of cost and productivity metrics (see Figure 18.13). In the case of HVM iCVD equipment, chemical cost, and facilities requirements (floor space, exhaust, air, water), quality metrics such as film uniformity or step coverage, and productivity metrics such as throughput will determine COO.

18.5
Summary

The relatively mild deposition conditions and the range of monomer precursors that can be used for iCVD make it an interesting technique for semiconductor device manufacturing. The technique has been demonstrated to be compatible

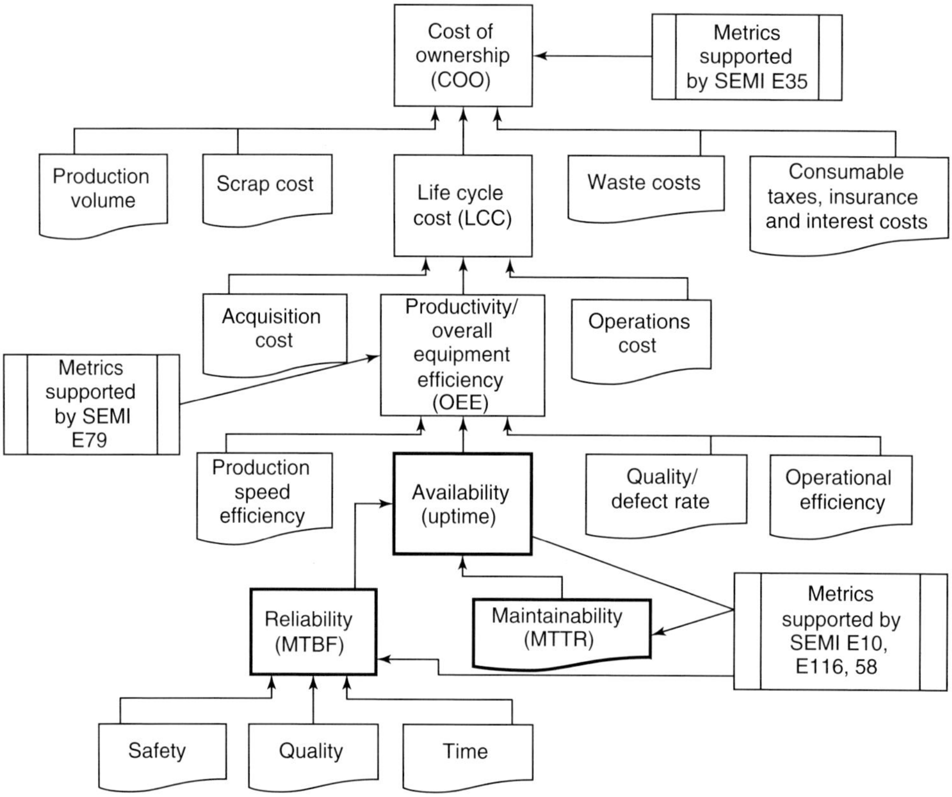

Figure 18.13 Hierarchy of equipment performance metrics. (Adapted from Ref. [89].)

with 300-mm wafer processing, and the technical know-how to build and integrate a production-ready iCVD system in a fabrication line is there. However, the COO for an industrial scale tool needs to meet application needs. Considering the various applications demonstrated on 300-mm wafer scale thus far, iCVD can fill a critical need in the tool box for air-gap engineering. iCVD also has potential to provide unique capability for adhesion and device packaging applications; however, additional research and evaluation is needed.

References

1. Capote, M.A., Zhou, Z., Zhu, X., and Zhou, L. (2000) Semiconductor flip-chip package and method for the fabrication thereof. United States, Patent: US20020014703 A1.
2. Wallraff, G.M. and Hinsberg, W.D. (1999) Lithographic imaging techniques for the formation of nanoscopic features. *Chem. Rev.*, **99**, 1801–1822.
3. de Sousa, I., Belanger, L., Dufort, C., and Chenier, S.D. (2011) A manufacturing approach to reducing underfill voiding on large die (>18 mm) flip chip organic laminate packaging. Paper presented at

61st Electronic Components and Technology Conference (ECTC), IEEE, Lake Buena Vista, FL.
4. Niklaus, F., Enoksson, P., Edvard, K., and Stemme, G. (2001) Low-temperature full wafer adhesive bonding. *J. Micromech. Microeng.*, **11**, 100 107.
5. Xu, J. and Gleason, K.K. (2010) Conformal, amine-functionalized thin films by initiated chemical vapor deposition (iCVD) for hydrolytically stable microfluidic devices. *Chem. Mater.*, **22**, 1732–1738.
6. Baxamusa, S.H. and Gleason, K.K. (2008) Thin polymer films with high step coverage in microtrenches by initiated CVD. *Chem. Vap. Deposition*, **14**, 313–318.
7. Chan, K. and Gleason, K.K. (2005) Initiated chemical vapor deposition of linear and cross-linked poly(2-hydroxyethyl methacrylate) for use as thin-film hydrogels. *Langmuir*, **21**, 8930–8939.
8. Ozaydin-Ince, G. and Gleason, K.K. (2010) Tunable conformality of polymer coatings on high aspect ratio features. *Chem. Vap. Deposition*, **16**, 100–105.
9. Chavan, A.V. and Wise, K.D. (2001) Batch-processed vacuum-sealed capacitive pressure sensors. *J. Microelectromech. Syst.*, **10**, 580–588.
10. Eaton, W.P. and Smith, J.H. (1997) Micromachined pressure sensors: review and recent developments. *Smart Mater. Struct.*, **6**, 530.
11. Ko, W.H. (2007) Trends and frontiers of MEMS. *Sens. Actuators, A: Phys.*, **136**, 62–67.
12. van Kessel, P.F., Hornbeck, L.J., Meier, R.E., and Douglass, M.R. (1998) A MEMS-based projection display. *Proc. IEEE*, **86**, 1687–1704.
13. Yazdi, N., Ayazi, F., and Najafi, K. (1998) Micromachined inertial sensors. *Proc. IEEE*, **86**, 1640–1659.
14. Rothschild, M. (2010) A roadmap for optical lithography. *Opt. Photon. News*, **21**, 26–31.
15. Altissimo, M. (2010) E-beam lithography for micro-/nanofabrication. *Biomicrofluidics*, **4**, 3–6.
16. Mao, Y., Felix, N.M., Nguyen, P.T., Ober, C.K., and Gleason, K.K. (2004) Towards all-dry lithography: electron-beam patternable poly(glycidyl methacrylate) thin films from hot filament chemical vapor deposition. *J. Vac. Sci. Technol., B*, **22**, 2473–2478.
17. Mao, Y., Felix, N.M., Nguyen, P.T., Ober, C.K., and Gleason, K.K. (2006) Positive- and negative-tone CVD polyacrylic electron-beam resists developable by supercritical CO2. *Chem. Vap. Deposition*, **12**, 259–262.
18. Moreau, W., Merritt, D., Moyer, W., Hatzakis, M., Johnson, D., and Pederson, L. (1979) Speed enhancement of PMMA resist. *J. Vac. Sci. Technol.*, **16**, 1989–1991.
19. Mao, Y. and Gleason, K.K. (2006) Positive-tone nanopatterning of chemical vapor deposited polyacrylic thin films. *Langmuir*, **22**, 1795–1799.
20. Petruczok, C.D. and Gleason, K.K. (2012) Initiated chemical vapor deposition-based method for patterning polymer and metal microstructures on curved substrates. *Adv. Mater.*, **24**, 6445–6450.
21. Bohr, M.T. (1995) Interconnect scaling-the real limiter to high performance ULSI. International Electron Devices Meeting, pp. 241–244.
22. Grill, A. (2003) Plasma enhanced chemical vapor deposited SiCOH dielectrics: from low-k to extreme low-k interconnect materials. *J. Appl. Phys.*, **93**, 1785.
23. Si, J.J., Ono, H., Uchida, K., Nozaki, S., Morisaki, H., and Itoh, N. (2001) Correlation between the dielectric constant and porosity of nanoporous silica thin films deposited by the gas evaporation technique. *Appl. Phys. Lett.*, **79**, 3140.
24. ITRS (2001) International Technology Roadmap for Semiconductors.
25. ITRS (2013) International Technology Roadmap for Semiconductors.
26. Baklanov, M. and O'Dwyer, D. (2006) Moisture induced degradation of porous low-k materials. Paper presented at MRS Spring Meeting, San Fransisco, CA.
27. Cheng, Y.-L., Leon, K.-W., Huang, J.-F., Chang, W.-Y., Chang, Y.-M., and Leu, J. (2014) Effect of moisture on electrical properties and reliability of low dielectric constant materials. *Microelectron. Eng.*, **114**, 12–16.

28. Iacopi, F., Brongersma, S.H., Vandevelde, B., O'Toole, M., Degryse, D., Travaly, Y., and Maex, K. (2004) Challenges for structural stability of ultra-low-k-based interconnects. *Microelectron. Eng.*, **75**, 54–62.
29. Volinsky, A.A., Vella, J.B., and Gerberich, W.W. (2003) Fracture toughness, adhesion and mechanical properties of low-K dielectric thin films measured by nanoindentation. *Thin Solid Films*, **429**, 201–210.
30. Anand, M.B., Yamada, M., and Shibata, H. (1996) NURA: a feasible, gas-dielectric interconnect process. Paper presented at Symposium on VLSI Technology, Widerkehr & Associates, Honolulu, HI.
31. Daamen, R., Bancken, P.H.L., Nguyen, V.H., Humbert, A., Verheijden, G.J.A.M., and Hoofman, R.J.O.M. (2007) The evolution of multi-level air gap integration towards 32 nm node interconnects. *Microelectron. Eng.*, **84**, 2177–2183.
32. Hoofman, R., Daamen, R., Nguyenhoang, V., Michelon, J., Gosset, L.G., Arnal, V., de Pontcharra, J., Gaillard, F., Caluwaerts, R., Bruynseraede, C. *et al.* (2006) Benefits and trade-offs in multilevel air gap integration, *Proc. MRS*, 914, F0910-0901.
33. Havemann, R.H. and Jeng, S. (1995) Multilevel interconnect structure with air gaps formed between metal leads. US Patent US5461003 A, Texas Instruments Incorporated.
34. Ryu, S.-K., Im, J., Ho, P.S., and Huang, R. (2014) A kinetic decomposition process for air-gap interconnects and induced deformation instability of a low-k dielectric cap layer. *J. Mech. Sci. Technol.*, **28**, 255–261.
35. Zhang, X., Ryu, S.-K., Huang, R., Ho, P.S., Liu, J., and Toma, D. (2008) Mechanical stability of air-gap interconnects. *Future Fab Int.*, **27**, 81–87.
36. Lee, E., Faguet, J., Brcka, J., Akiyama, O., Liu, J., and Toma, D. (2011) Single-chamber filament-assisted chemical vapor deposition of polymer and organosilicate films for air gap interconnect formation. *Thin Solid Films*, **519**, 4571–4573.
37. Chan, K. and Gleason, K.K. (2006) Air-gap fabrication using a sacrificial polymeric thin film synthesized via initiated chemical vapor deposition. *J. Electrochem. Soc.*, **153**, C223.
38. Ozaydin-Ince, G. and Gleason, K.K. (2010) Thermal stability of acrylic/methacrylic sacrificial copolymers fabricated by initiated chemical vapor deposition. *J. Electrochem. Soc.*, **157**, D41–D45.
39. Loo, L.S. and Gleason, K.K. (2001) Hot filament chemical vapor deposition of polyoxymethylene as a sacrificial layer for fabricating air gaps. *Electrochem. Solid-State Lett.*, **4**, G81.
40. Trujillo, N.J., Wu, Q., and Gleason, K.K. (2010) Ultralow dielectric constant tetravinyltetramethylcyclotetrasiloxane films deposited by initiated chemical vapor deposition (iCVD). *Adv. Funct. Mater.*, **20**, 607–616.
41. Chen, Q., Huang, C., and Wang, Z. (2012) Benzocyclobutene polymer filling of high aspect-ratio annular trenches for fabrication of Through-Silicon-Vias (TSVs). *Microelectron. Reliab.*, **52**, 2670–2676.
42. Chen, Q., Huang, C., Wu, D., Tan, Z., and Wang, Z. (2013) Ultralow-capacitance through-silicon vias with annular air-gap insulation layers. *IEEE Trans. Electron Devices*, **60**, 1421–1426.
43. Huang, C., Chen, Q., and Wang, Z. (2013) Polymer liner formation in high aspect ratio through-silicon-vias for 3-D integration. *IEEE Trans. Compon. Packag. Manuf. Technol.*, **3**, 1107–1113.
44. Huang, C., Chen, Q., Wu, D., and Wang, Z. (2013) Implementation of air-gap through-silicon-vias (TSVs) using sacrificial technology. *Packag. Manuf. Technol.*, **3**, 1430–1438.
45. Niklaus, F., Stemme, G., Lu, J.Q., and Gutmann, R.J. (2006) Adhesive wafer bonding. *J. Appl. Phys.*, **99**, 031101.
46. Makihata, M., Tanaka, S., Muroyama, M., Matsuzaki, S., Yamada, H., Nakayama, T., Yamaguchi, U., Mima, K., Nonomura, Y., Fujiyoshi, M. *et al* (2011) Adhesive wafer bonding using a molded thick benzocyclobutene layer for wafer-level integration of MEMS and LSI. *J. Micromech. Microeng.*, **21**, 085002.

47. Teixeira, R.C., De Munck, K., De Moor, P., Baert, K., Swinnen, B., Van Hoof, C., and Kuttel, A. (2007) Stress analysis on ultra thin ground wafers. *ECS Trans.*, **9**, 113–121.
48. Gupta, M., Kapur, V., Pinkerton, N.M., and Gleason, K.K. (2008) Initiated chemical vapor deposition (iCVD) of conformal polymeric nanocoatings for the surface modification of high-aspect-ratio pores. *Chem. Mater.*, **20**, 1646–1651.
49. Chen, H.-Y., McClelland, A.A., Chen, Z., and Lahann, J. (2008) Solventless adhesive bonding using reactive polymer coatings. *Anal. Chem.*, **80**, 4119–4124.
50. Im, S.G., Bong, K.W., Lee, C.-H., Doyle, P.S., and Gleason, K.K. (2009) A conformal nano-adhesive via initiated chemical vapor deposition for microfluidic devices. *Lab Chip*, **9**, 411–416.
51. Bharamaiah Jeevendrakumar, V.J., Altemus, B., Gildea, A., and Bergkvist, M. (2013) Wafer bonding via iCVD PGMA films and characterization of bond strength properties. Paper presented at TECHON, Semiconductor Research Corporation, Austin, TX.
52. Bharamaiah Jeevendrakumar, V.J., Altemus, B.A., Gildea, A.J., and Bergkvist, M. (2014) Influence of surface treatment on adhesion of iCVD PGMA thin films for wafer-level bonding. Paper presented at 2013 MRS Fall Meeting, Boston, MA.
53. Pascual, D.N. (2004) Effects of soft-bake parameters on a Benzocyclobutene (BCB) adhesive wafer bond. Paper presented at 2003 MRS Fall Meeting, Boston, MA.
54. Choy, K.L. (2003) Chemical vapour deposition of coatings. *Prog. Mater. Sci.*, **48**, 57–170.
55. Dobkin, D.M. and Zuraw, M.K. (2003) *Principles of Chemical Vapor Deposition: What's Going on Inside the Reactor*, Kluwer Academic Publishers, Boston, MA.
56. Jensen, K.F., Einset, E.O., and Fotiadis, D.I. (1991) Flow phenomena in chemical vapor deposition of thin films. *Annu. Rev. Fluid Mech.*, **23**, 197–232.
57. Jones, A.C. and Hitchman, M.L. (2009) *Chemical Vapour Deposition: Precursors, Processes and Applications*, Royal Society of Chemistry, Camebridge.
58. Martin, P.M. (2010) *Handbook of Deposition Technologies for Films and Coatings: Science, Applications and Technology*, 3rd edn, Elsevier Inc., Oxford.
59. Pauleau, Y. (2002) *Chemical Physics of Thin Film Deposition Processes for Micro- and Nano-Technologies*, Kluwer Academic Publishers, Boston, MA.
60. Pierson, H.O. (1999) *Handbook of Chemical Vapor Deposition: Principles, Technology and Applications*, 2nd edn, Noyes Publications, Norwich, NY.
61. Yan, X.-T. and Xu, Y. (2010) *Chemical Vapour Deposition: An Integrated Engineering Design for Advanced Materials*, Springer-Verlag, London.
62. Nishi, Y. and Doering, R. (2000) *Handbook of Semiconductor Manufacturing Technology*, 2nd edn, Marcel Dekker, New York.
63. Lau, K.K.S. and Gleason, K.K. (2006) Initiated chemical vapor deposition (iCVD) of poly(alkyl acrylates): an experimental study. *Macromolecules*, **39**, 3688–3694.
64. Mao, Y. and Gleason, K.K. (2004) Hot filament chemical vapor deposition of poly(glycidyl methacrylate) thin films using tert-butyl peroxide as an initiator. *Langmuir*, **20**, 2484–2488.
65. Martin, T.P., Chan, K., and Gleason, K.K. (2008) Combinatorial initiated chemical vapor deposition (iCVD) for polymer thin film discovery. *Thin Solid Films*, **516**, 681–683.
66. Martin, T.P., Lau, K.K.S., Chan, K., Mao, Y., Gupta, M., Shannan O'Shaughnessy, W., and Gleason, K.K. (2007) Initiated chemical vapor deposition (iCVD) of polymeric nanocoatings. *Surf. Coat. Technol.*, **201**, 9400–9405.
67. Economou, D.J., Panagopoulos, T.L., and Meyyappan, M. (1998) Examining scale-up and computer simulation in tool design for 300-mm wafer processing. *Micro*, **16**, 101–113.
68. Kee, R.J., Rupley, F.M., Meeks, E., and Miller, J.A. (1996) Chemkin-III: A Fortran Chemical Kinetics Package for the Analysis of Gas-Phase Chemical and Plasma Kinetics. Technical Report,

SAND96-8216, Sandia National Labs, Livermore, CA, p. 166.

69. Kleijn, C.R. (2002) Simulation of chemical vapor deposition processes, in *Characterization of Materials*, John Wiley & Sons, Inc.

70. Kleijn, C.R., Dorsman, R., Kuijlaars, K.J., Okkerse, M., and van Santen, H. (2007) Multi-scale modeling of chemical vapor deposition processes for thin film technology. *J. Cryst. Growth*, **303**, 362–380.

71. Krishnan, A., Lowry, S.A., and Harrand, V.J. (1995) *Virtual Prototyping Tools for Semiconductor Fabrication Equipment*, CFD Research Corporation, Huntsville, AL.

72. Rader, D.J. and Geller, A.S. (1994) Particle transport modelling in semiconductor process environments. *Plasma Sources Sci. Technol.*, **3**, 426.

73. Ozaydin-Ince, G. and Gleason, K.K. (2009) Transition between kinetic and mass transfer regimes in the initiated chemical vapor deposition from ethylene glycol diacrylate. *J. Vac. Sci. Technol., A*, **27**, 1135–1143.

74. Bakker, R., Verlaan, V., Verkerk, A.D., van der Werf, C.H.M., van Dijk, L., Rudolph, H., Rath, J.K., and Schropp, R.E.I. (2009) Heat transfer model of an iCVD reactor. *Thin Solid Films*, **517**, 3555–3558.

75. Brcka, J. (2009) Model of a filament assisted CVD reactor. Paper presented at COMSOL Conference, Boston, MA.

76. Chan, K. and Gleason, K.K. (2006) A mechanistic study of initiated chemical vapor deposition of polymers: analyses of deposition rate and molecular weight. *Macromolecules*, **39**, 3890–3894.

77. Honda, K., Ohdaira, K., and Matsumura, H. (2008) Study of silicidation process of tungsten catalyzer during silicon film deposition in catalytic chemical vapor deposition. *Jpn. J. Appl. Phys.*, **47**, 3692.

78. Hansen, S. (2013) Delivering the vapors. *Vac. Technol. Coat.*, **14**, 14–19.

79. May, G.S. and Spanos, C.J. (2006) *Fundamentals of Semiconductor Manufacturing and Process Control*, John Wiley & Sons, Inc., Hoboken, NJ.

80. Herman, I.P. (1996) *Optical Diagnostics for Thin Film Processing*, Academic Press Ltd, London.

81. Pulker, H.K. and Decosterd, J.P. (1984) Applications of quartz crystal microbalances for thin film deposition process control, in *Applications of Piezoelectric Quartz Crystal Microbalances* (eds C. Lu and A.W. Czanderna), Elsevier Science Publishers, Amsterdam, Holland.

82. Relihan, K., Geraghty, S., and O'Dwyer, A. (2007) Some aspects of process control in semiconductor manufacturing. Paper presented at Proceedings of the IMC-24, Waterford Institute of Technology, Ireland.

83. Cleary, T.J. and Wing, J.C. (2001) Exclusion of polymer film from semiconductor wafer edge and backside during film (CVD) deposition. US Patent US6281144 B1, Novellus Systems, Inc.

84. Berman, A. (1992) *Vacuum Engineering Calculations, Formulas, and Solved Exercises*, Academic Press, Inc, San Diego, CA.

85. Hablanian, M.H. (1997) *High-Vacuum Technology: A Practical Guide*, 2nd edn, Marcel Dekker, Inc., New York.

86. O'Hanlon, J.F. (2003) *A User's Guide to Vacuum Technology*, John Wiley & Sons, Inc, Hoboken, NY.

87. SEMI SEMI Standard F5: 90. (2001) *Guide for Gaseous Effluent Handling*, Semiconductor Equipment and Materials International, *www.semi.org* (accessed 18 November 2014).

88. SEMI SEMI Standard E35: 0307. (2007) *Guide to Calculate Cost of Ownership (COO) Metrics for Semiconductor Manufacturing Equipment*, Semiconductor Equipment and Materials International, *www.semi.org* (accessed 18 November 2014).

89. Dhudshia, V.H. (2008) *Hi-Tech Equipment Reliability: A Practical Guide for Engineers and Managers*, iUniverse, Inc..

Part IV
Reactors and Commercialization

CVD Polymers: Fabrication of Organic Surfaces and Devices, First Edition. Edited by Karen K. Gleason.

19
Commercialization of CVD Polymer Coatings

W. Shannan O'Shaughnessy

19.1
Introduction

The breadth and flexibility of CVD polymers makes them a tool applicable not just to materials research but also to the commercial deposition of thin polymer coatings. As previously mentioned, CVD polymer deposition allows the controlled, vacuum-based formation of a wide range of polymeric formulations. There are many reasons that CVD polymer deposition may be selected as a commercial coating process. In some instances, conformal coating of micro- to nano-scale features may be required. As mentioned previously, CVD polymer deposition can provide conformal coatings while allowing for the same polymer chemical specificity provided by solution polymerization. In other instances, CVD polymer coatings provide material properties not possible through solution polymerization. Alternately, the desired polymer coating may be impossible to form by a wet process either due to the insolubility, morphology, or very low required thickness of the desired polymer coating.

Despite these advantages, individual applications must be carefully evaluated for their suitability as commercial CVD polymer coating targets due to the complexity of vacuum-based processing. There are also many market, safety, and process variables to consider. What size of deposition chamber is needed to satisfy the required coating throughput? How will process by-products be handled safely and in what volumes? Where will coating occur, in-house by the end-user or at the facility of a specialty provider? What coating solutions currently exist to solve the targeted technical problem? All of these questions and many more must be answered before a new commercial CVD polymer deposition process can be conceived. Table 19.1, lists these and additional factors that must be considered to fully evaluate a new CVD polymer deposition opportunity. These considerations will be detailed in this chapter followed by a case study of the commercial application of CVD-deposited polytetrafluoroethylene (PTFE) for surface lubricity. A discussion of application-specific CVD polymer deposition chamber design will conclude the examination of commercial CVD polymer deposition.

CVD Polymers: Fabrication of Organic Surfaces and Devices, First Edition. Edited by Karen K. Gleason.

Table 19.1 Factors to consider when evaluating the commercial viability of a CVD polymer.

Commercialization factor	Considerations
CVD precursor(s)	Cost
	Vapor pressure
	Toxicity
	Ease of supply
Process	Consumables
	Control parameters
	Safety issues
	By-product abatement
Coating application	Material properties
	Size of parts
	Volume of parts
	Thickness required
	Pre- and post-treatments
	Testing
Market	Size and growth
	Business model
	Incumbent coatings
	Geographic spread
	Longevity

Although CVD polymer deposition provides many unique coating properties, the process complexity requires that thorough analysis be performed prior to commercialization.

19.1.1 Precursor Considerations

Of the many factors that must be considered in the commercialization of a new CVD polymer deposition application, primary among them are the precursors to the material itself. For efficient and cost-effective coating, these precursors must be delivered as a stable, uniform vapor throughout the deposition chamber. Given the wide variability of possible precursor materials for CVD polymer deposition, many potential issues arise. Low vapor pressure precursors can present a significant challenge. Although delivery of small volumes of vaporized liquid or sublimed solid monomer for the research-scale investigation of a new material can generally be achieved through conventional heating approaches, creation of the volumes of vaporous monomer necessary to feed a large-scale commercial process can require different technology entirely. The use of a diluent carrier gas can assist in the vaporization and distribution of low-volatility monomers. This can be achieved through a simple apparatus, such as a bubbler, wherein the carrier gas is simply passed through a low-volatility liquid while it is being heated. This approach increases the flux of monomer through the solubility of the monomer into the carrier gas and the constant convection of vaporized monomer from the heating vessel, increasing the driving force for further monomer vaporization. If larger volumes of low vapor pressure monomer are required, more technologically advanced solutions are available for vaporization. One of the most effective is an ultrasonic vaporizer. This device provides for vaporization of monomer at a precisely controlled rate. Utilizing a liquid mass flow controller, monomer is pumped to a vaporization tip where it is ultrasonically sheared into tiny droplets. These droplets present a very high surface area for the rapid absorption of energy and are then easily converted to vapor. An inert carrier gas, similar to that utilized in a bubbler, is then used to convectively transfer the vaporized monomer into the deposition system. While quite effective, these systems can also be a significant expense, adding 5–20% to the cost of the deposition system per vaporized monomer.

Another set of significant considerations relating to CVD polymer precursors are cost and utilization. High-cost specialty chemicals are easily within reach for small-scale research applications and proof-of-concept demonstrations. However, when 10–1000 m^2 of coating is required per day, the cost of the precursor chemicals in the deposition of said coating can quickly become a limiting factor in its utilization. Early awareness of the possible impact of this issue can allow for focus on applications that fit with the relative costs. Although it would always be preferable to have low-cost precursors, high-value precursor processes can still find application in areas requiring lower volume or throughput such as sensors or medical devices. Large volume commodity products, such as textiles or disposable consumer products, should be avoided for these chemistries.

In addition to outright precursor cost, the utilization efficiency of the precursor can also have a significant impact on the operational cost of the process. The chemistry of the CVD polymer will largely dictate here. Precursors forming stable, long-lived radicals can be utilized at high efficiency as process conditions can be tailored to allow for deposition of the vast majority of the monomer entering the deposition chamber. However, more reactive precursors, especially systems possessing significant gas-phase polymerization in competition with surface film growth, can possess surprisingly low precursor utilization, with the majority of the monomer flowing unreacted through the system. For high-value precursors, this loss of raw material is unacceptable and will require the separation and recycle of the process effluent to allow for cost-effective operation. These additional process steps can add significant complexity and capital cost to the overall CVD polymer coating process.

The safety and handling issues associated with the precursors to a CVD polymer must also be considered. Are the precursors extremely toxic? Do they need to be isolated from oxygen during transfer to vaporization vessels? Do the precursors break down over time when stored? What are the safety considerations associated with these decomposition products? All of these questions must be answered, not just for the safety of those workers involved in running the process but also for the local, regional, and national regulatory authorities in the area in which you intend to do business. Often additional process equipment may be required, such as gas leak detectors, to ensure that no operator precursor exposure is occurring. Special handling stations to prepare precursors for the process use may also be necessary, especially for precursors such as oxidative CVD (oCVD) oxidants necessitating oxygen-free handling.

One last consideration for process precursors is supply. How much of the raw materials will be required for the proposed process and application? Will you be able to buy in bulk to reduce costs? In fortunate situations, the process precursors may be commodity chemicals utilized widely for other applications. In these cases, large volume supply can be relatively straightforward though regulatory considerations associated with shipping should be considered early to avoid significant delays. However, in situations where the new CVD polymer process will be establishing a demand for the volume production of a precursor chemical, additional considerations apply. Early interaction with potential suppliers and understanding of the time that they will require to scale up production may be necessary.

19.1.2 Process Considerations

Precursor vaporization is far from the only parameter necessary for control of a commercial-scale CVD polymer deposition process. While rudimentary vaporization and flow rate control hardware, such as heated crucibles and needle valves, may be appropriate for bench-scale experiments, the required repeatability and uniformity of a commercial process necessitate significantly greater control and automation. Residence time and gas-phase composition are the two most crucial control parameters for any commercial CVD polymer process. Longer residence times generally result in higher precursor conversion rates, which is obviously favorable with respect to process cost. However, depending on the particular CVD polymer process, longer residence time can also lead to increased by-product concentrations or faster gas-phase reaction rates. These factors can cause undesirable changes in film morphology due to gas-phase particulate growth or film contamination. Variations in process vapor composition, due to inconsistent precursor inlet flows or process temperature, can lead to changes in film properties over the course of the deposition. Therefore, the precise control of gas-phase composition through the use of mass flow controllers, along with detailed tuning of the process residence time, is required.

In additional to precise control of process vapor composition, energy and heat management play a crucial role in the uniformity and repeatability of the process. Many of the CVD polymer processes discussed in this text require some thermal input to drive them. Vaporization and sublimation of monomers, thermal cracking of precursors in reactive furnaces to produce radicals, and heated filaments to drive initiation are all examples of significant heat inputs into CVD polymer processes. These must be managed in an automated manner to ensure that process temperatures, heating profiles, and reaction rates are uniform throughout the deposition chamber and do not vary significantly from deposition to deposition. Thermal management of the substrates onto which the polymer is deposited is also of crucial importance. As mentioned in Chapter 2, the surface concentration of adsorbed monomer on the deposition substrate greatly affects reaction rate and is controlled by the ratio of the partial pressure of the precursor to the saturation pressure of said precursor *at the temperature of the substrate*. For research-scale experiments, polymer deposition is often performed on uniform, thermally conductive substrates whose temperature can easily be controlled simply by setting them on a cooled stage. In contrast, commercial substrates are often composed of heat-retaining materials in highly non-uniform shapes. Ensuring consistent surface temperature for precise control of deposited polymer material properties can be very challenging and may require intricately machined cooling systems, shielding of parts from radiated heat inputs, and even gas shuttering to maintain.

All of these factors are of crucial importance to the primary goal of the process, the consistent, repeatable, uniform deposition of a polymer coating possessing defined thickness and material properties. The success of a given process will depend on the ability to provide this control over a deposition area, allowing for the necessary throughput of coated commercial parts. Often, due to the particular demands of a given CVD polymer deposition process for a selected application, custom chamber design is necessary. Further discussion of factors driving custom chamber designs and their attributes, including commercially deployed examples, will be covered at the end of the chapter.

One special case worth mention is the CVD polymer deposition of copolymers. These systems require two or more monomeric precursors, often along with initiating species and one or more inert carrier gasses. Precise process control for these systems can be quite challenging as standard control issues create exacerbated effects in the more complex system. Changes in substrate temperature vary not only the concentration of monomer on the surface but also the ratio of the two monomers due to differences in the change in their vapor pressures with temperature. This effect can quickly lead to undesirable material properties and even deposition of a homopolymer when the copolymer is desired. Another combined effect can be that of carrier gas dilution. If multiple low-volatility precursors are utilized, the combined volume of inert carrier gas can require an increased total process pressure in order to achieve the desired partial pressures of the reactants. Increased gas-phase molecular collisions can then lead to particle growth and poor film properties. Great care must be taken in process design for these systems, especially at scale.

All of these control considerations must also be contemplated with an understanding of who will be running the process. For a commercial process, it is unlikely that personnel will have a high level of technical training with respect to CVD polymer deposition. Understanding of drivers such as substrate temperature effects, thermal management and carrier gas dilution, and the ability to adjust the process to avoid issues, will likely be beyond the skills of the process operator. A high degree of automation is therefore required to ensure process repeatability. The process sequence, temperature ramps, flow rates, and shutdown must all be controlled independent of user input to avoid variations based on operator. Ideally, a push button, computer-driven system should be utilized.

Lastly, and most importantly, the health, safety, and environmental impact of the process must be analyzed and managed. Apart from handling issues associated with the precursors themselves, there are many additional factors to consider. What by-products are produced during the deposition? How will unreacted process gasses be handled? Can the process effluent be released untreated or must it be captured or abated in some way? How will process leaks be detected? If hazardous by-products are produced, what safety measures must be in place for process operators? All of these considerations must be evaluated for the safety of personnel, local environmental impact, and regulatory requirements of the process location.

19.1.3 Application Considerations

When considering the use of a CVD-deposited polymer for particular application, foremost among the concerns must be the material properties required of the coating. Are there special attributes necessary for the application that only CVD deposition can provide? Is conformality over micron-scale features necessary? Or does the solvent sensitivity of the substrate necessitate a dry application process? Perhaps the thickness of coating required is only in the tens to hundreds of nanometers, or a novel, insoluble material is needed. All of these are possible application requirements that may create good candidates for commercial CVD polymer coating. As previously mentioned, the complexity of implementing CVD polymer coatings commercially will necessitate finding applications which require the unique properties that these materials bring to the table. Also, be aware that even with one or more of the above needs, certain application requirements will involve great creativity to work around. Exceptionally large substrates can be very challenging to coat with CVD polymers. In addition, highly reactive CVD polymer systems can have difficulty in coving very high aspect ratio substrates.

Despite these complications, CVD polymers have demonstrated utility on a wide range of commercial applications. Figure 19.1 shows just some of the commercial applications for which CVD polymers present unique advantages in coating. Many of these products including mold release, seals and gaskets, filters, and electronics encapsulation have CVD polymer coating solutions commercialized for them. Many others have CVD coatings in development.

Once the need for a CVD polymer coating has been established, a range of additional factors should be considered. Evaluation of necessary coating thickness is important, especially relative to the deposition rate achieved during process exploration. CVD polymer coatings of 10 nm to 10 μm can readily be achieved, although the upper end of this range is more easily made commercially viable with deposition rates on the order of 100 nm min^{-1} or higher. While commercial processes do exist that take many hours to deposit a coating (parylene, for instance), a deposition time of a few minutes is highly preferable. Size and throughput of the parts to be coated for the application are also an important consideration. Small, high-value parts are obviously the best candidates for

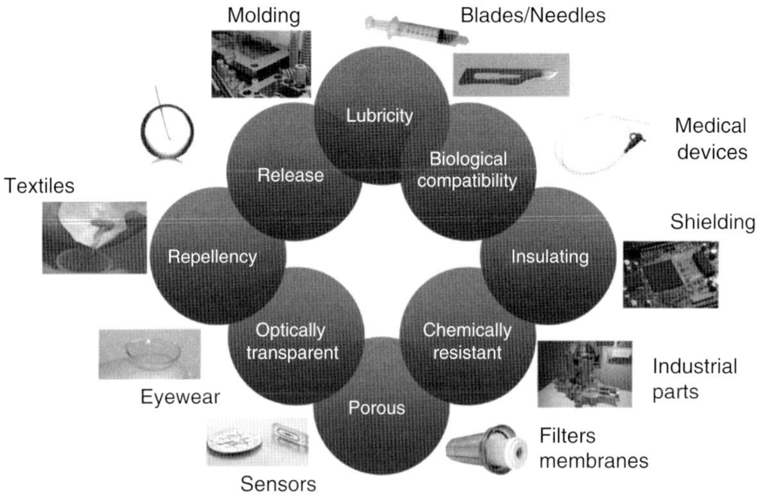

Figure 19.1 Application areas for the commercialization of iCVD coatings. High-value/high-margin products are best, with major commercial success in molding, industrial parts, and filter membranes to date.

CVD polymer coating, although larger industrial parts can present excellent opportunities if the application need is great enough. Anticipated yearly volumes of parts will help with decisions related to deposition chamber size and how many chambers will be required.

In addition to these considerations, many applications will require pre- or post-processing of the parts to ensure the technical success of the CVD polymer coating. Preparation of the surface prior to coating may be needed to ensure good adhesion; this can take many forms. Cleaning may be required to remove contaminants. Surface roughening may be used to increase surface area for beneficial energetic interactions such as van der Waals forces or hydrogen bonding. Chemical modification may be used to change the characteristics of the substrate and improve non-covalent interactions. It may also be utilized to present reactive groups for incorporation in growing polymer chains to provide for covalent attachment to the substrate surface. Energetic treatments, such as plasma, can also be utilized to create reactive groups and covalent bonding. Real-world examples include pre-cleaning to remove lubricant contaminants and silane treatment of metals to present surface groups advantageous to bonding.

One application-specific area that should not be neglected is the required testing of the coating necessary for commercial use. Some industrial and consumer products require no more for commercial approval than the customer-specific testing used to demonstrate the utility of the coating in the application. However, many industries have rigorous established testing standards through which the coating must be qualified before it can be sold. These can take the form of regulatory requirements, established government or customer standards, or standards developed by industry trade groups to ensure coating quality.

Table 19.2 Applicable testing standards for some common conformal coating applications.

Application	Industry	Testing standards
Conformal coating of electronics	Defense	MIL-I-46058C IPC-CC-830
	Aerospace	NASA-STD-8739.1 IPC-CC-830
	Consumer	IPC-CC-830[a]
Biocompatible coatings	Medical	FDA testing – varies by implant type and duration
Lubricious coatings	Industrial manufacturing	Application-specific testing
	Consumer products	Application-specific testing
Hydrophilic coatings	Sensors and microfluidics	Application-specific testing

Highly regulated industries such as medical or aerospace fields can have extremely rigorous testing requirements requiring months, if not years, of qualification before commercial use.
a) If required.

Table 19.2 shows a list of these standards for a few CVD polymer applications. Qualifying a novel material to an established testing standard can take anywhere from a few weeks to years (as in the case of Food and Drug Administration (FDA) requirements) and can be a significant expense in the development of a new CVD polymer coating product.

19.1.4 Market Considerations

Finally, we come to the least controllable of all factors in the evaluation of a new commercial CVD polymer application: the potential market. The size of the market, as well as the projected growth for said market, will play perhaps the largest role in determining the feasibility of commercialization. Sales data related to incumbent coating solutions can be invaluable in deciding if a novel, technically superior solution can be brought to market. An example of such data is shown in Figure 19.2. Many different technologies are currently utilized for conformal coating of electronics for total current materials' sales of greater than \$1.7 billion, as shown in the figure. In addition, the projected growth of this market is extremely health, making it an excellent target for a novel technology solution to enter. Even a limited initial market penetration of a new technology could justify significant capital investment.

When data are not available on the size of the coatings market associated with a potential CVD polymer application, such as when the market is small or highly fragmented, market data on the sales of the product to be coated should be

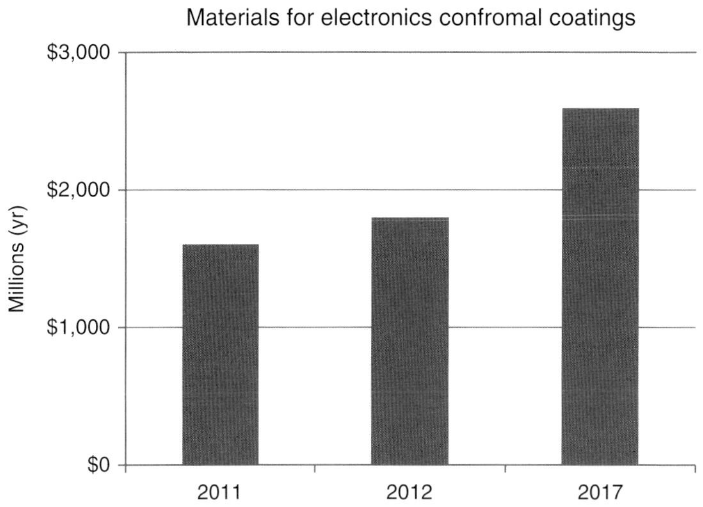

Figure 19.2 Market growth of materials for the conformal coating of electronics. Growth is projected to be steady at over 7% until beyond 2017 [1]. Potential market size and growth are two of the most important factors in selecting a new material to commercialize.

analyzed instead. A good rule of thumb for potential coating sales based on this data is that the coating should cost no more than 5–10% of the item to which it is applied. Obviously, there will be exceptions to this rule. CVD polymer coatings that provide enabling technological attributes to a product may command a significantly higher percentage of the sales price, whereas applications where incumbent coating technologies must be displaced may require smaller margins to gain market penetration.

In addition to the existing market potential, the business model necessary to tap this potential must also be evaluated. Where will coating of the product take place? Both the coating of a few high dollar parts at a central location and the coating of millions of parts at many locations could result in the same total market size, although the former will obviously be much more profitable. Will parts be sent to the coating supplier for processing or will coating systems need to be placed within the customer's manufacturing line? Will the system be utilized by three shifts of operators, 24 h per day, or will it be run once a week? These factors will determine not only how many CVD coaters, and of what size, are required, but also will provide input on how robust a process must be developed before first commercial sale. The manufacture and support of a new semiconductor CVD polymer deposition tool would require hundreds of millions, if not billions of dollars in Capitol, whereas the development of a coating system for use in contract coating small volumes of parts may only be tens of thousands. Knowing the scale of the opportunity, and ensuring the development effort is properly matched to that scale, is the key to commercialization success.

19.2 Case Study: CVD Deposited PTFE for Lubricity Applications

The application of these considerations is best demonstrated through a case study of a commercially applied CVD polymer coating. For this purpose, we examine the CVD deposition of PTFE for lubricity applications by GVD Corporation of Cambridge, MA. GVD Corporation, founded in 2001, is a spin-out from CVD polymer deposition pioneer Prof. Karen Gleason's labs at MIT and is focused on the commercial application of CVD polymer technology across a wide range of industries. PTFE deposition was one of the earliest and most successful products for GVD and continues to be a successful commercial product 13 years on.

19.2.1 PTFE Precursor and Process Considerations

As detailed in Chapter 2, PTFE deposition occurs from the thermal initiation of a gaseous precursor. Both gas phase and surface growth reaction mechanisms are possible, and a significant reaction side product is created during deposition, which must be managed in the process effluent. Based on these data, concerns related to precursor delivery are minimal. Health, Safety, and Environment (HSE) issues, however, cannot be overlooked. Degradation products of the PTFE process precursor can be hazardous, especially with long-term exposure. As such, GVD elected to install parts per million level gas detectors in all locations at which PTFE is deposited. These detectors alarm upon any release of process gas to allow operators to evacuate the active area. A control system can then be utilized to shut down the process remotely. GVD also elected to place all stored process gas and effluent pumping in a separate area, under negative Heating, Ventilating, and Air Conditioning (HVAC) pressure, to further minimize the risk of personnel exposure.

Unlike precursor delivery, process design for the CVD deposition of PTFE presents significant challenges. The competing gas and surface reactions greatly limit the available process envelope for depositing high-quality polymer films. Film morphology, adhesion, and material properties are extremely sensitive to the balance of these two mechanisms and many of the factors described in Section 19.1.2, including residence time, surface temperature, precursor partial pressure, and process energy input, must be properly accounted for to ensure that uniform repeatable coatings are produced. However, once a first principles understanding of process requirements was established, CVD deposition of PTFE has proven to be a remarkably scalable technology. Partially owing to the ease of delivery of the gaseous precursor, GVD has successfully scaled up the process more than 60-fold from early experimental depositions. The low cost of the precursor, a commodity chemical used widely in other industries, has allowed per-area production cost to drop dramatically in larger deposition systems, as shown in Figure 19.3.

Effluent abatement is another significant concern in the PTFE deposition process. As the single gas-phase precursor is cracked, an active radical species is formed that generates the CVD-deposited polymer film. However, for each radical

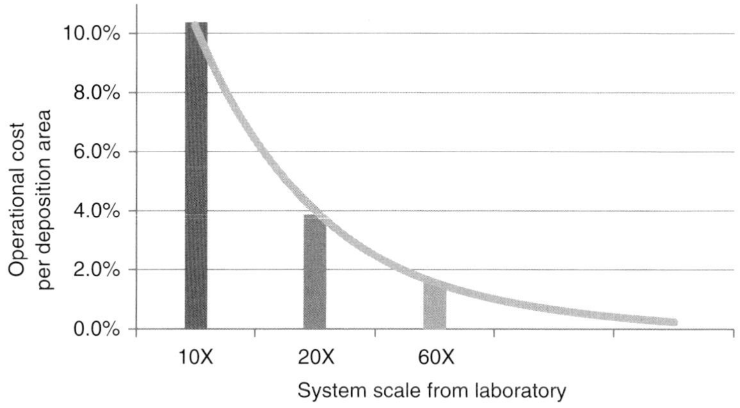

Figure 19.3 Comparison of operation cost per deposition area for iCVD PTFE commercial deposition systems. As chamber deposition area is increased, operating cost drops dramatically versus laboratory scale equipment.

produced, an additional molecule of an undesired side product is also formed. In order to minimize the environmental impact of the PTFE deposition process, GVD installed a scrubber downstream of the CVD polymerization chamber for the treatment of this waste. Through efficient mass transfer and solution-phase reaction, the side product is converted into a non-toxic salt that can then easily be disposed of at a wastewater treatment facility.

19.2.2
Lubricious CVD PTFE Application and Market Considerations

Relative to conventional PTFE, the CVD-deposited polymer possesses certain unique surface properties, while retaining the majority of the bulk characteristics that make the material so appealing. A detailed listing of the relative properties can be seen in Table 19.3.

Table 19.3 Material properties of iCVD-deposited PTFE versus the bulk material.

Material property	Bulk PTFE [2–5]	iCVD PTFE [2, 6]
Coefficient of friction	0.07–0.30	0.03–0.2
Water contact angle	110°	120–135°
Refractive index	1.35–1.38	1.35–1.38
Dielectric constant	2.1	1.7–2.1
Dielectric strength (V/mil)	4500	4000–7000
Density ($g\,cm^{-3}$)	2.2	1.5–2.2

As can be seen from the data, the iCVD polymer retains many of the desirable bulk physical properties of standard PTFE while providing enhanced surface lubricity and hydrophobicity.

For initial commercial deployment, GVD focused on the utilization of CVD PTFE's extremely low coefficient of friction (COF) to provide lubrication across a wide range of industries. PTFE coatings are traditionally sprayed in place using a surfactant/solvent system and then baked on. However, due to the high melting temperature (>350 °C) of PTFE, the substrates to which conventional coatings can be applied are primarily limited to inorganic materials. By contrast, CVD-deposited PTFE can be applied to parts held at room temperature, opening up a large untapped market. By providing coatings that effectively reduce stiction force to zero and frictional force by >90%, CVD PTFE's technical value proposition allowed for significant margins as an enabling technology in industries such as semiconductor manufacturing and industrial molding. GVD has utilized a coating services business model, focusing on high value-added parts in the selection of commercial applications. This in turn allowed for low-cost deposition systems as they are to be operated only by relatively skilled, CVD-trained personnel and do not require the level of development necessary for customer-deployed coating systems.

Figure 19.4 A range of built-to-purpose commercial CVD polymer deposition chambers. These systems are designed to maximize throughput and monomer use for industry-specific coating applications such as seals and gaskets, circuit boards, and mold release.

19.3 Commercial CVD Polymer Coating Systems

The deployment of CVD polymer coatings for commercial applications has necessitated the development of a wide range of coating systems. Although large-scale industrial applications can allow for the utilization of modular systems such as those described elsewhere in this text, niche applications often require systems built to the specification in order to optimize coating quality, repeatability, and cost. Even the deposition of PTFE, despite the cost per area reductions shown in Figure 19.3, can benefit from right-sized chamber designs; running a large chamber to coat a few small parts is still more expensive on the whole. Figure 19.4 shows a number of different CVD polymer deposition system used for applications ranging from coating of circuit boards to industrial molds. As can be seen, chamber geometry, degree of automation, and overall scale vary greatly depending on the process needs and application throughput.

One system of particular note is the roll-to-roll coating system shown in Figure 19.5. This CVD polymer deposition chamber allows for the semicontinuous coating of rolls of membrane or fabric. The substrate membrane is held in tension between the two rolls and slowly drawn across a cooled stage to maintain optimal monomer saturation during deposition. The rate of substrate motion, combined with the deposition rate, allows for a controlled, repeatable thickness of polymer to be applied to the full length of the membrane. If a vacuum load lock system were to be integrated with the rollers, full continuous deposition of CVD polymer on a membrane could be achieved. This would allow for the high volume

Figure 19.5 Prototype roll-to-roll CVD polymer deposition chamber. Fabric or membrane material is kept in tension between tow rollers and drawn at controlled speed across a cooled stage. Coating thickness is controlled by substrate residence time within the deposition area.

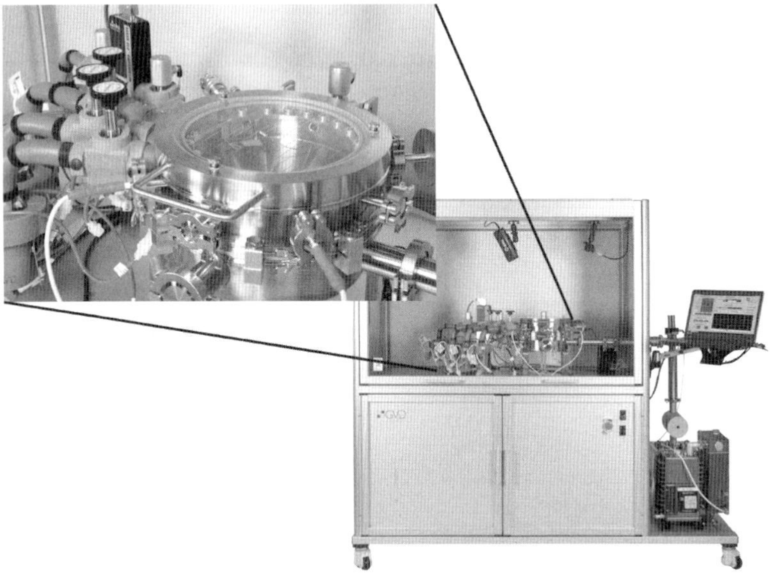

Figure 19.6 GVD's iLab deposition system. This self-contained, computer-controlled, bench-scale system is ideal for the development of novel iCVD polymers. Versions of this system have been installed at six universities throughout the United States and Europe, as well as multiple commercial and national laboratories.

production of CVD polymer-coated membranes or textiles, markets which have been difficult for the technology to access to date.

A final system of note is GVD's iLab polymer deposition system. Focused on the development of novel initiated CVD (iCVD) materials, the iLab is a fully integrated, computer-controlled system available commercially. It allows for novel experimentation in the CVD polymer space by those without a background in vacuum processing or hardware. By providing a flexible, computer-controlled apparatus on which to experiment, this system aims to expand the field of CVD polymer research and increase the future commercial deployment of this technology (Figure 19.6).

References

1. Electronics.ca Research Network (2012) Market report Global Markets for Conformal Coatings in Electronics, July 2012.
2. Kanarth, M.A., White, A.J., and Müftü, S. (2007) STLE/ASME International Joint Tribology Conference, Proceedings of the STLE/ASME International Joint Tribology Conference, IJTC2008, Miami, FL, October 20–22, 2007.
3. French, R.H. (2009) Optical properties of materials for concentrator photovoltaic systems. IEEE Photovoltaic Specialists Conference.
4. Dupont Corporation (1996) Teflon PTFE Properties Handbook, Technical Report H-37051-3, Dupont Corporation.
5. Haynes, W.M. (2012) *CRC Handbook of Chemistry and Physics*, CRC Press.

20
Carrier Gas-Enhanced Polymer Vapor-Phase Deposition (PVPD): Industrialized Solutions by Example of Deposition of Parylene Films for Large-Area Applications

Peter Baumann, Markus Gersdorff, Juergen Kreis, Martin Kunat, and Markus Schwambera

20.1
Motivation and Targets (Customer Requirements)

The formation of polymer films has seen increasing interest, as this approach of forming polymer thin films offers a number of advantages compared to the deposition of polymer thin films through solution-based wet-process approaches.

In recent years, a lot of interesting thin films have been demonstrated as possible candidates for gas-phase deposition with their specific advantages depending on the function of the respective material used.

While many people are quite familiar with the more common wet-process coating approach for polymers, deposition from gas phase and formation of a polymer thin film *in situ* by having the respective chemical building blocks being supplied from vapor phase and then being bonded to polymer chains by initiating a chemical reaction is rather new. A more commonly known approach is the formation of polymers by plasma enhanced chemical vapor deposition (PECVD) processes, where layers consisting of short molecular chains are exposed to a plasma source, and thus exposing them to energy that will create open molecular bonds. Then the polymer film is created by the rebonding of smaller molecules in forming longer molecular chains.

Various research institutes and companies have been working in the field of PECVD for the deposition of polymer thin films. A considerably smaller number has been investigating alternative approaches to form polymer thin films from vapor phase.

The Gleason group at MIT has reported interesting results for the formation of various functional polymers by using processes that are referred to as iCVD (initiated chemical vapor-phase deposition) or oCVD (for oxidative chemical vapor-phase deposition), depending on the formation mechanism (i.e., the respective polymerization path) [see Chapter 1].

Given the interest of researcher teams, the respective possibilities were mostly demonstrated on R&D level using smaller proof-of-principle implementations, thus allowing investigating the underlying processes at reasonable efforts.

CVD Polymers: Fabrication of Organic Surfaces and Devices, First Edition. Edited by Karen K. Gleason.

In order to take the respective applications to industrialized solutions, the processes need to be scaled, and, more importantly, the process technology being employed needs to be engineered for industrial use and throughput.

Organic precursor materials, due to their high reactivity and good absorption by organic tissue, must be considered as being potentially hazardous. This will directly lead to enhanced safety precautions to protect people being involved in such production processes as well as the respective environment, in which such processes are planned to be used.

Also, it should be emphasized that for industrial application of a process, it cannot be expected that the typical operator must have a university degree in chemistry. Hence, the requirements for a process solution on industrial scale are considerably more stringent than what typically is used if a system is used for proof-of-principle demonstrations in dedicated research facilities.

Requirements for ease-of-use, economies of scale, efficient utilization of resources, operator safety, intrinsic safety features such as design-in of fail-safe operation procedures, serviceability, and, most importantly, high productivity and reliability of a process are becoming of paramount importance for a production-ready implementation.

Therefore, taking a certain process implementation from lab scale to industrial relevance requires complete shift of priorities for the design criteria being used.

In the following sections, we explain and discuss these requirements, and, exemplarily illustrating the approach with the example of an industrial solution for Parylene deposition with an outlook to other applications and processes following the same logics.

20.2 Requirements for Industrial Solutions

Trying to take a process from an R&D level to an implementation of industrial relevance – marketing people often like to call this "from lab to fab" – requirements and design criteria for the respective equipment are significantly different.

For an R&D system, the focus mostly is on general demonstration capability, and experienced users with in-depth understanding of the processes happening in side of the tool are the more standard than an exception. For industrial realizations, other properties such as

- Ease-of-use
- Uptime
- Throughput
- Scaling effects
- Total cost of ownership (which includes operation costs, maintenance, and service requirements, material consumption as well as tool costs)
- Safety issues such as fail-safe operation
- Flawless process flow integration.

do become of paramount importance.

Table 20.1 Market requirements for production systems.

Market requirements for industrialized solutions			
Item	**Description**	**Impact**	**Comments**
Scalability	Technology scalable to enable mass manufacturing	Cost of ownership	Scaling must maintain key process parameters
Resource utilization	Efficient resource utilization. High output conversion factor	Cost of ownership	Consumables can easily exceed the cost of a certain system. This becomes specifically relevant, if output is scaled up to levels of industrial relevance
Ease-of-use	Easy and safe to use	User and fab manager acceptance	Systems should be easy to use to allow higher flexibility with respect to using different operators without extensive shop-floor training
	Low efforts to teach new operators	Cost of ownership	
Modularity	Modular system design	Lead times	Modular system design addresses shorter delivery times and scales of economy. Extensibility increases versatile utilization of an equipment
		Extensibility	
Maintenance	Long maintenance cycles and robust machine design to address high uptimes	Cost of ownership	—

In Table 20.1, we have listed some of the market requirements that production-type process equipment generally must fulfill.

While for an R&D system, ease-of-use is not necessarily an issue of major concern, user-friendly operation has a strong impact on the acceptance at operator as well as fab manager level. Training costs are a non-negligible part of the operation costs, and an easy-to-operate tool is much better suited to assign non-specialist operators. User-friendliness goes along with reduced complexity of maintenance, and the easier a system is to maintain at production-ready conditions, the more likely it will be accepted as an enabling process step.

Most times, as material consumption on R&D level is of no concern – the size of the tool as well as the potential throughput will not present a root cause of not utilizing a process on this level – for a production tool the material consumption easily can be the major cost driver, especially as for sophisticated processes the respective materials usually also come at a considerable price. Inefficient utilization of the material then directly translates into higher operation costs and ultimately elevated total cost for the respective process step. On top, inefficiency directly contributes to waste generation and, in the case of hazardous materials, in serious overhead costs for respective safety measures and post-treatment requirements.

For fabrication of sophisticated products, a manufacturing line usually integrates a number of process steps. Therefore, the flawless material flow from process step to process step is the key for efficient manufacturing with respect to cost, manpower required, and throughput. Hence, any subprocess being part of a manufacturing process is ideally perfectly embedded so that components move from process step to process step without any operator interference required and at processing times that avoid speed variations for the material flow.

This requirement adds the need for automation to the overall complexity of a production tool. For embedded process chains, the throughput of the production process is determined by the slowest process step. Hence, while specific processing times can be ignored as long as process steps are decoupled, for integrated manufacturing processes each subprocess must operate at highest possible throughput as otherwise it will become the intrinsic bottleneck. Different process times, of course, can be compensated by doubling certain process steps. This, on the other hand, would directly translate into higher investment costs, and therefore does not help to lower the total production cost per unit.

Finally, cost of ownership also is strongly influenced by the uptime of a tool – the longer it can operate at expected productivity levels, the lower the respective cost share of this specific process step will be. This understood, production-ready solutions must be designed with long-maintenance cycle times and high uptime being part of the core functional requirements.

Parylene is a good example to discuss the differences between a proof-of-principle solution and an industrialized solution, allowing cost-efficient mass manufacturing with high throughput.

20.2.1
State-of-the-Art Solutions for Parylene Deposition

Parylene is the generic trade name for a wide range of poly (*p*-xylylene) polymers deposited by CVD methods. The polymer was discovered in the 1940s, and, due to its versatile properties, these days are used in various fields, for instance, in the fabrication of electronic devices.

Depending on the specific derivatives of parylene, which mostly differ in their polymer side chains, properties of interest are

- good barrier characteristics;
- good dielectric properties;
- hydrophobic surface characteristics;
- chemical resistivity against fluorinated hydrogens (e.g., gasoline and diesel);
- biocompatibility;
- oxidative and UV stability;
- good adhesion to a wide range of substrate materials, and others.

Parylene films, in general, do show excellent contour conformity, thus making them ideal for protective overcoats.

The most widely used process to form parylene thin films is the Gorham process, named after its inventor Gorham who proposed this approach in the 1960s [1].

With this approach, parylene is formed in a two-step deposition process on an object put under vacuum. To start with, dipara-xylylene is vaporized under vacuum and exposed to a pyrolysis process between 450 and 700 °C. This pyrolysis step cracks the double-ring structure of the xylylene molecule and forms a *p*-xylylene intermediate product, which is able to react with siblings and from there to form the respective parylene polymer chains on surfaces.

The deposition is driven by physisorption, and therefore the resulting rate depends on the substrate temperature of the surface to be coated on. The reaction process is highly selective with respect to well-balanced combination of deposition temperature and pressure, which makes a good process control key for efficient fabrication methods.

For Parylene coating, most solutions are straightforward derivatives from the Gorham process and are either "homemade" vacuum systems that may fit for small scale lab test volumes or pilot run capacities at the best, or, as commercial type lab-coaters implemented as larger vacuum chambers available from various suppliers.

In both cases, these standalone solutions will require an experienced operator to deal with a number of manual interactions required for proper operation. Some of these interactions are material and substrate loading and unloading, manual selection of processing steps (going to vacuum level and venting the system, activation of furnace and vaporizer, cleaning of system components, and so on), how to employ detailed cleaning procedure at end-of-use, and so forth.

For R&D systems, usually processed devices are small. For a proof-of-principle of a certain process, small substrates are totally sufficient to show the general properties of a new material and the underlying process.

These boundary conditions will change when it comes to production requirements, where lowering manufacturing cost through scale of economies is one of the key requirements. Hence, process equipment always has to be designed with potential cost savings through larger scale implementation being a core criteria. At the same time, scaling must be possible without compromising on process features and performance of the devices being produced. Moreover, an easy transfer from system to system is a key requirement to allow processes to be developed or improved on smaller scale pilot lines and then being transferred to mass manufacturing without significant changes on the overall process control.

As described before, parylene is formed in a rather simple reaction involving only one singular precursor material; for other chemistries, complexity is added by the need of combining different components during the reaction process and the need for a controlling start and termination of the reaction process. While examples are explained in detail in other chapters, general implementation requirements for production-type solutions shall be mentioned here.

If different materials need to be introduced into the process space through separate channels to avoid unwanted mixing or recombination away from the desired process location, an implementation on small scale still can be considered easy to make as the diameter of the feeding tubes can be considered as relatively large when compared to the total deposition area, and hence uniformity of the precursor distribution can be obtained by simple means of symmetric design. If the deposition area gets larger, the technical challenges of maintaining good uniformity are increasing. Sophisticated gas distribution systems, which can ensure proper feeding as well as good balancing of the respective precursors will add to design complexity.

If gases have to be distributed over larger areas, the need for well-engineered gas distribution solutions recommends comprehensive and systematic design optimization through computational flow dynamic (CFD) simulation and other computer-aided design tools.

Furthermore, controlled initiation and termination of chemical reactions, in particular the need to introduce reaction energy in a controlled manner, adds to the design complexity of a solution when processes are being scaled.

A common method to introduce reaction energy for polymerization processes is the application of plasma. Plasma has the advantage of using high-energy densities without increasing the process temperature too much, which, especially for organic materials, always must stay under the specific decomposition limits of the respective component.

A plasma source can be embedded by relatively simple designs as long as the processed devices and, with that, the reaction chamber remains small. If the target deposition area becomes larger, scaling of a plasma process, too, requires for sophisticated design optimization for the plasma generator itself as well as uniform plasma distribution.

The scaling challenge is present for other technical means of energy infusion as well. If, for instance, a hot filament shall be used as it is the case of iCVD processes (refer to chapter 1.4.1, page 8ff), this feature, too, must be scaled with the process chamber without compromising its performance.

20.2.2
Impacts of Process and Chemistry on the Design of an Implementation

There are different types of parylene dependent on the substitution of hydrogen. The different parylene variants differ in physical and chemical properties and are

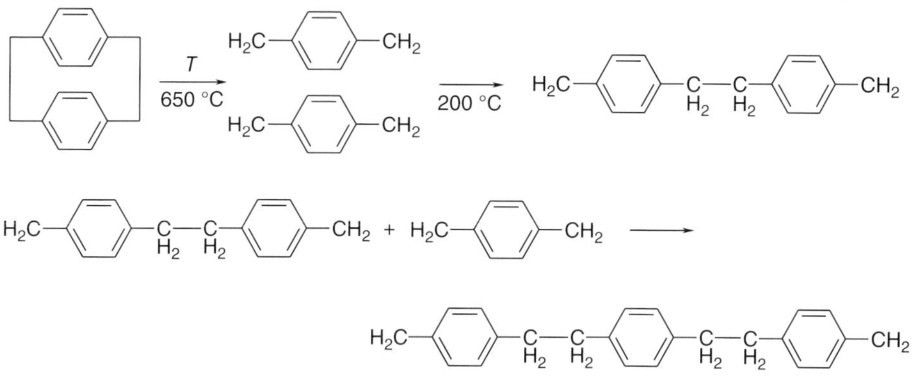

Figure 20.1 Reaction mechanism of parylene.

selected depending on the individual application. An overview can be found here [2].

The polymerization of parylene is based on cracking of the dimer at elevated temperatures of about 650 °C. By this pyrolysis step, one dimer molecule is transformed into two di-radicals. At lower temperatures below 200 °C, these radicals will react with each other again and form the next greater di-radical and in the end parylene polymers. The reaction equation is noted in Figure 20.1. The mechanism itself is a simple recombination of radicals. No further activator molecule is required.

Critical parameters are the temperature and the process pressure during the pyrolysis and within the polymerization chamber.

The thermal energy provided by the pyrolysis step is required to break the C-C bonds to form the di-radicals. The pressure during this reaction defines the mean free path of the molecules.

At too high pressure, the pyrolysis step will not be complete, as the di-radicals start to recombine again. At too low pressure levels, the residence time within the pyrolysis step will be too short for an efficient cleavage of all dimers. At low pressure the concentration, too, will be reduced, which then impacts the achievable deposition rate as well.

Apart from these process aspects, the technical effort to generate and maintain low pressure at given flow rates also increases with the size of an implementation. Once again, the engineering task is to achieve an optimum in order to obtain the most efficient process solution.

Due to the reaction kinetics, lower temperatures promote the polymerization process of Parylene. Above 200 °C, there will be almost no polymerization happening. This effect can be positively utilized to avoid unintended deposition on the process chamber walls by heating them above certain levels. By cooling the substrate, the polymerization rate, on the other hand, can be strongly accelerated. One major challenge in the design of the respective process system is to maintain a uniform temperature profile across the substrate.

An additional challenge lies in preventing the unwanted premature formation of polymers already within the gas phase instead of building polymers only on the substrate. This effect has to be avoided by proper concentration levels as otherwise it would lead to particle formation and rough film morphologies.

An important aspect that cannot be overestimated is the need for proper selection of the materials that are used to build a process solution. Compatibility of system components themselves with the process chemicals is of paramount importance to reduce undesirable side effects.

While this topic is present on R&D scale as well, the total amount of material is usually much smaller, and therefore the potential impact is much less pronounced as well. Therefore, there might be cases where, on R&D scale, no negative effect will be observed, but during scaling of the process to mass production level with 24 h/7 day operation, chemical reactions with surfaces such as the chamber walls of the process system will cause significant problems. Examples for such unwanted side effects could be memory or tailing effects, which then would reduce the controllability of the process to great length or might cause the unwanted maintenance need of regular exchanges of components, this dragging down the possible uptime of a system.

20.2.3 From Process Engineering to System Engineering

As mentioned earlier, understanding the driving forces of a specific process is of paramount importance for proper system design. The process parameters that influence the chemical reaction must be understood to an extend that implementation variants can be compared, and the design is optimized to yield maximum system performance. Hence, knowing these process parameters is the basis for designing and optimizing the respective process solution.

For gas-phase process solutions, key parameters to be understood are temperature, pressure, and concentration of the involved chemicals.

Generally, in a gas-phase deposition process, the film formation can be divided into three distinguishable steps.

First, chemicals need to be transferred into gas phase in a desired amount and concentration. Second, the chemicals need to be transported to the target deposition area. And third, the chemicals need to react and form a uniform film on the substrate.

Depending on the process, either between first and second step, or between second and third step, there is an activation impulse needed, which usually means the injection of additional energy to start the reaction process.

With respect to an industrial solution, which is easy to operate and maintain for continuous production, a modular realization of each step is a fundamental design principle.

The first step – the provision of chemicals by a source system – does not only need to deal with the evaporation of chemicals by means of controlling temperature and pressure, but also must reflect aspects such as stable deposition rates, avoidance of destructive material handling, for example, by overheating, and material feeding and provision. The later aspect is of great importance for 24 h/7 day operation as every outage in process readiness drags on uptime and ultimately increases the cost of operation.

In the worst case, not only the respective process is coming to a halt, but the whole production line is affected, if such a halt also impacts preceding or succeeding production steps as it is usually the case in an automated environment.

A stable deposition rate might be considered as not that important if only one chemical is to be processed and if only the final film thickness is of interest. Even then, as mentioned earlier, different deposition rates still can have an impact on the morphology of the film. If the functionality of the thin film is influenced by the morphology, then improper rates might have a direct impact on functionality of the fabricated device itself. This often will be the case where such thin films do have an interface function, and their functionality is defined by the interaction with other adjacent layers or surfaces. Unprecise rate control then makes quality control difficult if not impossible.

Rate control becomes of crucial importance when two or more different chemicals are to be mixed with precise concentration ratios. Then, well-defined stable rate of each injected precursor chemical is mandatory. If this rate control cannot be maintained, concentration ratios will change and may affect the desired polymer functionality or prevent the proper formation of the functional film right away.

One solution for proper rate control during the first stage of the process, the feeding of precursor materials by the source system, can be the mixing of the respective chemicals with an inert (non-reactive) carrier gas. Then the partial flow of the respective carrier gas concentrations can be used for controlling gas-phase transport and the respective concentrations.

While this carrier gas itself may influence the transition of the respective chemical from liquid or solid state to gas phase and therefore must be taken into consideration when designing a system, it allows for precise flow control by using mass flow controllers. This approach enables the amount of transported active species being precisely controlled by the flow of the carrier gas.

The underlying control mechanism of using carrier gases is much easier and much faster than controlling the temperature and simplifies the scaling of the process. However, even with this enhanced control feature, temperature and pressure still need to be kept as stable as possible.

Basically, by setting process temperature and process pressure within predefined ranges, a certain process window is defined, while the carrier gas flow itself then defines actual deposition rates for each precursor. Having all these parameters under control, the final deposition rate can be precisely controlled.

The second aspect for the system design, referred to as step 2 earlier, is the transportation of the process gases to the desired location for the actual process reaction. This aspect needs to be optimized as well.

In cases where thermal energy is required to evaporate the chemicals, the whole transport lines need to be heated in order to avoid any condensation by means of cold spots. Here, eventual design flaws would lead to reduced or instable deposition rates or tailing and memory effects for materials being absorbed and desorbed on the surfaces depending on the partial material concentration in the gas flow over time. Even in cases, where it is not possible to fully avoid these side effects, the carrier gas approach can help to minimize the resulting performance drop such as slower than desired system response time by introduction of run-vent stacks into the supply lines. Using a run-vent stack, which essentially allows switching the gas stream to bypass the deposition chamber, dedicated valve operation ensures that only a stable material flow reaches the deposition area. Start-up or tailing effects can then be routed to the vent line. With the carrier gas being used as transport medium it also reduces residence time of the active species within the tubing. Interactions with walls are significantly reduced, and sharp interfaces between different materials are possible to be formed.

The distribution of the precursors at the desired location of film formation, formerly referred to as third optimization step, decides about uniform and efficient material distribution over the target substrate.

In the literature, several approaches to achieve uniform distribution with minimized material loss are described. One solution is the utilization of close coupled showerheads. The showerhead, essentially a 2D assembly consisting of many holes and gas flow dynamics optimized for uniform gas velocities across all exit ports, is mounted in direct and close sight of the substrate being exposed to the process gases. By computer-aided design, using the principles of CFDs, the geometry of the nozzles, the showerhead distance to the substrate, and the gas flow itself are optimized to realize the best possible combination. In order to achieve the formerly listed performance criteria, but in particular a uniform gas distribution, the underlying computations must take process pressure, gas concentrations, and temperature into account. In particular, the flow field, the velocity distribution, the pressure drop over each component, and the temperature profile of each component are of paramount interest and need to be optimized. The ideal design represents a multidimensional optimization task as optimization of one item might directly affect the other items in a negative manner. Hence, iterative optimizations are required to address the design goals according to their priority. Examples for such iterative analysis and optimization runs are given in Figure 20.2a,b.

In Figure 20.2b, the uniform pressure drop over all nozzles can be seen. An additional design target could be to achieve a uniform flow pattern without leading to recirculations of the gases or areas forming turbulences.

Here, once again, the carrier gas-enhanced process provides certain advantages. In comparison, in a deposition process that is purely effusion driven, i.e. deposition rates are determined by temperature and concentration gradients only, obtaining uniform flow patterns of the active specifies and with that uniform film properties (thickness and morphology) should become increasingly difficult with larger substrate sizes.

As mentioned earlier, the activation of the reactants – if required – can be done between each of the aforementioned three steps. However, due to the chemical

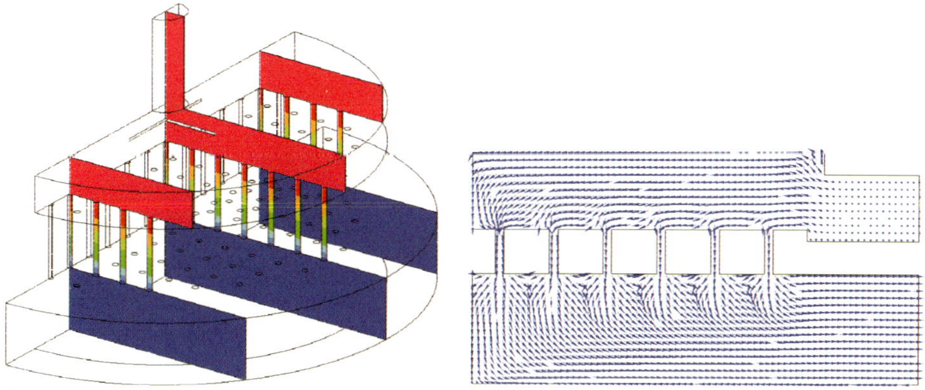

(a) (b)

Figure 20.2 Example for iterative numerical analysis and design optimization by CFD simulation. (a) Pressure distribution within a circular showerhead (blue area belongs to the process chamber). (b) Flow field analysis of a Gen1 showerhead.

reactivity and/or the lifetime of the active species, a position of the activating mechanism close to the target is usually preferred.

It should also be noted that the substrate itself can play an important role with respect to deposition rate and film uniformity. Especially the temperature often has a strong impact on these parameters, depending on the reaction kinetics. Therefore, understanding the reaction kinetics decides whether a substrate needs to be actively cooled or heated and how important this parameter might be. Then the quality of the substrate handling also determines the film quality. The better the uniformity of the temperature profile over a substrate area can be maintained, the better the uniformity of the deposited film then will be.

For polymer vapor-phase deposition (PVPD) processes, the deposition pressure is in the range of 0.1 mbar. This needs to be considered for the temperature management of the PVPD system. The surfaces along the gas flow path including the evaporation source, the showerhead, and the walls of the process chamber need to be heated in such a way that the precursors remain in the gas phase and do not cause parasitic deposition. Only the substrate needs to be kept at a relatively low temperature to enable deposition.

Thermal models have been used to calculate the temperature distribution in the process chamber that includes the hot showerhead and process chamber surfaces, as well as the cold susceptor with the substrate. For this purpose, heat distribution by forced convection (natural convection does not play a role at the low pressures), heat conduction, and heat radiation has been considered. Special attention has been given to heat transfer at contact areas and narrow gaps between different parts. If the mean free path is in the range of the size of the gap, the heat conduction of the gas in the gap becomes insignificant because of the low pressures. This will lead to effective thermal insulation in areas of low thermal contact. Figure 20.3 shows the calculated temperature distribution for a cross section of the

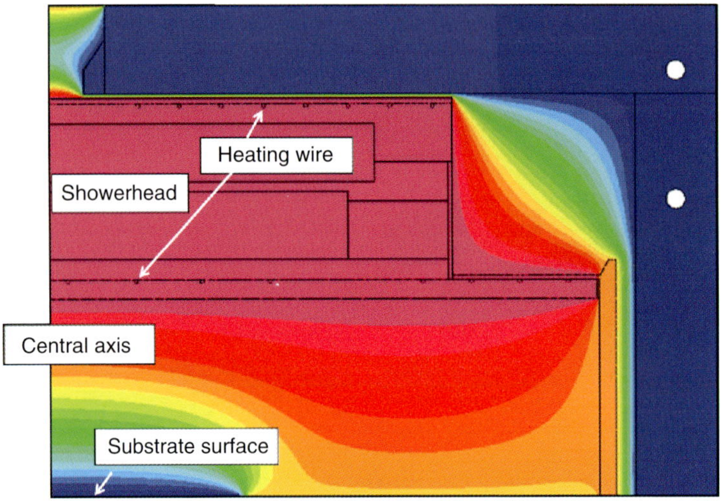

Figure 20.3 Calculated temperature distribution for a cross section of a PVPD process chamber.

PVPD process chamber. The heating power was adjusted to achieve the necessary temperature distribution.

The simulation shows that the hot showerhead can be thermally insulated very well from other surfaces in direct vicinity. In particular, it can be shown that the temperature distribution on the substrate is uniform.

In case of a parylene process, again, this parameter has major influence on film thickness uniformity.

For an industrial system design, parasitic deposition at unwanted locations in the process needs to be considered. Although parasitic deposition obviously reduces material utilization efficiency, it might also increase the need for regular maintenance. If, in short intervals, the process chamber and/or the vacuum system does require such maintenance cycles, it adds significantly to manufacturing costs. While additional design features such as the integration filtration system can help to protect sensitive devices such as system pumps, it should be noted that filters generally reduce the efficiency of these devices itself.

Parasitic deposition can be minimized by choosing proper process parameters and optimized deposition chamber design. In cases where these measures are not sufficient on their own, extra cleaning runs by chemical or physical etching, for example, by plasma cleaning, are to be foreseen, which then requires technical solutions to implement these capabilities. Such additions, on the other hand, must avoid negative impacts on the core process to be realized.

In the worst case, where self-cleaning approaches are not possible, parts must be foreseen for replacement during regular maintenance. In these cases, an upfront cost-optimization task for these components must be part of the system design.

Finally, all these aspects need to be considered from the beginning before a system is released for production.

Otherwise, countermeasures to overcome such oversights have to be implemented either at the last minute or even after a system has been put into the field will mean significant costs – on the supplier side and on customer side alike.

In the context of production solutions, the aspect of scalability is of enormous importance. From a customer point of view, even small-scale R&D equipment or pilot equipment already should obey the same design rules as a scaled-up version will use.

Otherwise, a certain process that can be proven on lab scale fails to be transferred to production scale because of different design principles. This would turn the developed process into a non-useable process.

This prerequisite also asks for the concept of a modular design approach. By this process, each component can be developed fairly independent from the others, and the replacement of an individual subsystem or detail does not affect other components. During the design phase, the requirement for iterative optimization of the overall system design is rather likely. Following the modular optimization approach, the risk of designing a system with unsatisfactorily performance or not being upgradable can be reduced.

It was highlighted earlier that, in comparison to small proof-of-principle implementations, production systems require a high level of integration and automation. To prove the feasibility of a certain process on small scale, on the one hand, and also to serve as convincing ready-for-production concept to customers, essentially the same capabilities for automation and integration need to be demonstrated on small-scale systems. In particular, this asks for computer-controlled process automation with integrated safety logics and a modular system interface to automation components. The human–machine interface itself needs to be operator friendly and fail-safe to receive acceptance with potential users.

While the control system allows to set and monitor all process parameters such as temperatures, pressures, flows, and the status of valves and any other actuators, all signals are evaluated by the safety system, which is capable to prevent or interlock critical operations. As an example, a safety system should independently recognize that opening one source valve while the process chamber is open due to maintenance is a condition to be prevented and must prevent such operation modes. Such measures are crucial to prevent potential misoperation and potential risk to operator and facilities.

In the design of typical safety systems for industrial solutions, there are at least two safety levels. The lower level handles machine safety aspects, and the higher level, often called performance level, takes care of human safety aspects.

Integrating such subsystems in a production environment is of paramount importance, as operators with basic training levels might be in command of a system where potentially hazardous materials, under pressure, over pressure, electricity, hot surfaces, and other dangers might be present.

Apart from increased safety regulations, also ergonomic aspects are becoming more important while approaching mass production. These aspects have no direct effect on the process itself but have an impact on system design as well as on operator acceptance.

20.2.4 Design Principles – Modularity as Enabling Criteria for Industrial Solutions

It has been explained that process equipment for CVD processes consists of a number of building blocks, which need to be designed in concerted efforts to obtain the desired performance and user-friendliness of a solution.

Subcomponents such as the reaction chamber, the gas delivery and distribution systems, energy injection methods (plasma, hot filament, UV sources, and so on), substrate handling, temperature control, safety features, just to name a few, all need to play together to obtain a working solution.

For industrial scaling, each of the functional blocks follow different scaling rules.

If a clear separation between the functional blocks can be maintained, industrialization tasks such as scaling, realizing scales of economies to reduce overall operation costs, performance optimization, and so on, all can be addressed item by item. The application of a modular system designs is clearly beneficial for achieving these goals.

On an abstract level, the functional blocks of a deposition system for gas-phase polymerization can be identified as shown in Figure 20.4. The respective clustering of functionalities can be used for optimizing each functional block independently.

Moreover, functional blocks that follow different scaling rules may be utilized throughout a range of different implementations.

As one example, a source delivery system, which, due to its specific design offers a wide dynamic range of possible feed rates, can be utilized throughout different implementations with respect to size. Hence, without adding additional costs of adapting this specific component, cost savings in the manufacturing of different production solutions can already be realized by utilizing the same component for different system implementations.

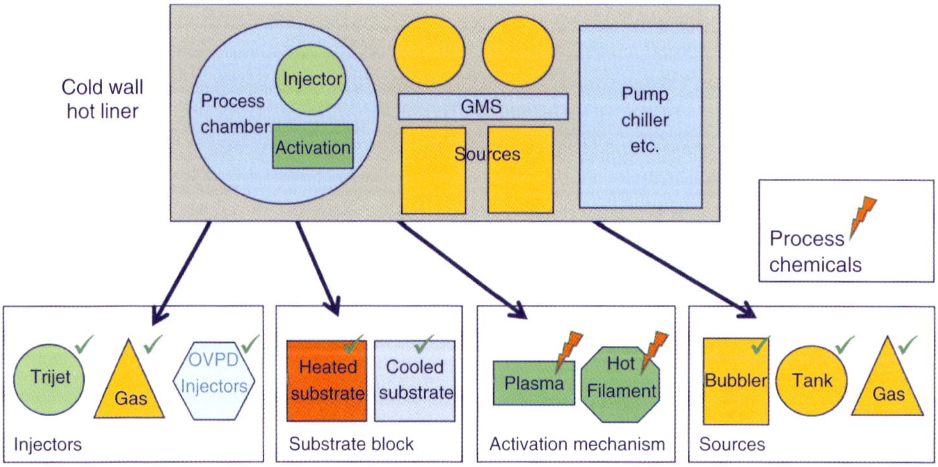

Figure 20.4 Conceptual representation of functional groups for an industrialized CVD system.

The conceptual separation of functional groups also simplifies the assignment of design tasks to independent engineering teams, thus potentially helping to shorten the overall design phase of a system by decoupling tasks.

Modified implementations then only ask to re-employ those teams that are responsible for the specific component due for adaptation.

A modular system concept is also reflected in a modular software concept, where the hardware modules are dubbed by the corresponding software subsystems. The user interface then can again be designed as an independent yet integrating part of the overall concept while system control units are designed to interact through well-defined interfaces between the different software modules. Scaled versions can utilize same interface concepts and building blocks independent from the specific realization.

Safety concepts can be derived in the same modular manner, thus ensuring that safety principles are ensured independently of the specific implementation. A small system then offers the same overall safety features as the "big brother" on the production floor.

As similarities in operation concept as well as look and feel usually simplify the operator training, additional cost savings can be realized by shorter ramp-up times.

20.2.5 Building Blocks – A Closer Look

20.2.5.1 Process Gas Supply and Mixing

Figure 20.5 shows a typical gas mixing schematic of a production system for Parylene. Similar schematics are usually designed for all system scales. Without diving into the details, the schematic shows the routing of tubes, position of valves, mass flow controllers, pressure controllers, pumps, and other gas flow devices and essentially visualizes the gas flow logics. In addition, interfaces to the surrounding lab/fab environment are shown and named.

The schematic can be divided into three functional sections. The upper section is related to the gas in-feeding and the source systems. In the middle part, the process chamber can be found and on the bottom there is the vacuum and exhaust system, which was not explicitly mentioned earlier.

As can be seen in this example, there were four identical sources attached in parallel. This feature was used in this specific implementation to avoid downtime by switching to alternate sources with same precursors while refilling one of the sources. So, while one source is under maintenance or is being refilled, another source can keep the process running. Of course, two or more sources could also be used at the same time to increase the deposition rate.

20.2.5.2 Gas Distribution – Showerhead Concept

A showerhead concept is useful whenever it comes to the requirement of uniform distribution over a flat surface area.

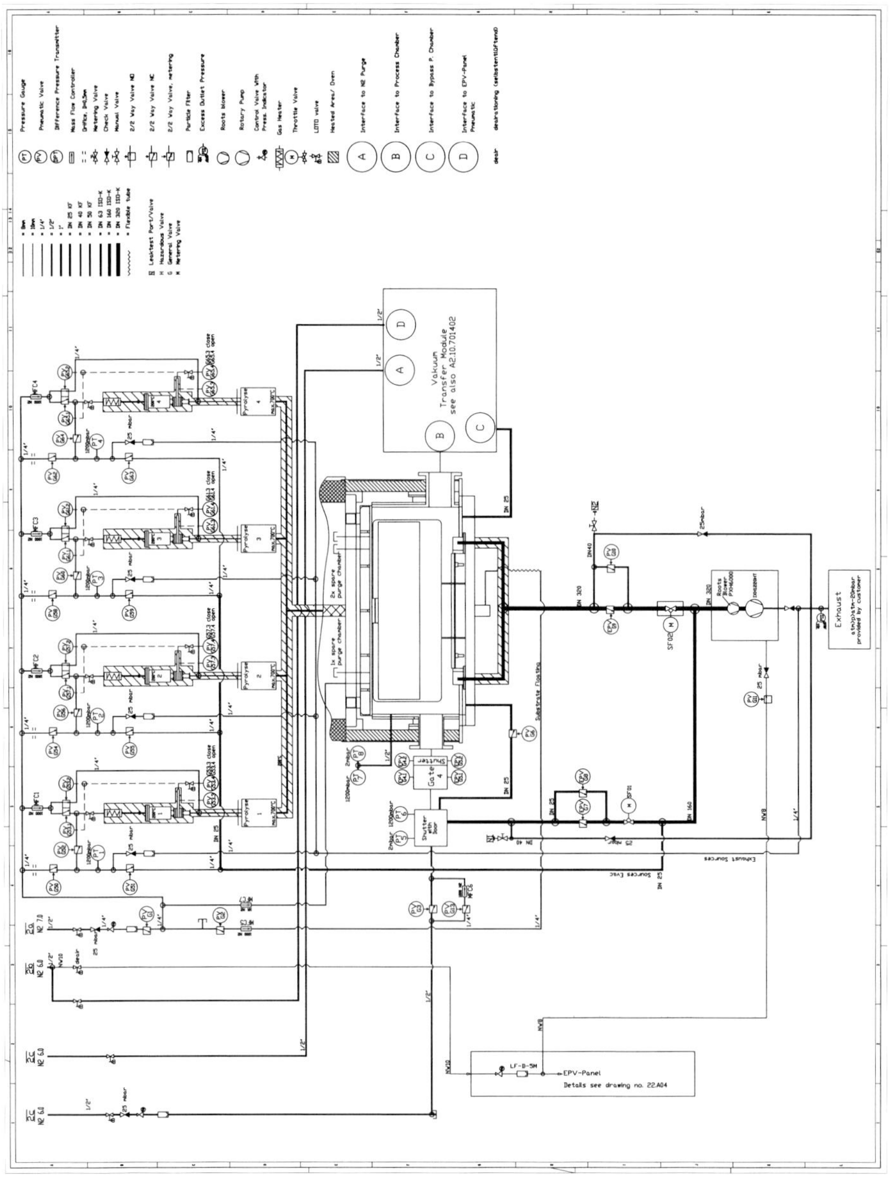

Figure 20.5 Gas flow schematic of PRODOS-Gen3.5 production system here: configured for parylene deposition.

Hence, for thin-film deposition processes, showerheads can simplify the scaling challenge by offering a method to transfer a process qualified on small area to a larger.

20.2.5.3 Modular System Design

Figure 20.6 Example for modular system design: gas supply cabinet (left) and reactor (right) implemented as modular trolleys.

Table 20.2 Example for modular system design.

Source system	Distribution system	Substrate stage	Post-treatment	Activation
For solid materials	Single-channel showerhead	Heated stage	None	Hot filament
For liquid materials	Dual-channel showerhead	Cooled stage	Filter	Plasma
For gaseous materials		Passive	Scrubber	UV light
		Roll-to-roll	Cold trap	Temperature

Functional building blocks on a PRODOS PVPD^{TM} system.

20.2.6 Results Example High-Throughput Deposition (e.g., Parylene)

Deposition of parylene-C has been performed in a PRODOS Gen 3.5 PVPD™ production system for substrates of up to 610 mm × 720 mm.

To verify the stability of the PVPD PRODOS system, the temperature stability of several parts of the reactor has been tracked over 24 h. The measured data included critical components of the system such as the evaporation source, the pyrolysis unit, the showerhead, and the chiller.

PVPD PRODOS system temperature stability measured over 24 h:

$$\Delta T \text{ Source} : \pm 0.1\ ^{\circ}\text{C}$$

$$\Delta T \text{ Pyrolysis} : \pm 0.1\ ^{\circ}\text{C}$$

$$\Delta T \text{ SH} : \pm 0.1\ ^{\circ}\text{C}$$

$$\Delta T \text{ Chiller} : \pm 0.35\ ^{\circ}\text{C}$$

Excellent system control and temperature stability could be shown, which is a prerequisite for reliable performance of the PVPD system and reproducible deposition runs under production conditions.

For a series of 195 deposition runs, the material utilization for parylene-C has been determined on a 590 mm × 710 mm surface area. Typical values of about **27%** have been achieved with good run-to-run reproducibility, as can be seen in (Table 20.3) below.

To check the deposition quality within one substrate, the thickness uniformity was measured on a 590 mm × 710 mm surface area with 12-mm edge exclusion. The resulting thickness of the measured film (**81** measurement points) showed a deviation of $\sigma = \mathbf{2.5\%}$.

The morphology of the substrate and the parylene-C films were measured by atomic force microscopy (AFM) (see Figure 20.7).

Using 27 measurement points, values of the roughness of 0.93 and 1.70 nm have been obtained for the substrate and film surfaces, respectively. Overall, it could be demonstrated that smooth film surfaces could be achieved, in which functional thin films relying on their interface properties is crucial.

Table 20.3 Material utilization parylene deposition.

Run number	Deposition rate time (s)	Measured average thickness (nm)	Dep. rate (nm s^{-1})	Layer volume (mm^3)	Layer weight (g)	Effective material utilization (%)
R001	14.80	541.3	36.57	226.7	0.2923	26.81
R095	14.80	560.7	37.88	234.9	0.3027	27.77
R195	14.80	534.6	36.12	223.9	0.2886	26.48

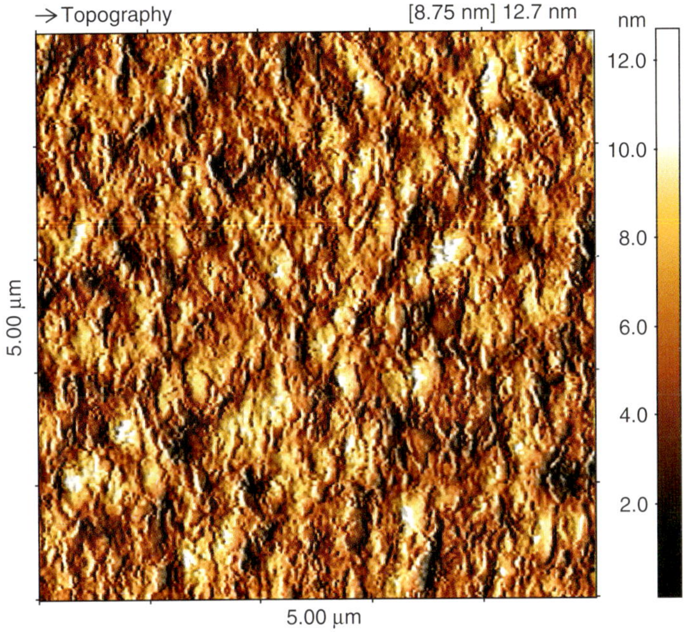

Figure 20.7 PVPD parylene-C deposition: AFM measurement. Roughness Ra (film) = 1.7 nm (for Ra substrate = 0.93 nm) 27 measurement points.

An average parylene-C film thickness of 550 nm was measured over the series of 195 runs, with a standard deviation of 1.48%, indicating good run-to-run reproducibility over the deposition series (see also Figure 20.8).

20.3 Conclusion

In the previous sections we explained that, on the one hand, a proper understanding of the underlying process as such as well as the chemical reactions involved is crucial and needs to be demonstrated with economically feasible effort.

On the other hand, the application requirements and the customer targets must be reflected in a system design, which usually exceeds the needs of a proof-of-principle demonstration setup.

Considering these two aspects, it is possible to design a system along design rules, which allow for easy scalability of a certain process and ultimately to unify both aspects. With the deposition system being the bridge between chemistry and application, an improper design that does not reflect both aspects will result in major failure regarding its intended application.

Different applications will usually yield different requirements. As such, a polymer film that works under one application might require to be tuned for

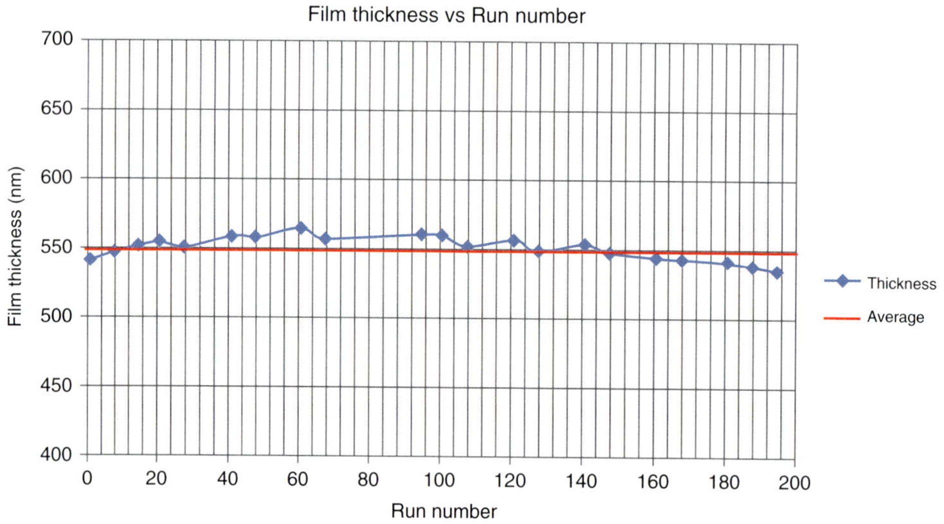

Figure 20.8 PVPD parylene-C deposition: run-to-run reproducibility of film thickness. Average film thickness 550 nm, standard deviation 1.48%.

another application with respect to its key properties. If parylene, to stick with this example, is used as a dielectric film, it needs to be pin hole free with good covering properties and excellent uniformity. If the material is used as a de-bonding film instead, apart from the precise thickness of the film, the interface quality, here especially a smooth surface morphology, is of paramount importance. While generally a system design should consider all aspects, the specific application might ask for special attention being given to the optimization of one or the other property. Also this flexibility should be reflected in the underlying design principles of a tool.

20.3.1
Outlook – Building Blocks to Create Systems and Variants Addressing a Variety of Polymer CVD Applications, For Example, Initiated CVD, Oxidative CVD

Following the modular concept described earlier, the universal platform concept then allows to replace the source system by another concept, add an activation mechanism, and with that to utilize the underlying system design for other processes such as iCVD (see chapter 1.4.1). To stick with this idea, very much in the same way an oCVD type of implementation can be derived if a feeding mechanism for the required catalyst material is provided in parallel to the monomer building blocks. This then would primarily ask for a modified showerhead design, allowing to inject several materials independently and without interfering with each other.

The required flexibility is well maintained if the design sticks with the building blocks principle described earlier, as this will allow to offer solutions according

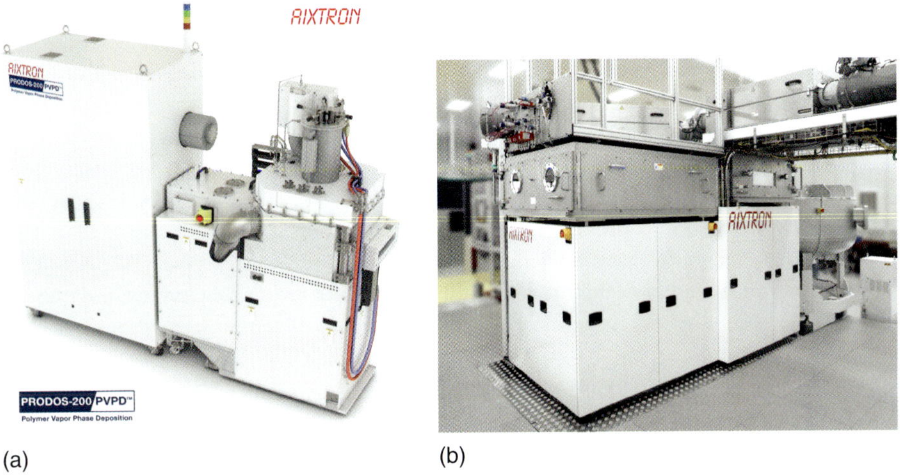

Figure 20.9 PRODOS series: industrial implementations of the PVPD^{TM} process technology. (a) PRODOS-200: designed as modular R&D platform for 200 mm × 200 mm substrate size. (b) PRODOS-Gen3.5 implementation of specific polymer process on industrial scale (Gen3.5 = 650 mm × 750 mm).

to customer demands and the application being addressed. An example for such implementation is shown in Figure 20.9.

Depending on process requirements, different kind of distribution systems might also be required with the aforementioned dual-channel showerhead where reacting species meet only right in front of the substrate and not already within the tubing of the gas mixing system being just one component that needs to be adapted. With flexible distribution systems, applications such as the deposition on flexible substrates, for example, by means of roll-to-roll, are possible and can be realized without too much of effort.

20.3.2
Scaling Polymer Film Fabrication from R&D Toward Large-Area Production

As said earlier, it is a fundamental design goal to have the same design rules being applied for small- and large-scale systems alike.

For larger implementations, the source systems need to deliver larger amounts of material per time. With the modular concept, this can be achieved either by integrating larger sources or by operating sources in parallel, thus allowing to reuse source designs proven on smaller areas without major modifications. In any case, the tubing must be capable to transport the higher material flows, and the showerhead needs to be larger to cover the whole substrate, which requires partial adaption of the system design.

From this top-level view, scaling seems to be quite simple. Here, it should be cautioned that each component still needs to be reviewed and verified regarding its intended use. At least the conceptual design needs to be reviewed, and, in some parts concepts that are required to ensure certain functionality need to be adapted. Nevertheless, as long as the underlying principles are not changed, the process itself can be claimed as already proven.

For the design review and concept evaluation, CFD and finite element model (FEM) simulations provide essential input.

With such numerical optimization methods critical design features can be analyzed with reasonable efforts and, if modifications or optimizations are required, respective impacts on critical properties such as thermal expansion or overall weight of the mechanical components can be determined upfront.

As an example, two implementations of the PRODOS product family are shown in Figure 20.10. On the left side, the PRODOS-200 system is depicted, which, being designed for substrates of 200 mm × 200 mm, addresses R&D and pilot setups. In comparison, the right side of this picture shows a PRODOS-Gen3.5 system with handling system on the rear side. Within this system, substrates of 650 mm × 750 mm can be processed.

It is apparent that depending on the size, differences in the implementation of a tool are required. As one example, a platform on top of the Gen3.5 system was required to allow for safe access of the sources (see Figure 20.9 , part b): platform on top of the system). on the right; upper side of the tool). Such efforts naturally are obsolete for smaller implementations.

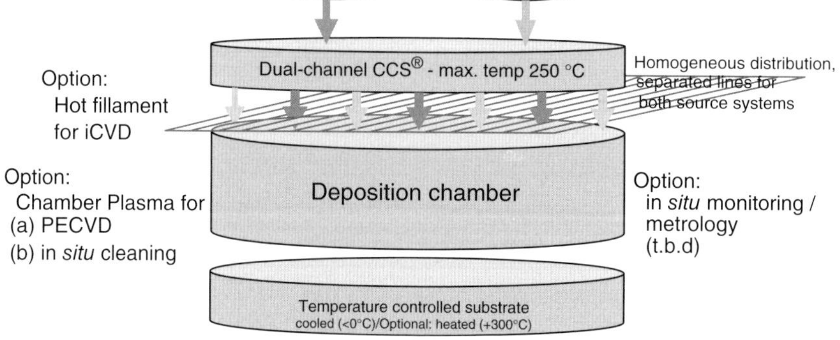

Figure 20.10 Modular system design allowing to address various polymerization processes.

Reference

1. Gorham, W.F. (1966) a new, general synthetic method for the preparation of linear poly-p-xylylenes. *J. Polym. Sci. A*, **4** (12), 3027. doi: 10.1002/pol.1966.150041209

2. Fortin, J.B. and Lu, T.-M. (2004) *Chemical Vapor Deposition Polymerization*, Kluver Academic Publishers, The Netherlands, p. 4ff.

Index

CVD Polymers: Fabrication of Organic Surfaces and Devices, First Edition. Edited by Karen K. Gleason.